Studies in Systems, Decision and Control

Volume 247

The series "Studies in Systems, Decision and Control" (SSDC) covers both new developments and advances, as well as the state of the art, in the various areas of broadly perceived systems, decision making and control–quickly, up to date and with a high quality. The intent is to cover the theory, applications, and perspectives on the state of the art and future developments relevant to systems, decision making, control, complex processes and related areas, as embedded in the fields of engineering, computer science, physics, economics, social and life sciences, as well as the paradigms and methodologies behind them. The series contains monographs, textbooks, lecture notes and edited volumes in systems, decision making and control spanning the areas of Cyber-Physical Systems, Autonomous Systems, Sensor Networks, Control Systems, Energy Systems, Automotive Systems, Biological Systems, Vehicular Networking and Connected Vehicles, Aerospace Systems, Automation, Manufacturing, Smart Grids, Nonlinear Systems, Power Systems, Robotics, Social Systems, Economic Systems and other. Of particular value to both the contributors and the readership are the short publication timeframe and the world-wide distribution and exposure which enable both a wide and rapid dissemination of research output.

** Indexing: The books of this series are submitted to ISI, SCOPUS, DBLP, Ulrichs, MathSciNet, Current Mathematical Publications, Mathematical Reviews, Zentralblatt Math: MetaPress and Springerlink.

More information about this series at http://www.springer.com/series/13304

Daniel Flaut · Šárka Hošková-Mayerová ·
Cristina Ispas · Fabrizio Maturo ·
Cristina Flaut
Editors

Decision Making in Social Sciences: Between Traditions and Innovations

Springer

Editors
Daniel Flaut
Faculty of History and Political Science
Ovidius University of Constanţa
Constanţa, Romania

Cristina Ispas
Universitatea "Eftimie Murgu" din Reşiţa
(UEMR)
Resita, Romania

Cristina Flaut
Faculty of Mathematics
and Computer Science
Ovidius University of Constanţa
Constanţa, Romania

Šárka Hošková-Mayerová
Department of Mathematics and Physics
Faculty of Military Technology
University of Defence
Brno, Czech Republic

Fabrizio Maturo
Department of Mathematics and Physics
University of Campania Luigi Vanvitelli
Caserta, Italy

ISSN 2198-4182 ISSN 2198-4190 (electronic)
Studies in Systems, Decision and Control
ISBN 978-3-030-30658-8 ISBN 978-3-030-30659-5 (eBook)
https://doi.org/10.1007/978-3-030-30659-5

This Springer imprint is published by the registered company Springer Nature Switzerland AG
The registered company address is: Gewerbestrasse 11, 6330 Cham, Switzerland

Preface

The book *Decision Making in Social Sciences: Between Traditions and Innovations* is part of the series *Studies in Systems, Decision and Control* published by Springer. This is the result of a scientific collaboration in the fields of Social Science among Prof. Daniel Flaut from the Ovidius University of Constanţa (Romania), Prof. Šárka Hošková-Mayerová from the University of Defence of Brno (Czech Republic), Prof. Fabrizio Maturo from the University of Campania "Luigi Vanvitelli" (Caserta, Italy), Prof. Cristina Ispas from the Eftimie Murgu University of Reşiţa (Romania) and Prof. Cristina Flaut from the Ovidius University of Constanţa (Romania).

Decision-making processes always bring to a final decision that can implies or not a prompt action. The decision-making process in social contexts is the result of analysis of founded alternatives using evaluative criteria. Therefore, this process is a process which can be rational or irrational and can be based on some knowledge and beliefs. Always a decision–making process always produces a final decision, which is implies or not a prompt action. This book explores some social science branches and their perspectives regarding their relations with decision-making process: computer science, education, linguistics, sociology and management. The concepts covered in the first part, the "Computer science" section, refer to how algorithms and the growth of the computing power during the years can control decision-making processes. In the second part, some traditional and innovative ideas and methods involved in various sectors of the educational process are presented: compulsory schooling, inclusive school, higher education, etc. The third part, the "sociology" section, includes some aspects regarding how society can be influenced by social norms, human interactions, culture, and religion. The last part of this book explores management, regarded as part of decision-making processes, some linguistic aspects, being examined the manner in which progress is manifested in a language, and finally some recent social perspectives. Moreover, also strategies of some organizations, their objectives and their resources, as for example, human resources, financial resources and technological resources, are presented.

The different studies included in this volume, selected after a peer-review process, are collected into above-mentioned four parts, as follows.

Part I, *Computer Science Between Traditions and Innovations*, has two chapters and is opened by the contribution of Syamal Kumar Sen and Ravi P. Agarwal, titled “Computing: Birth, Growth, Exaflops Computation and Beyond”. This contribution is considered as a brief history of birth, growth, exponentially increasing of computational power, specifically related to the computer mathematics/science and beyond.

In the second chapter of this part, “Methods and Algorithms for Creating and Reconfiguring Virtual Organizations”, Anata-Flavia Ionescu presents how virtual organizations (VOs) have gradually become a topic of interest among researchers representing numerous fields, ranging from technical domains such as optimization and soft computing to work psychology and organization studies.

Part II, *Some Remarks Regarding Traditions and Innovations in the Educational Process*, is composed by ten chapters. The first chapter, Mirela Minică, “Knowledge and Intellectual Capital Management in Higher Education”, underlines the main challenges of the Romanian higher education, in the context of an EU strategy to increase the quality of tertiary education. Moreover, is revealed the essential role of universities in creating skills and competencies required by the labour market in the twenty-first century.

Cristina Ispas in her chapter “Decision-Making Training of Teachers for Inclusive School”, considers that the successful implementation of inclusive education involves training for teachers. Inclusive school develops collaborative relationships with parents and communities through communication, participation and consultative decision-making. These things involve a series of substantial changes in terms of educational content, pedagogical and didactic approach.

Chapter “School Teaching in Germany Between Tradition and Innovation: The Concept of the Outdoor School” by Sarah Sahrakhiz, Marius Harring, Matthias D. Witte presents the idea that learning and educational processes can take place in different settings, and offer other experiences that the traditional classroom with its specific conditions largely neglected. The dimensions of space, time and body are re-evaluated from the perspective of outdoor education which faces children (and teachers) with new, sometimes challenging and adventurous experiences that contrast with an everyday routine of school instruction.

Dan Vătăman in his chapter “Reforms and Developments in Romanian Higher Education as a Result of the Bologna Process Implementation” presents the developments in Romanian higher education in the light of the Bologna Process evolution, with emphasis on reforms undertaken by Romanian authorities for renewing the legal, institutional and methodological frameworks applicable in higher education.

Daniela Căprioară, Annie Savard, Alexandre Cavalcante in their chapter “Empowering Future Citizens in Making Financial Decisions: A Study of Elementary School Mathematics Textbooks from Romania” present a qualitative study on the mathematics textbooks approved by the Romanian Ministry of Education and used during the school year 2018–2019. The authors investigated the learning situations (tasks) proposed in the textbooks whose content had a connection to financial education. Their results show that the tasks having a potential decision and those having a simple decision reflect practices of representational measurement.

In the chapter "Necessities and Expectations in Regard to Dobrudja's Rural Youth Access to Higher Education" by Rodica Gabriela Enache, Mariana Floricica Călin, Enache Tuşa, are analysed the particularities of the psycho-educational profile of young people from rural communities in the Romania's Dobrudja region, their professional values and interests as well as the factors that determine their access to studies through university education or professional orientation.

Afterwards, Jiří Jánský and Radovan Potůček, in the chapter "The Tradition and New Approach to the Mathematical Education of Officers in the Czech Army", discuss the tradition of teaching mathematics at the Faculty of Military Technology at the University of Defence in Brno. The thematic areas of mathematics teaching and the focus of relevant study programmes are compared in both bachelor's and master's degree programmes.

Irena Tuser, in her chapter "The Development of Education in Emergency Management", outlines the gradual development of the education system in crisis management, in Czech Republic, and offers a comparison of educational concepts, emphasizing their positive and negative aspects.

This part finishes with two papers of Engin Baysen and his co-authors. The co-author of the first one titled "Waste and Waste Management Awareness Among Teachers: A Phenomenographic Approach" is Fatma Baysen. It deals with a critical problem throughout the world that is waste. The success of waste management primarily depends on education and waste management which is the main topic of this chapter.

The last chapter of this part "Phenomenographic Approach to Teachers' Wait-Time Use: Reasons and Consequences" was prepared with Engin Baysen, Fatma Bayen and Nermin Çakmak. The purpose of this study is to inquire wait-time use of primary school teachers for the present. Phenomenography is the research approach adopted for this study.

Part III, *Some Aspects of Decision-Making Processes in Social Contexts*, has ten chapters. The first chapter, Gabriela-Felicia Georgevici, "The Family: Culture Traditional Keeper and Carrier. Study from the Belint Village", presents a general image of the manner in which traditions are kept and held in Belint, starting from the hypothesis: if the family from Belint keeps the old traditions and customs of this place, by maintaining a correct manner of their manifest. The family is a special community type of social will, which puts into motion all social frames and events. It is considered the traditional keeper and carrier of culture.

Lavinia Elisabeta Popp, Felicia Andrioni and Ion Petrică in their chapter, "The Migration Decision on Economic Level in the Contemporary Social Context" analyse the factors and consequences of the migration decision-making process on economic level, which may be correlated to a series of socio-economic indicators.

Cristian Rudolf in his chapter "Associations and Groups of the Germans from the Banat Highland 1918–1940", emphasizes the capacity of people to overcome difficulties, their power of continuous adaptation, steady progress and evolution, all come from the fact that they are social beings, they form communities, associating themselves based on different thematic groups.

The chapter "Asceticism for Society—*Integrating Body, Soul and Society. In Searching of the 'Real Self' that Actually Is 'Clothed with Christ'*" by Nichifor Tanase presents the greatest challenge which is facing Christian civilization and culture today is the following: the atrophy of spiritual senses.

Afterwards, Daniel Flaut and Cristina Flaut, in their paper "The Rural Population of Muntenia in the 1930s: A Brief Analysis Based on Statistical Data", feature a brief analysis of the structure and evolution of the rural population of Muntenia in the 1930s, as they emerge from statistical data. The period selected, albeit marked by social issues and crises preceding the beginning of the Second World War, was one in which the population was able to face up to all the challenges entailed by holding on to traditions as well as by accepting and harmoniously assimilating elements of novelty.

Enache Tușa and Ilie Iulian Mitran in their chapter "Vernacular and Urban-Influenced Housing Types in Constanța County in the 20th Century", present the evolution of housing in Constanța County, beginning with the last decade of the twentieth century, up to the late 1960s. The study highlights the way in which many rural communities suffered various social changes that greatly affected the mindset that was previously in place when it came to the housing design and household management.

The following chapter, by Elena-Ana Iancu, Ioana-Raluca Rusu and Nicolaie Iancu, "The Dimension of Criminality in the Context of the Variation of Migration Floods Trends" , focuses on some reflections on international migration in a context where there are variations in migration flows in the period 2015—first half of 2018 according to official statistics. To this end, are provided theoretical references concerning the notion of the state border and are explained its significance in international relations.

The chapter "Fuzzy Reasoning and Decisions in Primary School" by Luciana Delli Rocili and Antonio Maturo is based on the idea that the natural instrument of communication in a class is constituted by linguistic propositions, called fuzzy propositions, which are grammatically and syntactically correct statements, but which often cannot be classified either completely true or completely false. The truth value of a linguistic proposition depends on the opinions of a decision-maker that are conditioned by the circumstances that occur.

The objective of the next chapter, "Life Expectancy at Birth: A Causal Analysis of the Health Sector in Spain", by Pedro Antonio Martín Cervantes, Nuria Rueda López, Salvador Cruz Rambaud, is to determine the indicators which exhibit a unilateral or bilateral causal relationship with life expectancy, using the principal components analysis. Since the life expectancy at birth has become the main indicator of the performance of a country's health system, from a causal point of view, there is a series of socio-economic indicators which affect life expectancy.

The research of Annamaria Porreca "Assessing Diversity and Inclusion in the Context of the U.S. Federal Department of Health and Human Services" proposes an empirical application of diversity profiles in the organizational context; in particular, diversity profiles are applied for analysing diversity and inclusion of minorities in a United States Federal Agency, i.e. the Department of Health and Human Services.

Part IV, *Recent Social Perspectives Between Traditions and Innovations*, has ten chapters. This part is opened by the contribution of Cosmin Căprioară, titled "Romanian Language, Between Tradition and Innovation" opens this section. In this chapter is proposed an outline of the general characteristics of the Romanian language and the way in which tradition blends with innovation in this Romance language. Moreover, are considered aspects related to the manner in which progress is manifested in language, similar to a permanent interplay between the necessity of keeping the fundamental coordinates of the linguistic system and the need to adapt communication to the new challenges of reality.

In the second chapter of this part "Durkheim and Weber: Sociological Theory and Social Research", Vincenzo Corsi analyses the method in the sociology of Émile Durkheim and Max Weber. The basis of a research about the method that will have a fundamental value for sociology and for the empirical and theoretical development of the discipline in the twentieth century are identified.

The pedagogical aim of Fiorella Paone chapter "'Our Idea of School': Democracy Workshops in the Classroom" is to make students' point of view related to their school everyday life emerging in order to support their active role and participation to the cultural evolution process, in line with the latest perspectives of research of the movement Student Voice.

The contribution of Angela-Anca Dobre, "Aspects on the Evolution of the Press of Dobruja at the Turn of the 19th and 20th Centuries" . The chapter deals with the development of the press of Dobruja in close connection with the stages of the social-political and economic evolution of the Romanian provinces returned to Romania in 1878.

Nousheen Tariq Bhutta, Biagio Simonetti, Viviana Ventre in their work "Does Islamic Capital Asset Pricing Model Outperform Conventional Capital Asset Pricing Model?" present the difference between conventional and Islamic capital asset pricing model in order to ensure the efficient investment management of products and markets.

The chapter "Evaluative Aspects in Decision-Making Processes" by E. Serena Sanseviero deals with evaluative aspects in decision-making processes. Evaluative activity, however complex, precedes, accompanies and follows human actions. The evaluative activity, contrary to what could be thought in a simplistic and practical way, is instead expressed through a series of operations that are distinct but mutually connected, all of great importance.

Afterwards, Cornelia-Ecaterina Cornean in her chapter "Study on the Public Sector Personnel Motivation in the Caras-Severin County" highlights the motivational factors with a high percentage used by public institution from the Caras-Severin County, which help managers in motivating persons to choose and keep a job in the public sector.

Ferdinando Casolaro and Mario Cristiani in their chapter "The Volkswagen Case: Analysis of a 'Scandalous' Growth" present some aspects regarding the climb to the top of the Automotive sector of the Volkswagen group, between 2001 and 2015.

In the chapter "Experience-Based Course of Social Generativity in Villaggio Celdit, Chieti. History of Challenge for Popular Quarter's Rebirth After Big Factory's Closure" by Anna Crisant, Fabiola Nucci, Mario Perrucci, it is related about experience-based course of social generativity in Villaggio Celdit in Chieti, its name due to "Cellulosa d'Italia CEL.D.IT.", later named "Cartiera CIR".

Authors of the last chapter titled "Selected Aspects of Barrier Materials Assessment as a Part of the Reaction on Threats and Risks Connected with CBRN Problems" are Pavel Otřísal and Šárka Hošková-Mayerová. It deals with problems of barrier materials used in the branch of individual protection.

Constanţa, Romania — Daniel Flaut
Brno, Czech Republic — Šárka Hošková-Mayerová
Caserta, Italy — Fabrizio Maturo
Resita, Romania — Cristina Ispas
Constanţa, Romania — Cristina Flaut
May 2019

About This Book

The book discusses some aspects of social sciences related to their relations with the decision-making process which involve computer science, education, linguistics, sociology and management. First, the discussion focuses on how developments in algorithms and an extraordinary increase of the computing power over the years can influence and control the decision-making process. Second, some traditional and innovative ideas and methods involved in various sectors of the educational process: compulsory schooling, inclusive school, higher education, etc., are dealt with. Third, issues related to natural language are considered in the context of the volume. Fourth, an analysis of various social aspects is presented, notably how the society can be influenced by social norms, human interactions, culture and religion. The management, viewed as a science of the decision-making process, is explored in the last part of this book including analyses of strategies of organizations, their objectives and their resources exemplified by human resources, financial resources and technological resources, etc.

Contents

Computer Science Between Traditions and Innovations

Computing: Birth, Growth, Exaflops Computation and Beyond

Syamal Kumar Sen and Ravi P. Agarwal

Abstract Computing is a generic term implying activities such as design, development, and construction of hardware, firmware, and software systems using a computer or benefitting from it or creating it. Besides, structuring, processing, and managing different types of information, performing scientific/engineering work/research on and with the systems, making the systems behave intelligently, and creating and using communications and entertainment media are some of the applied aspects of computing. Computing comprises the subject areas such as computer science, computer/computational mathematics, and artificial intelligence. Mathematics—traditional/usual, computer/computational, and natural—did exist for millennia. These mathematics are connected intimately with computing explicitly or implicitly. We attempt to record a brief history of birth, growth, exponentially increasing computational power of over quintillion flops related to computing, specifically related to the computer mathematics/science, and beyond. In the process we stress the real-world importance and gigantic differences of computer mathematics over the usual mathematics as well as natural mathematics.

Keywords Computer/computational mathematics · Control flow architecture · Dataflow computer · Exaflops computation · Finite and infinite precisions of computation · Hardware, firmware, and software systems · History of computing · Natural mathematics · Quantum computing · Traditional mathematics · Systolic architecture

S. K. Sen
GVP-Prof. V. Lakshmikantham Institute for Advanced Studies, Visakhapatnam, India
e-mail: sksenfit@gmail.com

Department of Computer Science and Engineering, GVP College of Engineering, Visakhapatnam, India

R. P. Agarwal (✉)
Department of Mathematics, Texas A&M University-Kingsville, Kingsville, TX, USA
e-mail: Ravi.Agarwal@tamuk.edu

D. Flaut et al. (eds.), *Decision Making in Social Sciences: Between Traditions and Innovations*, Studies in Systems, Decision and Control 247,
https://doi.org/10.1007/978-3-030-30659-5_1

1 Introduction

God exists since mathematics is consistent and
Devil exists since its consistency cannot be proved.
Morris Kline (1908–1992)

In the same way we may say

God exists since computer mathematics is consistent and
Devil does not exist since we can prove its consistency.

1.1 Computer Mathematics Is Eternally Finite-Precision

Computation has become absolutely unavoidable for anybody including an engineer, a technologist, a scientist, a linguistics person, and even a mathematician if he/she has anything to do with a real-world application. Computational/computer mathematics (CM) provides the justification for any computation that we carry out using a computer—a finite precision machine. That is, the precision of computation (PoC) in any man-made computer is finite.

In other words, the CM is the only gateway for everybody whenever he does computation—numerical/semi-numerical/non-numerical to solve a problem derived out of nature/physical world; this computation is always finite precision (finite word length)

1.2 Usual, Computer, and Natural Mathematics: Comparison

We compare the usual/traditional mathematics (UM) and CM keeping in view the natural mathematics (NM) viz. the mathematics that nature uses all the time eternally parallelly in the background. The computation in the NM is always performed by nature only and it is **out of bound of human sphere** of activities. Further it is infinite precision and exact (completely error-free). Even the inputs for the computation are exact and cannot be captured error-freely by any means known to humans.

The computation in the UM is usually performed by humans occasionally assisted by a digital computer. The PoC in the UM is mainly infinite and occasionally finite. It is within the domain of human activity. Always an infinite precision quantity is written symbolically such as $\sqrt{2}$ since it can never be written numerically exactly by humans. However, such a symbolic representation would be only useful to an engineer when its sufficiently accurate numerical value is used.

The comparison focuses specifically on the vital differences between the UM and the CM from real-world applications point in view. It stresses on the fact that CM can never be bypassed since the engineers/applied scientists will be totally handicapped without CM in the exponentially fast-progressing computer world of today. Such a comparison among the CM, the UM, and the NM is useful for better appreciation of the scope of the CM.

1.3 Hardware, Software, Firmware and CM

The progress of the CM is directly linked with the advancement of hardware (h/w) digital computer, the related software (s/w), and the firmware (f/w). The f/w is a software program *and* is typically stored in the FROM (flash read-only memory) of a hardware device, that can be erased and rewritten unlike that in the ROM.

The f/w can provide a **standardized** operating environment for the device's more complex software (allowing more h/w-independence), or, for less complex devices, act as the device's complete operating system, performing all control, monitoring, and data manipulation functions. Devices containing f/w, for instance, are embedded systems, consumer appliances, computer peripherals, and others. Almost all electronic devices beyond the simplest ones contain some f/w.

Thus while presenting the CM over centuries we will necessarily talk about the birth, the growth, and the computing power of a computing device.

1.4 Finite and Infinite Precisions of Computation (PoC): Vital Differences

Since middle school days, students are taught mathematics (e.g. arithmetic, geometry of Euclid, and algebra) irrespective of the future disciplines to be adopted by a student. There is an implicit (or even explicit) feeling in the minds of mathematicians among others that the computational mathematics or, equivalently, the CM is a proper subset of the UM. There is nothing great about it. We will see below that this is not true from the real-world usage point of view.

The vital differences crop up due to primarily the difference in the PoCs or, equivalently, the word-lengths (used) of the computer (Krishnamurthy and Sen 2009; Lakshmikantham and Sen 2005; Sen and Agarwal 2016; Agarwal and Sen 2014). In **UM and NM, PoC is implicitly/explicitly known to be infinite** while **that in CM it is eternally finite**.

A finite precision (integer) is a proper part of infinite precision. Hence it might appear that the set of all properties due to a finite precision should be a proper subset of the set of all properties due to an infinite precision.

However, the reality is that this need not be. An nth degree (n is a positive integer) polynomial has exactly n zeros counting multiplicities when n is finite. If n is infinite (e.g. in a transcendental/trigonometric function), then the number of zeros can be anything viz. nil or 1 or 2 or … or a finite positive integer or infinity!

One may check the functions (infinite degree polynomials) $x - e^x$ (absolutely no real zero), $x - e^{-x}$ (exactly 1 zero), $x^3 - e^x x^2 - x + e^x$ (exactly 2 zeros), ..., $2\sin x - \cos x$ (infinity of zeros). One may execute the Matlab command **≫X=fzero(@(x) x-exp(-x), 5)** as well as similar other Matlab commands for other functions and check to convince oneself about the validity of the statement in the foregoing paragraph.

An input real number is always finite (in decimal representation) in the number of digits in CM which has always a finite PoC. The real number is not only always countable, and rational (never irrational unlike that in UM) in CM but also finite.

Consider a straight line (in a physical 1 dimensional (1-D) space). It will hold infinity of points whether its length is 10^{-1000} cm or 10^{1000} miles without any space between any 2 successive points. A point is defined as something which has no length, no width, and no height in 3-D space (no length in 1-D space), yet it exists.

Although this is a definition with inherent contradiction, there is no other simpler way to tell the audience exactly what a point (which is necessarily required to be physically geometrically represented in a 2D paper for conveying the true meaning/message of the figure) is.

With the approximate (erroneous) representation of the input data within the PoC in CM, the outputs produced due to the computation are also rational having exactly the same characteristics including errors (usually more pronounced) as those of the input data.

On the other hand, these are not so in UM which entertains a rational number with (i) exactly a finite number of digits in the decimal representation e.g. (1/4) and/or (ii) an infinite number of digits in the decimal representation (e.g. 1/7), and (iii) an irrational number (e.g. $3^{1.051}$). Most often the representation is symbolic or partially symbolic (such as $\surd 3$).

A complex number has a real part and an imaginary part where both the parts are real. The concerned arithmetic operations viz. complex add, subtract, multiply, and divide operations are the operations with these parts, which we have learnt in a high school. From both UM and CM points in view, the operations are similar. In UM, each of the real and imaginary parts of an operation could be rational or irrational while, in CM each of these parts is necessarily rational and never irrational.

In NM, however, the input real numbers are exact, could be irrational, and cannot be captured by us, the humans. The decimal number system is usually incapable of exactly representing any real quantity (real number) **which always physically exists without error**.

The natural computer executing a natural algorithm based on the NM was existing **infinite years ago**, is existing now, and will continue to exist **infinite years hence**. It is **completely maintenance-free, completely error-free, never stoppable, and ever parallel**. Simulation (by humans) of the natural computer/computing exactly using a digital computer is impossible. Even the completely error-free data accepted by the natural computer will be usually accepted by a digital computer only approximately (i.e. with errors).

There exists **no number system that could represent any real quantity exactly and there will never be any in future** because one of 2 real numbers representing two adjacent (distinct) points cannot be represented in a finite number of digits.

The output in UM, called a solution, is mainly in symbolic form which is not useful to an engineer unless converted to numeric form. In CM, there is no such symbolic form as output result; it is in numeric form always, which is all that is required by an engineer for a real-world implementation (for purely numerical problems).

Majority of the computational problems, in CM, involve intense numerical computations (nearly 100% numerical, not exactly 100%). Further, there are semi-numerical problems (roughly 50% numerical and the rest are non-numerical) such as the problem of generating prime numbers or/and testing whether a number is prime or not, as well as non-numerical problems (nearly 100% non-numerical, not exactly 100%) such as the problem of searching and sorting. These problems are also solved by a digital computer and the solutions—usually non-numeric—are useful to concerned humans including an engineer and a scientist.

1.5 Computation Is Built-in in a Living Being

Computation has been inherent in a living being such as a bird, an animal, a human, and even a bacterium. The capability of performing computation varies from one species to another. Even in a single species, the capability varies from one individual to another. The fact that, underlying the computation, the CM exists may/may not be consciously known to the concerned individual.

In Sect. 2, we describe computing in NM and UM over centuries along with their intimate occurrences in science and engineering. In Sect. 3 we present, in brief, the CM including a comparison with UM keeping in view the NM in the background besides describing the devices involved in computation from antiquity till present (2019). The comparison focuses specifically on the vital differences between the UM and the CM from real-world applications point in view. Section 4 comprises conclusions.

2 Computing in NM and UM Over Centuries: Birth and Growth Along with Computing Devices

2.1 Natural Mathematics

Any activity in nature may be viewed as a computation—(i) numerical, (ii) semi-numerical, (iii) non-numerical, (iv) a combination of any 2 of the foregoing 3, or (v) the combination of all the first 3. The instrument that performs these computations is nature or, equivalently, **natural devices such as a material object/non-material (spiritual) one around us. Besides these devices, the live humans themselves comprise the most important device**. The mathematics underlying this computation is termed as natural mathematics (NM).

Thus the natural computation based on NM knows no birth and hence no death. It is eternal. We, the humans, could see/experience the best dynamism and the resulting outputs all the time as long as we exist in human bodies. The speed of computation (essentially parallel subject to the required resources/inputs) is always the best within the laws of nature (both spiritual and material), which can never be violated.

So far as the growth in speed and storage is concerned, it is eternally the best in absolute sense at any point in time. In a relative sense, it may appear to us growing at some point in time while shrinking/decaying at some other point in time. In the ever dynamical environment of nature, such a growth and a decay have been alternating eternally. The theory of multi-verses, rather the theory of infinity of (dynamic) universes, as well as the essence of the big bang theory (applied to only one universe) appears not to contradict in essence the concept of the eternal natural computer.

In the sense of total energy, only the form of energy alternates. The situation when there existed no visible/apparent activity in nature, the energy was fully potential. In other words, the nature was at her final state of pregnancy with complete potential energy. This state happens to be, like a pendulum, at one of its 2 extremes and seems to precariously exist with kinetic energy 0. It develops tremendous tendency to translate its potential energy into kinetic one. Consequently it does. The potential energy starts reducing and the kinetic energy starts increasing. At this junction of energy transformation appears the "big bang" and hence the theory. There could be not just one bang but multiple (rather infinite ones in the sense of human comprehension). At a non-extreme state which is changing continuously at any point in time, appears the "multi-verses" and hence the multi-verse theory.

In any science—materials and/or spiritual—**something can never be created out of nothing**. Nobody has any evidence that a thing has been created/generated from complete void. **Sir Fred Hoyle** (1915–2001 AD), British mathematician and astronomer, and **Jayant Vishnu Narlikar** (born 1938 AD), Indian astrophysicist, during early 1960s were the proponents of the **Steady State Theory** in Cosmology (alternative to **Big Bang Theory** of the universe's origin). According to their theory, new matter is uninterruptedly created as the universe expands, thus adhering to the cosmological principle, was not flawless and did not succeed. Consequently the theory is now obsolete (Sen and Agarwal 2016). This is true for both material and non-material (spiritual) worlds.

According to today's physics, **the void i.e. the universal exact void was never there, is not there, and will never be there**. Two aspects are important to be considered here. One aspect is that of matter while the other aspect is that of non-matter or, may be termed, Spirit or Nature or God (encompassing all knowledge) or Consciousness that is omnipotent (having unlimited power), omnipresent (present everywhere), and omniscient (knowing everything).

One's realization/experience is the proof of the existence of spirit, which is the best proof (better than even a mathematical proof). Nature (or God if you wish to call) pervades all matters, all spaces containing matters of varying density including numerically zero (not exactly zero) density. Matter with "exact zero" density i.e. completely empty space does not seem to be fathomable by a physicist or possibly by anybody within the realm of traditional science. To put it in another way, exact zero energy—fully potential or partly kinetic and partly potential or fully kinetic—all the time at any point in the unlimited space is non-existent.

"Quantum" implies the smallest possible **discrete** (non-continuous) unit of any physical property such as *energy* (or matter) This may not be understood that there are points in space that possess no energy (matter) at all permanently.

Is there a sharp boundary—may be static or dynamic—just beyond which matter has exact zero density and just within (may be closest to the boundary) which a nonzero density? Is there a discontinuity of density (in mathematical/physical term)?

Interestingly, when we attempt to create vacuum in a container, we successively reduce the density of the gas (say, air) but we will never be able to make the density exactly zero by any process that we know of in physics or in any other area of knowledge.

All these theories are just created by humans with the sole goal to explain the functioning of the eternal nature and its laws. There is absolutely no way to experiment by humans in any material plane as the magnitude of both space (including all matters as well as their equivalence) and time is many folds beyond any astronomical figures. Even the concerned physical quantities computed by a physicist based on certain assumptions are not only often too approximate but also controversial. Hence any computation based on these quantities invariably will inflate/reduce the super-astronomical figures to such an extent that the final results may not be acceptable.

A proof in this manner is not palatable/acceptable to a critical mind. The significance of the resulting statement is considerably low for any real-world activity. Rarely we could depend on the statement/result.

Although these figures do occur in physics as well as in ancient Hindu scriptures, these super-astronomical figures are beyond the domain of common human logic (rather critical logic) to really fathom them or strictly prove the validity of the figures. Probably these can be experienced/realized by a human from his/her ultimate state of mind (e.g. ***Nirvikalpa Samadhi State***).

The scale of time and space measurements is truly hyper-astronomical and also controversial since we are dealing not with finite space and time but with unlimited space and time (or the space and time whose measures are out of bound of any common human including any materials scientist). If anybody tells us these measurements, he/she has no way to prove these convincingly to a well-versed common human/scientist.

2.2 Usual Mathematics and Its Occurrence in Science and Engineering

Any activity of a mathematician using the traditional/usual mathematics (UM) may be viewed, in essence, as symbolic vis-a-vis numeric derivation/computation. Implicitly and sometimes explicitly the UM underlying this activity uses, in digital computer terminology, the infinite precision (infinite word-length) in general. This is in contrast to that in CM (both modern and ancient) in which all numerical computations are always finite-precision.

The history of UM or, simply, mathematics has been the topic of several mathematical historians (Agarwal and Sen 2014; Agarwal et al. 2013; Al-Daffa 1977; Al-Daffa and Stroyls 1984; Allen 1966; Allman 1976; Andrade et al. 1954; Anglin 1994, 1995; Anthony 1960; Apostle 1952; Archibald 1949; Artmann 1999; Srinivasa Ayyangar 1967; Bag 1979; Ball 1960; Barker 1964; Baron 1993; Barrow 1992; Baumgardt 1952; Beckman 1976; Beiler 1964; Belhoste 1990; Bell 1986, 1992; Benson 1999; Berggren 1986; Berlinski 1995; Berndt 1985; Berndt and Rankin 1997; Beth 1959, 1965; Bhaskar et al. 2007; Bishop 1936; Black 1965; Blatner 1997; Bose et al. 1971; Boyer 1959, 1991; Brewer 1981; Brody and Brody 1997; Brumbaugh 1954; Bunt et al. 1976; Burton 1999; Cajori 1923, 1925, 1985; Calinger 1976, 1995, 1999; Cantor 1952; Cantor et al. 1907; Cardano 1962, 1968; Chandrasekhar 1995; Ranga Charya 1912; Chattopadhyaya 1986; Clagett 1963; Clark 1930; Clawson 2000; Closs 1986; Cohen 1958; Cole 1995; Conant 1923; Cooke 1984; Coolidge 1963; Copernicus 1995; Courant et al. 1996; Crowther 1995; Cullen 1996). The word "mathematics" meaning "a subject of instruction" was due to Pythagoras (570–495 BC), a Greek philosopher of Ionian (1 of the 4 major tribes) origin and the eponymous founder of the Pythagoreanism movement.

The first part, 'math', originates from an old Indo–European root related to the English word 'mind'. The Pythagoreans classified mathematics as the area of study comprising the 4 subjects viz, arithmetic, astronomy, geometry, and music. This classification survived for several centuries.

Later this area of study has been diversified. It includes, in addition to the foregoing subjects, the following topics such as Topology, Combinatorics, Number Theory, Dynamical Systems and Differential Equations, Mathematical Physics, Computation, Information Theory and Signal Processing, Probability and Statistics, Game Theory, and Operations Research.

- **Topology** (the study of geometrical properties and spatial relations unaffected by the continuous change of shape or size of figures) which originates in early 20th century, but several scattered activities can be found before 20th century. One of these activities comprises certain questions in geometry probed by Leonhard Euler (1707–1783), a Swiss mathematician, physicist, astronomer, logician and engineer. His 1736 article on the *Seven Bridges of Königsberg* is regarded as one of the 1st practical applications of topology (Dunham 1999, 2007).
- **Combinatorics** (deals with combinations of objects in a finite set in accordance with certain constraints, such as those of graph theory), which is known to have been used more than 3600 years ego. The earliest recorded use of combinatorics comes from problem 79 of the Rhind papyrus, which is dated 16th century BC. The problem concerns a certain geometric series, and has similarities to Fibonacci's problem of counting the number of compositions of 1s and 2s that sum to a given total.

The problem 79 may be stated as follows. A piece of land comprises 7 houses; each house had 7 cats; each cat ate 7 mice; each mouse ate 7 heads of wheat; and each head of wheat was capable of yielding 7 hekat measures of grain: Houses, cats, mice, heads of wheat, and hekat measures of grain, how many of these (viz. houses, cats, mice, heads of wheat, and hekat measures) in all were in the land? The answer is 19607 (i.e. the sum of the 1st 5 powers of 7 viz. $7^1 + 7^2 + 7^3 + 7^4 + 7^5$).

- **Number theory** is concerned with properties of the positive integers, sometimes referred to as *higher arithmetic*, It is among the oldest and most natural of mathematical pursuits. The history of Number theory is over 5000 years old, the term "number theory" was coined by **Pierre de Fermat** (1607–1665), a French lawyer during 17th century AD.

However, Pythagoras and his disciples are generally given credit for initiating the actual theory of numbers. Pythagoras believed that all universal laws can be explained by numbers, and hence he declared that "Everything is Number".

The first systematic treatment of number theory was given by Euclid in books 7, 8, and 9 of his *Elements*. The ancient Greeks were fascinated by number theory, in fact, Plato wrote that "We must endeavour to persuade those who are to be the principal men of the state to go and learn arithmetic".

Since antiquity, number theory has captivated the best minds of every era. During 1600–1850, Chebyshev, Dirichlet, Euler, Fermat, Gauss, Lagrange, and Legendre greatly enriched number theory. In 1801, Gauss asserted that number is purely a psychic reality, a free creation of mind, and he crowned number theory as "The Queen of Mathematics". According to Bell, "number theory is the last great uncivilized continent of mathematics".

In recent years, number theory has been divided into: (1) elementary or classical number theory, (2) algebraic number theory, (3) analytical number theory, (4) probabilistic number theory, and (5) computational number theory, which finds diverse applications.

An important aspect of number theory is that challenging problems can be formulated in very simple terms; however, hidden within their simplicity is complexity. Some of these problems have been instrumental in the development of large parts of mathematics. Amateurs and professionals are on an almost equal footing in this field.

Foregoing topics viz. **arithmetic, astronomy, and geometry** are some of the subjects which come under **pure mathematics**. Topics such as **Dynamical systems and Differential equations, Mathematical physics, Computation, Information theory and Signal Processing, Probability and Statistics, Game theory, and Operations research** constitute **applied mathematics**.

- **Dynamical systems and differential equations** A mathematical model of a dynamical system gives rise to differential equations—ordinary/partial, in general. The dynamical systems theory or, equivalently, the nonlinear dynamics/chaos theory involves techniques for analyzing differential equations.

The theory draws on analysis, geometry, and topology—areas which are in turn rooted in Newtonian mechanics (17th century)—and so should probably be viewed as a natural development within mathematics, rather than a scientific revolution or paradigm shift (Holmes 2007; Hildebrand 1965).

- **Mathematical physics** Christiaan Huygens *(1629–1695)*, a Dutch physicist, mathematician, astronomer, and inventor was the first to use **mathematical** formulas to describe the laws of **physics**, and for that reason Huygens is regarded as the first theoretical physicist and the founder of **mathematical physics**.
- **Computation** Computation is any type of calculation that includes both numerical and non-numerical (arithmetical and non-arithmetical) steps following an algorithm (a formal method or, equivalently, an ordered set of instructions/commands that can be executed by a computer in a finite time). The study of computation is paramount not only to the discipline of computer science but also to practically all scientific and engineering disciplines as well as even to the formal study of languages and related activities. As a matter of fact, **the 21st century civilization will simply collapse if "computation" using a digital computer is cut off**.

Computation has not only been an inherent part of any living being but also an integral part of very "nature". Since the nature is eternal and all-pervading, so is the computation, may be termed as the natural computation (following NM).

However, we have the computation—both finite and infinite precisions—involving the UM. Also we have the computation—eternally finite precision—using a man-made digital computer (ever a finite precision machine).

We will deal further about the most important subject "computation" using a modern digital computer when we discuss/describe computation under CM. Our focus is essentially on this aspect as mentioned in the title of the present article.

- **Information theory and signal processing** Information theory deals with the quantification, storage, and communication of information. The theory was propounded by Claude Elwood Shannon (1916–2001), an American mathematician, electrical engineer, and cryptographer acclaimed as "the father of information theory", in 1948 to find fundamental limits on signal processing and communication operations such as the data compression in an article *A Mathematical Theory of Communication* of profound importance.

Applications of information theory comprise, among others, lossless data compression (e.g. ZIP files) as well as lossy data compression (e.g. MP3s and JPEGs) and channel coding (e.g. digital subscriber line (DSL)). Shannon's discovery crucially impacted the success of the Voyager missions to deep space, the invention of the compact disc, the feasibility of mobile phones, the development of the Internet, the study of linguistics and of human perception, the understanding of black holes.

- **Probability and statistics** Statistics is a branch of mathematics dealing with the collection, analysis, interpretation, and presentation of (usually large amount of) numerical data. It may be viewed as the science of collecting and analysing masses of numerical data, especially for the purpose of drawing inferences about a whole from those in a representative sample.

Probability, on the other hand, is the measure of the likelihood that an event will occur and is quantified as a number in [0, 1], where, 0 indicates impossibility and 1 indicates certainty. The higher the *probability* of an even is, the more likely is that the event will occur.

While statistics deals with data and inferences from it, probability deals with the random processes existing behind data or outcomes. **Sir Ronald Aylmer Fisher** (1890–1962) a British statistical scientist, is widely acknowledged as the father of the modern statistics.

The mathematical theory of probability has its roots in attempts to analyze games of chance by **Gerolamo Cardano** in the 16th century while a gambler's dispute in 17th century (precisely 1654 when a gambling problem posed by **Chevalier de Mere**) led to the development of a mathematical theory of probability by two legendary French mathematicians, **Blaise Pascal and Pierre de Fermat**.

In 18th century (precisely 1786) **William Playfair** (1759–1823) introduced the idea of graphical representation into statistics. He developed the line chart, the bar chart, and the histogram and incorporated them into his works on economics as well as on the commercial and political atlas.

- **Game theory** The mathematical modelling of conflict and cooperation between intelligent rational decision-makers comprises Game theory. Besides computer science, it is used in economics, political science, and psychology.

It addressed 0-sum games originally, in which 1 participant's gains result in losses for the other participants. Today, game theory applies to a wide range of behavioural relations, and is now an umbrella term for the science of logical decision making in humans, animals, and computers.

The mathematical **theory** of games was developed during the 2nd world war (1939–1945). It was proposed by John von Neumann (1903–1957), a Hungarian-born American mathematician and computer scientist, and Oskar Morgenstern (1902–1977), a German-born economist and published in the book *Theory of Games and Economic Behavior* in 1944.

- **Operations Research (OR)** is an analytical method of solving problems and making decisions useful in the management of organizations. It encompasses a wide range of techniques applied in the pursuit of improved decision-making and efficiency, such as simulation, mathematical optimization, queuing theory, other stochastic (random) process models, Markov decision processes, and econometric methods.

Philip McCord Morse (1903–1985), an American physicist, administrator, and pioneer/father of OR in the U.S. during the 2nd World War.

There are several other areas in pure and applied mathematics, which have been invented/developed at different times by mathematicians/scientists. These developments along with the above-mentioned areas have contributed greatly in changing the humanity to one that we are a part of today (2019). We omit discussing/presenting these developments to conserve space.

2.3 *Historians May Not Always Be Unbiased Presenters of Facts*

However, it was found later that the study of the first 3 subjects viz. arithmetic, astronomy, and geometry mentioned earlier began long before Pythagoras. In fact, a majority of the significant breakthroughs in mathematics were due to the work of mathematicians other than those credited in the history books.

"But in science the credit goes to the man who convinces the world, not to the man to whom the idea first occurs" remarked Sir Francis "Frank" Darwin, (1848–1925), a British botanist and a son of the legendary British naturalist and scientist Charles Darwin (1809–1882).

In several instances, the fault rests on the historians themselves, who introduce their own opinion into their texts rather than reporting a fact in an unbiased way. Following these historians, we believe that mathematics originates from the **number sense**, which has played a pivotal role in the development of these subjects.

Number sense: The heart of CM The number sense can be viewed as an intuitive understanding of the integers, their magnitude, their patterns and relationships, and how they are affected by the arithmetic operations. This simple number sense, the ability to distinguish between 'plenty' and 'few' without counting, is a useful tool for a living being and a fundamental ability of humans. Mathematics is simply a comparison of small collections for children and ancient humans.

Outside the human domain, there are recorded incidences of living beings such as animals, birds, insects, and aquatic creatures who demonstrate through their behaviour a certain number sense—comparing/sorting, for example. These living beings possibly possess mathematical abilities that could be beyond our knowledge. Were we to read their minds, we would probably be surprised at our ignorance about their intellectual and mathematical abilities!

The number sense is the centre of mathematics (UM) and more specifically, the heart of CM. In fact, UM and CM are intimately/inseparably connected.

The **devices used in computation, both numerical and non-numerical, using UM comprise the minds of mathematicians themselves along with the devices starting from abacus** (before 2400 BC) **to modern digital computers (21st century and after)**.

Arithmetic. Algebra, Trigonometry, Spherical Trigonometry, Geometry, and Astronomy are the essential constituents of UM. There is some evidence of basic arithmetic in the petroglyphs at the Knowth and Newgrange burial mounds in Ireland, dating from about 3500 BC and 3200 BC, respectively (Agarwal and Sen 2014).

The ancient people of the **Indus River Valley (Harappan) Civilisation**—a Bronze Age civilisation (3300–1300 BC; peak period 2600–1900 BC)—residing mainly in the north-western regions of South Asia, extending from today's north-east Afghanistan to Pakistan and north-west India had a highly developed knowledge of arithmetic and a sophisticated system of weights and measures.

We will deal with the (history of) growth of more practically important CM, rather the modern CM viz. the CM developed with the advent of modern digital computer—a subject area that draws enormous information from UM based on reasoning termed as the supreme reasoning in the domain of common human logic.

3 Computer Mathematics and Computing Tools

Computer mathematics or, equivalently, computational mathematics is the heart of all computations—numerical and non-numerical/semi-numerical—using the digital computer. The modern world civilization has become compulsively dependant on computation. If the computation comes to a halt permanently or is forced to a stop eternally, then the current civilization will go back to the Stone age (30000 BC–3000 BC). However the natural computation is eternal and it will continue to exist for ever.

Both human-created computer mathematics and the concerned computation are inseparably connected. We present here the birth, growth, exa-flops computation, and beyond in brief. During the pre-digital computer days (before early 20th century), our civilization was a dwarf compared to that in the 21st century, more precisely after 2019 (i.e. in less than 100 years).

The **methods invented in applied mathematics** well before the modern digital computer era (i.e. before 1940s) and after constitute the backbone of computer mathematics and consequently that of the modern computation (Krishnamurthy and Sen 2009; Holmes 2007; Hildebrand 1965; Richtmyer 1957; Filon 1928; Egervary 1931; Farkas 1901; Kuhn 1956; Forsythe 1952; Wilkinson 1965; Young 1954; Moore 1920; Halmos 1950; Hildebrand 1956; National Bureau of Standards 1944; Abramowitz and Stegun 1965; Aitken 1937; Ralston and Wilf 1967; Sokolnikoff and Redheffer 1958; Hitchcock 1944).

Most/all of these methods have been used mainly to **solve very small real-world/test problems** using a non-automatic calculator and/or simply by a combination of hand calculation with or without the assistance of mathematical tables such as multiplication tables, logarithmic tables, trigonometric function tables, Bessel function tables, hyper-geometric function tables, and Chebyshev polynomial tables (Abramowitz and Stegun 1965).

However, with the advent of automatic digital computers since 1940s. use of all the foregoing tables gradually became obsolete (over the next about 60 years). This finally became practically extinct since the very dawn of 21st century throughout the world.

The current (21st century) generation of students appear to have forgotten that there used to be logarithmic/trigonometric tables supplied during concerned examinations/class tests in the under-graduate as well as graduate classes in schools, colleges, and higher educational institutions/universities all over the globe. They became accustomed to use a pocket calculator (programmable/non-programmable) or even a regular laptop/desktop computer.

Even they have forgotten early modern computer days' usage of tools such as punched (standard) 80-column IBM cards, paper-tapes, card/paper-tape readers, card/paper-tape punch machines for data and program preparation.

All these were required for inputting the program and data through punched cards/punched paper tapes into the memory of the digital computer (through a punched card/paper tape reader) for automatic execution of the instructions of the computer program (a computer program is an ordered set of instructions) and outputting the results using a printer.

The whole computing hardware system were physically large and heavy occupying large areas in an air-conditioned hall. Some of these electronic computing systems can be found in a museum for the visitors. This provides them a glimpse of the early human creation and activities in the field of computing. Little did one imagined that successive innovations in computing will change the whole world into one that we live in today (2019).

Numerous utility devices such as smart phones, smart washing machines, smart fridges, smart cars, and the internet (for communication) have become a reality. These appliances are used by common humans throughout the world. Each of these devices has embedded (artificially) intelligent computers in a highly miniaturized form.

There exists practically nobody whose day-to-day life has not been deeply affected due to the foregoing creations/appliances. Just imagine the world 80 years ago viz. the world of pre-1940 and that of today. The difference could not be possibly foreseen/imagined even in a fiction!

Our present (2019) civilization has grown so tall that our earlier civilization over hundreds of millennia could be viewed as too dwarf in the field of computing and the consequent influence/effect in all other scientific and engineering areas.

The progress is **exponential** in terms of speed of computation, storage for computation, and band-width for communication.

3.1 Computing: Tools

CM is not stand-alone. A digital computer with requisite software, firmware, and hardware along with a live user having sufficient knowledge in CM as well as that in the use of the concerned computer and the problem are required for a meaningful/successful solution of the problem. Omission of anyone of the foregoing components is enough to halt the solution process.

3.2 What Is Computing

Computing implies activities such as design, development, and construction of h/w, f/w, and s/w systems needing a computer or benefitting from it or creating it. Besides, structuring, processing, and managing different types of information, performing scientific/engineering work/research on and with the systems, making the systems behave intelligently, and creating and using communications and entertainment media are some of the applied aspects of computing. Computing comprises the subject areas such as computer science/mathematics, computational science/mathematics, information technology, computer h/w and s/w engineering, and artificial intelligence.

3.3 What Are Computational Science/Mathematics and Computer Science

Computational science or, equivalently, scientific computing or scientific computation comprises a mathematical modelling of a science/engineering problem, analyzing the model implemented on a computer, solving it using the computer (as opposed to using an analytic method of applied mathematics), and applying quantitative analysis techniques.

In practice, computational science is typically the application of a computer simulation and other forms of computation from problems in numerical analysis and theoretical computer science to problems in scientific and engineering areas.

On the other hand, **Computational mathematics**, involves mathematical research in science and engineering where computing plays the pivotal role, stressing on a **numerical algorithm** (in which most of the computational work, say, over 90%, consists of arithmetic operations, like finding a root of a non-linear equation using the Newton method), a **semi-numerical algorithm** (consisting of roughly 50% numerical and 50% non-numerical operations, for instance, testing whether a given integer is prime or not), and/or a **non-numerical algorithm** (comprising mostly non-arithmetic operations, such as searching/sorting).

Computational mathematics emerged as a distinct field of applied mathematics in the mid-1950s. Currently, it includes subjects such as numerical linear algebra, numerical solutions of ordinary and partial differential equations, discrete mathematics (e.g. in its search for mathematical structures like groups), number theory (primarily testing and factorization), cryptography, and computational algebraic topology. Computation (using a computer) is vital in research and even in almost all our day-to-day activities.

Computer science comprises theory and laboratory experiments – the traditional forms of science and engineering. **Computational science attempts to gain insight mainly through a mathematical model implemented on a computer**. Although computational science and computer science are intimately related and even overlap (e.g. both involve the study of fundamental computational algorithms), these are remarkably different.

Computational science/mathematics may also be viewed as the interface between a realistic mathematical model and a scientific computing method. A scientist/an engineer, whose primary interest is the efficient and cost-effective nature of the solution (not in the subject "computational science") that it provides, develops a computer program (application software) that model a system under investigation and execute the program with different sets of input data (parameters). In most real-real-world problems, these models are very compute-intensive and are executed on a hyper-speed computing or a distributed computing platform or on a best available digital computer in a given environment.

3.4 Pre-mid-17th Century Computing Devices

Computing dates back to the prehistoric era when a human had the sense of comparing any 2 quantities to say the one which is larger. **The human hands/fingers/finger-knots counting are the original "calculating machines"**. These have been the tools most widely used at all times as an aid to counting/calculation. But by its very nature, **these tools cannot serve as recording devices**.

The earliest pre-modern tools for computing comprise hands, fingers/finger-knots), sand and pebble, chalk and slate, charcoal ink and palm leaves, pen and paper with or without the help of a table.

Abacus, an oldest known computing tool, developed possibly in Babylon over 4400 years ago (in 2400 BC) originally consisted of lines drawn in sand with pebbles. Horace (65–8 BC) refers to the **wax abacus**, a board covered with a thin layer of black wax on which columns and figures were inscribed using a stylus (Agarwal and Sen 2014).

The **Chinese Abacus** (**suanpan** meaning counting tray), typically 20 cm tall and with various widths, has over 7 rods. There are over 2 beads on each rod in the upper deck and 5 beads each in the bottom. It was made of parallel grooves or rods, with buttons or pierced pellets that could be slid along these, that was a practical and formidable instrument still may be seen in use in the Far East.

Earlier the Chinese used little ivory or **bamboo sticks, called chou** (literally, "calculating sticks") that they arranged on the squares of a tiled floor or on a table made like a chessboard. The **abacists** still use these abaci but of a more modern design.

The Chinese also employed **counting boards** similar in design to abaci as mechanical aids for calculation as early as the 4th century BC. These were usually made of wood or metal, and contained carved grooves in which mounted beads, pebbles, or metal discs were moved.

The **Greeks** developed several mechanical devices for solving their computational problems. A most famous one was the **Antikythera**, a machine to compute astronomical positions, was constructed around 87 BC and lost a few years later. It has the oldest known complex **gear mechanism** and is sometimes referred to as the **first known analog computer (mechanical)** (Agarwal and Sen 2014).

3.5 Information Storage

A must An unavoidable requirement in computing is **to store information for future use or further computation**. In this context, the **Incas success** can be explained due to their ingenious method of preserving accurate records using a detailed and complex system of **knotted string** known as the device **quipu** (Inca word meaning "knot"), developed during 3rd millennium BC.

It consisted of a main piece of cord about 2 ft long onto which thinner, coloured strings were knotted in groups, these pendant strings themselves being knotted in various ways at regular intervals. Quipus are so simple and so valuable that they continued to remain in use for several centuries in Peru, Bolivia, and Ecuador.

They served not only as a device for concrete numeration, but also as a **mnemotechnic tool** for jobs such as recoding numbers, maintaining administrative archives, keeping count of contracts, and calendars. For our purposes, these numbers on strings may be considered as a special form of written numbers (Agarwal and Sen 2014).

3.6 Mid-17th-Mid-20th Century Computing Devices

The pioneers viz. the early computer scientists/technologists, contributed to both computer science h/w and/or s/w and computational science to varying degrees. **William Oughtred** (1574–1660), an English mathematician and Anglican clergyman, after Napier's (**John Napier of Merchiston** (1550–1617), a Scottish landowner known as a mathematician, physicist, and astronomer) invention of logarithm and Gunter's (**Edmund Gunter** (1581–1626), an English clergyman, mathematician, geometer and astronomer of Welsh descent) creation of the logarithmic scales, invented two such scales in 1622 that were slid by one another to perform multiplication and division.

Pascal (**Blaise Pascal** (1623–1662), a French mathematician, writer, physicist, inventor, and Catholic theologian, was a child prodigy.) invented the mechanical calculator in 1642. He was prompted to this invention when he saw the tedious arithmetical labour involved in his father's official work. His device was first called the Arithmetic Machine, then Pascal's Calculator, and later **Pascaline**. It could add and subtract two numbers directly, and it could multiply and divide by repetition.

Pascal went through 50 prototypes before presenting his 1st machine to the public in 1645. 9 machines have survived that are now on display in European museums.

In 1671 Leibniz (**Gottfried Wilhelm (von) Leibniz** (1646–1716), a German polymath and philosopher) embarked on inventing a machine that could add, subtract, multiply, and divide, which he gradually improved over the next 2 years. It was not considered a complete success because it did not fully mechanize/automate the operation of carrying. In 1674, he anticipated a machine capable of performing some algebraic operations. In 1679, while pondering over his binary arithmetic, Leibniz imagined a machine in which binary numbers were represented by marbles, governed by some sort of punched cards.

Modern electronic digital computers have replaced Leibniz's gravity-driven marbles with shift registers, voltage gradients, and electron pulses, but otherwise they run roughly as Leibniz had visualized. He also wondered if a machine could be built that could determine whether particular mathematical statements are true or false.

Later, Babbage (**Charles Babbage** (1791–1871), an English mathematician, philosopher, inventor, and mechanical engineer) did pioneering work on the **Difference Engine**, a sophisticated calculator that could be used to generate mathematical tables. In 1834, he also designed the **Analytic Engine**, which was the world's 1st mechanical computer. However, the machine was not built during Babbage's lifetime.

Pehr Georg Scheutz (1785–1873), a Swedish lawyer, invented in 1837 and formally completed in 1843 his **Scheutzian calculation engine**. This machine, which he designed and developed with his son Edvard Scheutz (1821–1881), was based on Babbage's Difference Engine.

An improved model, approximately the size of a piano, was created in 1853 and demonstrated in 1855 in Paris at the World Fair. The machine was then sold to the British government in 1859. Scheutz invented yet another machine in 1860, selling it to the USA. These devices were used for creating logarithmic tables.

While Scheutz's machine was not perfect and could not generate complete tables, the Swedish scientist Martin Wiberg (1826–1905) reworked the construction from the ground up and in 1875 developed a compact device, the size of a sewing machine, that could print complete logarithmic tables. The tables that it produced were published in English, French, and German (Agarwal and Sen 2014).

During 1935–1938, **Konrad Zuse** (1910–1995) built the world's first **program-controlled computer**, called **Z1**. Despite certain mechanical engineering issues it had all the basic ingredients of modern computers, using the binary system and today's standard separation of storage and control.

Zuse's 1936 patent application (Z23139/GMD Nr. 005/021) also suggests a von Neumann architecture (which is the fundamental architecture used in modern computers, which von Neumann (**John von Neumann** (1903–1957), a Hungarian-American mathematician, a nuclear physicist, and a computer scientist) recreated in 1945) with program and data modifiable in storage.

In 1936, *Alan* Mathison *Turing* (1912–1954) published a key paper on computability in which he introduced the theoretical/hypothetical **Turing machine**, a mathematical machine that is capable of solving any conceivable mathematical problem that has an algorithm.

The Church-Turing thesis states that anything that may be computed may be computed by a Turing machine. Turing also designed the **Automatic Computing Engine** (ACE), which was the first complete design of a stored–program computer in Britain.

The American physicist John Vincent Atanaso (1903–1995) invented the first electronic digital computer, the **Atanaso-Berry computer**, in 1939 at Iowa State College. This special-purpose computer was neither programmable nor Turing–complete. Challenges to his claim were resolved in 1973 (after 34 years) when the

"Honeywell v. Sperry Rand" lawsuit ruled that Atanaso was indeed the inventor of the computer.

In 1941, Zuse built world's first fully functional programmable computer, **Z3**. Z3 predates **the IBM's Harvard Mark I** (built by IBM between 1939 and 1944) and the Manchester Mark I. The Harvard Mark 1 computer (also known as the IBM Automatic Sequence Controlled Calculator (ASCC)), called simply the Mark I by Harvard University, was an **electro-mechanical machine** and the brainchild of **Howard Hathaway Aiken** (1900–1973).

Aiken conceived the idea of a general purpose machine for scientific computation while working on his doctorate at Harvard. Mark I. It weighed over 4500 kg (4.5 tonne), contained 500 miles of wiring, and accepted operations/instructions that were punched into a roll of paper tape.

Its multiplication operation consisted of the addition of partial products, just like multiplication using Napiers bones. The machine was kept running 24 h a day, and could handle in a day calculations that would have taken a contemporary computer six months.

The **ENIAC** (Electronic Numerical Integrator and Computer), the **1st general-purpose Turing–complete digital electronic computer**, was built at the **Moore School of Electrical Engineering** of the University of Pennsylvania in Philadelphia during 1943–1945. Since it used vacuum tubes (electronic valves), it is known to be a 1st generation machine. It was first put to work in the summer of 1945, and the Moore School is now known as the **birth-place of the computer industry**.

ENIAC was formally dedicated at the University of Pennsylvania on 15th February, 1946 and **was heralded as a "Giant Brain" by the press**. It worked over 1000 times faster than electro-mechanical machines such as the IBM **Automatic Sequence Controlled Calculator** (**ASCC**), also known as **Harvard Mark I** (1944) discussed above.

This computational power, coupled with general-purpose programmability, excited scientists and industrialists alike. The combination of speed and programmability allowed for thousands more calculations for problems, as **ENIAC calculated a trajectory in 30 s that took a human 20 h** (1 ENIAC hour equivalent to 24 hundred human hours).

The logical design of the system was based on the ideas of **John William Mauchly** (1907–1980), an American physicist, while the h/w design and implementation was due to **John Adam Presper Eckert Jr**. (1919–1995), an American electrical engineer.

ENIAC's construction was a great act of technological courage on the part of the designers as well as the Moore School. ENIAC contained 17.468 thousand (electronic) vacuum tubes, 7200 crystal diodes, 1.5 thousand relays, 70 thousand resistors, 10 thousand capacitors, and approximately 5 million hand-soldered joints.

It weighed over 27 thousand kg (27 tonne), was around 2.4 m × 0.9 m × 30 m, took up 167 m^2, and consumed 150 kW of power. It cost almost $500,000 in 1945 (approximately $ 8 million today in 2019). An addition operation took 0.2 ms (millisecond) while a multiplication took 2.8 ms.

The computer carried out calculations over 1 thousand times faster than its nearest rival, the Mark I as stated above; π was computed to 2035 decimal places in 70 h. **Its 17, 468 vacuum tubes were a serious drawback, requiring too much power and producing too much heat**. (A legend is that every time the ENIAC was turned on, lights dimmed in all the houses in Western Philadelphia.)

The **EDVAC** (Electronic Discrete Variable Automatic Computer) was a direct successor of ENIAC. It was a binary serial computer with automatic addition, subtraction, and multiplication, as well as programmed division, and automatic checking with an ultrasonic serial memory of 1 thousand 44-bit words (later set to a 1024-word re-circulating mercury-delay line memory).

It had 4 thousand tubes and10 thousand crystal diodes. The arithmetic unit did both fixed- and floating-point operations. EDVAC's average addition time was 864 μs (microsecond) compared to that 0.864 ms in ENIAC and its average multiplication time was 2900 μs as against 2.9 ms in ENIAC. Input and output were done via punched paper tape and punched IBM cards. It was delivered to the **Ballistic Research Laboratories in Aberdeen, Maryland i**n 1949 and underwent extensive tests. EDVAC became fully functional in 1951 and was used until 1962.

EDVAC used stored programs, and was also designed by Mauchly and Eckert. **UNIVAC I** (UNIVersal Automatic Computer I) was the first commercial computer. It was designed by the same two men who designed **ENIAC** and **EDVAC**.

During 1847–1854, Boole (**George Boole** (1815–1864), an English mathematician, philosopher, and logician,) formalized Boolean algebra, the foundation for digital logic and computer science. Boole and Babbage are among the **fathers of computing**. **Shannon** was the first person to see the **applicability of Boolean algebra to the design of circuits** and telephone routing switches. Shannon's later work at Bell Labs laid the foundation of modern information theory.

EDSAC 1 (Electronic Delay Storage Automatic Calculator), a serial binary computer and the second usefully operational electronic digital stored-program computer, was a British machine. The computer was inspired by von Neumann's seminal First Draft of a Report on EDVAC of June 30, 1945 and was constructed by **Sir Maurice Vincent Wilkes**, (1913–2010) and his team at the **Cambridge University Mathematical Laboratory** in England.

It became operational in 1949 when it computed a table of squares and a list of prime numbers. The project was later supported by the British firm J. Lyons & Co, who were rewarded with the first commercially applied computer, **LEO I**, based on the EDSAC design. EDSAC 1 was replaced by **EDSAC 2** in 1958, which stayed in use until 1965.

Many computer people left the Moore School around 1946. Eckert and Mauchly resigned and launched the first computer company, the **Eckert-Mauchly Computer Corporation** (later the name was applied to a division of **Remington-Rand Company** and successor organizations). Their product **UNIVAC** (Universal Automatic Computer), built in 1951, was the 1st general purpose computer for commercial use.

This was preceded by the computer **BINAC** in 1949. UNIVAC needed only 5 thousand vacuum tubes and relied on magnetic tape data storage instead of punched cards. The machine was initially installed at the **United States Census Bureau** in

1951, where it was used continuously over the next 12 years before its retirement to the Smithsonian Institution.

UNIVAC received considerable publicity when the CBS network used it to provide early election-night predictions of the 1952 presidential race. Although the network's political pundits were sceptical of the machine's forecast, **Dwight David Eisenhower** (1890–1969) was elected President by almost exactly the same landslide voting margin that was predicted by UNIVAC.

One of the system engineers contributing to the development of the UNIVAC was **Hopper** (**Grace Brewster Murray Hopper** (1906–1992), an American computer scientist and United States Navy rear admiral). Several versions of UNIVAC were developed that were widely used during the 1950s and 1960s (Agarwal and Sen 2014).

3.7 *Mid-20th Century-2019 Computing*

Post-mid-20th century computing systems, may be termed as the next era of modern computing systems based on the use of transistors instead of vacuum tubes since late 1950s as well as VLSI (Very Large Scale Integration) chips. These include both h/w and s/w devices invented around 1950 and regularly developed as increasingly better systems to tackle progressively larger computational problems. We present these devices designed and developed during post mid 20th century-present (2019).

There had been a surge of computing activity all over the world since 1950s. The activity has been growing exponentially not only in computation in practically all areas of science, engineering, and linguistics but also in the development of h/w and s/w areas of digital computer. Computing systems and computation (using the systems) both are highly linked in terms of their mutual growth, rather exponential growth.

The 1st generation (vacuum tube based) machine **TIFRAC** (Tata Institute of Fundamental Research Automatic Calculator) was developed in India at the Tata Institute of Fundamental Research (TIFR) in Mumbai in the 1950s that became functional in 1955 and was commissioned in November 1956. The enhanced TIFRAC computer became operational in 1957 and was commissioned in February of 1960. It performed computation for scientific research between the early 1960s and 1965.

India's first 2nd generation (transistor-based) computer, **ISIJU-1**, was designed and developed jointly by the Indian Statistical Institute, Kolkata and Jadavpur University, Kolkata during the early 1960s. ISIJU–1 became functional in 1965 and remained in use until 1975.

- **Ternary computers** All the computers so far (till 2019) designed and developed use 2-valued (binary) logic. However, in 1958, under the leadership of **Sergei Lvovich Sobolev** (1908–1989) and **Nikolay Brusentsov** (1925–2014), the world's first balanced ternary (using 3-valued (ternary) logic instead of 2-valued logic) computer, **Setun**, was designed and built at **Moscow State University**.

Base-3 or more computers are not as stable as base 2 (binary) computers because of required stable and fast errorless switching problem (existing in silicon technology in the physical world). In 1970, a new ternary computer, the **Setun–70**, was designed.

The ternary computer (silicon) technology could not become popular because of the foregoing problem and hence computer scientists did not devote their mental energy in this direction viz. developing Base-n computers, where n > 2 is a finite integer, the physical storage space required for n increasingly greater than 2 will be increasingly smaller though.

In some non-silicon technology, it could perhaps be possible in future to develop machines working with higher radix than the conventional radix 2 if the technology allows comparatively fast stable switching. As of now, we are still in dark about existence of such a technology, nor can we emphatically say/demonstrate the infeasibility of this technology.

- **Current computer hardware** Current computers use tiny electronic components which can be etched onto the surface of a silicon chip.

The most common electronic component is the "transistor" which works as a sort of amplifying valve for a flow of electrons. The transistor is a "solid state" device, meaning it has no moving parts. It is a basic building block used to construct more complex electronic components.

Specifically, a "bit" can be built with an arrangement of 5 transistors. The transistor was invented in the early 1950s, replacing the vacuum tube. Since then, transistors have been made increasingly smaller, allowing increasingly more of them to be etched onto a silicon chip.

- **Physical size of a component becomes increasingly smaller** Every (about) 24 months a transistor becomes smaller by 50%. Thus twice as many transistors can fit in a chip due to increasingly improved chip etching technology. In other words, keeping the capacity constant, a chip gets smaller and cheaper every 24 months. Very large scale integration (VLSI) is dynamically increasing fast non-stop. This is why cars, aeroplanes, washing machines, thermostats, (mobile) smart phones and numerous other devices have computers in them.
- **Minimal computer system** A minimal modern hardware computer system must have implicitly or explicitly 1 central processing unit (CPU) P, 1 executable (main) random access memory (RAM) M, 1 input/output (IO) processor C, also called channel, 1 IO device controller K, 1 input device α_1, say a keyboard and 1 output device α_2, say a printer.

M is connected to P and C by data and control buses (paths). C is also connected to P directly by a control bus. K is connected to C, α_1, and α_2 by data and control buses directly. If any one of these 6 units is absent, then the machine cannot be termed as a modern digital computer.

- **A general computer system** A generalization of the minimal machine is straightforward. The general system will have p central processors $P_1, \ldots, P_p$; c IO processors $C_1, \ldots, C_c$; k IO device controller $K_1, \ldots, K_k$; and m main memories

$M_1, \ldots, M_m$. Associated (connected) with each of the k IO processors are input and output devices α_i' s. These devices are connected by data and control buses (paths) and control buses. For further details including a pictorial diagram refer (Lakshmikantham and Sen 2005).

- **Main-frame/super-machines versus laptop/desktop with User pattern** The widely used main frame (many users using a single multi-processing systems) machines (there were thousands of them globally) all over the globe have been mostly replaced over the years by computers (laptop/desktop computers in millions for mainly individual usages) based on Intel/other processors.

The multi-user main frame computers had been increasing in number globally until 1980s. The increase then halted and gradually the main frame systems had started shrinking in number. Their popularity waned giving way to that of laptop/desktop computers. Since 1990s a laptop/desktop system started becoming almost every common man's household computational (including almost always with internet facility) utility.

However, while once (during 1960s–1980s) the main frame systems that ruled the world, these did not become extinct now (2019). These are housed in a few isolated locations in the world and are mainly used for vital activities (mainly government) requiring enormous computing power of the order of thousands of trillions of operations (arithmetic and non-arithmetic) per second and beyond. As a matter of fact, **our computing requirement for speed, storage, and communication knows no limit**. **The more we get the more we want**. This human psychology is eternal.

As an example of the most recent (June, 2018) main frame system. we present briefly below the following super-computer (Deccan Herald Daily News Paper 2018). The American scientists have unveiled the world's most powerful scientific supercomputer called **Summit for use** at US Department of Energy's Oak Ridge National Laboratory (ORNL). **Summit** or **OLCF-4** is developed by IBM. **As of June 8, 2018 it is the fastest supercomputer in the world**.

The Summit can execute more than 2×10^{17} operations (i.e. 200,000 trillion operations = 200 petaflops) (arithmetic/non-arithmetic) per second—providing exascale computing power for research in energy, advanced materials, and artificial intelligence (AI). Summit will be 8 times more powerful than its previous top-ranked system, **Titan**. For certain scientific applications, Summit is also capable of over 3.3×10^{18} i.e. 3.3 billion billion mixed precision calculations per second or 3.3 exaflops. It is around **a million times faster than a high-end laptop**.

"Summit will empower scientists to address a wide range of new challenges, accelerate discovery, spur innovation and above all, benefit the American people," said Rick Perry (born 1950), US Secretary of Energy.

The architecture of Summit has the **IBM AC922** server system that consists of 4608 compute servers, each containing two 22-core IBM Power9 processors and six NVIDIA Tesla V100 graphics processing unit accelerators, interconnected with dual-rail Mellanox EDR 100 GB per second InfiniBand. Summit also possesses more than 10 PB (petabytes) of memory paired with fast, high-bandwidth pathways for efficient data movement.

The combination of cutting-edge hardware and robust data subsystems marks an evolution of the hybrid CPU-GPU architecture successfully pioneered by the 27-petaflops Titan in 2012.

GPU (Graphics Processing Unit) is an electronic circuit specifically designed to rapidly manipulate and alter memory to accelerate the image creation in a frame store (A frame store, also called a frame buffer) is a portion of RAM containing a bitmap that drives a video display. It is a memory buffer containing a complete frame of data. Modern video cards contain **frame-buffer** circuitry in their cores.) intended for output to a display device. GPUs are used in machines/tools such as personal computers, mobile phones, embedded systems, workstations, and game-consoles.

ORNL scientists have figured out how to harness the power and intelligence of Summit's state-of-the-art architecture to successfully run the world's first exascale scientific calculation. Scientists led by ORNL's Daniel A. Jacobson, Computational Biologist and Wayne Joubert, Computational Scientist have leveraged the intelligence of the machine to run a 1.88 exaops comparative genomics calculation relevant to research in bioenergy and human health.

The mixed precision exaops calculation produced identical results to more time-consuming 64-bit calculations previously run on Titan. In addition to scientific modeling and simulation, Summit offers unparalleled opportunities for the integration of AI and scientific discovery, enabling researchers to apply techniques like machine learning and deep learning to problems in human health, high-energy physics, materials discovery and other areas. **There are several other high-end supercomputers developed in different countries**.

Supercomputers use their super-computational speeds (flops) and super-efficiency to assist many sectors such as weather forecasting, missile simulation, innovating life-saving pharmaceutical drugs, accelerating energy production (both conventional and non-conventional), scientific research (e.g. hypersonic aircraft design), and understanding the origin of the universe in India.

A **cluster supercomputer** is developed by stacking computer processors in a giant box through parallel processing. Each individual computer in the cluster is termed as a node.

A **grid supercomputer**, on the other hand, has the processors stacked in a way similar to that in a cluster system with the difference that the nodes can be located geographically at different places around the world and are connected through wide area network (WAN) or the internet.

India has 11 of the top 500 supercomputers around the globe. Some of the top supercomputers that serves India are as follows.

SahasraT (Cray XC40) supercomputer, a product from OEM Cray, is located in the Supercomputer Education and Research Centre (SERC), Indian Institute of Science, Bangalore.

This supercomputer comprises Intel Haswell Xeon E5-2680v3 processors, NVIDIA K40 GPU accelerators, Intel Xeon Phi 5120D coprocessors, and a gigantic storage of 2.1 PB i.e. 2.1×10^{15} bytes. There are around 1500 processors and coprocessors and 44 GPUs for complex tasks in the system.

The system implements Cray Aries interconnect routing using Dragonfly topology. The operating system works in Cray's Linux environment. The speed of computation in SahasraT is around 901.54 teraflops rated the highest among all supercomputers in India.

SahasraT is used in the areas such as aerospace engineering, meteorology predictions, molecular and materials research, and mapping climate condition of a specified region through simulation. It was used to simulate overlapping of supernovae.

Aaditya Indian Institute of Tropical Meteorology, Pune uses the Aaditya, an IBM X system supercomputer for research and development. This system consists of Intel Haswell E5-2670 2.6 GHz processors and a total of 15 Terabyte random access memory storage. The system has 2384 compute nodes, connected through infiniband interconnect technology, with a speed of 719.2 teraflops. The operating system environment used is Red Hat Enterprise Linux.

Aaditya assists the institute in operating and providing information regarding India's weather conditions, simulating weather models, forecasting monsoon rainfall cycles, air quality, and climate changes. Most of India's weather forecasting is carried out by Aaditya. Farmers depend on the information generated by it regularly.

TIFR Colour Boson The supercomputer TIFR Colour Boson, a Cray XC-30 2014 model located at the old Tata Institute of Fundamental Research (TIFR) campus in Hyderabad was installed in March, 2015. The Cray XC30 is a massively parallel supercomputer based on Intel Xeon processors and Nvidia Kepler GPUs connected by the Cray's Aries interconnect routing technology with Dragonfly topology.

The Colour Boson comprises 4760 nodes of Intel Xeon E5-2680 processors with NVIDIA Tesla K20x GPU. It has total 1.1 PB storage. The operating system for the system works in a Linus environment and can attain 558.7 teraflops speed.

The researchers of TIFR use the Colour Boson under Indian Lattice Gauge Theory Initiative program. They carry out extensive computational research in theoretical physics and quantum chromo-dynamics QCD (QCD involves the theory of study of elementary particles such as quarks, leptons, and gluons) on this supercomputer. The focus is on the research of quark-boson—a phase of matter—that, many scientists think/believe, holds the **mystery of the creation/origin of the universe**.

IIT-D HPC High Performance Computing (HPC) at the Indian Institute of Technology, Delhi (IIT-D) has a supercomputer based on GPU Tesla platform. The system consists of HP ProLiant XL230a server and HP ProLiant XL250a Gen9 server with 322 NVIDIA K40 M GPU cards. Total storage of the system is 1.5 PB. Infiniband interconnect technology is used for network connections. The performance is at 524.4 teraflops. Preferred programming platforms are CUDA and OpenACC.

The system is used for research in atmospheric science, bioinformatics, biology, and nano systems. In addition, it provides services for all IIT-D departments for computational study and research in various fields such as deep learning, computational physics, data analytics, materials science, and computational fluid dynamics.

C-DAC Param Yuva 2 Param Yuva 2 is the supercomputer housed at the Centre for Development of Advanced Computing (C-DAC) in Pune. It has 221 Intel Xeon E5-2670 nodes which also consist of Intel Xeon Phi 51 10P supported by Linux 64-bit (CentOS 6.2) and a total storage of 200 terabyte. It has the computational speed of 388.44 teraflops. It has totally 3536 CPU cores and 26520 coprocessor cores, The system is interconnected using Mellanox FDR Infiniband primary interconnect (for interconnection of nodes).

The system can be interconnected with Indian Institute of Technology and National Institute of Technology through National Knowledge Network—a multi-grid backbone network for high-speed data transfer among several educational institutions in India.

It is used in areas such as seismic data analysis, aerospace engineering, bioinformatics, weather forecasting, and pharmaceutical development.

Home-developed Super-machines The Param series of supercomputers are designed and assembled in India by C-DAC, Pune though several supercomputers in India have been purchased from abroad from manufacturers such as Cray Inc. from USA. C-DAC was established in 1988 when the United States denied the import of a supercomputer to India.

C-DAC successfully built the India's 1st supercomputer **PARAM 8000 in 1991**. Param Yuva 2, the latest C-DAC super-machine, was built in 3 months at the cost of only 16 crore Indian Rupees (i.e. around US$ 2.7 million).

The network, **PARAMnet,** allows all PARAM supermachines in India to be connected via 2.5 GB/s bandwidth in full-duplex mode.

- **What is in store for us** India, a largest country in terms of population in the world, has only 11 super-machines featured in the top 500 supercomputers in the world in 2018. Evidently, we, the Indians, need to do better. Consequently the Indian government initiated the project called **National Supercomputing Mission** with a target of 70 supercomputing facilities spread across India. These super-machines will be connected by the **National Knowledge Network**.

This project has several diverse goals such as (i) to accelerate scientific research and development specifically in the areas concerning cutting edge technologies, (ii) to boost the quality of higher education in science and engineering, (iii) to forecast more accurately the weather and climate changes, with timely warning for natural disasters, (iv) to design rockets/missiles, and (v) to discover life-saving pharmaceutical drugs,

Due to a gigantic youthful (perhaps the largest in the world) manpower availability in India, the foregoing ambitious goals will most likely be achieved with our satisfaction within the specified time-frame.

- **Supercomputing activities in other countries** Besides the USA and India as stated earlier, there are several countries around the globe, which use/build supercomputers for their respective countries' need and/or for the use of other countries. The scope of a supercomputing device is primarily for solving highly compute- and storage-intensive problems—not tractable in a laptop/desktop—needing maybe over a million hours of computational time on a high-end laptop/desktop.

The more the computing power of a system is, the more exploration capability it provides in an affordable time frame. Our need for exploration capability is never limited. So we continue to grow in all scientific and technological areas (with computation as the backbone of every area) without any stop in any foreseeable future.

- ***China*** A Chinese supercomputer is among the world's fastest super-machines built using indigenous chips (not US chips). The **Sunway TaihuLight** consisting of around 41000 chips each with 260 processor cores (i.e. 10.66 million cores) takes the top spot replacing the earlier record-holder Tiahne-2 (also belonging to China). The computational speed of TaihuLight viz. 93 petaflops (1 petaflops = 10^{15} floating-point operations) is over 3 times that of Tianhe-2. It uses Sunway Raise OS 2.0.5 operating system

As of March 2018, TaihuLight costing around US$ 273 million occupied the top spot among the top 500 super-machines in the world. As of June 2018, it is ranked second in the TOP500 list with a LINPACK benchmark rating of 93 petaflops. It uses Sunway Raise OS 2.0.5 operating system (based on Linux).

- ***France*** **Tera 100** of France was the fastest supercomputer in Europe in June 2011. It was the 1st petascale super-machine at that time.
- ***Switzerland*** An upgraded version of **Piz Daint** became the fastest supercomputer in Europe in October 2016 with a peak speed of over 25 petaflops.
- ***HPC Europa Project*** Distributed access to several supercomputers located at different centers across Europe is coordinated by **European initiatives** to facilitate supercomputing. The **HPC Europa** project, a one such initiative established in 2002 as a consortium of 11 supercomputer centers from 7 European countries operating with CORDIS framework, fits within the **Distributed European Infrastructure for Supercomputing Applications** (DEISA). The aim of the project is to provide access to supercomputers across Europe.
- ***High-performance Computing in other continents*** Several countries in other continents such as Australia, Africa, and Latin America have their share of supercomputing. We omit presenting them to conserve space. The reader may refer the concerned literature for further details.

However, the march for increasingly higher processing speed for both computation and communication along with increasingly higher storage space is going on in many locations around the globe. These computing activities are having enormous impact in all areas of science and engineering resulting in amazing discoveries for humanity besides progressively more digital convenience in our day-to-day interactions. The progress is neither linear nor (finite-degree) polynomial non-linear. This is exponential i.e. ∞-degree polynomial non-linear.

In 2018 we have mostly a super-computer (such as a high end/super-laptop/desktop) having a processor viz. (i) Intel Core i7-8700K processor (CPU) with boost clock 4.7 GHz and 12 MB L3 cache or (ii) AMD Ryzen Threadripper 1950X with boost clock 4.0 GHz and 32 MB L3 cache or some other processor.

For further details, refer the concerned literature. The number of such laptops/desktops are currently (2019) tens of millions in the world and is increasing. Usually each laptop/desktop is used exclusively by a single user.

- **Past computer programming languages** Numerous computer programming (formal) languages (totally devoid of ambiguity unlike natural languages) were created from the early days of computing based on the requirements of the computational problems at hand vis-à-vis the then available (hardware) computers. This is because writing a machine language (lowest level computer-specific programming language) program including debugging is not only too tedious, but also too much time consuming. Even writing a program in the assembly language (mnemonics) akin to the concerned machine language is also tedious (slightly less though).

The high-level programming language **FORTRAN** was developed in the mid-1950s. This was followed by many versions of FORTRAN, including **FORTRAN II** (1958), **FORTRAN IV, FORTRAN 66, FORTRAN 77,** and **FORTRAN 90**.

Besides FORTRAN, which (in its highly improved version keeping in view the then continuously updated computer hardware availability) is still in use, many other programming languages such as **ALGOL** (developed in the late 1950s and 1960s) and its many variants such as **ALGOL 60** and **ALGOL 68, COBOL** (designed in the late 1950s), and **PASCAL** (developed in the early 1970s) were widely used when they were developed but now (2019) we see practically no use of these languages.

Other programming languages that have been developed after the 1960s include **BASIC** (designed in 1963 as a teaching tool), **visual BASIC**, the **C** language (introduced in the early 1970s and has become popular in industry), and **C++** (an extension of C implemented in 1985).

Besides **Miranda** (mid-1980s) and **Java** (1996) programming languages, various functional programming languages such as **HOPE** (1970s) and **REXX** (1979), and logic programming languages have been designed and developed.

Many of the earlier programming languages have become practically **history** in computing, **although many of their structures can be found in modern programming languages in some form or the other**.

- **Currently used programming languages** Currently (2019) for general scientific and engineering computations, a most widely used high-level programming language in the scientific world is **Matlab**. **Mathematica, Mathcad, Python, and Maple** are other high-level programming languages which are also in use.

(i) **Matlab**—a high-level programming language—with **Simulink**, **Variable precision arithmetic** (vpa), and **fixed-point tool box** is a most used user-friendly software developed and updated periodically by its maker **The MathWorks Ltd**. It is not, in general, a free software.

While MATLAB and **Simulink** are well-suited for prototyping **signal processing and control systems**, the **fixed point toolbox** is useful if one needs to develop **real time embedded algorithms**.

(ii) **Mathematica**—a widely appreciated programming language created by **Stephen Wolfram**: Founder & CEO of *Wolfram* Research (creator of Mathematica, *Wolfram* Alpha & the *Wolfram* Languages) for both technical computation and elegant ease of use—provides a single integrated, continually expanding system that covers the breadth and depth of technical computing.

With Mathematica Online, it is now seamlessly available in the **cloud** through any web browser, as well as natively on all modern desktop systems.

(iii) **Mathcad**, PTC's (Parametric Technology Corporation's) engineering Math Software allows one to perform, analyze, and share one's most vital calculations. It is used by engineers and scientists in various disciplines—most often those of mechanical, chemical, electrical, and civil engineering.

Originally conceived and written by **Allen Razdow** (of MIT, co-founder of Mathsoft), Mathcad is currently (2019) owned by PTC and is widely accepted as the 1st computer application to automatically compute and check consistency of engineering units such as the International System of Units (SI), throughout the entire set of calculations.

Mathcad today includes some of the capabilities of a computer algebra system, but remains oriented toward ease of use and simultaneous documentation of numerical engineering applications.

(iv) **Python** is an interpreted high-level programming language created by **Guido van Rossum** for general-purpose programming. First released in 1991, Python has a design philosophy that emphasizes code readability, and a syntax that allows programmers to express concepts in fewer lines of code, notably using significant whitespace (Python is famous among programming languages for its fairly unique syntax: rather than being delimited by curly braces or "begin/end" keywords, blocks are delimited by indentation. Indenting a line is like adding an opening curly brace, and de-denting is like a closing curly brace).

It provides constructs that enable clear programming on both small and large scales. Python also features a dynamic type system and automatic memory management. It supports multiple programming paradigms—including object-oriented, imperative, functional, and procedural—and has a large and comprehensive standard library.

(v) **Maple** is a symbolic and numeric computing environment, and is also a multi-paradigm programming language. Developed by **Maplesoft**, Maple also covers other aspects of technical computing, including visualization, data analysis, matrix computation, and connectivity. A toolbox, **MapleSim**, adds functionality for multi-domain physical modeling and code generation.

- **Special-purpose programming lang*uages*** There are numerous special-purpose programming languages. A special-purpose programming language is a **programming language** designed to solve a specific class of problems. **LISP** and **Prolog**, for instance, are designed for and used extensively in **AI** (Artificial Intelligence) applications. Even more specific are languages such as **COGO**, for **civil engineering problems**, and **APT** for directing machine tools.

Still many more specific special-purpose programming languages—physically much more concise than the foregoing languages—for performing very limited activities are used in various utility devices not only for our day-to-day usage but also for occasional usage.

These languages are usually hard-wired and remain hidden from the users. The users are primarily concerned in issuing a command of a language to obtain the desired output/result of the execution out of the concerned device.

For further details in the development of computer hardware and software, reader may refer (Agarwal and Sen 2014) as well as other relevant literature.

- **Writing a program for a virtual computer and not for the real machine** A user (scientist/engineer/technologist) writes a high-level program (an ordered set of instructions to a computer) in general. He does not write a program for the real hardware machine which knows only its own machine language—the lowest level language—and nothing else.

He writes a program for a virtual machine such as a **Matlab machine** and a **Mathematica machine**. A Matlab machine, for example, is a virtual machine which is the machine realized when the Matlab programming language with the requisite software including the operating system is superimposed on the hardware (physical) machine. The Matlab machine understands the Matlab commands and executes them.

The user need not know/learn the **machine language, the writing of which is a wearisome, and highly time consuming task with tedious debugging effort**. A machine language program could be usually at least 7 times as large as a Matlab (or a Mathematica) program in physical size.

The numerical computations are done **always with finite precision** (finite word-length) and never infinite precision. The precision used in a Matlab machine is about 15 digits (good enough for most real-world problems). All the programming languages—general-purpose as well as special-purpose—use finite precisions only and would continue to do so for ever.

- **Analog computing** During 1930–2000, the world experienced extensive and rapid refinements in (electronic) machine computation. **Vannevar Bush** (1890–1974), an American electrical engineer, inventor, and science administrator, was an **analog computing** pioneer. He introduced the **Memex** concept in 1930, which led to the development of **Hypertext**.

An **analog computer** is based on realizing arithmetic operations by reading/measuring physical quantities such as current, voltage, and resistance in a carefully designed electrical circuit. Unlike a digital computer, analog computers are very limited in their accuracy. Analog computers might give at most 0.01% accuracy and have very limited use as general-purpose computers.

- **Miniaturization of electronic circuitry** By the late 1950s, vacuum tubes were superseded by transistors, which generated little heat and had long service lives. The next decade saw the **miniaturization of electronic circuitry**; soon a million transistors were replaced by a single silicon chip.

Computers accordingly became smaller, more powerful, and lower cost; instead of filling rooms, machines now sat on desks as well as on laps. Just as mechanical devices led to the **Industrial Revolution**, computing machines have become the engines that power the **modern information revolution**.

- **Mathematical conjectures, computational mathematics and intractability** Rapid advances in computer technology have led to an intriguing interplay between a mathematician, specifically a computational mathematician, and the machine. The computer has been an invaluable tool in furnishing counterexamples to **outstanding conjectures** or in verifying conjectures up to a specific numerical bound.

There have been some remarkable successes with previously intractable problems (problems such as **practical travelling salesman** and **chess** problems that cannot be solved in an affordable time frame).

Further, computation of irrational (as well as rational) numbers such as the exponential function e^x for $x = 1$ (say) and π to thusands (or beyond) of digits have been achieved. For example, π has been computed to 10 trillion digits (Agarwal et al. 2013), and the 48thMersenne Prime, 2 raised to the 57,885,161 power minus 1 (which is 17,425,170 digits long), was discovered in 2013.

- **Internet: Email communication and world wide web** The English computer scientist **Sir Timothy John Berners-Lee** (born 1955), invented the world wide web (**www**) that has revolutionized global communication. He evolved an information management system in March 1989, and he implemented the first successful communication between a hypertext transfer protocol (**http**) client and a server via the Internet in November, 1989.

Email communication, an important development in computer communication, was followed by the invention of the world wide web.

Raymond Samuel Tomlinson (born 1941) developed an email system in 1971 on the **ARPANET,** one of the world's first operational packet switching networks that implemented **TCP/IP** (Transmission Control Protocol/Internet Protocol), the basic language of the Internet. ARPANET was the progenitor of what was to become the global internet.

Tomlinson sent several test messages from one computer/terminal to the other, then he sent a message to the rest of his group explaining how to send messages over the network. The first use of network email announced its own existence.

- **Mathematically satisfactory breakthroughs in computational mathematics** In computational science, several breakthroughs were achieved. Some of them are as follows. **James William Cooley** (1926–2016) **and John Wilder Tukey** (1915–2000) **developed the Fast Fourier Transform (FFT) algorithm to compute discrete Fourier Transform and its inverse in 1965**.

Computing the DFT of n points in the naive way, using the definition, takes $O(n^2)$ arithmetic operations, while an FFT can compute the same DFT in only $O\left(n \log_2 n\right)$ operations—a significant reduction in computation and hence in computational error (for large n); For small n, the computation of DFT is never a problem; the FFT algorithm may be used but not at all necessary). An improved version of the FFT, developed later, results in significantly less computation and also less error.

Narendra Karmarkar (born 1957) developed a polynomial time $O(n^{3.5})$ projective transformation iterative deterministic algorithm to compute a solution of a linear program Max $c^t x$ subject to $Ax \leq b$, $x \geq 0$ (null column vector) in 1984, where n is the number of variables. This algorithm obtains the next feasible direction toward optimality and scales back by a factor $0 < \alpha \leq 1$.

It has undergone improvements over the years and a variant of it is now more widely used than the simplex algorithm (which ruled the linear optimization area for over 3 decades – 1950s to 1980s) which is an exponential-time method in the worst case.

In 2002, **Manindra Agrawal** (born 1966)**, Neeraj Kayal** (born 1979)**, and Nitin Saxena** (born 1981) of the Indian Institute of Technology, Kanpur, India, developed the **AKS** (Agarwal-Kayal-Saxena) **primality test algorithm**—a deterministic primality testing method—which is general, polynomial-time, deterministic, and unconditional all at the same time.

Earlier algorithms developed over the centuries achieved at most 3 of these properties but not all 4 concurrently. The algorithm is based upon the following theorem: An integer n($\geq$2) is prime if and only if the polynomial congruence $(x - \alpha)^n \equiv (x^n - \alpha) \pmod{n}$ holds for all integers α coprime to n (or even just for some such integer α, in particular for $\alpha = 1$), where x is a free variable.

The foregoing (breakthrough) algorithms have had a tremendous positive impact on cost-effective computation in solving many real-world engineering and scientific problems. We emphasize the fact that these algorithms are the most–desired **mathematically proved** ones and are valid for any infinite (as well as any finite precision computation) precision computation.

- **Mathematically unsatisfactory but computationally satisfactory breakthroughs in computational mathematics** (**4-colour conjecture** (Agarwal and Sen 2014)) The **4 colour conjecture** (1852) is: Given any separation of a plane into contiguous regions, producing a figure called a map, 4 colours are enough to colour the regions of the map so that no 2 adjacent regions have the same colour.

In the following map (Fig. 1), for instance, we must make regions A, B, and C of different colours since they share pairwise boundaries, but then it would be impossible to colour region D unless a 4th colour was introduced.

Fig. 1 4-color conjecture

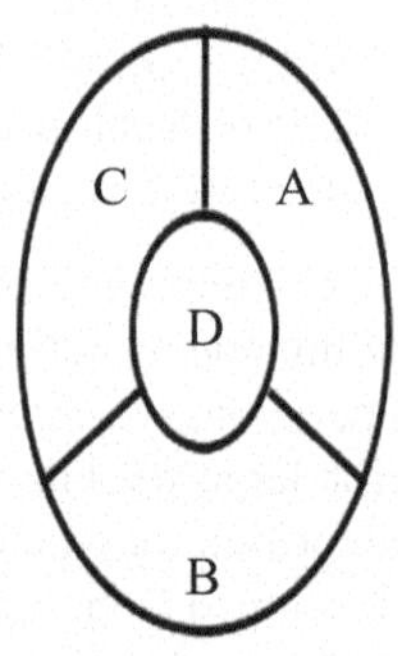

Several false proofs and counter-examples appeared during the 20th century. It has been shown during 1890s that 5 colours suffice to colour a map while 3 colours are insufficient.

Kenneth Appel (1932–2013) and Wolfgang Haken (born 1928) of the University of Illinois (USA) demonstrated in 1976 that there is a particular set of 1936 maps, each of which cannot be part of a smallest-sized counter-example to the 4 colour theorem.

This was followed by a special-purpose computer program, written to confirm that each of these maps has this property. Additionally, any map (including a counter-example) must have a portion that looks like 1 of these 1936 maps. This demonstration needed 100s of pages of manual analysis (almost a humanly intractable proposition). Appel and Haken inferred that no smallest counter-example existed because any must contain, yet not contain, 1 of these 1936 maps.

This contradiction implies that there is no counter-example at all and that the 4 colour conjecture is indeed a theorem. Their proof was rejected by mathematicians because the computer-assisted proof was infeasible/intractable for a human to check by hand.

Ronald Lewis Graham (born 1935), a mathematician, asked when mulling over this complicated issue, "The real question is this: if no human being can ever hope to check a proof, is it really a proof?" The question has no definite answer as of now (2019). However. as computer proofs become more common, mathematicians will probably feel more at ease with them.

All mathematicians would feel most comfortable if the 4 colour theorem were shown valid with a 2-page (mathematical) proof – short and elegant, rather than with the brute-force machinations of a computer. Traditional mathematicians long for the good old days of mathematics unplugged.

Finally, we do emphasize that the proofs of the Karmarkar and AKS algorithms (mentioned earlier) were mathematical and not computer-assisted, hence these are accepted whole-heartedly by all mathematicians and others (including computational scientists) without any doubt (Agarwal and Sen 2014).

- **Outstanding computational problems** We still have several outstanding problems for which a deterministic polynomial-time algorithm is yet to be developed. For example, the **Travelling Salesman Problem** (TSP—Suppose we are given n cities and the distances between each pair of cities. Determine the shortest path that passes through each city only once and returns to the city of origin.).

A deterministic solution needs the evaluation of $(n-1)!$ paths. If $n = 1000$, then 999! paths need to be evaluated. The evaluation is computationally intractable even using the world's most powerful computer of today and also of tomorrow (future).

Both the **solution of the TSP and a verification of the solution** viz. the shortest path (if determined by some way/method) are **NP-hard** and hence intractable (for large n). However, if n = 11, then the TSP can be deterministically solved in less than a minute on a modern laptop!

- **Useful solution of an intractable problem by a randomized algorithm** Intractable problems (intractable problems are those which cannot be solved in an acceptable time frame deterministically) are usually solved using an **evolutionary approach/probabilistic method**. Such a method is a randomized algorithm in a broader sense. The solution (result) produced by such a method usually may not be **the best** solution, but a solution usually reasonably close to the best one in terms of accepting it in a real-world application.

- **Computers in all walks and exponentially growing in computing power** Computers are everywhere, not just on our desks. Computers are hidden in our stereos and televisions, in our cars, in our wrist watches, in our smart mobile phones and laptops, and in our microwave ovens and other household appliances. Computers make our telephone connections.

Computers are used to make our airline reservations. Many of the images we see on the cinema screen are produced on a computer. In fact, the exponential growth that the world has seen during the last 6 decades in architecture, engineering, and technology, specifically information technology, has been made possible by fantastic progress in both computing devices (hardware) and software.

Every 18 months processor speed doubles. Every 12 months band–width doubles, and every 9 months hard disk space doubles. Behind all this exponential growth is the computational mathematics that is superimposed on the hardware.

- **Computations change it all** All views—both negative and positive – of mathematics need to be modified when we land in the world of computational mathematics. Here we can see the **real power** of mathematics/mathematical reasoning through its computational arm. In conclusion, mathematics and computational science/mathematics are intimately connected and inseparable, and this union has presented modern mathematicians with new challenges viz. finding hidden patterns.

All of the science and engineering that we know today would be dwarfed without this computational arm of mathematics. At the root of science and technology is this arm. If we cut this arm, the whole of today's scientific world would simply collapse.

Through computations one develops **tremendous confidence** in all that one does in science, engineering, and in all other areas (including mathematics). But for these, the power of mathematical reasoning would remain mainly of academic interest and without the unimaginable advancement of human civilization that has taken place during the past few decades.

- **Beyond: Can we increase the speed of computing indefinitely** We have seen over years the exponential growth in the speed of computing. We have already stated that every 18 months the processor speed in doubling, every12 months the band-width is doubling, and every 9 months the hard disk space is doubling. All these growths have been possible due to the relentless effort of computer h/w and s/w scientists of the world. In fact the growths are truly astounding unlike any other field of research.

These innovations have been profoundly affecting all the scientific and engineering fields readily and positively resulting in the continuous upliftment of the society all over the globe.

However, can this exponential growth be sustained eternally (without limit)? The answer is "no" according to physics. Following are the 3 barriers posed by laws of physics, that set the limit beyond which computation cannot be speeded up even if the whole universe (as viewed by the Big Bang theory) is dedicated to information processing.

- *Speed of light barrier.* Electrical pulses cannot propagate faster than the speed of light viz. 3×10^{10} cm/s. A random access memory used to 10^9 cycles/s (1 GHtz) will deliver information at 0.01 ns (0.01×10^{-9} s) speed if it has a diameter of 0.3 cm.
- *Thermal efficiency barrier.* Information processing enhances the entropy of the system. Hence the amount of heat that is absorbed is kT $\log_e 2$/bit, where k = Boltzmann constant = $1.38064852 \times 10^{-16}$ erg K^{-1} and T is the absolute temperature (taken as room temperature) viz. 300 K (K denotes Kelvin). It is not possible to economize any further on this. If we want to process 10^{30} bits/s, the amount of power that we need is $p = 10^{30} \times k \times 300 \times \log_e 2$ = 28696293275181728 erg/s obtained using the Matlab vpa command **≫p=vpa(10^30*1.38064852*10^-16*300*log (2), 40)** This produces p = 28709778869467836 erg/s = 2870977886.9467836 W ≈ 2.87×10^9 W, where 1 W = 10^7 erg/s.
- *Quantum barrier.* Associated with every moving particle, there is a wave which is quantified so that the energy of 1 quantum E = hν, where h = Plank's constant and ν = frequency of the wave. The maximum frequency $\nu_{max} = mc^2/h$, where m = mass of the system and c = velocity of the light. Hence the frequency band that can be used for signalling is limited to the maximum frequency ν_{max}.

Following the Information theory of **Claude Elwood Shannon** (1916–2001), an American electrical engineer, mathematician, and cryptographer widely proclaimed as the **father of information theory,** the number of information that can be processed per second cannot exceed ν_{max}. The mass of hydrogen atom is 1.67×10^{-24} gm, c = 3×10^{10} cm/s, and h = 6×10^{-27}. Thus per mass of hydrogen atom, maximum $1.67 \times 10^{-24} \times 3^2 \times 10^{20}/(6 \times 10^{-27}) = 2.5050 \times 10^{23}$ bits/s can be transmitted.

The number of atoms in the observable universe (continuously expanding) is estimated to be around 10^{82}. Thus if the whole universe is dedicated to information processing i.e. if all the atoms are employed to information processing, then maximum

2.5050*10^23*10^82 = 2.5050×10^{105} bits/s
=2.5050*10^105*60*60*24*365
=7.8998e + 112 = **7.8998 x 10^{112} bits/year**

can be processed. This is the **theoretical limit** (for massively parallel processing) set by the current (2019) laws of physics **assuming the validity of the Big Bang theory and we are still hyper-astronomically far behind it!**

Hence the exponential growth in digital computation and communication will apparently continue not only for decades but also for centuries before it comes to a halt. Maybe the humanity will never have the scope to reach the halting stage.

However, as of now **there are many surprises for us in store in terms of innovations/discoveries in practically all sciences and engineering areas**. These are due to human ingenuity combined with increasingly more powerful computing facilities (both in terms of h/w and s/w) available to humans.

- **Computational error and complexity: Importance** CM denoting computer/computational mathematics (as stated earlier) essentially refers to the methods—numerical/arithmetical. non-numerical/non-arithmetical, and a combination of them—oriented toward computer. Most often these methods are computer-based numerical methods. These differ from the pre-1940 computational/applied mathematics methods precisely in stressing more on (i) computational error due to eternal finite precision of the computer as well as the developed mathematical methods—iterative and non-iterative and (ii) computational complexity. Both the error and the complexity determine the quality of the solution and the cost of the solution (in terms of amount of computation), respectively.

Pre-1940 computation was carried out for **very small problems** using a mathematical table, mechanical/electromechanical calculator that could do mainly simple arithmetic operations using a few storage locations for operands and results and was manual (non-automatic).

The **post-1940** computation was performed for relatively large problems using an automatic digital computer (ADC)/main-frame computer. However, the foregoing pre-1940 computing devices continue to remain active for around 3 more decades till 1970s since ADC's were yet to be widely available to a common man and were too expensive to be afforded by most of the people (individually) in the society.

In 1970s the computational error and complexity concern was yet to be deeply rooted among people. They did the computation and got the solution/result. If they like the result, they take it as valid/correct, use/publish them, and remain satisfied with the quality of the result (without a comparison to know its relative merit). The computational complexity issue is still a lesser issue and yet to gain enough momentum.

Since 1980s the **error and complexity issues start gaining importance**. **Today these issues are considered vital, need to be computed and also need to accompany any (computer-produced) solution to justify (scientifically) its quality (error) and cost (amount of computation) of the solution**.

Exact error in a quantity is never known and can never be known. However, numerical error implying relative error is/can be known in terms of an error bound in which the exact error lies (Krishnamurthy and Sen 2009; Lakshmikantham and Sen 2005).

If the error implying the accuracy (quality) of the solution is not computed/known, then using such a solution could be catastrophic. For instance an apparent precision (or surgical) strike on a target could destroy some other target causing perhaps enormous loss of lives of people and property (defeating the very purpose of the strike).

If, on the other hand, computational complexity is not known, then we will not be able to compare 2 or more algorithms and choose that one which is the/a best. This may result in an undesirable increased computational cost.

4 Conclusions

- ***Advent and unimaginable impact of computing on humans*** We have briefly described the birth, the growth, the exaflops computation and beyond for the area of "computing". The motive is to stress the facts that (i) there exists no other area/field of activity in the world, which has been growing exponentially (so fast) and that (ii) there is no other area that has immensely and uniquely impacted all other areas in terms of gigantic innovations/discoveries. Computing has unimaginably changed the globe to the one today (2019), that could not have been imagined 8 decades ago even in a fiction!
- ***The glory of silicon technology underlying the h/w*** During the past 80 years the only technology that has made modern computing so prodigiously tall is the silicon technology. It is going strong and is the life-line of the world of computing. Were we to halt silicon technology permanently, our civilization will simply collapse. Our dependence on silicon technology and hence on computing has gone so deep that, without our knowledge, we have entered into a realm of the environment in which it (computing) is almost as important as air and water needed for our very survival.

The banking system, the air-communication system (involving hundreds of flights in the sky at any point in time), the smart phones, and many more systems and devices will come to a grinding halt once the silicon technology becomes extinct. As we rarely realize the gigantic importance of air and water as long as these continue to remain serving us continuously behind the scene, so do we seldom realize the immeasurable importance of computing.

- ***Dataflow computer*** A dataflow computer is a computer based on dataflow architecture using silicon technology. The architecture contrasts the classical control flow architecture (also called von Neumann architecture). Unlike the control flow architecture of a classical computer, the dataflow architecture is without a program-counter (or at least conceptually). The execution of an instruction is solely determined based on the availability of input arguments to the instructions. Thus the order of the execution of an instruction is unpredictable. Or, in other words, the behaviour is non-deterministic.

Hardware architectures for dataflow was a major topic in computer architecture research in 1970s and 1980s. **Jack Bonnell Dennis,** an emeritus professor of computer science and engineering at Massachusetts Institute of Technology (MIT), USA pioneered the field of static dataflow architectures while the Manchester Dataflow Machine and MIT Tagged Token architecture were major projects in dynamic dataflow (Manchester Dataflow Research Project 1997; Dennis et al. 1972).

However, the dataflow architecture has been successfully implemented in a special-purpose h/w such as in network routing, digital signal processing, telemetry, graphics processing, and data warehousing, no commercially successful general-purpose computer h/w has used a dataflow architecture though.

- ***Systolic computer*** **Hsiang Tsung Kung** (born 1945), a Chinese American professor of computer science and electrical engineering in Harvard University, USA (known for his work in machine learning, signal processing, and parallel computing), and **Charles Eric Leierson** (born 1953), a Norwagian-American professor of computer science in Massachusetts Institute of Technology, USA published the first paper describing systolic arrays in 1978 (Kung and Leiserson 1979; Kung 1988; Petkov et al. 1992). Systolic computers are a new class of pipelined array architecture also based on silicon technology. It comprises an array of processing elements called **cells**. Each cell is connected to a small number of nearest neighbours in a mesh-like topology, has same operations, and performs the operations (sequentially) on each data item that flows between them, and passes it to its neighbour. Systolic arrays compute "lock-step" with each cell undertaking alternate compute/communicate phases.

Features of a systolic array (a computing network) are

- **Synchrony** (the data is rhythmically computed (timed by a global clock) and passed through the network),
- **Modularity** (the array—finite/infinite—consists of modular processing units),
- **Regularity** (the modular processing units are interconnected homogeneously),
- **Spatial locality** (each cell has a local communication interconnection),
- **Temporal locality** (each cell transmits the signals from one cell to other needing at least 1 unit time delay), and
- **Pipelinability** (the array can achieve a high speed).
- ***Difference between SIMD and Systolic arrays*** The classical SIMD (Single Instruction Multiple Data) array usually loads data into its local memories before starting the computation. It is a synchronous array of processing elements (PEs) supervised by 1 control unit. All PEs receive the same instruction broadcast from the control unit but operate on different data sets from distinct data streams. Systolic arrays, on the other hand, usually pipe data from an outside host and pipe the result back to the host. These are almost like classical multiple-instruction single-data (MISD) arrays.
- ***Application of Systolic arrays*** Systolic arrays are usually hard-wired for specific operations such as *multiply-and-accumulate* to perform tasks such as massively parallel integration, matrix multiplication, data sorting, convolution, correlation, and dynamic programming (used in DNA and protein sequence analysis besides other activities).
- ***Quantum computing*** is computing performed by a quantum computer. It is based on theoretical quantum bits that, unlike classical bits of a binary digital electronic computer based on transistors, can accept quantum values. While conventional digital computing needs that the data be encoded into bits, each of which is either 0 or 1, quantum computing uses qubits (quantum bits) which can be in superpositions of states.

The field of quantum computing was initiated by the work of **Paul A. Benioff** (born 1930), an American physicist and **Yuri Ivanovitch Manin** (born 1937), a

Russian mathematician, (known for work in algebraic geometry and diophantine geometry, and many expository works ranging from mathematical logic to theoretical physics), were among the pioneers to propose the idea of a quantum computer in 1980. The others were **Richard Phillips Feynman** (1918–1988), an American theoretical physicist and physics Noble Laureate in 1965, known for his work in the path integral formulation of quantum mechanics (the theory of quantum electrodynamics), in 1982 and **David Elieser Deutsch** (born 1953), an Israeli-born British physicist known for formulating a description for a Quantum Turing Machine and specifying an algorithm designed to run on a quantum computer in 1985.

The tasks of quantum computing include (i) building devices with a specialized behaviour and (ii) designing and developing an algorithm to use the behaviour. The use of quantum phenomena may allow us to perform computational tasks that are not otherwise possible/efficient.

A classical computational model provide a level of abstraction with discrete systems. Devices are increasingly becoming smaller. **The only description that work on an extremely small scale are quantum.**

- ***Classical versus quantum computing*** Classical computing relies, at its ultimate level, on principles expressed by Yes-or-no logic (Boolean algebra), operating with a 7-mode (and, or, xor, not, nand, nor, xnor) logic gate (a logic gate is an elementary building block of a digital circuit) principle, it is possible to exist with only 3 modes (which are and, not, and copy) though.

Data processing at any instant must be exclusively in a binary state. Each of millions of transistors and capacitors at the heart of a computer can only be in one state at any point in time. As we increasingly tend to develop smaller and faster circuits, we begin to reach the physical limits of materials and the threshold for laws of classical physics to apply. Beyond this, the quantum world takes over, which opens up a potential as great as the challenges that are faced.

While the time that each transistor or capacitor needs be either in 0 or 1 before switching states is now measurable in billionths of a second, there is still a limit as to how quickly these devices can be made to switch state as we progress to smaller and faster circuits.

The quantum computer, on the other hand, works with a 2-mode (xor, qo1 (the ability to change 0 into a superposition of 0 and 1—a logic gate non-existent in classical computing)) logic gate principle.

In a quantum computer, an elemental particle such as an electron or a photon can be used, with their charge or polarization acting as a representation of 0 and/or 1. Each of these particles is known as a qubit (or just qbit), the nature and behaviour of these particles form the basis of quantum computing. The two most relevant aspects of quantum physics are the principles of superposition and entanglement.

Quantum superposition, a fundamental principle of quantum mechanics, states that, much like waves in classical physics, any two (or more) quantum states can be added ('superposed') together and the output will be another valid quantum state; conversely, that every quantum state can be expressed as a sum of 2 or more other distinct states. The superposition implies that each qubit can represent both a 1 and a 0 at the same time.

Quantum entanglement is a quantum mechanical phenomenon in which the quantum states of 2 or more objects have to be described with reference to each other, even though the individual objects may be spatially separated. This leads to correlations between observable physical properties of the systems. A quantum state must be described for the system as a whole (Finilla et al. 1994; Feynman 1982; Nielsen 2000; Mermin et al. 2007; Kitaev and Shen 2002; Hirvensalo et al. 2004).

As of now (2019) there exist negligibly few commercially available quantum computers meant more for research than for real-world commercial use. The design and development of an actual quantum computer is in its infancy even after 39 years (from 1980 to 2019). However, laboratory experiments have been carried out in which quantum computational operations were executed on a very small number of qubits. Both practical and theoretical research have been continuing since 1980.

Governments of several countries are funding quantum computing research in an effort to develop quantum computers for trade, civilian, environmental, and national security purposes.

A small 20-qubit quantum computer exists and is available for experiments through the *IBM Quantum Experience* project *D-Wave Systems* (a quantum computing company, located in Burnaby, British Columbia, Canada. It is the world's first company to sell quantum computers.) D-Wave Systems has been developing their own version of a quantum computer that uses **quantum annealing** (a metaheuristic to find the global optimum of a specified objective function over a given set of candidate solutions (states) by a process using quantum fluctuations).

Large-scale quantum computers, if and when developed, would be able to solve certain problems much faster than any classical computer that use even the best currently known algorithms such as the integer factorization using **Shor's algorithm** (a quantum algorithm) and the simulation of quantum many-body systems (a highly compute intensive complex problem). There exist quantum algorithms, such as the **Simon's algorithm**, that run faster than any possible randomized classical algorithm/evolutionary approach.

A classical computer could, in principle (with exponential resources), simulate a quantum algorithm, as quantum computation does not violate the **Church-Turing thesis**. A quantum computer, on the other hand, may efficiently solve a problem which is *practically* infeasible (or intractable) on a classical computer.

- ***Differences Between UM and CM*** We describe below the differences, rather striking differences existing between UM and CM from the point of view of the real-world applications (Sen 2014).

(i) ***Proof*** In UM most statements cannot be proved mathematically while in CM all statements can be proved or disproved. This is because PoC (Precision of Computation) in UM is infinite while it is finite in CM.

(ii) ***PoC: Finite versus infinite*** Within a specified PoC which corresponds to non-intractable computation, there exists no statement which cannot be proved or disproved in CM. In fact there is **no conjecture in CM** unlike that in UM.

The literature including those in the internet is considerably crowded with statements such as the 4-color conjecture.

(iii) ***Problems easy mathematically but involved computationally*** For example, an n-sided regular polygon inscribed in a given circle: Determining the shortest and the longest diameters can be mathematically (geometrically) easy while it is involved computationally ($n \rightarrow \infty$). A centre of a polytope (useful in linear optimization) can be relatively easily obtained geometrically while it is more involved computationally.

(iv) ***Problems easy computationally but involved mathematically*** Relative error computation, for instance, is a must for any computation to logically justify the quality of the result. It is easy in numerical computation in real-world situation while it is either not possible or more involved and not often useful in physical world application, mathematics does provide us the insight though.

(v) ***Mathematical solutions are too limited/impossible*** Most real-world problems cannot be solved mathematically. For example, partial differential equations involving **Navier-Stokes** equations or **Laplace** equations or any other partial differential equations can only be solved for some well-defined boundaries with some geometrical shapes.

There could be indefinitely large number of possible non-geometrical shapes such as the shape of the map of India and of New Zealand, mathematical solutions are difficult or impossible. However, all real-world problems including partial differential equations with irregular boundaries can be solved computationally and these solutions are all-important for engineers/applied researchers.

- ***Advances in memory technology*** A vital requirement of any computing system is the storage (memory) for data and programs. **But for the storage modern computing collapses**. There has been lots of activities in advancing memory technologies. IBM is an organization in the forefront of this venture.

These technologies, among others, include MRAM (magnetic RAM) and ReRAM (resistive RAM), and PCM (phase change memory), where RAM denotes random access memory. IBM continues to innovate and drive advances in memory technology.

These multifarious technologies have diverse applications across both existing and emerging technology sectors. With a unique combination of speed, non-volatility, and power consumption, emerging memory technologies have the potential to replace DRAM (dynamic RAM), SRAM (static RAM), or flash memory in certain applications such as ultra-low-power devices for IoTs (**Internet-of-Things**) applications. MRAM could eventually serve as a fast, dense, cache memory. ReRAM and PCM show potential as storage class memories. The near-analog range of states available through ReRAM and PCM are also promising candidates for incorporation into complex neuromorphic computers (James 2013; Prall and Parat 2010).

History of computing: ***Refer literature cited for details*** This article depicts brief, rather very brief, history of computing. For further details, the reader may refer the concerned literature cited in "References".

Acknowledgements Certainly an article of this type cannot be written without deriving many valuable ideas from several sources. We express our indebtedness to all authors, too numerous to acknowledge individually, from whose specialized knowledge we have been benefitted. We have also been immensely benefitted from several websites such as Wikipedia.org.

References

Abramowitz, M., Stegun, I.A. (ed.): Handbook of Mathematical Functions: with Formulas, Graphs, and Mathematical Tables. Dover Books on Mathematics (1965)
Agarwal, R.P., Sen, S.K.: Creators of Mathematical and Computational Sciences. Springer, New York (2014)
Agarwal, R.P., Agarwal, H., Sen, S.K.: Birth, growth and computation of Pi to ten trillion digits. Adv. Differ. Equ. **2013**(2013), 100 (2013)
Aitken, A.C.: Studies in practical mathematics, Part II: the evaluation of the latent roots and latent vectors of a matrix. Pro. Roy. Soc. (Edinburgh) **57**, 269 (1937)
Al-Daffa, A.A.: The Muslim Contribution to Mathematics. Humanities Press, Atlantic Highlands (1977)
Al-Daffa, A.A., Stroyls, J.J.: Studies in the Exact Sciences in Medieval Islam. Wiley, New York (1984)
Allen, R.E.: Greek Philosophy: Thales to Aristotle. The Free Press, New York (1966)
Allman, G.J.: Greek Geometry from Thales to Euclid. Arno Press, New York (1976)
Andrade, E.N.C.: Sir Issac Newton, His Life and Work. Doubleday & Co., New York (1954)
Anglin, W.S.: Mathematics: A Concise History and Philosophy. Springer, New York (1994)
Anglin, W.S.: The Queen of Mathematics. Kluwer, Dordrecht (1995)
Anthony, H.D.: Sir Isaac Newton. Abelard-Schuman, New York (1960)
Apostle, H.G.: Aristotle's Philosophy of Mathematics. The University of Chicago Press, Chicago (1952)
Archibald, R.C.: Outline of the history of mathematics. Amer. Math. Monthly **56** (1949)
Artmann, B.: Euclid—The Creation of Mathematics. Springer, New York (1999)
Bag, A.K.: Mathematics in Ancient and Medieval India. Chaukhambha Orientalia, Varanasi (1979)
Ball, W.W.R.: A short Account of the History of Mathematics. Dover, New York (1960)
Barker, S.F.: Philosophy of Mathematics. Prentice-Hall, Englewood Cliffs (1964)
Baron, M.E.: The Origin of Infinitesimal Calculus. Dover, New York (1993)
Barrow, J.D.: Pi in the Sky: Counting, Thinking, and Being. The Clarendon Press, New York (1992)
Baumgardt, C.: Johannes Kepler: Life and Letters. Victor Gollancz, London (1952)
Beckman, P.: A History of π. St. Martin's Griffn, New York (1976)
Beiler, A.: Recreations in the Theory of Numbers. Dover, New York (1964)
Belhoste, B.: Augustin–Louis Cauchy: A Biography (trans. by F. Ragland). Springer, New York (1990)
Bell, E.T.: Men of Mathematics. Touchstone Books, New York (1986)
Bell, E.T.: The Development of Mathematics, 2nd edn. Dover, New York (1992)
Benson, D.C.: The Moment of Proof: Mathematical Epiphanies. Oxford University Press, New York (1999)
Berggren, J.L.: Episodes in the Mathematics of Medieval Islam. Springer, New York (1986)
Berlinski, D.: A Tour of Calculus. Vintage, New York (1995)
Berndt, B.E.: Notebooks of S. Ramanujan. Springer, New York (1985)
Berndt, B.C., Rankin, R.A.: Ramanujan Letters and Commentary. Affiliated East West Press, New Delhi (1997)
Beth, E.W.: The Foundations of Mathematics. North-Holland, Amsterdam (1959)

Beth, E.W.: Mathematical Thought: An Introduction to the Philosophy of Mathematics. D. Reidel, Dordrecht (1965)

Bhaskar, T.G., Kovach, D., Lakshmikantham, V.: The hybrid set theory. Nonlinear Anal. Hybrid Syst. **1**, 414–416 (2007)

Bishop, M.G.: Pascal, The Life of Genius. Reynal & Hitchcock, New York (1936)

Black, M.: The Nature of Mathematics: A Critical Survey. Routledge & Kegan Paul, London (1965)

Blatner, D.: The Joy of Pi. Walker Publishing Co., New York (1997)

Bose, D.M., Sen, S.N., Subbarayappa, B.V.: A Concise History of Science in India. Indian National Science Academy, New Delhi (1971)

Boyer, C.: The History of the Calculus and its Conceptual Development. Dover, New York (1959)

Boyer, C.: A History of Mathematics, 2nd edn. Wiley, New York (1991)

Brewer, J.W., Smith, M.: Emmy Noether, A Tribute to Her Life and Work. Marcel Dekker, Inc., New York (1981)

Brody, D.E., Brody, A.R.: The Science Class You Wish You Had: The Seven Greatest Scientific Discoveries in History and the People Who Made Them. Berkeley Publishing Group, New York (1997)

Brumbaugh, R.S.: Plato's Mathematical Imagination. Indiana University Press, Bloomington (1954)

Bunt, L.N.H., Jones, P.S., Bedient, J.D.: The Historical Roots of Elementary Mathematics. Prentice-Hall, Englewood Cliffs (1976)

Burton, D.M.: The History of Mathematics: An Introduction, 4th ed. McGraw-Hill, New York (1999)

Cajori, F.: The history of the notations of calculus. Ann. Math. **25**, 1–46 (1923)

Cajori, F.: Leibniz, the master builder of notations. Isis **7**, 412–429 (1925)

Cajori, F.: A History of Mathematics, 4th edn. Chelsea Publishing, New York (1985)

Calinger, R.: Gottfried Wilhelm Leibniz. Rensselaer Polytechnic Institute, Troy, N.Y. (1976)

Calinger, R. (ed.): Classics of Mathematics. Prentice-Hall, Englewood Cliffs (1995)

Calinger, R.: A Contextual History of Mathematics: To Euler. Prentice Hall, Upper Saddle River (1999)

Cantor, G.: Contributions to the Founding of the Theory of Transfinite Numbers (trans. by P.E.B. Jourdain). Dover, New York (1952)

Cantor, M.: Vorlesungen uber¨ Geschichte der Mathematik, Leipzig (1907)

Cardano, G.: The Book of My Life (trans. by J. Stoner). Dover, New York (1962)

Cardano, G.: The Great Art (trans. by T.R. Witmer). MIT Press, Cambridge (1968)

Chandrasekhar, S.: Newton's Principia for the Common Reader. Oxford University Press, Oxford (1995)

Chattopadhyaya, D.: History of Science and Technology in Ancient India: The Beginnings. Firma KLM PVT., Calcutta (1986)

Clagett, M.: Greek Science in Antiquity. Collier, New York (1963)

Clark, W.E. (ed.): The Aryabhatia of Aryabhata. The University of Chicago Press, Chicago (1930)

Clawson, C.C.: Mathematical Mysteries: The Beauty and Magic of Numbers. Plenum Press, New York (2000)

Closs, M.P. (ed.): Native American Mathematics. University of Texas Press, Austin (1986)

Cohen, I.B.: Sir Issac Newton's Papers and Letters on Natural Philosophy. Harvard University Press, Cambridge (1958)

Cole, J.R.: Pascal: The Man and His Two Loves. New York University Press, New York (1995)

Conant, L.: The Number Concept. Its origin and Development. Macmillan, New York (1923)

Cooke, R.: The Mathematics of Sonya Kovalevskaya. Springer, New York (1984)

Coolidge, J.L.: The Mathematics of Great Amateurs. Dover, New York (1963)

Copernicus, N.: On the Revolutions of Heavenly Spheres, Great Mind Series. Prometheus Books, New York (1995)

Courant, R., Robbin, H., Stewart, I.: What is Mathematics?. Oxford University Press, New York (1996)

Crowther, J.G.: Six Great Scientists. Barnes & Noble, New York (1995)

Cullen, C.: Astronomy and Mathematics in Ancient China: The Zhou Bi Suanjing. Cambridge University Press, Cambridge (1996)
Dennis, J.B., Fosseen, J.P., Linderman, J.P.: Data flow schemes. In: Symposium on Theoretical Programming, Novosibirsk, pp. 187–216 (1972)
Dunham, W.: Euler: The Master of Us All. Mathematical Association of America (1999)
Dunham, W.: The genius of Euler: reflections on his life and work. Mathematical Association of America (2007)
Egervary, E.: On combinatorial properties of matrices (trans. by H.W. Kuhn). Office of Maval Research Logistic Project Report, Department of Mathematics, Princeton University, Princeton (1931)
Farkas, J.: Uber die theorie der einfachen ungleichungen. Jr.und angew. Math. **124**(1–24), 1901–1902
Feynman, R.P.: Simulating physics with computers (PDF). Int. J. Theor. Phys. **21**(6), 467–488 (1982)
Filon, L.N.G.: On a quadrature formula for trigonometric integrals. Proc. Roy. Soc. (Edinburgh) **49**, 38–47 (1928)
Finilla, A.B., Gomez, M.A., Sebenik, C., Doll, J.D.: Quantum annealing: a new method for minimizing multidimensional functions. Chem. Phys. Lett. **219**, 343 (1994)
Forsythe, G.E.: Escalator method for latent roots. Q. J. Mech. **5**, 178–190 (1952)
Halmos, P.R.: Measure Theory. D. Van Nostrand, New York (1950)
Hildebrand, R.B.: Introduction to Numerical Analysis. McGraw-Hill, New York (1956)
Hildebrand, F.B.: Methods of Applied Mathematics. Dover Books on Mathematics (1965)
Hirvensalo, M.: Quantum Computing. Springer, Berlin (2004)
Hitchcock, F.L.: An improvement on the G.C.D. method for complex roots. J. Math. Phys. **23**, 69–74 (1944)
Holmes, P.: History of Dynamical System. Princeton University, Princeton (2007)
James, D.: Recent advances in memory technology. In: Advanced Semiconductor Manufacturing Conference (ASMC) (2013). http://dx.doi.org/10.1109/ASMC20136552766
Kitaev, A.Y., Shen, A.H., Vyalyi, M.N.: Classical and Quantum Computation. AMS (2002)
Krishnamurthy, E.V., Sen, S.K.: Numerical Algorithms: Computations in Science and Engineering. Affiliated East West Press, New Delhi (2009)
Kuhn, H.W.: Variants of the Hungarian method for assignment problems. Nav. Res. Logist. Q. **3**, 253–258 (1956)
Kung, H.T., Leiserson, C.E.: Algorithms for VLSI processor arrays. In: Mead, C., Conway, L. (eds.) Introduction to VLSI Systems. Addison-Wesley, New York (1979)
Kung, S.Y.: VLSI Array Processors; Prentice-Hall, Inc., Upper Saddle River (1988)
Lakshmikantham, V., Sen, S.K.: Computational Error and Complexity in Science and Engineering. Elsevier, Amsterdam (2005)
Manchester Dataflow Research Project: Research Reports: Abstracts, Sept 1997
Mermin, N.D.: Quantum Computer Science. Cambridge University Press, Cambridge (2007)
Moore, E.H.: On the reciprocal of the general algebraic matrix (abs.). Bull. Amer. Math. Soc. **26**, 394–95 (1920)
National Bureau of Standards, Tables of Lagrangian Interpolation Coefficients, Washington (1944)
Nielsen, M.A., Chuang, I.L.: Quantum Computation and Quantum Information. Cambridge University Press, Cambridge (2000)
Petkov, N.: Systolic Parallel Processing. North Holland Publishing Co., Amsterdam (1992)
Prall, K., Parat, K.: 25 nm 64 Gb MLC NAND technology and scaling challenges. Trans. IEDM 102–105 (2010)
Ralston, A., Wilf, H.S.: Mathematical Methods for Digital Computers, vol. 2. Wiley, New York (1967)
Ranga Charya, M. (ed.and trans.): Mahavira (author), Ganita Sara Sangraha, Madras (1912)
Richtmyer, R.D.: Difference Methods for Initial-Value Problems. Wiley-Interscience, New York (1957)

Sen, S.K.: Natural mathematics, computer mathematics, and mathematics: scope in engineering computation. Nonlinear Stud. **21**(2), 309–318 (2014)

Sen, S.K., Agarwal, R.P.: Zero: A Landmark Discovery, The Dreadful Void, and the Ultimate Mind. Academic Press, New York (2016)

Sokolnikoff, I.S., Redheffer, R.M.: Mathematics of Physics and Modern Engineering. McGraw-Hill, New York (1958)

Srinivasa Ayyangar, C.N.: The History of Ancient Indian Mathematics. World Press Private Ltd., Calcutta (1967)

Wilkinson, J.H.: The Algebraic Eigenvalue Problem. Clarendon Press, Oxford (1965)

World's most powerful supercomputer unveiled in US, p. 9. Deccan Herald Daily News Paper, Tuesday, 19 June 2018

Young, D.: Iterative methods for solving partial difference equations of elliptic type. Trans. Amer. Math. Soc. **76**, 92–111 (1954)

Methods and Algorithms for Creating and Reconfiguring Virtual Organizations

Anata-Flavia Ionescu

Abstract As organizations tend to specialize in ever narrower and more diverse activities, virtual organizations (VOs) have gradually become a topic of interest among researchers representing numerous fields, ranging from technical domains such as optimization and soft computing to work psychology and organization studies. Due to the ad-hoc, temporary nature of VOs, most of the research attention has been devoted to optimizing the selection of partners for the strategic alliance, and this concern still holds a significant share of the current research agenda in the VO literature. This chapter reviews the most prominent approaches to solving partner selection problems. We present and discuss some of the most documented methods and algorithms for VO creation and reconfiguration, as well as a number of example implementations in applied research. Gaps and future research directions are identified.

Keywords Virtual organization · Virtual enterprise · Partner selection problem · Multi-objective optimization · Heuristics

1 Introduction

Virtual Enterprises (VEs) and the more general Virtual Organizations (VOs) are temporary alliances of organizations formed to serve one specific purpose or goal, i.e., develop one specific product (Camarinha-Matos and Afsarmanesh 2001). The common goal serves as the main binder for the VO and, once the goal disappears, the VO dissolves. Examples of VO purposes include event management, large-scale construction projects, media and entertainment projects, and consulting services (Von Danwitz 2017). Generally speaking, most forms of project-based work usually lend themselves to being carried out by VOs.

A.-F. Ionescu (✉)
Faculty of Mathematics and Informatics, Ovidius University of Constanta, 124 Mamaia Bd, 900527 Constanta, Romania
e-mail: anata.ionescu@univ-ovidius.ro

D. Flaut et al. (eds.), *Decision Making in Social Sciences: Between Traditions and Innovations*, Studies in Systems, Decision and Control 247,
https://doi.org/10.1007/978-3-030-30659-5_2

The general rationale for the existence of a VO is that no one single enterprise or organization holds all the resources necessary to attain the specific goal. One way to tackle this challenge is by having organizations team up to share resources. In this context, one of the main issues that arises concerns finding the optimal team for the specific purpose the VO is meant to serve.

The growing research interest in the VE/VO paradigm has been stimulated by researchers (e.g., Mladineo et al. 2017) diagnosing an ascending trend of replacing large enterprises with networks of small and medium-sized enterprises in the production of new products. In order to build and sustain competitive advantage, firms tend to engage in highly specialized, narrow activities meant to be integrated in complex chains or networks, thus generating an inter-firm specialization trend (Cunha and Putnik 2006).

Due to the inherent separation between form and function in VOs (Zato et al. 2012), integrating the contributions of the participating organizations such that the strategic alliance appears to the outside world as one seamless organization can be rather challenging. Though substantial bodies of research in different fields such as organization studies, project management, network studies, and optimization have addressed temporary alliances between organizations, their contributions remain disconnected (Von Danwitz 2017), giving VO researchers a lot of space to develop methodologies and applications that capitalize on results exogenous to their research area.

The chapter is structured as follows: in Sect. 2, we describe VO creation/reconfiguration in the multi-objective optimization framework and offer a review methods for creating and reconfiguring VOs; in Sect. 3 we present two example applications of VO creation methods; relevant conclusions and perspectives for future research are discussed in the final section.

2 Theory

Both VO creation and reconfiguration involve an issue that has consequently become one of the most prominent topics of interest in VO research: the ***Partner Selection Problem (PSP)***.

2.1 *The Partner Selection Problem (PSP) in VOs*

PSP loosely refers to the optimal choice of VO members, or more specifically to optimally assigning parts of the production process of the intended product to individual organizations. PSP has been proven to be an NP-hard problem (Ko et al. 2001). Because it involves simultaneous optimization of generally conflicting goals such as cost and time, PSP can be (though it may not always explicitly be) formalized as a multi-objective optimization problem.

Preliminaries. Multi-Objective Optimization Problems (MOPs). A MOP is a problem that involves the optimization of two or more criteria. These criteria are usually conflicting and trade-off solutions are therefore sought, governed by a partial order called Pareto dominance. The standard form of a MOP is the following (Elarbi et al. 2017):

$$\begin{cases} \min f(x) = [f_1(x), f_2(x), \ldots, f_m(x)]^t, \; x \in \mathbb{R}^n \\ g_j(x) \geq 0, \; j = 1, \ldots, p \\ h_k(x) = 0, \; k = 1, \ldots, q \\ {x_i}^L < x_i < {x_i}^U, \; i = 1, \ldots, n \end{cases}$$

where x_i, $i \in \{1, \ldots, n\}$ are **decision variables** (that influence the criteria), $f_l : \mathbb{R}^n \to \mathbb{R}$, $l \in \{1, \ldots, m\}$ are the **objective functions (criteria)**, g_j, $j \in \{1, \ldots, p\}$ are inequality constraints, and h_k, $k \in \{1, \ldots, q\}$ are equality constraints. We denote by $\Omega = \{x \in \mathbb{R}^n | {x_i}^L < x_i < {x_i}^U, x$ satisfies the $(p+q)$ contraints$\}$ the set of **feasible solutions**, or the **feasible search space**.

One solution $x \in \Omega$ is said to **Pareto-dominate** another solution $y \in \Omega$ if and only if $f_l(x) \leq f_l(y), \forall l \in \{1, \ldots, m\}$ and $\exists l \in l\{1, \ldots, m\}$ such that $f_l(x) < f_l(y)$ ($f(x)$ is partially better than $f(y)$, and we note $f(x) \prec f(y)$).

One solution $x^* \in \Omega$ is said to be **Pareto-optimal** if and only if $\forall x \in \Omega$, either $f_l(x) = f_l(x^*)$, $\forall l \in \{1, \ldots, m\}$, or $\exists l \in \{1, \ldots, m\}$ such that $f_l(x) > f_l(x^*)$ (there is no feasible solution that is better with respect to one criterion without being worse at other criterion, i.e., there is no feasible solution that Pareto dominates it). For a given MOP, the **Pareto optimal set** is $P^* = \{x \in \Omega | \nexists x' \in \Omega, f(x') \preceq f(x)\}$. For a given MOP, with $f(x)$ and the Pareto optimal set P^*, the set $\{f(x) | x \in P^*\}$ is said the **Pareto front**.

The **ideal point** $Z^I = (Z_1^I, \ldots, Z_m^I)$ is the vector composed of the best values of the objective functions over the search space:

$$Z_l^I = \min_{x \in \Omega} f_l(x), \; l \in \{1, \ldots, m\}$$

The **nadir point** $Z^N = (Z_1^N, \ldots, Z_m^N)$ is the vector composed of the worst values of the objective functions over the Pareto optimal set:

$$Z_l^N = \max_{x \in P^*} f_l(x), \; l \in \{1, \ldots, m\}$$

There is no absolute consensus regarding what solving a MOP means. Ideally, the entire Pareto front should be determined. In practice, however, most approaches confine themselves to either finding one or a few Pareto-optimal solutions, or generating an approximation of the Pareto front.

Methods for solving MOPs can be classified according to the involvement of a (human) decision maker (DM) (Coello et al. 2002; Miettinen 1999) in the following categories: (1) no-preference methods, and (2) preference-based methods, which are in turn divided into a priori, a posteriori, and interactive methods.

In the case of **no-preference methods**, no decision-maker is required to specify any preferences for the optimal solution sought. E.g., the ***method of the global criterion*** consists of minimizing the distance to the ideal point:

$$\min_{x \in \Omega} ||f(x) - Z^I||_p, \; 1 \leq p \leq \infty$$

In **a priori methods**, decision-makers specify their preferences *before* running the algorithm. Decision-making thus precedes the search for solutions. This category includes the weighted sum method; ϵ-restriction method, utility function method, lexicographic method, goal programming. For example, goal programming seeks to minimize weighted sum of deviational variables (deviations of criteria with respect to their target/aspiration levels, i.e., to the goals set for the respective criteria).

$$\min_{x \in \Omega} \sum_{l=1}^{m} w_l |f_l(x) - y_l^*|, \; w_l \in (0, 1) \; \forall l \in \{1, \ldots, m\}, \; \sum_{l=1}^{m} w_l = 1$$

In the case of **a posteriori methods**, decision makers specify their preferences *after* running the algorithm (they choose the preferred solution from the set of (sub)optimal solutions generated by the algorithm). This time, the search precedes the decision-making process. Examples are mathematical programming techniques such as Directed Search Domain (DSD; Erfani and Utyuzhnikov 2011), or evolutionary algorithms. Among the evolutionary algorithms, genetic algorithms (GAs) are the most popular.

GAs for MOPs can be either elitist or non-elitist depending on whether or not they store an "elite", i.e., a set of the best solutions so far (based on the Pareto dominance relation). In the following, we list some of the most prominent **non-elitist GAs**:

- ***Vector Evaluated Genetic Algorithm (VEGA; Schaffer** 1985)* divides the population of feasible solutions in m subpopulations and for subpopulation l, it optimizes (through selection) the objective l, $l \in \{1, \ldots, m\}$. Speciation is avoided by mating restrictions within the same subpopulation.
- ***Multiobjective GA (MOGA; Fonseca and Fleming** 1993)* assigns to every individual a rank equal to the number of individuals that Pareto dominate it. MOGA diversifies solutions (avoids overcrowding and thus early convergence) through a fitness sharing (niching) function and mating restriction mechanisms.
- ***Niched Pareto GA (NPGA; Horn et al.** 1994)* uses a binary tournament selection procedure: two individuals from the population are selected and compared based on Pareto dominance to the individuals in a randomly selected sample from the population. If neither or both competing individuals are dominated by at least one solution in the random sample, a niching function is used to determine the winner (the least crowded of the two).
- ***Non-Dominated Sorting Genetic Algorithm (NSGA; Srinivas and Deb** 1994)* builds successive Pareto fronts by assigning a rank to non-dominated individuals

and then excluding them from the population. Non-dominated individuals from the initial population are therefore assigned the rank of one, non-dominated individuals from the remaining population are assigned the rank of two, etc. A fitness sharing mechanism is then used to differentiate between individuals of the same rank.

Elitist GAs usually implement an archive mechanism that updates the set of Pareto dominant solutions at every generation. Several noteworthy examples are listed below:

- ***NSGA-II (Deb et al.*** 2000***)***, an improved version of NSGA, is currently one of the most cited and competitive GAs for solving MOPs. It uses an optimized ranking algorithm called fast non-dominated sorting, it introduces a crowding factor to differentiate individuals with equal ranks. Being an elitist algorithm, it keeps an archive of Pareto dominant solutions.
- ***Indicator-Based Evolutionary Algorithm (IBEA; Zitzler and Künzli*** 2004***)*** is a general algorithm that uses what is termed a *quality indicator* as a fitness function. More specifically, the fitness of an individual is given by the loss in quality of the population if that particular individual were to be eliminated. At each generation, the individual that minimizes the loss in quality is removed from the population. In fact, rather than an algorithm, IBEA is a class of as many algorithms as proposed valid quality indicator functions. For example, in the ***S Metric Selection-based Evolutionary Multi-objective Algorithm (SMS-EMOA; Beume et al.*** 2007***)***, the quality indicator is the hypervolume determined by a reference point and the worst Pareto front given by the Deb et al.'s fast non-dominated sorting algorithm.
- ***Multi-objective Evolutionary Algorithm Based on Decomposition (MOEA/D; Zhang and Li*** 2007***)*** is another recent innovative approach that consists of decomposing the MOP in N single-objective sub-problems; each sub-problem corresponds to a weight vector for the criteria. Each individual in the population is assigned a weight vector and should solve the MOP by optimizing the scalarizing function based on the weight vector. Mating is restricted to weight vectors in the same neighborhood.

Finally, in **interactive methods** decision makers specify their preferences *during* algorithm execution (by describing how the optimal solutions generated at each iteration could be improved). In these cases, search and decision processes interweave. For example, in the ***Light Beam Search (LBS)*** method, the DM specifies an aspiration point and a reservation point (by default, the ideal and the nadir point, respectively), and local preference information. Then, the algorithm iteratively projects the direction determined by the two points on the Pareto front, calculates an outranking neighborhood of the projection based on the DMs preferences, and prompts the DM to either stop the algorithm, choose one of the calculated neighbors, update the preference information, or define other aspiration and reservation points for the next iteration.

Nevertheless, please note that most methods currently classified as a priori or a posteriori can be turned into interactive methods. For example, iterative methods can generally be adapted to allow DM feedback after every iteration. It is also worth mentioning that another classification of methods for solving MOPs (Elarbi et al. 2017) opposes **classical (aggregative) methods**, that turn the MOP into a single-criterion optimization problem by combining all of the objective functions in one, and at each run of the algorithm it generates a single element of the Pareto optimal set, to **population-based heuristics** (such as the evolutionary methods) that simultaneously generate more elements of the Pareto optimal set at each run.

2.2 PSP Criteria

A minimal set of objectives in the great majority of PSPs comprises cost (price), quality of the product (service), and time. Other criteria, reflecting modern business concerns such as safety and environmental issues, have ulteriorly been added to the list. Please note, however, that these objectives are rather general, and applied research has derived many more specific criteria. Huang and Keskar (2007) proposed a taxonomy of objectives that offers a fine-grained view of PSP criteria. E.g., cost-related metrics such as inventory turns, payment terms, return policy, and warranty cost were classified by the authors under the "Cost and financial" category, while published delivery cycle time, order fulfillment lead time, and return product velocity were subsumed under "Responsiveness".

Another aspect worth highlighting is the fact that PSP is usually a multi-objective combinatorial optimization (MOCO) problem, because more than one partner is regularly needed, and choosing the best partner for each part of the decomposed production process does not necessarily guarantee the best team or vice-versa. VO goals should be decomposable in tasks, activities, etc. susceptible of being organized in input-output chains, generating an activity network of precedence relationships (Crispim and de Sousa 2010) and therefore a high degree of interdependence between VO partners.

Therefore, optimizing the criteria for each individual partner may not lead to the optimal partnership. Instead, in these cases, common criteria oftentimes apply only to the team as a whole. For example, the quality of a product produced by a VO may depend not only on each partners performance in executing its share of the production process, but also on how these processes are integrated. Similarly, the delivery time depends on how the activities are scheduled, which in turn depends on the composition of the team. Price can be also subject to variation due to integration in a specific network of partners. Task precedence constraints are among the most common factors that make objectives in a PSP depend on the VO as a whole (Huang et al. 2017). Furthermore, additional network-specific objectives such as, e.g., minimizing transportation distances may arise.

Consequently, Wang et al. (2009) group PSP-specific objectives into three major classes:

- **Independent goals**, pertaining to the individual partners or their individual contributions (e.g., the quality of a specific service provided by one of the partners, its cost, or the risks associated to it).
- **Association goals**, pertaining to the VO as a whole, such as collaboration costs.
- **Constraint goals**, usually time constraints.

2.3 Approaches to PSP

Early PSP research had treated it as a single-objective optimization problem. Most early approaches to supplier selection were chiefly cost-based, i.e., they focused on minimizing an overall cost-related objective, given time-related constraints (Huang et al. 2017). However, multi-criteria decision making has gradually come to far out-represent such approaches, both in the literature and in practice (Ho et al. 2010).

PSP approaches cover both exact and heuristic algorithms. Optimization algorithms are said to be exact when they provide optimal solutions, and heuristic when they generate approximate (suboptimal) solutions. The latter are often preferred for many real-world applications because, rather than finding an exact solution, applied research oftentimes seeks to generate approximate solutions to complex problems (e.g., NP-hard problems) in a reasonable computational time. The number of Pareto-optimal solutions for a given MOP can be very large (and this can be thè case with PSP, too). Therefore, a method that can provide a diverse and evenly-distributed approximation of the actual Pareto front in a reasonable time is preferred (Elarbi et al. 2017).

Aggregative Exact Methods. According to reviews by Ho et al. (2010) and Wu and Barnes (2011), mathematical programming based approaches to PSP are among the most popular. One of the oldest established method is Data Envelopment Analysis (DEA). DEA classifies criteria as inputs and outputs and computes the efficiency of a solution based on a ratio of the weighted sum of outputs and the weighted sum of inputs. DEA is appreciated for its robustness, but the required classification of criteria under inputs and outputs is sometimes invoked as a limitation. Other MP approaches are often derived from goal programming (e.g., Karpak et al. 2018; Ravindran et al. 2010).

Other widespread approaches to PSP are based on the Analytic Hierarchy Process (AHP) method. The decision problem is organized in a hierarchical structure with the goal of the VO at the top, the criteria (objectives) to be optimized at the middle, and the alternatives (in our case, potential partnerships) at the bottom level. One or more decision makers fill in a decision table in which each alternative is evaluated (quantitatively of qualitatively) with respect to each criterion. Then, a pairwise comparison matrix for the criteria is created, containing the decision makers judgements of the relative importance of criteria with respect to each other. This judgement matrix is then used to derive criteria weights. The alternatives can finally be evaluated and ranked as weighted sums of their criteria evaluations (Rao 2007).

Most of these approaches have the advantage of guaranteeing an optimal solution. However, one general disadvantage of this category of methods is their reliance on human decision makers for the assignment of weights or priorities to criteria, which may be a rather challenging task, all the more so as many of these methods are highly sensitive to criteria priorities. Additionally, they often require many conditions for the objective functions, and are therefore not suitable for any MOP. The approaches presented in the following two subsections are heuristics, and are applicable to a wider range of MOPs, though they do not guarantee optimal solutions.

Neighborhood Search (Local Convergence-Based) Algorithms. For Neighborhood Search Algorithms (NSA), the solutions are generated sequentially along the Pareto front, producing a local convergence to it (Ehrgott and Gandibleux 2008).

Multi-Objective Simulated Annealing (MOSA) is a generalization of the simulated annealing method (suitable for single-objective optimization) to MOPs. An initial solution is randomly generated and a set of criteria weights is used to determine the search direction. At each iteration, a neighboring solution on the search trajectory is compared to the current solution. The neighbor always replaces the current solution if it Pareto dominates it. Conversely, if the neighbor is not Pareto dominant, it can still replace the current solution with a probability that decreases with the "temperature" of the system. The system "cools" as a function of the weighted distance between the current and new solution. Each accepted solution is added to the set of potentially optimal solutions. The algorithm stops at a predetermined number of iterations or temperature.

Multi-Objective Tabu Search (MOTS) is an extension of the tabu search to MOPs. Tabu search is an iterative method of neighborhood exploration whose main particularity is it that stores a history of the visited items (in our case, potential partnerships) in order to avoid cycling through solutions. Within each iteration of a "tabu process", MOTS guides the search of a dominant neighboring solution according to a reference point and a weighted scalarizing function of the objectives. Sequential tabu processes are run in order to vary weights. Two types of "memory" are considered in MOTS: one on the decision space, to avoid cycling, and one on the objective space, to guide the search in the direction of improving objectives and update weights for the next tabu process. An archive of the best solutions so far is updated after each iteration.

Population-Based (Global Convergence-Based) Algorithms. For these methods, the generation mechanism is parallel along the Pareto front, thus they converge globally to the front (Ehrgott and Gandibleux 2008).

Particle Swarm Optimization (PSO) is a population based heuristic inspired by natural "swarm intelligence" found in animal groupings such as bird flocks or fish schools. A population of individuals ("particles") is generated with the purpose of exploring the search space. At each iteration, a particle has three vectors stored in memory: one representing the current position of the particle (i.e., a feasible solution), one representing the particles velocity, and one for the best solution encountered so far. The particle updates its velocity based on a function of these three vectors and an additional vector representing the best position (solution) found by the entire swarm so far. The position is then updated according to the new velocity.

Ant Colony Optimization (ACO) methods mimic the ant foraging behavior and have recently been adapted to MOPs. A colony of artificial ants is generated in random points of the feasible space. The trajectory of each ant can be determined, for example, by a random assignment of criterion weights. Finding a good solution in its way determines an ant to mark the path to the solution with a pheromone trail to let other ants know of the solution and follow that trail with a greater probability, reinforcing it. The pheromone strength depends on the quality of the solutions.

Evolutionary approaches are dominated by GAs. VO researchers adopting this approach are either inspired by influential genetic algorithms for solving MOPs such as NSGA (e.g., Chuang et al. 2009), or NSGA-II (Ding et al. 2005), or develop their own GAs (e.g., Zhang et al. 2013). Criticism of GAs for PSP includes feasible solutions that cannot be generated by the crossover operator (Wu and Barnes 2011).

Other Methods and Techniques. It is worth mentioning that most of the aforementioned methods have been adapted or hybridized so as to accommodate uncertainty. The qualitative nature of certain criteria (such as quality of service), as well as the vagueness of some common criteria (such as cost or time) in some PSPs, or the vagueness of their relative importance (priorities) calls for integration with models of uncertainty. The most common approach to representing and operating with uncertainty in PSP is, by far, fuzzy set theory. Vague criteria and their hierarchies can be expressed through linguistic variables by decision makers and treated as fuzzy numbers. Various fuzzy aggregation operators are used to aggregate the decision makers fuzzy preference information.

Other artificial intelligence techniques are sometimes employed, usually with the purpose of enabling the system to learn from experience. Expert systems using case-based reasoning and artificial neural networks are typical examples. Solutions proposed by the system are inferred using performance evaluations of decisions in previous similar situations (Wu and Barnes 2011).

Hybrid Approaches. The combinatorial nature of PSP has induced a growing interest in population-based algorithms among PSP researchers. Nevertheless, scholars have noticed that these meta-heuristics have limitations that can be compensated by their integration with NSA or exact methods. The first hybrid algorithms for MOPs combined local and global convergence based heuristics to neutralize each other's disadvantages. The former converged more quickly, but to a single Pareto optimal solution, whilst the latter converged to the entire frontier, but slowly. Metaheuristics can be hybridized among themselves for reasons such as making a population-based method more aggressive (e.g., adding a neighborhood search exploitation mechanism to advance more quickly from individuals generated with population based heuristics towards the Pareto front), or "driving" a NSA method (i.e., using a population-based algorithm to generate search directions for a better coverage of the Pareto front). Similarly, they can be hybridized with exact methods by either providing a rather diverse set of good quality solutions as starting points for the exact method (in order to find a diversity of Pareto optimal solutions) or by including a set of exact solutions in the initial population in order to improve the performance of a population-based algorithm. Last but not least, integrating preference information from a decision-maker can also be considered a form of hybridization (Ehrgott and Gandibleux 2008). Current approaches to PSP are dominated by hybrids of population-based heuristics and classical methods such as DEA, AHP, or PROMETHEE (see, e.g., Chuang et al. 2009; Goyal and Kaushal 2018; Mladineo et al. 2017), or with other artificial intelligence techniques (e.g., Zato et al. 2012).

2.4 Reconfiguration-Specific Issues

Creating and reconfiguring VOs raise similar concerns, but reconfiguration, inasmuch as it is frequent and unpredictable, introduces additional challenges. Be it triggered by changes in product specifications or in partners contracts, reconfiguration requires what VO researchers have termed “agility”. Agility subsumes dynamism, proactivity, flexibility, and context-specific behavior. Above all, it implies rapidness (Cunha and Putnik 2006). An agile VO should listen and quickly react to changes in the dynamic environment.

Reconfiguration incurs re-assigning tasks or roles, as well as rescheduling activities. It therefore adds objectives to the classical creation MOP: besides minimizing the time and cost of the production process, the time and cost of the transition to the new configuration should also be minimized; needless to say, the robustness of the product should not be affected by this transition. Moreover, a turbulent environment pressures product design to consider reusability, reconfigurability, scalability, and other principles that allow easy transitions from one configuration to another (Cunha and Putnik 2006).

Relatively few researchers have been concerned with VO reconfiguration. An exception is provided by Zato et al. (2012), who presented an approach for dealing with reconfiguration issues using a neural network, queueing theory and case based reasoning to dynamically assign roles to agents in an e-government multi-agent system.

3 Results

In this section, we chose to describe two applications of the reviewed PSP methodologies in industry. Our choice was guided, besides the number of citations, recentness, and impact of the outlet, by representativeness for PSP approaches.

In their paper on production optimization in a defective supply chain system (with production losses in the multi-stage production process, along the way from suppliers to customers through manufacturers and warehouses), Yeh and Chuang (2011) tested multi-objective GAs. The four objectives to be optimized were cost (a sum of product cost and transportation cost, to be minimized), time (a sum of product time and transportation time, to be minimized), average product quality (to be maximized), and green appraisal score (to be maximized). Nine constraints were formulated for the problem. Their methodology consisted of identifying quantitative and qualitative parameters (decision variables) for the problem. A database of potential partners in each stage was created, storing parameter values for all possible combinations with a partner in the next stage of the chain. This information was used as input for two multi-objective GAs the authors compared in terms of average number of Pareto-optimal solutions and CPU time. The algorithm that performed best from both points of view had been proposed by Murata et al. (1996). This algorithm starts from an initial population of individuals satisfying all specified constraints. The fitness evaluation function was a weighted sum of objective functions, with weights randomly generated for each generation. The elitist strategy of the algorithm consisted of replacing, at each generation, a given number of less efficient solutions with a set of Pareto optimal solutions in an archive gene pool. The algorithm terminated at a predetermined number of iterations. The experimental study was based on a scenario built for an electronics company in Taiwan. A (4, 4, 4, 4) design was tested, with four potential partners for each stage of the chain.

A more recent study by Ghadimi et al. (2018) focused on sustainable supplier selection and order allocation. Their approach starts with a manufacturer defining Triple Bottom Line (TBL) sustainability attributes (environmental, economic, social) for the potential suppliers considered for manufacturing a given product. In a Supplier Evaluation sub-model, the manufacturer should then specify criterion weights and provide criterion scores for potential suppliers. These scores are inputs for a Fuzzy Inference System (FIS)-based algorithm that provides the final global evaluations of the potential partners (the TBL criteria scores). In turn, these evaluations are inputs for an Order Allocation sub-model, wherein a fuzzy max-min operator algorithm is used to determine the optimal order quantities. Calculating order quantities was approached as a bi-objective optimization problem, the two objectives being total purchasing cost (to be minimized) and supplier sustainability performance value (to be maximized). The authors proposed a Multi-Agent System (MAS) model consisting of four FIPA-compliant agents: a Database Agent (responsible for storing supplier evaluation data), a Supplier Agent (handles sustainability performance and order allocation results), a Decision Maker Agent (responsible for supplier evaluation) and an Order Allocator Agent (that calculates optimal order quantities). The authors implemented the model using the JADE platform and the open-source JFuzzyLogic library for the FIS, and GAMS with GAMSJavaAPI for the order allocation problem. The authors conducted practical experiments using a real-world scenario from the medical device industry and demonstrated that use of their application for a 24-week period facilitated information exchange with suppliers and generated an increase in profitability for the manufacturer.

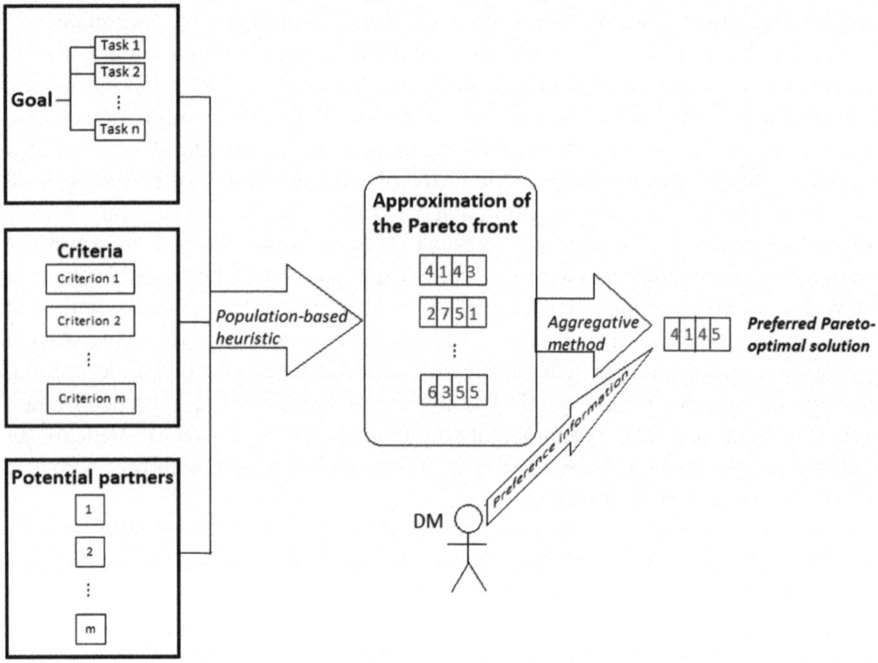

Fig. 1 A general scheme of an integrated PSP methodology

4 Conclusion and Perspectives

Our review of VO creation and reconfiguration theory and practice corroborates earlier observations by Ehrgott and Gandibleux (2008) that methods are becoming more and more: (1) hybridized—hybridizing metaheuristics between themselves and with other methodologies is becoming increasingly popular in solving MOCOs; and (2) specific to the problem they are meant to solve. This explains the combinatorial proliferation of PSP methodologies we are currently witnessing. Despite the diversity of integrated approaches, we can abstract some general scheme or pattern, like the one depicted in Fig. 1: after the main goal is decomposed into tasks and, in parallel, criteria and decision variables (and their contributions as influencing factors for the criteria) are established using some sources of expert knowledge, while a database provides the set of potential partners, a population-based algorithm is regularly used to generate an approximation of the Pareto front, and then some aggregative method uses some input preference information from a decision-maker to finally generate the desired Pareto-optimal solution.

Our review also suggests that VO methodologies can successfully be applied in a plethora of organizational settings, and we expect significant contributions for less explored scenarios. One cannot help noticing that most applications of PSP methodologies come from the manufacturing area, while some other organizational research areas have yet to full-heartedly embrace the VO paradigm, though applicable. Public VOs, for example, have garnered limited research attention, though papers such as the Zato et al. (2012) study demonstrate the suitability and usefulness of VO methodologies in such cases. Applied VO research has also demonstrated the appropriateness of Agent-Based Modeling (ABM) to VO problems. ABM is a handy approach whenever the problem domain is distributed, complex or large-sized (consisting of many variables, parameters, and constraints), heterogeneous, modular, and dynamic/changeable (Barbati et al. 2012). The organizations in the virtual market can be modeled as agents and negotiate to form alliances to respond to a specific request. ABM can, on the one hand, help automate processes such as partner selection, and on the other hand, provide information and feedback to decision makers and facilitate a deeper—and desirably interdisciplinary—understanding of these processes through simulation.

Last but not least, given the dynamic competitive economic environment, PSP should perhaps start to be almost exclusively regarded as a reconfiguration problem, rather than a creation problem. Despite the calls for more research attention to VO agility, VO reconfiguration research still appears to leave much room for advancements. A certain amount of research effort should perhaps be devoted to adapting some established PSP methods to accommodate environment dynamics issues, and even to proposing innovative reconfiguration-centered methodologies.

References

Barbati, M., Bruno, G., Genovese, A.: Applications of agent-based models for optimization problems: a literature review. Expert Syst. Appl. **39**(5), 6020–6028 (2012)

Beume, N., Naujoks, B., Emmerich, M.: SMS-EMOA: multiobjective selection based on dominated hypervolume. Eur. J. Oper. Res. **181**(3), 1653–1669 (2007)

Camarinha-Matos, L.M., Afsarmanesh, H.: Virtual enterprise modeling and support infrastructures: applying multi-agent system approaches. In: Luck, M., Marik, V., Stpankova, O., Trappl, R. (eds.) LNAI, vol. 2086, pp. 335–364. Springer (2001)

Chuang, C.L., Chiang, T.A., Che, Z.H., Wang, H.S.: Using DEA and GA algorithm for finding an optimal design chain partner combination. In: Global Perspective for Competitive Enterprise, Economy and Ecology (pp. 117–127). Springer, London (2009)

Coello, C.A., Lamont, G.B., Van, V.: Evolutionary Algorithms for Solving Multi-objective Problems. Springer (2002)

Crispim, J.A., de Sousa, J.P.: Partner selection in virtual enterprises. Int. J. Prod. Res. **48**(3), 683–707 (2010)

Cunha, M.M., Putnik, G.: Agile Virtual Enterprises: Implementation and Management Support. IGI Global, Hershey, New York (2006)

Deb, K., Agrawal, S., Pratap, A., Meyarivan, T.: A fast elitist non-dominated sorting genetic algorithm for multi-objective optimization: NSGA-II. In: Schoenauer, M., Deb, K., Rudolph, G., Yao, X., Lutton, E., Merelo, J.J., Schwefel, H.-P. (eds.) Proceedings of the Parallel Problem Solving from Nature VI Conference. Lecture Notes in Computer Science, vol. 1917. Springer, Paris, France, pp. 849–858 (2000)

Ding, H., Benyoucef, L., Xie, X.: A simulation-based multi-objective genetic algorithm approach for networked enterprises optimization. Eng. Appl. Artif. Intell. **19**(6), 609–623 (2006)

Ehrgott, M., Gandibleux, X.: Hybrid metaheuristics for multi-objective combinatorial optimization. In: Blum, C., et al. (eds.) Hybrid Metaheuristics—An Emerging Approach to Optimization, pp. 221–260. Springer (2008)

Elarbi, M., Bechikh, S., Ben Said, L., Datta, R.: Multi-objective Optimization: classical and evolutionary approaches. In: Bechikh, S., Datta, R., Gupta, A. (eds.) Adaptation, Learning and Optimization. Recent Advances in Evolutionary Multi-objective Optimization, vol. 20, pp. 1–30. Springer (2017)

Erfani, T., Utyuzhnikov, S.: Directed search domain: a method for even generation of the Pareto frontier in multiobjective optimization. Eng. Optim. **43**(5), 467–484 (2011)

Fonseca, C.M., Fleming, P.J.: Genetic algorithms for multiobjective optimization: formulation discussion and generalization. In: ICGA, vol. 93, pp. 416–423. Citeseer (1993)

Ghadimi, P., Toosi, F.G., Heavey, C.: A multi-agent systems approach for sustainable supplier selection and order allocation in a partnership supply chain. Eur. J. Oper. Res. **269**(1), 286–301 (2018)

Goyal, R.K., Kaushal, S.: Deriving crisp and consistent priorities for fuzzy AHP-based multicriteria systems using non-linear constrained optimization. Fuzzy Optim. Decis. Making **17**(2), 195–209 (2018)

Ho, W., Xu, X., Dey, P.K.: Multi-criteria decision making approaches for supplier evaluation and selection: a literature review. Eur. J. Oper. Res. **202**(1), 16–24 (2010)

Horn, J., Nafpliotis, N., Goldberg, D.E.: A niched Pareto genetic algorithm for multiobjective optimization. In: Proceedings of the First IEEE Conference on Evolutionary Computation, 1994. IEEE World Congress on Computational Intelligence, pp. 82–87. IEEE (1994)

Huang, B., Bai, L., Roy, A., Ma, N.: A multi-criterion partner selection problem for virtual manufacturing enterprises under uncertainty. Int. J. Prod. Econ. (2017). https://doi.org/10.1016/j.ijpe.2017.08.024

Huang, S.H., Keskar, H.: Comprehensive and configurable metrics for supplier selection. Int. J. Prod. Econ. **105**(2), 510–523 (2007)

Karpak, B., Kumcu, E., Kasuganti, R.: An application of visual interactive goal programming: a case in vendor selection decisions. J. MultiCriteria Decis. Anal. **8**(2), 93–105 (1999)

Ko, C.S., Kim, T., Hwang, H.: External partner selection using tabu search heuristics in distributed manufacturing. Int. J. Prod. Res. **39**(17), 3959–3974 (2001)

Miettinen, K.: Nonlinear multiobjective optimization, international series in operations research and management. Science **12** (1999)

Mladineo, M., Veza, I., Gjeldum, N.: Solving partner selection problem in cyber-physical production networks using the HUMANT algorithm. Int. J. Prod. Res. **55**(9), 2506–2521 (2017). https://doi.org/10.1080/00207543.2016.1234084

Murata, T., Ishibuchi, H., Tanaka, H.: Multi-objective genetic algorithm and its applications to flowshop scheduling. Comput. Ind. Eng. **30**(4), 957–968 (1996)

Rao, R.V.: Decision Making in the Manufacturing Environment Using Graph Theory and Fuzzy Multiple Attribute Decision Making Methods. Springer, London (2007)

Ravindran, A.R., Ufuk Bilsel, R., Wadhwa, V., Yang, T.: Risk adjusted multicriteria supplier selection models with applications. Int. J. Prod. Res. **48**(2), 405–424 (2010)

Schaffer, J.D.: Multiple objective optimization with vector evaluated genetic algorithms. In: Proceedings of the 1st International Conference on Genetic Algorithms, pp. 93–100. L. Erlbaum Associates Inc. (1985)

Srinivas, N., Deb, K.: Muiltiobjective optimization using nondominated sorting in genetic algorithms. Evol. Comput. **2**(3), 221–248 (1994)

Von Danwitz, S.: Managing inter-firm projects: a systematic review and directions for future research. Int. J. Project Manag. (2017). https://doi.org/10.1016/j.ijproman.2017.11.004

Wang, Z.J., Xu, X.F., Zhan, D.C.: Genetic algorithm for collaboration cost optimization-oriented partner selection in virtual enterprises. Int. J. Prod. Res. **47**(4), 859–881 (2009)

Wu, C., Barnes, D.: A literature review of decision-making models and approaches for partner selection in agile supply chains. J. Purch. Supply Manag. **17**(4), 256–274 (2011)

Yeh, W.C., Chuang, M.C.: Using multi-objective genetic algorithm for partner selection in green supply chain problems. Expert Syst. Appl. **38**(4), 4244–4253 (2011)

Zato, C., De Paz, J.F., de Luis, A., Bajo, J., Corchado, J.M.: Model for assigning roles automatically in egovernment virtual organizations. Expert Syst. Appl. **39**(12), 10389–10401 (2012)

Zhang, Q., Li, H.: MOEA/D: a multiobjective evolutionary algorithm based on decomposition. IEEE Trans. Evol. Comput. **11**(6), 712–731 (2007)

Zhang, Y., Tao, F., Laili, Y., Hou, B., Lv, L., Zhang, L.: Green partner selection in virtual enterprise based on Pareto genetic algorithms. Int. J. Adv. Manuf. Technol. **1–17** (2013)

Zitzler, E., Künzli, S.: Indicator-based selection in multiobjective search. In: Parallel Problem Solving from Nature-PPSN VIII, pp. 832–842. Springer (2004)

Some Remarks Regarding Traditions and Innovations in the Educational Process

Knowledge and Intellectual Capital Management in Higher Education

Mirela Minică

Abstract This chapter underlines the main challenges of the Romanian higher education, in the context of the EU strategy to increase the quality of tertiary education. By underlining the three basic and strongly interrelated components of Intellectual Capital (Human Capital, Structural Capital and Relational Capital), this article reveals the essential role of universities in creating skills and competencies required by the labour market in the 21st century. The case study outlines the teaching-learning-assessment based on the aims of education: ideal, purpose, objectives, implemented by the author at the Faculty of Economics.

Keywords Intellectual capital · Competencies · Tertiary education · Aims of education · Communication · Labor market · Educational design

1 Introduction

In order to correctly comprehend the present economic phenomenon, it is necessary to have a precise identification of the main society coordinates, of existing dogmas, of social archaeology, where the individual and the community manifest themselves. This phenomenon concerns postmodernism, whose pregnant manifestation of which seems to be "the knowledge-based society" (Bretcu 2016, p. 78).

The rapid and continuous changes produced in almost all fields of knowledge, thanks to intense concerns, in the last decades, for scientific research and innovation, lead to the rethinking of training and education, giving them a more relevant role in the social, economic, cultural development. The assuming of this role gives the new school new quality standards, which can be realized by streamlining and optimizing all available resources for the organization and the educational process. At the beginning of the third millennium, the Romanian school is connected, on one hand, to local

M. Minică (✉)
Faculty of Economic Sciences, "Eftimie Murgu" University of Resita, Piaţa Traian Vuia, no. 1–4, 320060 Reşiţa, Romania
e-mail: m.minica@uem.ro

D. Flaut et al. (eds.), *Decision Making in Social Sciences: Between Traditions and Innovations*, Studies in Systems, Decision and Control 247,
https://doi.org/10.1007/978-3-030-30659-5_3

and national traditions and realities, and on the other hand, to European trends which encourage and value research and innovation in education (Ispas 2013a, p. 42).

The correlation between education quality, economic performances (economic growth) and institutional environment are shown in numerous papers: Faruq and Taylor (2011), Kawamoto (2007), Wang and Wong (2011), Năstase and Hodoroabă (2010), whereas Khataybeh et al. (2010) consider that Education for Sustainable Development is absolutely vital for helping humanity to escape the present system crisis (Minică 2016, p. 17).

In order to correctly analyse communication features in this epoch, the paradigm of knowledge and its implications in post-modern society must be underlined.

Post-modern communication technology produces a revolution in knowledge, both due to its unprecedented vastness, allowing knowledge in a largely disseminated manner, and as a form of manifestation—without limits of interpretation—which sketches fluid and fractured perspectives. In fact, the analysis of knowledge triggers postmodernism—its core idea being that the "absolute" definitive and immutable truth does not exist, and thus indeterminacy replaces eternal and immutable values (Cărtărescu 1999, p. 19).

In the paper: *Implementation model of the student focused education in a university in Romania* (Roşca 2015) the main categories of educational beneficiaries: the student, their family, the employers, the society and the employees of the High Educational Institution (HEI) are highlighted (Minică 2016, p. 19).

2 Characteristics of Intellectual Capital at the Level of Tertiary Education

Tertiary Education (TE) is a key asset in knowledge-based economies: tertiary educated workers stimulate economy-wide productivity and growth, and are crucial for innovation and the use of new technologies (Aghion and Cohen 2004; Vandenbussche et al. 2006; Minică 2016, p. 19).

The academic environment must become one of the main vectors of implementation and development for the present phase of human civilization: the knowledge society.

European universities must face a large frame of challenge:

- A strong competition at an international level between public and private university, between traditional and "corporation" universities, developed by transnational companies;
- Internationalisation, transferability and education and research mobility;
- The need to adapt the national system to the demands of an European pace of higher education;
- The expectations of stakeholders and the need to insure transparency by using public funds from the perspective of costs-benefits analysis.

In this context, the management of intellectual capital becomes extremely important for universities, because knowledge is at the same time their main input and output.

The term of intellectual capital has been introduced for the first time by Kenneth Galbraith in 1969, who considered that it doesn't refer only to corporate intelligence, but more to a level of knowledge, abilities, aptitudes held by employees, thus representing the manner in which these are valued in concordance to the organization mission and objectives.

Human capital is thus defined as "a dynamic nexus of the human and the social capital of a company and its knowledge management" (Murale et al. 2010, p. 285).

The tri-dimension structure of intellectual capital has the following attributes at the university level.

Human capital is made of the educational capital (abilities gained by the didactic staff and by researchers in the educational process, but also outside it) and the biological capital (the individuals' physical abilities, most often synthesized through their health state).

William Petty, was the first to try to evaluate the necessary expenses concerning human education—as a source of qualifications and development—idea which occupies a central place in Adam Smith's economy theory. The theory of human capital was created during the decade 1961–1970, its main creators being considered the American economists: Theodor Schultz, Gary Becker and Robert Mincer.

The structural capital refers to communication and management processes of technical and scientific knowledge from an organization, manifested through organizational culture and the valuing degree of a technological potential (software data bases, patented inventions, licenses).

The relational capital aims a set of economic, social and political relations developed by universities in partnerships with their main stakeholders.

The Intellectual Capital Management (ICM) is a set of managerial activities aimed at identifying and valuing knowledge assets of organisations, leveraging these assets through knowledge sharing and creating new knowledge (Easterby-Smith and Lyles 2003; Holsapple 2003; Easterby-Smith 2003, p. 1).

The concept of IC and its implementation calls for a new perspective of strategic management, a new taxonomy, new management methods and new measurement systems (Intellectual Capital Center Croatia 2007, p. 35).

2.1 Universities—Learning Organisation

The Bologna Declaration issued in 1999 can be considered the starting point for a paradigm change in the European Higher Education Area. The European Qualification Framework (EQF) is the backbone of the paradigm change in Europe. It covers all aspects of qualification, including, but in no way limiting, the three cycles of the Bologna process, agreed on in 2008. Its implementation has started all over

Europe, it "*acts as a translation device to make national qualifications more readable across Europe, promoting workers' and learners' mobility between countries and facilitating their lifelong learning*" (European Commission 2013; Minică 2015, p. 135).

An increasing complexity of organization also determined the apparition of a new managerial orientation, inspired from psycho-social theories on organizations (the theory of human relations, the theory of human relations). New aspects of the organization were approached: employees' motivation, labour conditions, groups' dynamics, and informal dimension of relations between the organization members, etc. In addition, organizational development accentuates in an excessive manner the psychological and the sociologic aspects. Among its presuppositions the following can be mentioned (Păun 1999):

- changes in the human resources domain (organisation) in order to increase employees' life quality;
- flexible organizations which adapt rapidly to environment changes, to appearing conflicts;
- change in organization is not realized at once, but gradually, through change and development programs.

In the tertiary education domain, organizational development may be considered an *educational strategy* through which an analysis of internal orientation of the theory can be realized, that is the *learning organisation*.

The performance of a university is expressed not only by adapting to the new, but especially through its capacity to anticipate and produce change. A real change is realized from the inside towards the exterior, when its didactic staff proves initiative and creativity during courses and seminaries and as members of the organization. But in order to obtain a creative activism from university personnel, an organization culture must also be present. The university has its specific, this is the reason why the application of an organization development theory must keep in mind the values and autonomy of the institution actors (students, didactic and administrative personnel).

Change is realized among humans and through resources used by the organization, this is the reason why people must intervene in an active manner in the organizational life, to overpass obstacles met. The individual is aware of that the organization destiny is the same as his.

The new managerial optics promoted by organizational development give us a management pyramidal vision in favour of a holistic one; the organization thus resembling an open system which has numerous relations with the social. In the organization management, all actors are trained by eliminating limits between the management function and the execution one, this management being characteristic to *participative management* "which also becomes a situational, adaptable and flexible management, in rapport to concrete situations, by eliminating prefabricated solutions with solutions elaborated according to concrete situations and date" (Păun 1999, p. 44).

The learning organisation is not just a concept, but a reality which can't be ignored, organizational progress being dependent on its learning capacities, change is realized by redefining, resizing and restructuring the groups' and the collective's values. Thus change isn't realized individually, but with the help of the collective; individual change is realized collectively. Specialty literature underlines two definitions for the term: the first referring to the organisation that offers learning and the second to organizations that promote and produce learning. The first definition, the organization that teaches is associated to *staff development*, organization change thus referring to conception, objectives and values, activities, administration and management. The second definition is centred on *individual change,* the students' education being put on individual learning.

The two definitions intertwine, but can also exist separately, thus altering the proper activity of the organization. For example, in each educational organization the second meaning is the predominant one for the concept of learning organisation, but the management's lack of interest realizing, creating conditions, stimulating staff development will lead to negative effects, the educational staff becoming a conservatory one.

2.2 *Learning Outcomes*

All programmes from the *Qualifications Framework for the European Higher Education Area (QF-EHEA)* need to be written in terms of **Learning Outcomes,** proposed as an increasingly complex set of **competence** levels to be acquired by successful students.

According to the EQF reference system, Learning Outcomes represent *knowledge, skills* and *attitudes* that identify with the learning process.

- to promote quality assurance;
- to increase mobility and comparability of job profiles, skills and learning transparency;
- to exemplify a methodological approach for the expression and description of the curriculum;
- to describe the student's learning progress in terms of the knowledge acquired, the comprehension of that knowledge and the capacity to apply it;
- summaries of essential areas of learning that result from a course of study, mostly explicit and clearly expressed as "you will be able to", limited in number, written with a level of learning in mind, must also exist (Fig. 1);

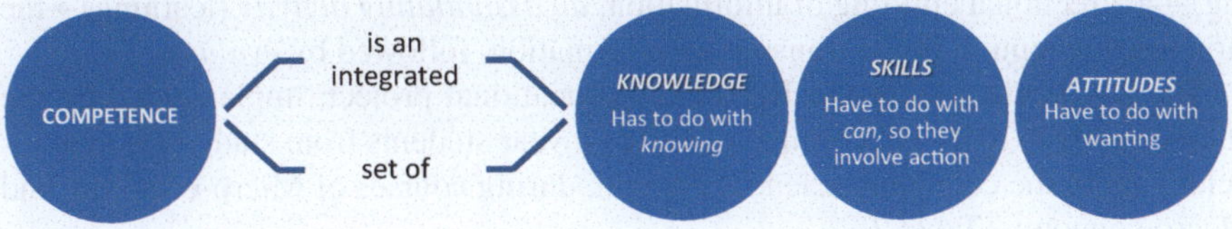

Fig. 1 How can competence be defined? *Source* Realized by the author

- Competences are contextual: someone is competent if he/she can use certain knowledge, skills and attitudes in a particular context and act appropriately;
- One of the first steps while planning a study programme is to define the group of competences that students should achieve during their studies;
- Competences must be taken into consideration by the curriculum or study programme development process (planning, development/fulfilment and assessment results).

The success of the educational process depends on all the factors involved in teaching. Educational psychology studies the learning processes carried out in school (and beyond school), the factors that facilitate learning, that motivate and contribute to the establishment of the physical, social and mental well-being (Ispas 2013b, p. 54).

According to Atkinson's achievement and motivation theory (Atkinson and Feather 1966):

$$Motivation = function\ of\ (Reason \times Expectancy \times Stimulant)$$

From the student's perspective the reason for choosing a certain study programme is correlated to his expectations for the future and the environment of teaching-learning-evaluation in which the university activity is realized.

In order to develop a curriculum through a *constructive alignment* (Biggs 1996, 2003) a congruence between intended learning outcomes, teaching and learning activities and assessment must be achieved.

Curriculum, as an ongoing social activity, where change, both in practice and the curriculum context is engendered through "day-to-day" interactions of students, teachers, knowledge, also appears as a response to influences of an increasingly changing society (Table 1).

3 The "RESPECT" Educational Project

The main purpose of this chapter is to highlight the educational strategy applied to economic sciences, representing a methodological model of teaching—learning—evaluation, and the foundation of an experimental project to identify the best teaching methods tailored according to students' learning styles (Minică 2016, p. 19).

Learning represents a cognitive process that supposes *a tutoring activity* (teaching)—a directional emitting of information, *an assimilating activity* (learning)—the structured reception of the transmitted information, followed by *evaluation*.

The key landmarks of the RESPECT educational project, implemented by the author in 2014–2016, on a sample of 180 first year students from study programmes belonging to the economic sciences domain, during courses of Micro-economy and Macro-economy (Table 2).

Table 1 Factors of a bi-factorial motivation model at the level of tertiary education

Intrinsic factors (*content factors*)	Extrinsic factors (*context factors)*
The Realization, or a continuous fulfilment of interesting learning tasks and the perception of one's effort results	*Technical supervision* or the educator's competence
Recognition, as appreciation of well done work through a final grade obtained	*The University politics and administration*
Responsibility for one's work and of others, control possibilities over your work	*Earnings:* a high level of future income
The Work itself, the content of work (varied or monotonous)	*European recognition* of one's diploma
Development, learning new things, new abilities	*Safety*: a real chance of integrating to the labour market
Promotion, a hierarchical change of the graduation status	

Sources Adapted by the author to the specific of university education, from Mamali (1981, p. 144)

After long discussion with educational experts and with representatives of the business environment, three learning strategy aces were identified and a number of seven specific attributes (Table 3).

Each attribute was evaluated during and at the end of the semester with the help of auto-evaluation questionnaires.

The anterior curricular and methodological design was realized by identifying students' characteristics after the application of a set of 3 tests and the adaptation of teaching methods to the dominant learning style (visual, auditory or kinaesthetic).

During the two semesters of the academic year, different individual and group working methods were tested, by putting an accent on the realization of projects, case studies and internship stages.

In establishing the content and objectives of courses, the necessity of fulfilling all 7 attributes of the model proposed was kept in mind, together with the recommendation of business experts, so after graduation, students could possess a knowledge and understanding mechanism that would allow them integrate on the labour market.

Communication was realized with the help of the Edmodo educational platform, fact which allowed transparency and operability during all the phases of the educational project.

In the design of courses materials and of seminary and applications notebooks, what Toohey (1999, p. 21) presents as the typical model of course design was taken into consideration (Fig. 2).

Table 2 The structure of the RESPECT educational project

Phase	Activity	Task	
Project initiation	Development of an educational model	Identification of concepts Defining elements of a learning strategy Analysis of the RESPECT model with the help of specialists from the academic community Specialists round table: Discussion with the business and educational environments Establishing of educational goals	
Project planning	Projection of deliverable materials and of didactic resources	Curricular design	Modular projecting of courses materials
		Methodological design	Realization of methodological worksheets: Time-sheet model Lecture sheet model Brainstorming sheet SINELG lecture sheet model Students' scientific paper model Schema model LOTUS Technique sheet Mosaic method Thinking Hats methods Pyramid method Stellar explosion method Cube method Aquarium method Creative controversies Brunch method Q&A thematic sheets Key terms method T table Little Words Intensified speech Discussions networks Gallery tour Value Line Text comment model Team project model Power point presentation method Reflexive paper method

(continued)

Table 2 (continued)

Phase	Activity	Task
	Initial evaluation	Communicational profiled of the D.I.S.C. MODEL (Dominance-Influence-Steadiness-Compliance) Questionnaire regarding learning styles BELBIN test for the establishment of team roles
	Realisation of individual and team plan tasks and for students	Planning of individual activities and of team projects Performance statement establishing a minimal accepted success level
Project execution	Ongoing evaluation	Weekly evaluation—Reflexion of the day Seminary evaluation Module of critical evaluation Knowledge evaluation—Edmodo Platform Quiz Evaluation of objectives realization
	Sustainment and evaluation of team projects	Sustainment and evaluation of team projects (collegial evaluation + evaluation realized by the educator)
	Additional activities	Work visits Invitation of specialists during didactic activities Internship stages
	Students' involvement in extracurricular activities	Educational projects in partnerships with schools and the business environment Voluntary activities
	Realisation of scientific papers	Students' communication sessions Future Economists' Olympics organized by the Economics Faculties Association in Romania (AFER)
Project closure	Final evaluation	Evaluation wheel Self-evaluation of objectives realized
	Analysis	Analysis of results obtained in comparison to intermediary evaluations data gathering and results analyses of satisfaction tests on modules Realisation of a synthesis report
	Educational model improvement	Reviewing and completing the model proposed
	Results multiplication	Application of an improved model to other specialties

Source Made by the author

Table 3 Attributes of the RESPECT educational model

Cognitive axis	**R***esponsibility, rationality*
	E*fficiency, balance*
Social behaviour axis	**S***ustainability, social abilities*
	P*ersonalization, practical character*
	E*quity, empathy*
Competences and attitudes axis	**C***ommunication, creativity*
	T*ransdisciplinarity, technology*

Source Made by the author

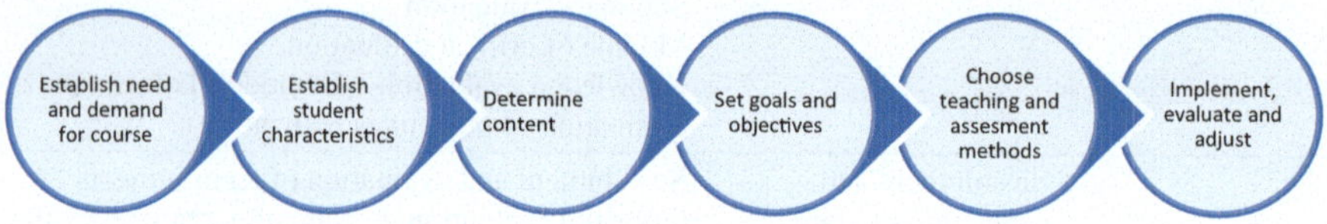

Fig. 2 University course design. *Source* Made by the author

4 Conclusions

This article has as main objective the underlining of the need to adapt superior education to particularities of the new students' generation, to challenges of a knowledge society, to the internalization of education and research, and to the important task of developing useful competencies for integrating graduated on the labour market.

Competences are always associated with a particular meaningful context. One is indeed competent if he/she can use certain knowledge, skills and attitudes in a particular context and if he/she act appropriately. Throughout the study programme development, three moments can be clearly described: the curriculum planning, its development or fulfilment, and the assessment results. Within all of them, competences must be taken into account (Caggiano and Gordon 2016).

The author's endeavour of designing and implementing an educational project in the economics sciences domains enlists in this goal, but it is not complete without implementing a strategy at the level of the entire university, which could be permanently reviewed and adapted to the challenges of an European Space of Higher Education.

A very important objective was the easy realisation of a transition from high school to universities of students and later, their integration on the labour market, from an affective and motivational perspective.

Starting with the Affective objectives taxonomy described by D. Krathwohl through the RESPECT education projects, the following could be realized:

1. *Reception*—being aware of a message by differentiating, accepting, accumulating, combining, choosing, listening, controlling and assimilating concepts specific to economic sciences;
2. *Reaction/answer*—generates a feeling through discussions, practice, leisure time activities that involve specialty literature and the analysis of economic shows;
3. *Valuing*—a preference for value—it refers to encouraging, offering assistance, arguing, debating with colleagues during team working activities;
4. *Organization* of a value system was underlined by putting into theory of themes, discussion, conclusions formulated and the harmonization of different points of view and creative controversies;
5. *Characterisation*—a generalized organization—was realized at the end of each semester by every student, while making valuable appreciations on his/her activity and that of his/her educator, assuming more numerous and more complex working tasks during the second semester and collaborating, with supplementary hours, during the educational project.

Frequent modification in the Romanian educational system most of the times modified on political grounds, without an proper analysis of impact measures, lead to a very difficult situation in universities.

A decrease of school population for demographic external migration reasons, the imposing of extremely difficult to realize conditions by authorizing and conducting study programmes in the lack of correlations with finances and to the need to adapt to international demands from the perspective of this research results, and in the conditions of didactic norms, of high bureaucracy, low incomes etc., all these being problems confronted by the Romanian superior education.

In these condition, the daily struggle for survival has demoralized and demotivated most educators belonging to higher education and students, and as a result the quality of the teaching-learning-evaluation process had to suffer.

Among the criteria sets evaluated for the promotion of didactic personnel, the pedagogic component and the working with students component are missing, the results of the research being the only ones of importance, because the lack of internal funds and the access to international ones is conditioned by experience and restricted by the membership of internationally recognized research teams.

Individual approaches may offer an example of good practice only by calling, but these are not sufficient if they are appreciated and stimulated on a systematic basis.

The adaptation of a teaching-learning-evaluation process to the challenges of a knowledge society and to the specificity of a digital general (which according to a Hart Consulting study, are narcissistic, lacking attention during processes, rules and working procedures and have a low degree of responsibility) is an extremely difficult task, that needs a participative organizational culture at the universities level. The author's effort will continue with the publishing of this chapter in hope that it will attract the colleagues' attention, at an international education level, thus leading to common partnership projects.

References

Aghion, P., Cohen, E.: Education ET Croissance. La Documentation Francaise, Paris (2004)

Atkinson, W.J., Feather, N.T.: A Theory of Achievement Motivation. Wiley, New York (1966)

Biggs, J.: Enhancing teaching through constructive alignment. Higher Educ. **32**(3) (1996)

Biggs, J.: Teaching for Quality Learning at University, 2nd ed. SRHE and Open University Press, New York (2003)

Bretcu, A.: Importance of the human factor in the knowledge-based society. Ecoforum J. **5**(1) (2016)

Caggiano, V., Gordon, J.: Curriculum design process of university education. European Project UNIVERSITARIA, POSDRU/157/1.3/S/135590 (2016)

Cărtărescu, M.: The Romanian postmodernism. Humanitas, Bucharest (1999)

Easterby-Smith, M., Lyles M.A.: Introduction: watersheds of organisational learning and knowledge management. In: Easterby-Smith, M., Lyles M.A. (eds.) The Blackwell Handbook of Organisational Learning and Knowledge Management. Blackwell Publications (2003)

Faruq, H.A., Taylor, A.C.: Quality of education, economic performance and institutional environment. Int. Adv. Econ. Res. **17** (2011)

Holsapple, C.W.: Knowledge and its attributes. In: Holsapple, W. (ed.) Handbook on Knowledge Management 1: Knowledge Matters. Springer, Berlin (2003)

Kawamoto, K.: Preferences for educational status, human capital accumulation and growth. J. Econ. **91**(1) (2007)

Khataybeh, A.M., Subbarini, S., Shurman, S.: Education for sustainable development, an international perspective. Procedia Social Behav. Sci. **5** (2010)

Intellectual Capital Center Croatia: Intellectual Capital: Handbook of IC Management in Companies, Zagreb (2007)

Ispas, C.: School through the eyes of the pupils in the fourth and the eighth grade. J. Educ. Sci. **1**(27), An XV (2013a)

Ispas, C.: The quality of the educational system in the vision of the Romanian teachers. Ann. "Eftimie Murgu" Univ Resita. 2nd Fascicle. Econ. Stud. **XX** (2013)

Mamali, C.: Motivational Balance and Co-evolution. Scientific and Encyclopaedic Publishing House, Bucharest (1981)

Minică, M.: Proposal for a modular approach to discipline microeconomics—fundamental landmark in designing learning strategies. "RESPECT", Ann. "Eftimie Murgu" Univ. Resita. 2nd Fascicle. Econ. Stud. **XXIII** (2015)

Minică, M.: Education—the foundation for sustainable economic development. The "RESPECT" learning strategy principles implemented in higher economic education. Ecoforum J. **5**(2), 9 (2016)

Murale, V., Jayaraj, R., Ashrafali: Impact of intellectual capital on firm performance: a resource based view using VAIC approach. Int. J. Bus. Manage. Econ. Inform. Technolo. **2**(2) (2010)

Năstase, C., Hodoroabă, A.: Education—an economic growth factor. Ann "Stefan cel Mare" Univ. Suceava Fascicle of the Faculty Econ. Public Administr. **10**(2), 12 (2010)

Păun, E.: School Socio-pedagogical Approach. Polirom, Iasi (1999)

Toohey, S.: Designing Courses for Higher Education. Open University Press, Philadelphia (1999)

Vandenbussche, J., Aghion, P., Meghir, C.:, Growth, distance to frontier and composition of human capital. J. Econ. Growth **2**(11) (2006)

Wang, M., Wong, M.C.S.: FDI, education, and economic growth: quality matters. Atl. Econ. J. **39** (2011)

Decision-Making Training of Teachers for Inclusive School

Cristina Ispas

Abstract Every child has the right to a quality education. All children need access to quality education in an inclusive environment. Inclusive education is a process by which school respond to diverse needs of children through increasing participation in learning and reduce exclusion. This involves a series of substantial changes in terms of educational content, pedagogical and didactic approach, the structure of the education system and educational strategies. The successful implementation of inclusive education involves training for teachers. In the inclusive school, the students and their parents participate in setting learning goals and take part in decisions that affect them. The relationship between a school, students and their families is dynamic and interactive. The inclusive school develops collaborative relationships with parents and communities through communication, participation and consultative decision-making.

Keywords Inclusive school · Training of teachers · Decision-making

1 Introduction

> A school for everyone is not only a right for everybody but a responsibility for all. Inclusive education is a civil and human responsibility act. (http://ictito.gov.it/index.php?option=com_content&view=article&id=587&Itemid=1522)

In the specialised literature, the ideas and policies of inclusive education appeared more clearly outlined in the latest years, as compared to the previous decades (Engelbrecht 2013, pp. 115–118; Rouse 2008, p. 16; Waitoller and Artiles 2013, pp. 319–356).

Unfortunately, there are not many researches regarding the way in which the teachers are supported in order to become promoters of inclusive education (Rouse 2008 apud Isosomppi and Leivo 2015, pp. 686–694).

C. Ispas (✉)
"Eftimie Murgu" University of Resita, 1-4 Train Vuia Square, Resita, Caras-Severin County, Romania
e-mail: c.ispas@uem.ro

D. Flaut et al. (eds.), *Decision Making in Social Sciences: Between Traditions and Innovations*, Studies in Systems, Decision and Control 247,
https://doi.org/10.1007/978-3-030-30659-5_4

Inclusive school also called *school for all* promotes learning and participation in the school activity of all pupils, regardless if they face learning difficulties or not due to some disabilities or to a different pace and styles of learning or of certain different cultural identities or of some socio-economic problems which affect the pupils' results in learning. The culture of inclusion supposes co-responsibility of everyone towards all, where the individual interest is in synergy with the others'; it represents holistic learning which involves interpersonal and emotional-affective interaction regarding learning, recognizing the multitude of intelligences, of different cognitive styles and the importance of social context. (http://ictito.gov.it/index.php?option=com_content&view=article&id=587&Itemid=1522) The inclusion aims the prevention of the learning barriers for all: children, young people or adults (Booth et al. 2003, p. 2). "*The school difficulties are many and of different types and often they are not the consequence of a single specific cause, but they are due to many factors which regard either the pupil, or the context in which he/she is found*" (Cornoldi 1999).

The training of teachers for inclusive school becomes a necessity of the present time. The complexity and diversity of educational needs of the pupils require, from the part of the school, adequate answers, meant to facilitate the access of all pupils to education, preventing and eliminating the risk of exclusion and/or social marginalisation. Inclusive school operates changes and modifications at the level of curriculum, the organisation of the didactic activity, the adopted educational strategies, the assumed responsibilities through the recognition of the right to education for all. The teachers represent the essential educational vectors in promoting the model of the inclusive school.

The debate regarding the importance of the inclusive school model's promotion in the community brings together the political, educational decision-makers, teachers, researchers, pedagogues, social workers, parents, children, representatives of the community etc. The inclusive education highlights the valorisation of the existing differences between pupils, the development of the maximal potential everyone disposes of.

2 The Research of the Teachers' Perception About the Importance of the Professional Formation in the Field of Inclusive Education

Purpose and objectives of the research—The identification of the teachers' perception about the importance of the professional formation in the field of inclusive education. The purpose was reached by the following objectives:

- The identification of the professional training activities in the field of inclusive education in which the investigated subjects participated in the last 5 years;

- Highlighting the main reasons and criteria which stay at the base of making the decision to participate in the professional training activities in the field of inclusive education;
- The identification of the teachers' perception regarding the use of professional training activities in the field of inclusive education.

The period of the research development—the research was achieved in the period May–September 2016.

***The investigated population*—the sample structure**—in the research there were 174 teachers from Caraş-Severin County, thus:

- 58 people (1/3 of the total sample, respectively 33.33%) are teachers for primary and pre-school education,
- 58 people (1/3 of the total sample, respectively 33.33%) are teachers from the secondary education system (gymnasium)
- 58 people (1/3 of the total sample, respectively 33.33%) are teachers from the secondary superior education (high school, professional schools).

If we analyse the investigated group from the placement perspective of the workplace, 87 people (1/2 of the total sample, respectively 50%) work in school institutions in the urban area and 87 people (1/2 of the total sample, respectively 50%) work in school institutions in the rural area.

Research methodology

The research was achieved with the help of the method *sociologic inquiry*, based on a questionnaire. For the accuracy of the data provided, the teachers involved in the research have anonymously completed the questionnaires. The items of the questionnaire explore the type of professional training activity in which the investigated subjects participated in the last 5 years; the main reasons which stayed at the base of the decision to participate in the training activities in the field of inclusive education; the use for the daily practical activity of the professional training activities in the field of inclusive education; the main criteria which contribute to the crystallization of the decision to participate in the professional training activities in the field of inclusive education.

Presentation and interpretation of results

In the presentation of data and their interpretation we have chosen the structuring of information according to the items used in the questionnaire.

According to the legislation in force, the continuous training of teachers "represents both a right and an OBLIGATION" (https://www.edu.ro/formare-continua). The teachers have the possibility to participate in a series of professional training activities. Given their interest towards the problems of teachers' training for the inclusive school, the first question explored the type of activities of professional development in which the interviewed teachers took part in the last 5 years (Table 1).

Table 1 The type of activities of professional development in which the teachers took part in the last 5 years

Answer variant	No. of answers	(%)
Initial training courses in the field of inclusive education	2	1.15
Continuous training courses in the field of inclusive education	52	29.89
Workshops in the field of inclusive education	112	64.37
Debates/round tables on the theme of inclusive education	127	72.99
Conferences and seminars on topics of inclusive education	72	41.38
Observation visits in other schools with programs for the integration of children with SEN	14	8.05
Observation visits in the school centres for inclusive education in the country or abroad	27	15.52
Participation in a network of teachers which act in the field of inclusive education	21	12.07
Individual research or in collaboration on topics of inclusive education	38	21.84
Thorough reading of bibliographic resources in the field of inclusive education	159	91.38
I have not participated in professional training activities in the field of inclusive education	0	0.00
Others. Which?	0	0.00

Q1. In the last 5 years, in which of the following professional training activities did you take part in? (Tick your answers in the box on the right)

The data collected have revealed the fact that all the teachers in the research have participated in the professional training activities in the field of inclusive education. From the total of 174 interviewed people 159 people (91.38%) have stated that they *have thoroughly read the bibliographic resources in the field of inclusive education* while 112 people (64.37%) have said that they have participated in "*workshops in the field of inclusive education*" and 127 people (72.99%) have taken part in "*Debates/round tables on the theme of inclusive education*". The results obtained in this item strengthen the fact that the *bibliographic research* is the most accessible way of teachers' training for those interested in this topic. Likewise, the organisation in the methodical committees at level of the school/town/county of some professional development activities of type *workshops, round tables/debates* on the topic of inclusive education have facilitated the participation of the teachers in this type of activities.

On the opposite pole there were variants such as "*initial training courses in the field of inclusive education*" (2 people—1.15%); "*observation visits in other schools with integration programs of children with SEN*" (14 people—8.05%); "*participation in a network of teachers which act in the field of inclusive education*" (21 people—12.07%).

Table 2 The main reasons which have determined the teachers to participate in the training courses in the field of inclusive education in the last 5 years

Answer variant	No. of answers	(%)
Desire of continuous improvement	162	93.10
The obligation of accumulating credits in the field of continuous training	131	75.29
The increase of the annual self-evaluations score	112	64.37
The need to acquire competencies in the field of inclusive education	132	75.86
The development of professional abilities of working with children with SEN	101	58.05
Perspectives of professional development	123	70.69
Others. Which?	0	0.00

The number relatively low of teachers who state that they have participated in the *continuous training courses in the field of inclusive education* (52 people—29.89%) can be explained also by the relatively reduced offer of courses on this theme. For example, in the school year 2015–2016 in Caraş-Severin there is only one course entitled "Education and inclusive school", course which did not benefit of professional credits but only the notice of the The House of the Didactic Staff Caraş-Severin.

Participation in the training activities in the field of inclusive education involves a process of analysis and decision from the part of the teacher at whose base stay a series of factors which influence the respective decision. The 2 items of the questionnaire explore the main reasons which have determined the investigated teachers to participate in the training courses in the field of inclusive education in the last 5 years (Table 2).

Q2. Which are the main reasons for which you participated in the training activity in the field of inclusive education in the last 5 years? (Tick the right box with the variants which correspond to your opinions).

All the variants of this item have obtained high scores (respectively over 50%). Analysing the answers received by the teachers participating in the research we observe that an overwhelming majority (93.1%) stated that one of the main reasons to participate in the training activities in the field of inclusive education in the last 5 years is represented by the desire of continuous *improvement*. We could understand that this desire represents the engine which puts in movement all the other components respectively it also trains other motivations in favour of the participation in other training activities in the field of inclusive education.

The European Agency for Development in Special Needs Education has achieved a study in the 25 member states of the European Union after which it outlined "the profile of the inclusive teachers" based on the professional competences regarding the inclusive education in the initial education of the teachers (European Agency for Development in Special Needs Education 2012). The consolidation of the professional competencies in inclusion require permanent connections in theory and practice, between the theoretical concepts and approaches present in the specialised literature and the practical experiences of inclusive education in schools (Heikkinen et al. 2012, pp. 3–30; Waitoller and Kozleski 2013, pp. 35–45).

The answer variant *The need to acquire competencies in the field of inclusive education* was indicated by 132 people, respectively 75.86% of the interviewed teachers. The inclusive education starts from the principle that each person has the right to a qualitative education regardless of the problems/difficulties they are confronted to. Being given the large nature of the students' deficiencies, of the difficulties met in the learning process, and the multiple causes which can generate these difficulties, there is a need of teachers who own and continuously develop the competencies in the field of inclusive education.

According to the legislation in force, the teachers are forced to participate periodically in programs of continuous training, thus, to accumulate, for every interval of 5 years, considered from the date of promoting the definitive exam in education, minimum 90 transferable professional credits. As a consequence there is a permanent preoccupation from the part of the teachers to acquire the number of credits foreseen by the law for continuous professional training. From the total of teachers interviewed 13 people, respectively 75.29% consider that the *Obligation to accumulate credits in the field of continuous training* represent an important reason for which they participate in the training activity in the field of inclusive education.

For 123 teachers, respectively 70.69% from the research sample, *the perspectives of professional development* created by the participation in the professional training activities in the field of inclusive education contribute in making decisions to participate in the activities of continuous training.

The annual self-evaluation is achieved based on the criteria where we can also found the continuous training activities. 112 interviewed teachers, meaning 64.37% from the total of the sample consider as being an important reason for *the increase of the annual self-evaluation score*.

The lowest score was registered by the variant *the development of the work abilities with the children with SEN* being indicated by 101 people, respectively 58.05%. A possible explanation of this result can be given by the fact that the activities for the development of the work abilities with the children with SEN are to be found within the activities meant to lead to the acquirement and development of competences in the field of inclusive education (Table 3).

Q3. In the activity with the pupils, how much did the professional training activities help you in the field of inclusive education you participated in the last 5 years?

Table 3 The degree to which the professional training activities in the field of inclusive education for the last 5 years have been useful in the activity with the students

	Not at all	Little	Average	Much	Very much	Other situations	Total
No. of answers	0	12	33	67	62	0	174
(%)	0.00	6.90	18.97	38.51	35.63	0.00	100.00

Other situations = I have not participated in any training activities in the field of inclusive education in the last 5 years.

The ultimate principle of inclusive education according to which all children should learn together whenever it is possible, eliminating any difficulty or difference with which they can be confronted (Mukhopadhyay 2012, p. 68), brings in the forefront the need of the teachers' participation in coherent and efficient programs of initial and continuous training for inclusive education.

Most of the answers received from the teachers participating in the research for this item are to be found distributed in the superior area of the grid. Thus, 129 people (74.14%) have chosen the variants of answer *much* and *very much* as follows: 67 people, respectively 38.51% from the total sample state that the professional training activities in the field of inclusive education used *much*, while 62 people, respectively 35.63% indicated the variant *very much*. 18.97% (33 people) appreciate that the training activities in the field of inclusive education were used at an average level and 6.9% (12 people) state that these activities of professional training were *little* used in their activity with the children.

The fact that most of the respondents have chosen answers from the median or superior part of the grid show the practical use of these training activities in the activities developed with the pupils. The answers which are found in the inferior part of the grid can have different explanations such as: ambiguous quality of the training activities, disinterest of the teachers for the training activities they participated in, the poor knowledge of the practical aspects from the teachers' part which can be applied in the activity with the pupils in class etc.

The teachers training for the inclusive education play a regulation and self-regulation part of the education process at functional level (regarding the reaching of the assumed educational finalities), structural (by valuing adequately all the available resources for education) and operational (by specific activities of design, achievement and evaluation of the didactic process and the obtained results). The evaluation by the teachers of their own needs of training in the field of inclusive education is reflected in their interest to participate in the present and future in specific activities of professional formation in this field (Table 4).

Q4. To what extent are you interested in participating in training activities in the field of inclusive education this year? What about the next 3 years?

Table 4 The training needs of teachers in the field of inclusive education in the present and in the next 3 years

	School years	Not at all	Little	Average	Much	Very much	Total
No. of answers	Current school year	0	3	27	65	79	174
(%)		0.00	1.72	15.52	37.36	45.40	100.00
No. of answers	Next 3 years	0	1	16	74	83	174
(%)		0.00	0.57	9.20	42.53	47.70	100.00

According to a study (Magrab 1999) achieved in the developed countries which are part of the organisation for Cooperation and Economic Development (OCDE) the teachers' training in the field of inclusive education constitute a powerful priority and a huge challenge having in view the generalisation of experiences and positive results.

Referring to inclusive education, Marchesi (1997) considers that teachers must prove their interest and competence towards inclusion. By interest towards inclusion, the author understands the teachers' attitude, their theories about the education of disabled pupils, difficulties and disadvantages completed by the teacher's desire to contribute to their education.

The data obtained in the framework of our research reveals the fact that the problems of inclusive education suscites the teachers' interest *much* and *very much*, this is why they are interested to participate in the activities of training in the field of inclusive education, both this year and also in the following 3 years. Surely inclusive education is not a trend, but a necessity in a world which tends toward globalization, in a world full of contradictions and searches which knows an explosion in the plan of technological development but it is also confronted to high rates of school abandonment or with low levels of instruction (Table 5).

Q5. Which are the main criteria in your decision to participate in training courses in the field of inclusive education?

The decision to participate in the training activities in the field of inclusive education is the fruit of the analysis of a series of criteria. It is obvious that such criteria are tightly connected by the reasons which determine the teacher to choose or not the participation in such activities of continuous training. Thus, the teachers participating in the research in an overwhelming number state that when the training activities are selected, they take into account *the topics of the training activities* (167 people, respectively 95.68%), *the credits number of continuous training* (157 people respectively 90.23%), *the nature of the training activities*: *training courses, conferences/seminars, other didactic activities* (154 people, respectively 88.51%), *the venue of the training activities* (142 people, respectively 81.61%). All the variants

Table 5 The main criteria to participate in training courses in the field of inclusive education

Criterion	No. of answers	(%)
The nature of the training activities (training courses, conferences/seminars, other didactic activities)	154	88.51
The themes of the training activities	167	95.98
The trainer's competence	121	69.54
Participation fees	135	77.59
The venue of the training activities	142	81.61
Period of development of the training activity	132	75.86
Number of credits in continuous training	157	90.23
Others. Which?	0	0.00

Table 6 The degree of importance of the training activities in the field of inclusive education

	Not at all important	Little important	Quite important	In a great extent important	Very important	Total
No. of answers	0	0	34	67	73	174
(%)	0.00	0.00	19.54	38.51	41.95	100.00

registered high scores, which confirm the importance that the teachers give to all the elements which represent the development in the best conditions of the training activities (Table 6).

Q6. How important are your professional training activities in the field of inclusive education?

> Inclusive schools have a belief that all children can learn and that all of them can benefit when that learning is done together. (Daniels and Garner 2012, p. 17)

The teachers participating in the research consider the professional training activities in the field of inclusive education being *quite important* (34 people, respectively 19.54%), *in a great extent important* (34 people, respectively 19.54%), *very important* (73 people, respectively 41.95%). The results for this item were predictable if we take into consideration the data obtained in Q3 and Q4.

3 Conclusions

The inclusion theories are relatively recent, they have appeared in the scientific literature at the middle of the 1990s.

> The United Nations Educational, Scientific and Cultural Organization (UNESCO) (2009) clearly indicate that inclusive education is a question of equity and is therefore a quality issue impacting upon all learners. (European Agency for Special Needs and Inclusive Education 2014)

The present research shows the opinion of the teachers in Caraş-Severin County regarding the importance of the professional training in the field of inclusive education. After the achieved research, the following conclusions were reached:

- The teachers interviewed have participated in professional training activities in the field of inclusive education in the last 5 years; consequently the theoretical-practical approaches in this field are known by them.
- The most accessed ways of continuous training in the field of inclusive education by the research sample were observed in "*reading thoroughly the bibliographic resources in the field of inclusive education)*" (159 people—91.38%), the participation in "*workshops in the field of inclusive education*" (112 people—64.37%), and the participation in "*debates/round tables on the topic of inclusive education*" (127 people—72.99%).

- The main reasons why the teachers participated in the training activities in the field of inclusive education in the last 5 years reflect "*the desire of continuous improvement*" (162 people) 93.1%, corroborated with "*the need to acquire competences in the field of inclusive education*" (132 people—75.86%), as well as "*the obligation of accumulating credits in the field of continuous training*" (127 people—72.99%).
- In the activity with the pupils, the professional training activities in the field of inclusive education in which the teachers took part in the last 5 years were used "*much*" (67 people—38.51%) and "*very much*" (62 people—35.63%).
- Both this year and also in the following 3 years, most of the didactic staff declare that they are "*much*" and "*very much*" interested in participating in the training activities in the field of inclusive education.
- The main criteria which influence the decision of the teachers to participate in the training courses in the field of inclusive education are represented by "*the topics of the training activities*" (167 people, respectively 95.68%), by the number of the credits of continuous training (157 people—90.23%), by the "*nature of the training activities: training courses, conferences/seminars, other didactic activities*" (154 people—88.51%), by "*the venue of the training activities*" (142 people—81.61%).
- For the interviewed teachers the professional training activities in the field of inclusive education are "*in a great extent important*" (67 people—38.51%), *very important* (73 people—41.95%).

The education and training in the field of inclusion remain present challenges to which the teachers are invited to adequately answer in harmony with the exigencies of the present and future time.

References

Booth, T., Nes, K., Strømstad, M.: Developing Inclusive Teacher Education. RoutledgeFalmer, London, New York (2003)

Cornoldi, C.: Le difficolta' di apprendimento a scuola. Il Mulino, Bologna (1999)

Daniels, H., Garner, P. (eds.).: Inclusive Education, IInd ed. Routledge, New York (2012)

Engelbrecht, P.: Teacher education for inclusion, international perspectives. Eur. J. Special Needs Educ. **28**(2) (2013)

Heikkinen, H.L.T., Jokinen, H., Tynjälä, P.: Teacher education and development as lifelong and lifewide learning. In: Heikkinen, H.L.T., Tynjälä, P. (eds.) Peer-Group Mentoring for Teacher Development, pp. 3–30. Routledge, Milton Park (2012)

Isosomppi, L., Leivo, M.: Becoming an inclusive teacher at the interface of school and teacher education. In: Procedia—Social and Behavioral Sciences. p. 171 (2015)

Magrab, P.: Training professionals to work in inclusive settings. In: OECD, Inclusive Education at Work: Including Students with Disabilities in Mainstream Schools, Paris (1999)

Marchesi, A.: Quality for all: some comments about inclusive schools from Spanish educational reform. In: OECD, Implementing Inclusive Education, Paris (1997)

Mukhopadhyay, S.: Rethinking inclusive education: action points for communities. În: Daniels, H., Garner, P.(eds.) Inclusive Education, IInd ed. Routledge, New York (2012)

Rouse, M.: Developing inclusive practice: a role for teachers and teacher education? Educ. North **16**, 1 (2008)
Waitoller, F.R., Artiles, A. J.: A decade of professional development research for inclusive education: a critical review and notes for a research program. Rev. Educ. Res. **83**(3) (2013)
Waitoller, F.R., Kozleski, E.B.: Working in boundary practices: identity development and learning in partnership for inclusive education. Teach. Teacher Educ. **31** (2013)
Waitoller, F.R., Kozleski, E.B.: European agency for development in special needs education (2012). Profile of Inclusive Teachers. European Agency for Development in Special Needs Education, Odense, Denmark (2016). [Online 12 Sept 2016]. Available online accessed on: https://www.european-agency.org/sites/default/files/Profile-of-Inclusive-Teachers.pdf
Waitoller, F.R., Kozleski, E.B.: European agency for special needs and inclusive education (2014). Five key messages for inclusive education Putting Theory into Practice (2016). [Online 12 Sept 2016]. Available online accessed on: https://www.european-agency.org/sites/default/files/Five%20Key%20Messages%20for%20Inclusive%20Education.pdf
Waitoller, F.R., Kozleski, E.B.: L'istituto comprensivo di Tito- Inclusione: DSA—BES (2016). [Online 25 Sept 2016]. Available online at: http://ictito.gov.it/index.php?option=com_content&view=article&id=587&Itemid=1522
Waitoller, F.R., Kozleski, E.B.: Ministerul Educației Naționale. Formare continuă (2016). [Online 10 Dec 2016]. Available online accessed on: www.edu.ro/formare-continua

School Teaching in Germany Between Tradition and Innovation: The Concept of the Outdoor School

Sarah Sahrakhiz, Marius Harring and Matthias D. Witte

Abstract It is a well known fact that school is seen as having a spatial characteristic, being associated to indoor learning. The aim of this chapter is to underlined the idea that learning and educational processes can take place in different settings, and to offer other experiences that the traditional classroom with its specific conditions largely neglected. The dimensions of space, time, and body are re-evaluated from the perspective of outdoor education which faces children (and teachers) with new, sometimes challenging, and adventurous experiences that contrast with an everyday routine of school instruction.

Keywords Children and adolescents · Pedagogical activities · Outdoor school

1 Introduction

All pedagogical activities take place in social spaces that both facilitate and place limitations on learning processes. This statement may appear to be a truism, yet it gives rise to far-reaching questions about (socio)spatial education that have long been neglected by educational science. As recently as the early 1990s, the field of school studies in Germany paid little or no attention to the issue of space; only during the last few years have scholars increasingly addressed the significance of space for learning and educational processes in the school context (see, for example, Böhme 2009; Böhme and Herrmann 2011; Göhlich 1993; Rittelmeyer 1994). Nevertheless, spatial research in the field of school and educational science has mostly limited its focus to the architecture of school buildings and their interior design. Pedagogical theories often remain beholden to the conventional spatial boundaries of school buildings,

S. Sahrakhiz · M. Harring · M. D. Witte (✉)
Johannes Gutenberg University of Mainz, Mainz, Germany
e-mail: matthias.witte@uni-mainz.de

S. Sahrakhiz
e-mail: s.sahrakhiz@outlook.de

M. Harring
e-mail: harring@uni-mainz.de

D. Flaut et al. (eds.), *Decision Making in Social Sciences: Between Traditions and Innovations*, Studies in Systems, Decision and Control 247,
https://doi.org/10.1007/978-3-030-30659-5_5

largely because school is traditionally regarded as a circumscribed and relatively coherent place. Spatial properties of public schools are a product of the national states of the nineteenth century (see Herrlitz et al. 2009; Oelkers 2009). School buildings and schoolyards are clearly delimited from the out-of-school environment. Children and adolescents spend most of their school time within a circumscribed school compound. Learning activities mostly take place indoors and pupils leave the school building only on special occasions, for example during excursions or school trips. The materialization of the school space manifests institutionalized values and standards. The school building and classrooms create a functional homogenization of space that meets the institution's needs in terms of safety, control, predictability and efficiency; teachers and students further more adopt and shape the school's spatial order and incorporate its norms.

Alongside this relative spatial coherence, however, the history of education in school has also seen repeated attempts to open up the school space towards its immediate surroundings. The dissolution of school boundaries was an expression of social developments such as urbanization and movements like the Enlightenment, Romanticism or the German Youth Movement. Even in the twenty-first century, schools are still tied to specific locations and keep being challenged to respond to social changes—in this case, globalization and growing mobility. "The most important thing that schools must impart is the ability to learn and to be open for the unknown. To develop this openness, it is necessary to experience various different places of education" (Oelkers 2009, p. 40). This includes both the immediate environment of school and more distant places and educational spaces.

2 Outdoor Education as a Way of Spatial Opening of School

Many European countries have a long tradition of school teaching and learning at out-of-school places. In the English-speaking world, such concepts, programs, and methods go by the name of *Outdoor Education* or *Outdoor Learning* (see, for example, Lynch 2009; Rickinson et al. 2004; Waite 2011), while in Scandinavia common terms are *Friluftsliv* (see, for example, Bentsen et al. 2009a; Brügge et al. 2002; Pedersen-Gurholt 2008), *Uteskole/Udeskole* (see, for example, Bentsen et al. 2009b, 2010; Jordet 2007), and *Utomhuspedagogik* (see, for example, Dahlgren et al. 2007). The exact definition of these terms and the practices denoted by them vary from one country to another or even within the countries themselves. The concepts exhibit different degrees of formalization and may refer to a separate school subject, to parts of the curricula of various everyday subjects, or to occasional activities as part of supplemental school activities. What they all have in common is an enlargement of space within school activities. The aim is often to initiate learning and educational processes through varied experiences in different natural and cultural settings—experiences that the traditional classroom with its specific conditions largely neglect. In Germany such approaches have been referred to as *Abenteuer- und Erlebnispädagogik* ("adventure and experience-based pedagogy") or sometimes as *Outdoor-Pädagogik* ("outdoor pedagogy"). Meanwhile terms like *Draußen-Pädagogik* ("outside pedagogy") and *Draußenschule* ("outdoor school") have very little currency. A systematic, theoretically and empirically backed study of school pedagogy in out-of-school learning places is still to be done. In the following, we will present the German *Draußenschule* (Outdoor School), a practice of teaching and instruction that remains largely unknown at elementary schools in Germany and that implies innovative forms of teaching and learning.

3 Outdoor School as an Innovative Concept

The outdoor school describes a pedagogical teaching concept that has become very widespread since the 1990s in Norway (*Uteskole*) and Denmark (*Udeskole*) as a grassroots movement (see Bentsen et al. 2009b, 2010; Jordet 2007; Witte and Gräfe 2010a, b). Used primarily for teaching the lower grades, it calls for the students to leave the classroom once a week throughout the whole school year. Together with their teacher, they discover natural and cultural spaces in the surroundings. Almost all school subjects are taught in an interdisciplinary way at these out-of-school locations: Math, Language, Art as well as Biology or Physical Education. The weekly "outdoor days" encourage social competences as well as academic ones. The students enter various social spaces, which often have not been prepared to serve as a setting for school learning and therefore exhibit a lower degree of standardization, for example a nearby forest. Thus, conditions in the outdoors are fundamentally different from those indoors. The curricular contents, that guides the instruction, however, remain the same. In Germany, outdoor school has found little resonance so far in theory and practice. With our outdoor school project (running from 2014 to 2016) for the first time a weekly outdoor day has been implemented systematically at three German elementary schools. Our assumption is that the outdoor school initiates processes of learning and education, which are important for the children's development.

4 Learning Potential of Outdoor School

Education is an ongoing process where an individual appropriates the world without a specific purpose. It is in the course of their engagement with the world that human beings form and develop every aspect of their nature and personality. Education changes the fundamental figuration of human's relationship with the world and with themselves, especially in the course of dealing with unfamiliar situations (Koller et al. 2007). Moments of irritation and disruptions of everyday routine can trigger educational processes. The outdoor school faces children (and teachers) with new, sometimes challenging, and adventurous experiences that contrast with an everyday routine of school instruction. By moving the teaching and learning activities out of the classroom into a more stimulating environment, it is possible to link easily physical-sensory and cognitive-abstract learning. In this context, learning is not a passive state, but "a process of increasingly self-regulated self-renewal" (Fournés 2008, p. 5). As a long-term and, by definition, open-ended process, education includes a multiplicity of different learning activities (Schulze 2007). In the following, we will deal in detail with space, time, and body (physicality) as the three fundamental dimensions of learning and education in the (outdoor) school.

4.1 The Dimension of Space

In general, schools are perceived as defined locations and enclosed spaces. This perception is reflected in phrases like "going to school" or "building a school". Conceptually, they are not geared towards mobility, but defined by their location-dependence. School buildings and classrooms are highly functionalized spaces that conform to the institutional order and its criteria of safety, control, and predictability. There are desks for students, the size of a classroom is limited, and the objects within are static. Although distraction from the outside is not completely avoidable, it is rare. The students "inhabit" their classrooms for several years, which means they develop certain habits and routines during this time, internalizing the spatial and social structures of school.

In the outdoor school, on the contrary, for one day of the school week unaccustomed experiences dominate. The appearance of natural space changes according to the season, and a forest, for example, is "living" and in a state of constant flux. The change of scene enables the children to discover and learn new things on each of their weekly outdoor days. The outdoor locations are less functionalized than the classroom. That allows an appropriation of space in various ways. The children can investigate pond life at natural locations, discover animal tracks, or be creative while doing handicrafts with natural materials. On visits to craft businesses, they can talk to "experts" and learn how products are manufactured. On the way to the out-of-school-places, they also learn road safety and how to find their way around their town (Armbrüster et al. 2016). These regular trips challenge not only the students, but their teachers as well, to engage with the character of the outdoor-learning-places. Too much standardization of these places (e.g. by "furnishing" the forest setting up fixed benches, tables, and blackboards) would reduce the place-based potential for triggering autonomous and spontaneous learning processes.

4.2 The Dimension of Time

In traditional school teaching, learning processes not only take place in enclosed, functional spaces, but are also subject to a fixed timetable to which both teachers and students must adhere. Like the spatial organization, the chronological structure of the school serves to habituate children to conforming to externally imposed rules. The school bell determines when lessons begin and when it is allowed to go out to recess. Despite the fact that schools occasionally break with the convention of 45-min teaching units, time remains a factor that structures and delimits the learning process. Teachers plan their lessons according to these time frames, and the pre-existing frameworks force them to ensure that the curriculum will be completed in the time available — these time frames can effect even outdoor days.

However, teaching and learning in the outdoor school allows time to be perceived differently than in regular classroom teaching. Outdoors, time is less a mathematical unit than a phenomenon to be experienced when temperatures, sounds, and lighting conditions change along with the seasons. The subjective perception of time thus gains priority over time as an objectively measurable unit. The ever-recurring sequence of spring, summer, fall, and winter allows the children to experience the natural cycle of growth and decay. The clock fades to irrelevance while pupils play, explore, make things, or run around outdoors. In the outdoor school, there is time to understand and think about ostensibly unimportant things. At the same time, however, openness for spontaneous events poses a challenge for teachers. Our observations indicate that the time pressure, which is frequently felt in everyday school teaching, can also carry over into teaching in natural spaces. Principals, teachers, and parents often worry that weekly outdoor days may come at the cost of having less time for covering the formal curriculum (Gräfe et al. 2015; Harring et al. 2015). Inevitably, the children cannot explore every intriguing thing that attracts their attention in the outdoors. A number of different curricular elements must be taken into account when planning an outdoor day. At the same time, however, over-emphasizing the temporal framework runs the risk of failing to recognize the learning potentials of the supposed distractions and irritations of the environment. In a world that is often perceived as fast-paced, in which time is a precious resource and acceleration has become the dominant guiding principle, deceleration has an intrinsic value as a contrasting and psychosocially significant experience. Schools should allow students to also satisfy their need for play and exploration by making time for them to engage in activities that have no specific purpose. The learning and educative potential of the outdoor school may come into its own precisely in the combination of structured phases, in which the children work on specific tasks on the one hand, and open phases, in which they can act freely and autonomously on the other (Witte 2015).

4.3 The Dimension of the Body

Closely related to the spatial and temporal dimensions is the category of the body, since human (and so corporeal) existence is constituted through space and time. Similarly, learning and education would be inconceivable without the corporeal nature of the human being. Nevertheless, the school's relationship to the body remains ambivalent to this day. In the tradition of Enlightenment Pedagogy, school learning is often regarded as a cognitive process based on the duality of body and mind. This perception reduces the body to a baser, impulse-driven nature of the human being that must be tamed and disciplined. It overlooks the fact that the body itself is an instrument of cognition and that one of the ways in which people learn is through physical, sensory experience.

The weekly outdoor days in different natural and cultural spaces lend added significance to the children's physical and sensory experiences. The outdoor school offers scope for a wide variety of ways for the children to experience their bodies in motion—ways that are fundamentally different from the options available in classroom, hallway and gym. The children appropriate the spaces through different modes of movement—balancing along a fallen tree trunk, jumping into a puddle, or taking the subway in a metropolitan city. In each case, children's physical experiences must be considered as being of holistic, immediate, and existential significance. In the outdoor school, the school subjects can be experienced with all senses. The body becomes the point of departure for new and old units of measurement. Geometrical shapes can be sought out in nature and reproduced. Experiences from the out-of-school learning places can be recalled during classroom teaching, such as when pupils are asked to write or talk about a birds nest, or to draw a picture of it. Additionally, the outdoor school allows children to experience and test their bodies: How can we cross this stream? Which of these branches can I carry on my own and which one will take several of us to lift? These physical experiences can go hand in hand with changes in their relationships to themselves and the world (Armbrüster et al. 2016). Finally, the outdoor school facilitates stronger physical communication both among the children and between students and teachers, and this can have beneficial effects on relationship structures.

5 Conclusion

The school is a social environment in which teachers and students naturally interact according to certain rules and norms. The historical hierarchies between the acting subjects as well as their roles and functions are inscribed into the spatial order of school. The pedagogical concept of outdoor school allows this order to be suspended, varied, and potentially changed for one day per week. We have illustrated the learning and educative potential of outdoor school focusing the three dimensions space, time, and body. However, the realization of this potential depends on a number of prerequisites at the levels of individual and institution, since the dissolution of spatial boundaries on the outdoor days poses special organizational and didactic challenges for teachers and school as such. Teachers who regularly leave the classroom with their students to visit natural spaces often have a special affinity for nature as a result of their biographies (Gräfe et al. 2015; Jordet 2007). Their motivation and their practical methodological knowledge is due in large part to their past socialization outside of their professional lives. To professionalize these activities, however, the intuitive and personally relevant actions of such teachers should be supplemented by a more in-depth theoretical study of out-of-school learning. Additionally, there is a need for more empirical studies that focus systematically on learning processes of elementary school students and teachers at out-of-school learning places (on this subject see, for example, Armbrüster et al. 2016; Sahrakhiz et al. 2016, 2017; Witte 2015). Finally, the outdoor school also touches on issues of school development in general.

References

Armbrüster, C., Gräfe, R., Harring, M., Sahrakhiz, S., Schenk, D., Witte, M.D.: Inside we learn, outside we explore the world—children's perception of a weekly outdoor day in German primary schools. J. Educ. Human Develop. **5**(2), 1–12 (2016)

Bentsen, P., Andkjær, S. & Ejbye-Ernst, N.: *Friluftsliv. Natur, samfund og pædagogik* [Natur, Community and Pedagogy]. Kopenhagen (2009a)

Bentsen, P., Mygind, E., Randrup, T.: Towards an understanding of udeskole: education outside the classroom in a Danish context. Education **37**(1), 29–44 (2009b)

Bentsen, P., Jensen, F.S., Mygind, E., Randrup, T.B.: The extent and dissemination of udeskole in Danish schools. Urban Forest. Urban Greening **9**(3), 235–243 (2010)

Böhme, J.: Raumwissenschaftliche Schul- und Bildungsforschung" [Spatial school and education research]. In: Böhme, J. (ed.) Schularchitektur im interdisziplinären Diskurs, pp. 13–24. Wiesbaden (2009)

Böhme, J., Herrmann, I.: Schule als pädagogischer Machtraum. Typologie schulischer Raumentwürfe [School as a Sphere of Pedagogical Power. Typology of School Spatial Concepts]. Wiesbaden (2011)

Brügge, B., Glantz, M., Sandell, K.: Friluftslivets pedagogik. För kunskap, känsla och livskvalitet [Friluftslivets pedagogy. For knowledge, self-esteem and quality of life]. Kristianstad (2002)

Dahlgren, L., Sjölander, S., Strid, J., Szczepanski, A.: *Utomhuspedagogik som kunskapskälla* [Utomhuspedagogik as a Source of Knowledge]. Pozkal (2007)

Fournés, A.: Lernen an außerschulischen Orten. Oder: Zur Verbindung innerschulischen Lernens mit der 'Welt da draußen' [Learning at out-of-school places, or: On the connections between in-school learning and the "world out there"]. Grundschule Sachunterricht **39**, 2–5 (2008)

Göhlich, M.: Die pädagogische Umgebung. Eine Geschichte des Schulraums seit dem Mittelalter [The Pedagogical Environment. A History of the School Space Since the Middle Ages]. Weinheim (1993)

Gräfe, R., Gillessen, C., Harring, M., Sahrakhiz, S., Witte, M.D.: Einmal wöchentlich draußen unterrichten?! Eine qualitativ-empirische Studie zur Draußenschule aus der Perspektive von Grundschullehrerinnen [Teaching outdoors once a week?! A qualitative empirical study of the outdoor school from the perspective of elementary school teachers]. In: Von Au, J. (ed.) Raus aus dem Klassenzimmer – Outdoor Education als Unterrichtskonzept, pp. 79–95. Weinheim and Munich (2015)

Harring, M., Sahrakhiz, S., Schenk, D.: (Draußen-)Schule aus der Perspektive von Eltern – Befunde einer empirisch-quantitativen Studie [(Outdoor) School from the perspective of the parents—findings of an empirical qualitative study]. In: Grundschulzeitschrift, 09/2015, pp. 28–32 (2015)

Herrlitz, H.G., Hopf, W., Titze, H., Cloer, E.: Deutsche Schulgeschichte von 1800 bis zur Gegenwart. Eine Einführung [History of Schools in Germany from 1800 to the Present. An Introduction]. Weinheim and Munich (2009)

Jordet, A.: Nærmiljøet som klasserom. En undersøkelse om uteskolens didaktikk i et danningsteoretisk og erfaringspedagogisk perspektiv [Local Community Classrooms. A Survey on Uteskole Education from a Theoretical and Experience-Educational Perspective]. Oslo (2007)

Koller, H.-C., Marotzki, W., Sanders, O. (eds.): Bildungsprozesse und Fremdheitserfahrung. Beiträge zu einer Theorie transformatorischer Bildungsprozesse [Educational Processes and the

Experience of Otherness. Contributions to a Theory of Transformative Educational Processes]. Bielefeld (2007)
Lynch, P.: Camping in the Curriculum. A History of Outdoor Education in New Zealand Schools. Canterbury (2009)
Oelkers, J.: Globalisierte Bildungsansprüche im lokalen Schulraum [Globalized education in the local school space"]. In: Böhme, J. (ed.) Schularchitektur im interdisziplinären Diskurs. Territorialisierungskrise und Gestaltungsperspektiven des schulischen Bildungsraums, pp. 25–41. Wiesbaden (2009)
Pedersen-Gurholt, K.: Norwegian Friluftsliv as Bildung. A critical review. In: Becker, P., Schirp, J. (eds.) Other ways of learning. The European Institute for Outdoor Adventure Education and Experiential Learning 1996–2006, Marburg, pp. 131–154 (2008)
Rickinson, M., Dillon, J., Teamey, K., Morris, M., Choi, M.Y., Sanders, D., Benefield, P.: A review of research on outdoor learning (2004). http://www.field-studies-council.org/documents/general/NFER/A_review_of_research_on_outdoor_learning.pdf. Accessed on 7 Dec 2015
Rittelmeyer, C.: Schulbauten positiv gestalten: wie Schüler Farben und Formen erleben [Positive School Building Design: How Students Experience Colors and Shapes]. Wiesbaden (1994)
Sahrakhiz, S., Harring, M., Witte, M.D.: Learning opportunities in the outdoor school – empirical results from the children's perspective. J. Advent. Educ. Outdoor Learn. **18**(3), 214–226 (2017)
Sahrakhiz, S., Witte, M.D.,Harring, M.: Nähe und Distanz in der Lehrersprache. Eine Konversationsanalyse am Beispiel von Draußenschule und Klassenzimmer [Closeness and Distance in the Teacher's Language. A Conversational Analysis in the Outdoor School and the Classroom]. Scheider-Verlag Hohengehren, Baltmannsweiler (2016)
Schulze, T.: Modi komplexer und längerfristiger Lernprozesse. Beobachtungen und Überlegungen zu einer Theorie des Lernens und der Bildung [Modes of complex and long-term learning processes. Observations and thoughts on a theory of learning and education]. In: Koller, H.-C., Marotzki, W., Sanders, O. (eds.) Bildungsprozesse und Fremdheitserfahrung. Beiträge zu einer Theorie transformatorischer Bildungsprozesse, pp. 141–159. Bielefeld (2007)
Waite, S. (ed.): Children Learning Outside the Classroom: from Birth to Eleven. London (2011)
Witte, M.D.: Draußenschule aus Kindersicht – Eine vernachlässigte Perspektive in der Udeskole-Forschung [The outdoor school in the child's view: a neglected perspective in Udeskole research]. Die Grundschulzeitschrift **29**(287), 22–27 (2015)
Witte, M.D., Gräfe, R.: Die skandinavische Draußenschule. Schulbildung jenseits des Klassenzimmers [The Scandinavian outdoor school. School education beyond the classroom]. erleben & lernen. Internationale Zeitschrift für handlungsorientiertes Lernen **18**(4), 20–22 (2010a)
Witte, M.D., Gräfe, R.: Schule unter freiem Himmel. Die norwegische Uteskole als Lern- und Bildungskonzept [School in the open air. The Norwegian Uteskole as a learning and educational concept]. Motorik. Zeitschrift für Motopädagogik und Mototherapie **33**(3), 119–124 (2010b)

Reforms and Developments in Romanian Higher Education as a Result of the Bologna Process Implementation

Dan Vătăman

Abstract This year we celebrate 20 years from the Bologna Declaration, a document that represents the starting point for a reforming process which changed not only the Higher Education systems in European countries, but also in other countries across the world. As in the case of others participating countries, Bologna Process meant for Romania profound reform of Higher Education system, marked by changes in legislation and reconsideration of the basic principles of the educational process. Therefore, this chapter is aimed to overview the developments in Romanian Higher Education in light of the Bologna Process evolution, with emphasis on reforms undertaken by Romanian authorities for renewing the legal, institutional and methodological frameworks applicable in Higher Education.

Keywords Bologna process implementation · European higher education area · Romanian higher education system · Structural reforms · Legal framework

1 Historical Overview on the Bologna Process in Light of the Ministerial Declarations and Communiqués

1.1 The Inception Phase of the Process

In connection with the celebrations of the 800th anniversary of the University of Paris-Sorbonne, on 25 May 1998, four ministers responsible for higher education in France, Germany, Italy and the United Kingdom adopted Joint declaration on harmonization of the architecture of the European higher education system (so-called Sorbonne Declaration), document that announced their intention to establish structural compatibility between European institutions of higher education, remove existing obstacles to mobility and establish the basis for improved European cooperation between institutions of higher education. With the same occasion, the four

D. Vătăman (✉)
Doctoral School of Humanities, "Ovidius" University of Constanţa, Constanţa, Romania
e-mail: vataman.dan@gmail.com

D. Flaut et al. (eds.), *Decision Making in Social Sciences: Between Traditions and Innovations*, Studies in Systems, Decision and Control 247,
https://doi.org/10.1007/978-3-030-30659-5_6

signatories of the Sorbonne Declaration extended an invitation to fellow Ministers in charge of higher education in other countries to engage in the endeavour to create a European area of higher education (Sorbonne Declaration 1998).

One year after the Sorbonne Declaration, Ministers responsible for higher education from 29 European countries (including Romania) met in Bologna on 18–19 June 1999. As part of the preparations for the planned Bologna Forum, the Confederation of European Union Rectors' Conferences undertook, in cooperation with the Association of European Universities (CRE), a project for the European Commission (DG XXII), on Trends in Learning Structures in Higher Education. The objective of the project has been to provide an outline and overview of learning structures in higher education and a comparative analysis of the different systems embodying these structures, thereby offering a tool to identify possible divergences and convergences in the national and institutional policies (Project Report 1999).

At the end of the meeting it was adopted the Joint declaration of the European Ministers of Education—The Bologna Declaration (hence the name of the whole Process), a document expressing the willingness of the participants to increase the international competitiveness of the European system of higher education. As it appears from Bologna Declaration, Ministers responsible for higher education affirmed their support for general principles laid down in the Sorbonne declaration and engaged that, by coordinating the national policies of the participating countries, to reach a number of six objectives which represent the essence of the Bologna process: adoption of a system of easily readable and comparable degrees; adoption of a system essentially based on two main cycles, undergraduate and graduate; establishment of a system of credits—such as in the ECTS; promotion of mobility by overcoming obstacles to the free movement of students, teachers, researchers and administrative staff; promotion of European co-operation in quality assurance; promotion of the necessary European dimensions in higher education (Bologna Declaration 1999).

1.2 The Implementation of the Bologna Process Throughout the First Decade

As shown in the Bologna Declaration, convinced that the establishment of the European area of higher education requires constant support, supervision and adaptation to the continuously evolving needs, the Ministers decided to meet again within two years in order to assess the progress achieved and the new steps to be taken.

Consequently, at the next meeting of European Ministers in charge of Higher Education, held in Prague on 19 May 2001, not only has the number of members increased to 33 (Croatia, Cyprus, Liechtenstein and Turkey joined the Bologna process), but there has also taken place an expansion of the objectives with another 3, in terms of lifelong learning, involving students as active partners and enhancing the attractiveness and competitiveness of the European Higher Education Area. Also, the participating Ministers committed themselves to ensure the further development of quality assurance and development of national qualification frameworks (Prague Communiqué 2001).

The following Conference of Ministers responsible for Higher Education took place in Berlin on 19 September 2003, occasion with the Ministers decided to accept the requests for membership of Albania, Andorra, Bosnia and Herzegovina, Holy See, Russia, Serbia and Montenegro, "the Former Yugoslav Republic of Macedonia" and to welcome these states as new members thus expanding the Bologna Process to 40 European Countries. The main provisions of the Berlin Communiqué dealt with an expansion of the objectives, in terms of promotion of linking European Higher Education Area to European Research Area, as well as the promotion of quality assurance. Another important aspect referred to establishing the follow-up structures supporting the process in-between two Ministerial meetings. This arrangement established the Bologna Follow-up Group, the Board and the Bologna Secretariat (Berlin Communiqué 2003).

The next Ministerial Conference, held in Bergen (Norway) on 19 May 2005, marked the accession of 5 new countries to the Bologna Process, namely: Armenia, Azerbaijan, Georgia, Moldova and Ukraine. As stated in the Bergen Communiqué, the Ministers responsible for Higher Education decided to enlarge the follow-up structure set up in Berlin, with the inclusion of the Education International (EI) Pan-European Structure, the European Association for Quality Assurance in Higher Education (ENQA), and the Union of Industrial and Employers' Confederations of Europe (UNICE) as new consultative members of the Follow-up Group. Also, the Communiqué underlined the importance of partnerships, including stakeholders—students, Higher Education Institutions (HEIs), academic staff and employers, together with the further enhancing of research, especially with regard to the third cycle—Doctoral Programmes. It was also highlighted the Ministers' desire to make further progress, in particular with regard to: implementation of the standards and guidelines for quality assurance as proposed in the ENQA report; implementation of the national frameworks for qualifications; the awarding and recognition of joint degrees, including at the doctorate level; creating opportunities for flexible learning paths in higher education, including procedures for the recognition of prior learning (Bergen Communiqué 2005).

At the London Conference, held on 18 May 2007, Montenegro was welcomed to the Bologna Process following its declaration of independence in 2006, bringing the number of participating countries to 46. As appears from the London Communiqué, the Ministers responsible for Higher Education reaffirmed their determination to seeing the process continue until its full implementation in 2010 by addressing important issues such as the recognition of higher education qualifications, prior learning, and the recognition of non-formal and informal learning. They also emphasized the importance of putting in place qualifications frameworks in each country and of pursuing the implementation of a European Register of Quality Assurance Agencies, social dimension of higher education and the need to promote the European Higher Education Area in a global context. As regards to Doctoral Studies, the Ministers appreciated that enhancing provision in the third cycle and improving the status, career prospects and funding for early stage researchers are essential preconditions for meeting Europe's objectives of strengthening research capacity and improving the quality and competitiveness of European higher education. Consequently, they invited HEIs to reinforce their efforts to embed Doctoral Programmes in institutional strategies and policies, and to develop appropriate career paths and opportunities for doctoral candidates and early stage researchers (London Communiqué 2007).

The 46 Ministers responsible for Higher Education met again in Leuven/Louvain-la-Neuve, Belgium, on 28 and 29 April 2009 to take stock of the achievements of the Bologna Process and to establish the priorities for the European Higher Education Area (EHEA) until 2020. As shown in the conference's communiqué, the main working areas set for the 2010–2020 decade are focused on: social dimension, lifelong learning, employability, student centred learning and the teaching mission of education, international openness, mobility, education, research and innovation, as well as data collection, funding of the Higher education and multidimensional transparency tools (Leuven/Louvain-la-Neuve Communiqué 2009).

As can be seen, these main working areas show a new orientation of the Bologna Process, towards a more in-depth approach of the reforms, thus ensuring the Bologna Process implementation in the decade up to 2020. At the same time, it must be emphasized that Leuven/Louvain-la-Neuve Conference represents a key moment in the evolution of the "external dimension" of the Bologna Process, that's because on this occasion took place the first Bologna Policy Forum that gathered high ranking representatives of Australia, Brazil, Canada, China, Egypt, Ethiopia, Israel, Japan, Kazakhstan, Kyrgyzstan, Mexico, Morocco, New Zealand, Tunisia, USA, along with the International Association of Universities (IAU) and other international organizations and NGOs (Statement by the Bologna Policy Forum 2009).

1.3 Launching the European Higher Education Area and the Beginning of a New Phase in the Bologna Process Evolution

The next meeting of the Ministers responsible for higher education in the countries participating in the Bologna Process was held in Budapest and Vienna on 11 and 12 March 2010, the main purpose being the launching the European Higher Education Area, as envisaged in the Bologna Declaration of 1999. It should be noted that on this occasion Kazakhstan officially joined the Bologna Process and became the 47th member of the European Higher Education Area. According to the Budapest-Vienna Declaration on the European Higher Education Area, the Ministers showed their commitment to the full and proper implementation of the agreed objectives and the agenda for the next decade set by the Leuven/Louvain-la Neuve Communiqué. At the same time, they asked the Bologna Follow-up Group (BFUG) to propose measures to facilitate the proper and full implementation of the agreed Bologna principles and action lines across the European Higher Education Area, especially at the national and institutional levels, among others by developing additional working methods, such as peer learning, study visits and other information sharing activities (Budapest-Vienna Declaration 2010).

A conference that marked a defining moment in the Bologna Process was the one held in Bucharest on 26 and 27 April 2012, a meeting considered as an opportunity to take stock of progress of the Bologna Process and set out the key policy issues for the future. As the Bucharest Communiqué shows, the Ministers responsible for higher education in the 47 countries of the European Higher Education Area (EHEA) took account of the achievements since the last Ministerial conference in Leuven in 2009 and set objectives that would be operationalised in the plan of work for the upcoming sequence of the Bologna Process (Bucharest Ministerial Communiqué 2012).

It is important to remind that in conjunction with the Bucharest Ministerial Conference it was organised the third Bologna Policy Forum with the theme: "Beyond the Bologna process: Creating and connecting national, regional and global higher education areas". Discussions were focused on four topics reflecting on future approaches for dialogue, more exactly: Public responsibility for and of higher education within national and regional context; Global academic mobility: Incentives and barriers, balances and imbalances; Global and regional approaches to quality enhancement of higher education; and the contribution of HE reforms to enhancing graduate employability (Statement of the Third Bologna Policy Forum 2012).

The Yerevan Conference, held on 14–15 May 2015, represented a milestone in evolution of the Bologna Process, a fact highlighted by the 47 participating Ministers who acknowledged that "the EHEA has come to a turning point where a new sense of direction is needed in order to move ahead". As is reflected from the Yerevan Communiqué, the Ministers adopted a renewed vision of the EHEA and sets goals that would be pursued in the plan of work for the upcoming period. Thus, they states that are determined to achieve an EHEA where the common goals are implemented in all member countries to ensure trust in each other's higher education systems; where automatic recognition of qualifications has become a reality so that students and graduates can move easily throughout it; where higher education is contributing effectively to build inclusive societies, founded on democratic values and human rights; and where educational opportunities provide the competences and skills required for European citizenship, innovation and employment. Also, they adopted the following three policy measures: the revised Standards and Guidelines for Quality Assurance in the European Higher Education Area (ESG); the European Approach for Quality Assurance of Joint Programmes; and the revised ECTS Users' Guide, as an official EHEA document (Yerevan Communiqué 2015).

Furthermore, during the Yerevan Conference the Ministers also welcomed Belarus as the 48th member of the EHEA, under condition that the Belarusian authorities and the BFUG to collaborate on developing and implementing a roadmap which includes a list of structural reforms pertaining to qualifications frameworks, quality assurance, recognition, transparency instruments, mobility of higher education staff and students and internationalization, lifelong learning and the social dimension of higher education (Belarus roadmap for higher education reform 2015).

In conjunction with the Yerevan Ministerial Conference it was organised the Fourth Bologna Policy Forum with declared objective to reinforce policy dialogue and cooperation between the EHEA and the countries of the Middle East, North Africa and Asia (MENAAS). For this purpose, the participants presented their priorities, which consist of: developing national qualifications frameworks within the EHEA and establishing compatibility with those developed by MENAAS countries; developing cooperation in quality assurance, thus encouraging mutual trust among education systems; improving the mutual recognition of qualifications; and developing and implementing a credit transfer system bearing in mind the ECTS and the newly revised Users' Guide (Statement of the Fourth Bologna Policy Forum 2015).

The most recent EHEA Conference was held in Paris on 24–25 May 2018, opportunity for Ministers not only to celebrate the progress made in building the European Higher Education Area over the past two decades after the Sorbonne Declaration, but also to make strong and ambitious commitments for its further development. As shown in the Paris Communiqué, in order to unlock the full potential of the EHEA and ensure the implementation of Bologna key commitments, the Ministers adopted a structured peer support approach based on solidarity, cooperation and mutual learning, so as to, in 2018–2020 period, thematic peer groups will focus on three key commitments crucial to reinforcing and supporting quality and cooperation inside the EHEA: a three-cycle system compatible with the overarching framework of qualifications of the EHEA and first and second cycle degrees scaled by ECTS; compliance with the Lisbon Recognition Convention; and quality assurance in compliance with the Standards and Guidelines for Quality Assurance in the European Higher Education Area (Paris Communiqué 2018).

The Paris Ministerial Conference included the Fifth Bologna Policy Forum which provided an important opportunity for a multilateral dialogue and exchange of ideas between members of the European Higher Education Area (EHEA), Ministers from other countries and a range of stakeholders. It should be mentioned that the participants committed to support the interregional dialogue on issues of common concern amongst policy makers, stakeholder organisations, students, staff and higher education institutions. Also, they proposed the establishment of a Global Working Group in the next 2018 -2020 Bologna work programme to take this agenda forward and invited countries to express their interest to hold high level workshops on a yearly basis to continue the dialogue on social inclusion and the wider role of higher education (Statement of the Fifth Bologna Policy Forum 2018).

2 The Implementation of the Bologna Process in Romania

As shown above, Romania was among the first 29 countries that signed the Bologna Declaration on 19 June 1999, this being a commitment freely taken to reform its own higher education system. The Bologna Declaration was not just a political statement, but a binding commitment to an action programme based on a clearly defined common goal, a deadline and a set of specified objectives.

Therefore, in line with the commitments made under the Bologna Process, Romanian authorities initiated the reform process for renewing the legal, institutional and methodological frameworks applicable in higher education.

2.1 *Structural Reforms Made by Romanian Authorities During 2004–2010 Period*

One of the first steps towards achieving the Bologna objectives was adoption of the Law no. 288/2004 concerning the organization of the higher education, a normative act which provide the legislative framework for the introduction of the three cycles beginning with the 2005–2006 university year, as follows: the first cycle, having a duration of 3–4 years and amounting to 180–240 credit points, grants the graduates the title of Bachelor; the second cycle, having a duration of 2 years and amounting to 120–160 credit points, is completed with the a title of Master; and the third cycle is represented by doctoral studies, has a duration of 3 years and grants the title of a doctor of science.

Pursuant to the provisions of Chapter II of Law no. 288/2004, the Minister of Education and Research adopted Order no. 3235/2005 concerning the organization of the first cycle (bachelor) of university studies thus ensuring that the reform of the education curricula of the first cycle programmes within the same field of study includes the clear definition of the knowledge and skills acquired by the graduate, in correlation with the master and doctorate studies. Also, a new structure of the first cycle (Bachelor) was adopted by the Government Decision no. 1175/2006 regarding the organization of undergraduate studies and the approval of the fields and specializations list within them. The decision establishes only 15 general fields of study and the correspondent specializations applicable in all public and private higher education institutions, thus allowing a better coordination between study programmes offered by the Romanian higher education institutions and the other higher education systems within the EHEA.

The legislative framework defining the structure of the second and third cycle was established through the Government Decision no. 404/2006 on organization and conduct of Master Studies and Government Decision no. 567/2005 on organization and conduct of Doctoral Studies. Furthermore, by Order of the Minister of Education and Research no. 3861/2005 have been also implemented the two-year post-doctorate programmes, their aim being to strengthen the research excellence and therefore increase the attractiveness of the scientific career for young researchers in Romania.

Among the main objectives to be achieved by countries participating in the Bologna Process is also found introducing widely the European Credit Transfer System (ECTS) as the only broadly tested credit system in Europe. For this purpose, the Minister of Education and Research adopted Order no. 3617/2005 which stipulates a general use of ECTS which stimulates students in their mobility inside the same HEI and/or between programmes offered by various Romanian universities, similar to the mobility freedom promoted by ECTS across the borders of the country.

In close relationship with ECTS, the Bologna Process also looked at the Diploma Supplement as one of the commitments within the European Higher Education Area. Consequently, the Minister of Education and Research adopted Order no. 4868/2006 concerning the implementation of the Diploma Supplement for certifying the graduation of one study cycle which stipulates that, starting from the 2005–2006 academic year, all HEIs will issue, free of charge, the Diploma Supplement providing a standardised description of the nature, level, context, content and status of the studies completed by its holder.

Since Bergen conference, an increased attention was paid also to the development of quality assurance at institutional, national and European level. One of the main achievements in this sense for the Romanian higher education was adoption of the Government's Emergency Ordinance no. 75/2005 concerning quality assurance in education, approved with modifications by Law no. 87/2006. Thus, besides a number of general provisions, the Government Emergency Ordinance on quality assurance in education includes: Quality Assurance Methodology in Education; Internal Quality Assurance in Education; External assessment of quality education; and Institutional arrangements involved in quality assurance. Also provides the establishment of the Romanian Agency for Quality Assurance in Higher Education (ARACIS in its Romanian acronym, which stands for *Agenția Română de Asigurare a Calității în Învățământul Superior*) as an autonomous institution of national interest, with competencies in accreditation, academic evaluation and quality assurance. According to the law provisions, ARACIS's mission is, to assess the quality of the study programs, master study domains and higher education institutions as well as to contribute, together with the HEIs, to a continuous enhancement of quality. In addition, ARACIS is to proceed to the external evaluation/accreditation of the third cycle of university studies (doctoral). The agency is also aiming to contribute at the development of an institutional culture of higher education quality (Articles 16–17 of Government Emergency Ordinance No. 75/2005, as approved by the Law no. 87/2006).

Despite all the reforms attempts in Romanian education system, according to the Report of Commission for the Analysis and Development of Policies in the fields of Education and Research (established by the Romanian president in January 2007), Romanian education was inefficient, irrelevant, inequitable and of poor quality. Consequently, the authors of the report recommended that, on the basis of rational discussions, "all responsible actors to reach a fundamental agreement, a true national pact for education and research, otherwise - they added - the future will not forgive us" (Report of the Presidential Commission 2007).

As a result of report, on 5 March 2008, a National Pact for Education was signed by the leaders of all parties represented in the Parliament, and subsequently by the Romanian President and representatives of other organizations interested in the development of education and research in Romania. Thus, the signatories set a series of eight objectives, including: Modernization of the education system and institutions in 2008–2013 period; to insure in 2008–2013 period, allocation of a minimum of 6% of GDP for education and a minimum of 1% for research; to make early education generally accessible to all, to achieve compulsory school education of 10 years, and to guarantee unrestricted access to free education until graduating the high-school; Comprehensive decentralization of financial, curricular and human resources by adapting the curriculum to children's specific personal development needs, to the requirements of the labour market and communities, based on the principle that issues should be resolved at the most local level possible; Adoption of the principle "the money follows the students" in precollege education, with respect to the principle "multi-annual financing on cycles, study and project programs" in university education; Adopting a charter for rights and liberties in education, to guarantee the access to a quality education; Defining priority education areas in order to close the gap which dramatically divides rural and urban regions and different social groups; Permanent education will become the base of the educational system from Romania and will be extended so until 2013 will include annually at least 12% from the active work force of the country (National Pact for Education 2008).

The objectives set out in the Pact were supposed to provide the foundation of a coherent and unitary strategy for the national education system and educational institutions, but it was not meant to be. The Pact was not followed by concrete actions.

As a result of disputes and lack of consensus among political decision-makers, on 28 October 2010, the Romanian Government assumed responsibility in the Parliament for the Education Law, which made that Opposition to notify the Constitutional Court of Romania (CCR).

After a series of debates and another notification to the Court, this time by the Government, on 4 January 2011, the Constitutional Court dismissed, as inadmissible, the objection of unconstitutionality referring to the Education Law (Decision 1/2011 of CCR).

It must be noted that, only few hours after the Constitutional Court's decision, the president of Romania issued the decree for the promulgation of the National Education Law no. 1/2011, which was published in Official Gazette, Part I, no. 18 of 10 January 2011 and entered into force on 9 February 2011.

2.2 *Romanian Higher Education Following the Reforms Made by the National Education Law no. 1/2011 with the Subsequent Modifications and Completions*

2.2.1 Reforms in the Field of Governance and Coordination of the Higher Education

With regard to the public authorities responsible for higher education, Article 121 of the Law no. 1/2011 states that Ministry of Education is a state authority and is qualified to follow up and control the enforcement and observance of the legal regulations in the field of higher education and, if necessary, to apply sanctions. As a novelty, in Article 217 of the Law no. 1/2011 were introduced some consultative bodies to support the Ministry of Education in fulfilling its mission, such as: National Council of Statistics and Forecast of Higher Education (Consiliul Naţional de Statistică şi Prognoză a Învăţământului Superior—CNSPIS); National Council for the Attestation of University Titles, Diplomas and Certificates (Consiliul Naţional de Atestare a Titlurilor, Diplomelor şi Certificatelor Universitare—CNATDCU); National Scientific Research Council (Consiliul Naţional al Cercetării Ştiinţifice—CNCS); Advisory Board for Research, Development and Innovation (Colegiul consultativ pentru cercetaredezvoltare şi inovare—CCCDI); National Council for Higher Education Financing (Consiliul Naţional pentru Finanţarea Învăţământului Superior—CNFIS); National Council of University Libraries (Consiliul Naţional al Bibliotecilor Universitare—CNBU); Council of ethics and university management (Consiliul de etică şi management universitar CEMU); National Council of Ethics for Research, Technological Development and Innovation (Consiliul Naţional de Etică a Cercetării Ştiinţifice, Dezvoltării Tehnologice şi Inovării—CNECSDTI).

Referring to the types of Higher Education Institutions (HEIs), the first version of the Law no. 1/2011 stated in Article 114 only that "higher education in Romania is ensured by universities, academies, institutes, higher education schools, and other such, hereinafter referred to as higher education institutions or universities. Higher education institutions may be public, private, or religious, they are legal persons of public utility and also non-profit and apolitical organisations". On 30 June 2014, Article 114 of the Law no. 1/2011 was amended by Article I, Point 34 of Government Emergency Ordinance No. 49/2014, adding the provision according to which "higher education institutions are education providers that carry out educational activities based on study programmes (authorised and, as applicable, accredited as required by law) of university-level initial and continuous education and training, programmes that operate based on the principle of ensuring the quality necessary to meet the trust of the direct and indirect beneficiaries form the society".

In terms of the HEIs mission, Article 117 of the Law no. 1/2011 provides that "the mission of the higher education is to generate and transfer knowledge to the society: by basic and continuous training at academic level, for personal development, professional insertion and in order to satisfy the need for competency of the socioeconomic environment; and by scientific research, innovation and technologic transfer, collective and individual creation, in the field of science and engineering, arts, literature and languages, by ensuring the sports and physical development and performance, as well as by putting to good use and disseminating their results mission".

Pursuant to Article 123 of the Law no. 1/2011, all HEIs are autonomous and have the right to establish and implement their own mission, institutional strategy, structure, activities, organization and operation, and to manage their own material and human resources in strict compliance with the legislation in force, the university autonomy is correlated with the principle of personal and public accountability for the quality of the entire teaching and scientific research activity accomplished by the higher education institution.

An element of novelty that sparked controversy was the introduction by Article 193 of the Law no. 1/2011 of Universities' assessment made with the purpose of temporary authorisation and accreditation, as well as for ranking the educational programs and classifying universities, as two distinct activities. According to the law provisions, the assessment for the ranking of the educational programs and classifying universities is made based on an assessment methodology proposed by the Ministry of Education and approved by Government decision, within 6 months from enacting the law. Based on the law provisions, the Romanian Government adopted Decision no. 789/2011 regarding the approval of the Assessment Methodology for University Classification and Ranking of the Study Programmes.

According to the Article 193 (4) of the Law no. 1/2011, Universities are classified in 3 categories, as follow: universities, mainly for education; universities for education and scientific research and artistic creation; universities of advanced research and education. Also, in compliance with Article 10 of the Assessment Methodology approved by the Government Decision No. 789/2011, the study programs are placed in 5 categories according to the classification position of the university that organizes them: A, B, C, D, and E, where A represents the category of the most performing study programmes, and E represents the category of the less performing study programmes. Following these legal provisions, the first classification results were made public through the Minister of Education Order No. 5.262/2011, these being immediately contested by some universities that saw their reputation damaged.

2.2.2 Reforms in the Field of Academic Educational Programs

Referring to the academic educational programs, the National Education Law no. 1/2011 states that a program of this type represents a group of teaching, learning, research, practice and evaluation curricular units planned so that they lead to an academic training certified by a diploma and a diploma supplement. Also, the law reiterates that the higher education educational programs are grouped by domains and organized in three study cycles: bachelor, master and doctorate programmes (Article 137).

The first (Bachelor) cycle includes a minimum of 180 and a maximum of 240 transferable study credits equivalent to ECTS, and lasts three to four years, depending on the field and area of specialisation (Article 150).

The second (Master) cycle includes a minimum of 60 and a maximum of 120 transferable study credits (in exceptional cases and depending on the length of the first cycle, the lower limit may be 60 transferable study credits), and lasts one to two years (Article 153).

In the case of third cycle, the law provides that doctoral programs usually lasts 3 years, with the possibility to be extended by 1–2 years with the approval of the university senate, at the proposal of the doctoral supervisor and within the limit of the available funds (Article 159). The third cycle include 180 or 240 ECTS depending on the specific field of the doctoral studies. The law establishes clearly 240 ECTS only for the medical fields which lasts for 4 years (Article 174). It should be noted that in initial form of Article 140 (4) of the Law no. 1/2011 it was provided that the doctoral programs can only be organized in the form of *full time* studies, fact which has changed by Government Emergency Ordinance no. 49/2014 according to that the doctoral programs can be organized in the form of *part time* studies.

In close relationship with the three study cycles it has been developed a national qualifications framework.

As a first step, based on Article 340 of the Law no. 1/2011, it was established the National Qualifications Authority (NQA) by Government Decision no. 556/2011, the purpose being to reorganize two institutions: the National Council for Adult training, in charge of VET (Vocational Education and Training) qualifications and National Agency for Qualifications in Higher Education. This single legal entity—under the coordination of Ministry of Education—has the following competences: proposes elements of national policies and strategies, draft legislation on the national qualifications framework; develops, implements and updates the NQF and manages the national qualifications register; develops and updates the methodologies necessarily for the implementation of NQFs; develops the instruments needed for monitoring, evaluation and control of the NQF; quality assures the implementation of the NQF (Article 4 of the Government Decision no. 556/2011).

The Romanian national qualifications framework it was established by Government Decision no. 918/2013, and includes 8 levels of qualifications that can be acquired through the formal education and training system in Romania and by recognition of learning outcomes acquired through non-formal and informal learning. This decision was amended by Government Decision no. 567/2015, which provides some updates and modifications in previous Annex no. 2, especially with regard to correspondence between the NQF levels, the study papers to be issued, the type of education and training programs in Romania through which the qualification levels can be obtained, the reference levels of the European Qualifications Framework (EQF), as well as the access conditions corresponding to each level of qualification.

2.2.3 Reforms in the Field of University Management

Some of the most controversial provisions concerning the reform of the University Management were introduced by former Article 215 of the Law no. 1/2011, according to that upon attaining the retirement age, holding leadership positions in public, private and confessional universities is prohibited, with the exception of the terms of offices in course when this law comes in effect". In addition, the law provided that persons who exercise a leading position or public dignity cannot exercise a rector position during the execution of the term of office, the rector position being also incompatible with holding high office in a political party. These provisions were corroborated with disposal of former Article 285 (2) of the Law no. 1/2011, which stated that "after retirement, it is forbidden to hold any managerial or administrative position, at any level of the state, private or confessional higher education institutions", and hence "the mandates of those who hold managerial and administrative positions at any level of the university are terminated de jure in the case of the persons who have reached the retirement age". It is self-evident that immediately after the change of government these provisions were abrogated by Article I, paras. 10 and 17 of

Government Emergency Ordinance no. 92/2012, thus disappear all incompatibilities and age limit.

2.2.4 Reforms Concerning Teaching and Research Staff Recruitment

As shown in a 2011 European Commission's communication, the reform and modernisation of Europe's higher education depends also on the competence of teachers and researchers, which requires transparent and fair recruitment procedures (COM (2011) 567 final).

In this sense, the National Education Law no. 1/2011 introduced in Article 219 (1) and Article 295 (1) some performance criteria for the award of teaching titles in higher education, professional research and development degrees, doctoral supervisor position and habilitation certificate.

Based on the provisions of the Law no. 1/2011, the Romanian Government adopted Decision no. 457/2011 regarding the approval of the Framework Competition Methodology for Filling in Vacant Teaching and Research Positions in Higher Education, according to which, within 60 days from the entry into force of the decision, the higher education institutions were obliged to elaborate their own Competition Methodology for filling vacant positions (Article 2 of the Government Decision no. 457/2011).

More than that, under the above-mentioned legal provisions, the Minister of Education adopted a series of orders on the approval of the necessary and required minimal standards for conferring academic titles in higher education, professional titles in research and development and habilitation certificate, as follow: Order no. 4478/2011 for the scientific fields related to the panels on fundamental areas P1—Mathematics and Natural Sciences, P2—Engineering Sciences and P3—Biomedical Sciences; Order no. 4692/2011 for the scientific fields related to the panel on fundamental areas P5—Arts and Humanities; Order no. 4691/2011 for the scientific fields related to the panel on fundamental areas P4—Social Sciences.

All these legal acts were in force for only a few months, their provisions being amended after the change of Government.

Thus, the necessary and required minimal standards for conferring academic titles in higher education were reformed by Order no. 6560/2012, a normative act that compressed all scientific fields within 33 Annexes. It should be noted that after almost 4 years, following the change of government of course, this order has been abrogated by Order no 6129/2016 regarding the approval of the necessary and compulsory minimum standards for the award of higher education teaching titles, professional research and development degrees, doctoral supervisor position and habilitation certificate. The new order set the necessary and compulsory minimum standards for all scientific fields within 35 Annexes (two more than the previous order) and provided that, within 90 days of the entry into force of the Order, higher education institutions shall amend and supplement their own standards.

The Framework Competition Methodology was in turn modified by Government Decision no. 36/2013, among the changes can be mentioned the elimination of the condition to hold the habilitation certificate in the case of the competition for a university professor position. Then, it was modified by Government Decision no. 883/2018, the most important changes being as follow: removal of some interdictions regarding members of the contest committee (associated with a candidate in commercial companies, remunerated through research projects to which a candidate has been the project manager, benefited from services or receive an advantage of any kind from a candidate over the past 5 years); elimination the word *only* in connection with public competition for filling the vacant positions (Article 1 of the Government Decision no. 883/2018).

It should be remembered that in the summer of 2018, it was adopted the Government Ordinance no. 9 which modified the Law 1/2011 by introducing of a new Article 294^1, according to that is introduced an innovative concept regarding academic career, respectively *promotion in the teaching career based on the exam*, not just through public competition as before (Article 1, para. 20 of the Ordinance no. 9/2018).

As is provided in the new Article 294^1 of the Law no. 1/2011, the applications for exam enrolment are approved by University Management Board, although among its duties established in Article 213 (13) is only competence *to approve the proposals made for organising contests for teaching and research positions*. Given the touchy issue and also vague provisions of the law, conflicts may again arise because when we say promotion, the spectrum of those who can participate in these competitions narrows only to teaching staff already employed in a certain university, without outside competition.

3 Strengths and Weaknesses of the Romanian Higher Education System as a Result of Reforms Over the Past 20 Years

3.1 Achievements

As regards the strengths of the Romanian higher education system, it is undeniable that in recent years there have been important developments in the direction of reforming and improving the quality of higher education in line with the commitments made under the Bologna Process. Thus, the Romanian authorities reviewed the legal, institutional and methodological frameworks for quality assurance according to the Standards and Guidelines for Quality Assurance in the European Higher Education Area.

In light of the Bologna process, various initiatives have been launched for achieving the assumed obligations and also to align Romanian higher education system with education systems in other European countries, such as: implementing of the three cycles of education (bachelor/master/doctorate); implementing ECTS and adopting the Diploma Supplement; promoting the mobility of students and teaching and administrative staff; development of Erasmus partnerships and issuing of joint degrees; establishment and development of the Romanian Agency for Quality Assurance in Higher Education (ARACIS) and also a National Qualifications Framework for Higher Education that is fully compatible with the European Qualifications Framework.

It can be said that, following these ongoing efforts, education programs organised by the Romanian universities and scientific achievements acquired by some of them can be compared to those of universities with a similar profile in Europe.

3.2 Deficiencies

As concerns the deficiencies of the Romanian higher education system, can be identified a number of weaknesses that are based on impaired or abnormal functioning, both in terms of the Ministry of National Education, but especially the higher education institutions.

Regarding the activity of the Ministry, the following issues may be mentioned:

- *Managerial incoherence*—that's because it has reached the "performance" of having 26 Ministers in the period 1989–2018, of course, each Minister with their own vision of the reform to be made in education (for example only in 2012 were 6 Ministers);
- *Legislative instability*—in just 8 years the National Education Law no. 1/2011 has been changed over 60 times, most often by Government Emergency Ordinance, although pursuant to Article 73 of the Romanian Constitution, the general organization of education is regulated by organic laws;
- *Poor administrative capacity*—part of the higher education system malfunctions are the result of the Ministry's low capacity to identify relevant issues, to conceive the right actions and to have a consistent and long term perspective, this in relation to the Article 216 (2) of Law 1/2011 which sets the Ministry's attribution to propose the national policies and strategies for higher education as part of the European Higher Education Area;
- *Inability to provide adequate funding for education activities*—although Article 8 of Law 1/2011 establishes that a minimum of 6% of the GDP of the year in question is allotted annually from the state budget and from the budgets of the local authorities in order to finance national education, Romania has the lowest general expenditure on education as a proportion of GDP in the EU, of 3.7% in 2017 (Education and Training Monitor 2018, p. 239).

The weaknesses in the activity of Romanian higher education institutions are numerous, fact reflected by the negative perceptions of public opinion on the higher education system, especially as a result of the scandal concerning plagiarism in the doctoral studies. Among the problems faced by higher education institutions in Romania, can be listed:

- *Uniformization of the mission and organizational structure*

All Romanian universities, whether state-owned or private, declare that are "Higher education and research institutions", although Article 193 (4) of Law 1/2011 establishes that Universities are classified in 3 categories: universities, mainly for education; universities for education and scientific research and artistic creation; universities of advanced research and education. As regard to internal organization, the universities have adopted the minimum model set by law, while Article 131 of Law 1/2011 leaves them free to set up other organizational components, such as: faculties, departments, institutes, centre or laboratories, design facilities, consultancy centres, university clinics, artistic studios and workshops, theatres, museums, centres for continuous learning of the human resources, services and micro-production facilities, experimental stations, and other entities for production and know-how and technologic transfer;

- *Improper mode of recruiting and motivating the teaching staff*

Despite a clear legislative framework that ensures conditions for respecting the principles transparency and equal opportunities in higher education (Article 3 of Law 1/2011), among the public opinion there is suspicion that the vast majority of vacant teaching positions are designed for certain people (on the basis of personal relationships or political criteria) and then are open for "competition", that's why usually competes only one person (anyone who is interested in finding out whether it's true or not, can access the portal of vacant positions in higher education http://jobs.edu.ro).

Another problem is the lack of a strategy for attracting, motivating and keeping of the competent young people in the higher education system, which led to their exodus to foreign universities and research institutes.

4 Conclusions

Twenty years ago 29 countries signed the Bologna Declaration, thus initiating a wave of coordinated reform which changed not only the European system of higher education, but also across other world regions.

This reforming process launched with the Bologna Declaration (hence the name of the whole Process), has led to the creation of the European Higher Education Area (EHEA) that includes 48 Member States, their common goal being the establishment of comparable and compatible systems of higher education in order to facilitate mobility, increase employability, allow equitable student access and progression and strengthen Europe's attractiveness and competiveness worldwide.

Bologna process meant for all 48 participating countries profound reform of higher education, marked by changes in legislation and reconsideration of the basic principles of the educational process, especially in terms of the three key commitments that underpin the EHEA and concern the implementation of the three cycle system (bachelor/master/doctorate), strengthened quality assurance and easier recognition of qualifications and periods of study.

As shown by the regular Bologna Process implementation reports, a lot of progress has been made in implementing the reforms, but implementation remains uneven, both between policy areas and between countries, which means that additional efforts are needed.

In order to comply the Bologna Process requirements, the Romanian authorities have undertaken a series of structural reforms since 1999, with a focus on revising the legal, institutional and methodological frameworks for implementation of the commitments set out in the Ministerial Declarations and Communiqués. All these measures have contributed to the strengthening of the Romanian higher education system, implementing high standards for modern and relevant educational provision.

It is undeniable that in recent years there have been important developments in the direction of reforming and improving the quality of the Romanian higher education, but further efforts need to be made in order to ensure the correct implementation of the Bologna tools and the EHEA principles. Thus, fulfilling of Romania's goals and commitments depends on the concerted efforts of national policy-makers, public authorities, institutions, staff, students and other stakeholders as well as coordination at EHEA level.

As an argument in this regard, can be quoted the European Commissioner for Education who, in the Foreword of the Bologna Process Implementation Report on 2018, have shown: "We should of course be proud of Bologna Process achievements, but we must not be complacent, we need to redouble our efforts to bring Europe's higher education institutions, researchers and students even closer together".

References

Berlin Communiqué—Realising the European Higher Education Area (2003, Sept 19). Retrieved from http://www.ehea.info/media.ehea.info/file/2003_Berlin/28/4/2003_Berlin_Communique_English_577284.pdf

Belarus roadmap for higher education reform (2015, May 21). Retrieved from http://www.ehea.info/media.ehea.info/file/2015_Yerevan/70/9/Roadmap_Belarus_21.05.2015_613709.pdf

Bergen Communiqué—The European Higher Education Area-Achieving the Goals (2005, May 20). Retrieved from http://www.ehea.info/media.ehea.info/file/2005_Bergen/52/0/2005_Bergen_Communique_english_580520.pdf

Bologna Declaration: Joint declaration of the European Ministers of Education (1999, June 19). Retrieved from http://www.ehea.info/media.ehea.info/file/Ministerial_conferences/02/8/1999_Bologna_Declaration_English_553028.pdf

Bucharest Communiqué—Making the Most of Our Potential: Consolidating the European Higher Education Area (2012, April 27). Retrieved from http://www.ehea.info/media.ehea.info/file/2012_Bucharest/67/3/Bucharest_Communique_2012_610673.pdf

Budapest-Vienna Declaration on the European Higher Education Area (2010, Mar 12). Retrieved from http://www.ehea.info/media.ehea.info/file/2010_Budapest_Vienna/64/0/Budapest-Vienna_Declaration_598640.pdf

Communication from the Commission to the European Parliament, the Council, the European Economic and Social Committee and the Committee of The Regions: Supporting growth and jobs—an agenda for the modernisation of Europe's higher education systems, COM (2011) 567 final (2011, Sept 20). Retrieved from https://eur-lex.europa.eu/LexUriServ/LexUriServ.do?uri=COM:2011:0567:FIN:EN:PDF

Education and Training Monitor 2018, Country analysis – Romania. Retrieved from https://ec.europa.eu/education/sites/education/files/document-library-docs/volume-2-2018-education-and-training-monitor-country-analysis.pdf

Leuven and Louvain-la-Neuve Communiqué—The Bologna Process 2020-The European Higher Education Area in the new decade (2009, April 29). Retrieved from http://www.ehea.info/media.ehea.info/file/2009_Leuven_Louvain-la-Neuve/06/1/Leuven_Louvain-la-Neuve_Communique_April_2009_595061.pdf

London Communiqué—Towards the European Higher Education Area: responding to challenges in a globalised world (2007, May 18). Retrieved from http://www.ehea.info/media.ehea.info/file/2007_London/69/7/2007_London_Communique_English_588697.pdf

National Pact for Education (2008, July 6). Retrieved from http://old.presidency.ro/static/ordine/Raport_CPaepdec_2007_.pdf

Paris Communiqué (2018, May 25). Retrieved from http://www.ehea.info/media.ehea.info/file/2018_Paris/77/1/EHEAParis2018_Communique_final_952771.pdf

Prague Communiqué—Towards the European Higher Education Area (2001, May 19). Retrieved from http://www.ehea.info/media.ehea.info/file/2001_Prague/44/2/2001_Prague_Communique_English_553442.pdf

Project Report: Trends in Learning Structures in Higher Education (1999, June 7). Retrieved from https://eua.eu/resources/publications/677:trends-in-learning-structures-in-higher-education.html

Report of the Presidential Commission for Analysis and Development policies in the fields of education and research (2007, July 6). Retrieved from http://old.presidency.ro/static/rapoarte/Raport_CPAEPDEC.pdf

Sorbonne Declaration: Joint Declaration on harmonisation of the architecture of the European higher education system (1998, May 25). Retrieved from http://www.ehea.info/media.ehea.info/file/1998_Sorbonne/61/2/1998_Sorbonne_Declaration_English_552612.pdf

Statement by the Bologna Policy Forum (2009, April 29). Retrieved from https://enqa.eu/wp-content/uploads/2013/03/Bologna_Policy_Forum_Statement_29April2009.pdf

Statement of the Third Bologna Policy Forum—Beyond the Bologna Process: Creating and connecting national, regional and global higher education areas (2012, April 27). Retrieved from http://www.ehea.info/media.ehea.info/file/2012_Bucharest/63/8/Bucharest_BPF_Statement_610638.pdf

Statement of the Fourth Bologna Policy Forum—Yerevan (2015, May 15). Retrieved from https://enqa.eu/wp-content/uploads/2015/06/4th-Bologna-Policy-Forum.pdf

Statement of the Fifth Bologna Policy Forum—Paris (2018, May 25). Retrieved from http://www.ehea.info/media.ehea.info/file/2018_Paris/36/8/BPFStatement_with_Annex_958368.pdf

Yerevan Communiqué (2015, May 15). Retrieved from http://www.ehea.info/media.ehea.info/file/2015_Yerevan/70/7/YerevanCommuniqueFinal_613707.pdf

Legislation

Constitutional Court Decision no. 1/2011 on the referral of unconstitutionality of National Education Law, published in Official Gazette, Part I, no. 135 of 23 February 2011

Government Decision no. 567/2005 on organization and conduct of Doctoral Studies, published in Official Gazette, Part I, no. 540 of 24 June 2005, abrogated by Government Decision no. 681/2011 regarding the approval of the Code of Doctoral Studies, published in Official Gazette, Part I, no. 551 of 3 August 2011

Government Decision no. 1175/2006 regarding the organization of undergraduate studies and the approval of the fields and specializations list within them, published in Official Gazette, Part I, no. 769 of 11 September 2006
Government Decision no. 404/2006 on organization and conduct of Master Studies, published in Official Gazette, Part I, no. 319 of 10 April 2006
Government Decision no. 457/2011 regarding the approval of the Framework Competition Methodology for Filling in Vacant Teaching and Research Positions in Higher Education, published in Official Gazette, Part I, no. 371 of 26 May 2011
Government Decision no. 556/2011 on the organization, structure and functioning of the National Authority for Qualifications, published in Official Gazette, Part I, no. 435 of 22 June 2011
Government Decision no. 789/2011 regarding the approval of the Assessment Methodology for University Classification and Ranking of the Study Programmes, published in Official Gazette, Part I, no. 569 of 10 August 2011
Government Decision no. 457/2011 regarding the approval of the Framework Competition Methodology for Filling in Vacant Teaching and Research Positions in Higher Education, published in the Official Gazette, Part I, no. 371 of 26 May 2011, as amended by Government Decision no. 36/2013, published in Official Gazette, Part I, no. 88 of 11 February 2011, in turn amended by Government Decision no. 883/2018, published in Official Gazette, Part I, no. 967 of 15 November 2018
Government's Emergency Ordinance no. 75/2005 concerning quality assurance in education, approved with modifications by Law no. 87/2006, published in Official Gazette, Part I, no. 334 of 13 April 2006
Government Emergency Ordinance no. 92/2012 on measures in the field of education and research, published in Official Gazette, Part I, no. 864 of 19 December 2012
Government Emergency Ordinance no. 49/2014 on the establishment of measures in the field of education, scientific research and for the modification of some normative acts, published in Official Gazette, Part I, no. 486 of 30 June 2014
Government Ordinance no. 9/2018 on amending and completing certain normative acts in the field of education, published in Official Gazette, Part I, no. 744 of 28 August 2018
Law no. 288/2004 concerning the organization of the higher education, published in Official Gazette, Part I, no. 614 of 7 July 2004
Law no. 1/2011 of the National Education, published in Official Gazette, Part I, no. 18 of 10 January 2011
Order of the Minister of Education and Research no. 3861/2005 on the establishment of postdoctoral research programs, published in Official Gazette, Part I, no. 407 of 13 May 2005
Order of the Minister of Education and Research no. 3235/2005 concerning the organization of the first cycle of university studies. Retrieved from http://www2.edu.ro/index.php/articles/31
Order of the Minister of Education and Research no. 3617/2005 on the generalized application of the European Credit Transfer System. Retrieved from http://oldsite.edu.ro/index.php/articles/37
Order of the Minister of Education and Research no. 4868/2006 concerning the implementation of the Diploma Supplement, published in Official Gazette, Part I, no. 705 of 7 August 2006
Order of the Minister of Education, Research, Youth and Sport no. 5.262/2011 regarding the results of the classification of universities, published in Official Gazette, Part I, no. 637 of 9 September 2011
Order of the Minister of Education, Research, Youth and Sport no. 4478/2011 approving the necessary and binding minimum standards for the award of teaching titles in higher education, professional research and development degrees and the habilitation certificate, for the scientific fields related to the panels on fundamental areas P1—Mathematics and Natural Sciences, P2—Engineering Sciences and P3—Biomedical Sciences, published in Official Gazette, Part I, no. 448 of 27 June 2011
Order of the Minister of Education, Research, Youth and Sport no. 4691/2011 approving the necessary and binding minimum standards for the award of teaching titles in higher education, professional research and development degrees and the habilitation certificate, for the scientific fields

related to the panels on fundamental areas P4—Social Sciences, published in Official Gazette, Part I, no. 535 of 28 July 2011

Order of the Minister of Education, Research, Youth and Sport no. 4692/2011 approving the necessary and binding minimum standards for the award of teaching titles in higher education, professional research and development degrees and the habilitation certificate, for the scientific fields related to the panels on fundamental areas P5—Arts and Humanities, published in Official Gazette, Part I, no. 540 of 29 July 2011

Order of the Minister of Education, Research, Youth and Sport no. 6560/2012 approving the necessary and binding minimum standards for the award of teaching titles in higher education and the professional degrees in research-development, published in Official Gazette, Part I, no. 890 of 27 December 2012, as amended by Order no. 4204/2013, published in Official Gazette, Part I, no. 499 of 8 August 2013

Order of the Minister of Education no. 6129/2016 regarding the approval of the necessary and compulsory minimum standards for the award of higher education teaching titles, professional research and development degrees, doctoral supervisor position and habilitation certificate, published in Official Gazette, Part I, no. 123 of 15 February 2017

Empowering Future Citizens in Making Financial Decisions: A Study of Elementary School Mathematics Textbooks from Romania

Daniela Căprioară, Annie Savard and Alexandre Cavalcante

Since the major financial crisis in 2008, Financial Education is getting an international recognition in Education. In Romania, the challenge to financially educate students is particularly important because the country is still adapting to a capitalist economy and most adults grew up under a communist economy. The elementary school national curriculum includes Financial Education embedded into the Mathematics Curriculum. In this book chapter, we present a qualitative study on the Mathematics textbooks approved by the Romanian Ministry of Education and used during the school year 2018–2019. We investigated the learning situations (tasks) proposed in the textbooks whose content had a connection to financial education. Three themes emerged from our analysis: no potential decision, potential decision and decision. In the third theme, decision, we sub-divided it into five categories: four of them related to simple decisions, and one related to complex decisions. Our results show that the tasks having a potential decision and those having a simple decision reflect practices of representational measurement. On the other hand, the complex decisions category, despite being rarer, reflects practices of pragmatic measurement. The tasks under this category are not only richer in regard to the making-decision opportunity, but also in regard to financial education.

Financial literacy · Decision making · Elementary school · Mathematics curriculum · Mathematics textbooks

---

D. Căprioară (✉)
Ovidius University of Constanța, Constanța, Romania
e-mail: ccaprioara@yahoo.fr

A. Savard · A. Cavalcante
McGill University, Montreal, Canada
e-mail: annie.savard@mcgill.ca

A. Cavalcante
e-mail: alexandre.cavalcante@mail.mcgill.ca

D. Flaut et al. (eds.), *Decision Making in Social Sciences: Between Traditions and Innovations*, Studies in Systems, Decision and Control 247,
https://doi.org/10.1007/978-3-030-30659-5_7

1 Financial Literacy as a Competency to Make Financial Decisions

Since the major financial crisis in 2008, Financial Literacy (FL) is getting an international recognition in Education. For instance, the Organisation for Economic Co-operation and Development (OECD) started to assess in 2012 in their Programme for International Student Assessment (PISA) 15 years-old students' competencies in regard of Financial Literacy. OECD is looking for more than knowledge and behaviors, OECD is looking for being competent with finances. At this end, the OECD (2016) definition of FL presents a competency to make effective decisions for individual and for their community:

> Financial literacy is knowledge and understanding of financial concepts and risks, and the skills, motivation and confidence to apply such knowledge and understanding in order to make effective decisions across a range of financial contexts, to improve the financial well-being of individuals and society, and to enable participation in economic life. (p. 85)

Making effective decisions implies to mobilize knowledge accordingly to different situations. For doing so, knowledge, attitudes and skills are required. Coming with the definition, OECD (2016) proposed four areas of knowledge and understanding that students should mobilize: *money and transaction, planning and managing finances, risk and reward, and financial landscape*. This definition shows a scope of well-being: from individual to the society. Thus, as pointed out by Arthur (2012), FL is more than creating better consumers; it is about citizenship and the development of civic disposition. Civic disposition includes going beyond self-determination by being critical toward choices made. Democratic participation in debates and social issues encourages members of society to contribute to it by making personal and collective choices (ten Dam and Volman 2004). In fact, making a decision might have individual and social responsibilities (Swartz and Perkins 1990). Making financial decisions might have strong implications for individuals and their communities, because it might affect the use of individual and collectives resources.

Making decision starts with the recognition of a need to make decision (Halpern 2003; Paul and Elder 2001) and why a decision should be made (Swartz and Perkins 1990). Recognizing the need to make a decision also implies knowing the short-term and long-term effects. A short-term effect leads to instant gratification or pleasure, while long-term effects requires a deeper thinking and some self-discipline, because is not always possible to predict the long-term effect (Paul and Elder 2001). After recognizing the need to make a decision, the generation and the exploration of different options leads to assess them in order to find the best choice (Swartz and Perkins 1990). This assessment includes using different criteria: personal feelings, values and knowledge, cognitive and cultural bias, and environmental variables (availability of an object for example) (Halpern 2003). This assessment of the options is critical in order to select one that can meet the criteria identified by the individual and should reflect the "pros" and "cons" of them. It involves taking into consideration the probabilities, the possible consequences and the possible risks or benefits involved, which involves considering the uncertainty of the decision chosen for self and others' well-being (Savard 2015). The selection of an option is followed by the actions to undertake with this option. This should lead to re-assessing the decision and finding other options, in a dynamic and iterative process (Halpern 2003). The re-assessment of an option also leads to making a decision and validating the choice afterward.

2 Financial Education in the Romanian Curriculum

In Romania, an European emerging country in a post-communism area (1990), the challenge to financially educate students is particularly important because the country is still adapting to a capitalist economy and most of the adults grew up under a communist economy. In one hand, the challenges coming from the economy of the country, and the other hand, the European context forced the country to adapt the curriculum. The curricula reform had many steps over the years, for all level of teaching and for all teaching domains. The last step started in 2013. The new curriculum pays attention to financial literacy of early grades students, and has also an optional course on Financial Education (from 9 to 11 years old) which aim students to develop a better understanding of the role of money, credit and banks in the context of the market economy.

In the elementary school national curriculum, Financial Literacy is embedded into the Mathematics Curriculum. For each competency which should be developed, some examples of learning activities are presented for the teacher to use. The mathematical content is also presented and the content related to Financial Literacy belongs to the mathematical domain Measurement. Table 1 presents, for each grade in elementary school, the competencies, some examples of learning activities related to them, along with the content.

The Romanian new curriculum focuses on problem solving and inquiry-based learning as teaching and learning approaches. Thus, high ability of thinking are targeted through creativity, heuristic, critical thinking and metacognition. Different strategies such as trials and errors, reflexive thinking, self-assessment, risk assessment and management of emotions are supposed to be developed by students. Collaboration, cooperation and communication are promoted in classrooms.

3 Methodology

A qualitative study was conducted on the Mathematics textbooks approved by the Romanian Ministry of Education and used during the school year 2018–2019. The electronic version of the Mathematics textbook could be find on this webpage: https://www.manuale.edu.ro/. Table 2 presents the textbooks used for this study.

We investigated all of the learning situations (tasks) proposed in the textbooks, whose content had a connection to financial education. Our inclusion criteria, upon

Table 1 Extract from the Primary School Mathematics Curriculum (Ministerul Educației Naționale 2013; Ministerul Educației Naționale 2014)

Preparatory class (6–7 years old)	
Competency	Realization of some value exchanges by using the non-conventional representations in problems-games of the type revenue-expenditure, with the numbers in the limits 0–31
Examples of learning activities	– The recognition of banknotes of 1 leu, 5 lei, 10 lei – The production of "silver" necessary for a game activity – Games for the use of money (e.g. "At store", "In the amusement park" etc.) – Budget games—for an excursion, museum visit, park walk, movie viewing, etc.
Content (Domain "Measurement")	Money: Leul (banknotes of 1 leu, 5 lei, 10 lei) Equivalent value exchanges within the limits 0–31
1st Grade (7–8 years old)	
Competency	Realization of some value exchanges using the standard and non-conventional representations in problem-games of the type revenue-expense, with the numbers in the limits 0–100
Examples of learning activities	– The recognition of the banknotes of 1 leu, 5 lei, 10 lei, 50 lei, 100 lei – Recognition of the coins of 1 ban, 5 bani, 10 bani, 50 bani – The exchange of a group of coins/banknotes with another of the same value – Matching: 1 leu → a loaf; 1 leu → a lollipop etc.; 5 lei → 1 juice; 5 lei → a magazine – The negotiation of certain objects equivalent in value (barter) – The equivalence in value of certain groups of objects with a single object (3 pencils ≈ 1 pen) – Games: Token exchange (e.g.: a token with animals is worth two with football players) – The calculation of the costs (and profit, in the case of sale) of certain objects made for children – The addition and subtraction within the 0–100 limits, based on the learned banknotes and

Table 1 (continued)

Content (Domain “Measurement”)	“Money Leul: banknotes of 200 lei, 500 lei Euros (1 € = 100 cents); coins and banknotes Equivalent value exchanges within the limits 0–1000”
2nd Grade (8–9 years old)	
Competency	Realization of some value exchanges using the standard and non-conventional representations in problem-games of the type revenue-expense, with the numbers in the limits 0–1000
Examples of learning activities	– The recognition of the banknotes of 1 leu, 5 lei, 10 lei, 50 lei, 100 lei, 200 lei, 500 lei – The exchange of a group of coins/banknotes with a singular banknote/other group of coins or banknotes of the same value – Addition and subtraction within the 0–1000 limits, based on the learned banknotes and coins – Involve children in experiences where they can decide alone whether or not they can buy an object with the sum of money available – Money-use games – Comparison of sums of money made up of different coin and banknotes – The resolution of certain expenditure/budget/purchasing problems, oral and written, using addition and/or subtraction, multiplication, division – Games: “Exchange of tokens”, “In the shopping center”, “On a trip”
Content (Domain “Measurement”)	Money Leul: banknotes of 200 lei, 500 lei Euros (1 € = 100 cents); coins and banknotes Equivalent value exchanges within the limits 0–1000
3rd Grade (9–10 years old)	
Competency	The use of certain standardized instruments and units of measurement in concrete

Table 1 (continued)

Content (Domain "Measurement")	Monetary measurement units – Units of measurement: leu and ban, euro and eurocent – Equivalent currency exchanges in the same currency unit
4th Grade (10–11 years old)	
Competency	The use of certain standardized instruments and units of measurement, in concrete situations, including to validate certain transformations; Operations with standardized measurement units, using transformations
Examples of learning activities	– Comparison of certain sums of money composed of different coins and banknotes; – Money-use games – The analysis and interpretation of the results obtained by the resolution of certain practical problems, with reference to the units of measurement studied – Perform some calculations based on monetary units
Content (Domain "Measurement")	Monetary measurement units – Units of measurement: leu and ban, euro and eurocent (coins and banknotes in use) – Equivalent currency exchanges in the same currency unit

Table 2 Distribution of textbooks by grade

Grade	Curricular area/domain	Number of textbooks available for use
Preparatory class	Mathematics and exploration of the environment	–
Grade I	Mathematics and exploration of the environment	2
Grade II	Mathematics and exploration of the environment	3
Grade III	Mathematics	3
Grade IV	Mathematics	3

which we based our analysis, was the degree of contribution that each task presented to the development of decision making for elementary school students.

The majority of tasks analyzed in this study are part of the learning unit dedicated to measurement (measures and units of measurement). Table 3 shows the distribution of these activities/tasks in each textbook.

4 Results

Three themes emerged from our analysis: no potential decision, potential decision and decision. The first theme, no potential decision, describes the tasks that did not present any potential decision to be made. It was just problems that touched upon basic knowledge to exercise mathematical operations. So there is not enough room for the teachers to discuss potential decisions, or even the real life context. We do not present this category in our results, because of the lack of potential in making financial decision.

In the third theme, decision, we sub-divided it into five categories: four of them related to simple decisions, and one related to complex decisions. For all simple decision categories, we specify the context of the decision to be made. For instance, simple decision could be made in a financial, a mathematical or a sociocultural context. In the mathematical context, the decision to be made is related to mathematics, not finance. Also, the decision to be made could be related to simple or complex mathematic.

We present our results by describing each theme and category, and we provide examples coming from our data collection. We also present a short analysis of the data presented.

Potential Decision

Some activities presented situational problems that could be exploited by the teachers. These problems did not require any decision to be made in order to solve them, but we understand that teachers could go beyond the simple context of these problems and exploit decisions that are rich and connected to real situations. The decision-making dimension could also be brought up by students in the development of the activity. For instance, the 4th grade students who participated in Savard (2015) study provided different arguments for justifying their decisions. For instance, they took in consideration the risks involved in participating in gambling activities to decide to participate or not.

Examples:

Table 3 Distribution of learning situations related to financial education in each textbook analyzed

Grade	Textbook	Learning unit(s)	Rate (%)[a]
Grade I	https://manuale.edu.ro/manuale/Clasa%20I/Matematica%20si%20explorarea%20mediului/EDP/A590.pdf	Units of measurement/Measurement of value: Leul, banul	2
	https://manuale.edu.ro/manuale/Clasa%20I/Matematica%20si%20explorarea%20mediului/EDP2/A591.pdf	Measurement, measure, instruments of measurement/Money. Leul: coins and bills; Equivalent exchanges in terms of value from 0 to 100	3
Grade II	https://manuale.edu.ro/manuale/Clasa%20a%20II-a/Matematica%20si%20explorarea%20mediului/EDP/A627.pdf	We and the world around us/Money. Nous et le monde autour de nous/Les argents. Equivalent exchanges in terms of value from 0 to 100	2
	https://manuale.edu.ro/manuale/Clasa%20a%20II-a/Matematica%20si%20explorarea%20mediului/EDP2/A628.pdf	Measurement (II). The planets of the solar system/Coins and bills	2
	https://manuale.edu.ro/manuale/Clasa%20a%20II-a/Matematica%20si%20explorarea%20mediului/EDP3/A629.pdf	Units of measurement/Money. Leul, euro. Coins and bills. The intensity of acoustic force	4
Grade III	Part I https://manuale.edu.ro/manuale/Clasa%20a%20III-a/Matematica/Intuitext/Partea%20I/A0611.pdf Part II	Part II, 2nd semester Units of measurement/Coins and bills. Leul and banul. Euro and eurocent	3

3. Administratorul unui cămin de copii cumpără fructe. Analizează tabelul și află:

Produsul	Cost al unui kilogram	Cantitatea achiziționată
mere	3 lei/kg	25 kg
căpșuni	9 lei/kg	20 kg

a) Cu cât este mai scump 1 kg de căpșuni decât 1 kg de mere?
b) De câte ori este mai ieftin 1 kg de mere decât 1 kg de căpșuni?
c) Câți lei a costat cantitatea achiziționată?
d) Ce rest primește administratorul dacă a folosit trei bancnote de 100 de lei?

Translation:

The administrator of a child's home buys fruit. Analyze the table and find out:
a) How much is more expensive 1 kg of strawberries than 1 kg of apples?
b) How many times is cheaper 1 kg of apple than 1 kg of strawberries?
c) How much does the whole purchase cost?
d) What else does the administrator receive if he has used three 100 lei banknotes?

In this task, students are presented with a context that, depending on the teacher's practices and approach, could raise discussions about financial decisions regarding pricing or products, quantity to be purchased, promotions and sales, etc. The financial context presented in such task is rich and provides an opportunity for students to experience a situation that is similar to real life. In other tasks under this category, students were asked to create a problem out of the information provided.

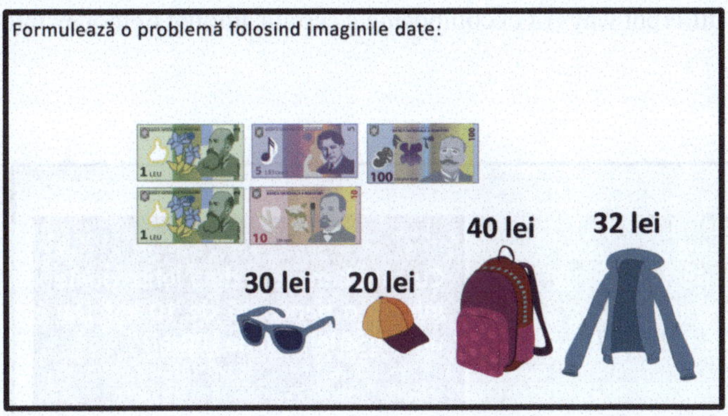

Translation:
Formulate a problem using the given images

Students might come up with a decision to be made regarding the amount of money and the items. This request in itself is complex for today's standards of education (instead of answering problems, students are being asked to create them, which subverts traditional practices in most pedagogical materials). In this activity, multiple decisions could potentially arise from the mathematical or the financial domains of the problem, depending on how teachers approach this kind of task (or the reaction of students and their questions).

Simple Decision—Financial Context

The decision to be made is simple and is situated in a financial context. Usually, the decision has to be made among some options presented, under certain constraints, such as the amount of money available.

Example:

6 Andrei va cumpăra un trening de 80 lei și încă două obiecte, dintre cele ilustrate alături. Știind că a economisit 200 de lei, numește cele două obiecte pe care le va cumpăra, astfel încât să îi ajungă banii. (Estimează, apoi verifică prin calcul.)

Translation:

Andrei will buy a training of 80 lei and two other objects, illustrated next. Knowing that he has saved 200 lei, calls the two items he will buy so he can get his money. (Estimate, then select by calculation.)

Simple Decision—Mathematical Context with Simple Mathematics

The decision in these problems was related to finding multiple mathematical solutions.

Example: Finding different ways of decomposing a certain amount of money into different bills.

Example:

3 Joc

Confecționați bani din carton. Elevii vor veni la catedră și vor cere casierului să le schimbe bancnote și monede, de valoare mai mare sau mai mică.

Cu câte bancnote de 5 lei se poate înlocui o bancnotă de 10 lei? Dar una de 50 lei? Dar de 100 lei?

Translation:

Make money using cardboard. Students will come to the chair and ask the cashier to change their banknotes and coins of higher or lower value.

[...]

How many 5 lei banknotes can replace a 10 lei banknote? But one of 50 lei? But 100 lei?

The decision to be made is simple: changing money into other bills and coins. We consider that the decision to be made is situated in a mathematical context, because the mathematical knowledge about the value of the money (bills and coins) is a simple knowledge needed to make the decision (addition, multiplication).

Simple Decision—Mathematical Context with Complex Mathematics
The decision in these problems were related to finding multiple mathematical solutions. However, these problems often involved more than one mathematical structure (additional/multiplicative) and mixed different currencies (leu, euro, cents, etc.). For instance, finding different ways of decomposing a certain amount of money into different bills and coins using the national currency (lei) as well as a different currency (euro).

Example:

7 **Calculează, în fiecare caz, numărul de bancnote din fiecare, pentru care se verifică egalitatea.**
a) $a \times 5$ **lei** $+ b \times 10$ **lei = 80 lei** **b)** $c \times 100$ **lei** $+ d \times 200$ **lei = 1400 lei**
Sunt mai multe soluţii? Explică!

Translation:
Calculate, in each case, the number of banknotes in each, for which equality is checked.
[...]
Are there multiple solutions? Explain!

4 **Mario a oferit un magnet pentru frigider, fiecăruia dintre cei 9 copii prezenţi la petrecere. Ştiind că preţul unui magnet este de 2 euro, află ce rest a primit Mario dacă a dat la casă o bancnotă de 50 de euro.**
- **Descoperă trei moduri în care putea primi restul, ştiind că a primit şi monede şi bancnote.**

Translation:
Mario offered a fridge magnet to each of the 9 children present at the party. Knowing that the price of a magnet is 2 euros, find out what the rest Mario has received if he handed a 50 euro banknote to cash register.
Discover three ways to get the rest, knowing he also received coins and banknotes.

In both cases, the decision to be made is simple: finding different ways to come up with a certain amount of money. However, the mathematical knowledge involved is complex, because an algebraic reasoning is required to make the decision in a mathematical context. Furthermore, the first task asks students to explain the existence of multiple solutions and utilizes a currency that students are not very familiar with at this moment (euro). Romania, which is part of the European Union, utilizes more and more the Euro in multiple financial transactions and will soon replace its currency.

Simple Decision—Sociocultural Context
The activities in this category are connected with the social sciences part of financial education, and involved decisions that pertain to the social aspect of money, bills, etc. Although these decisions are not strictly mathematical, we believe that, since they are presented in the manual for mathematics, they engender a special category of decision making that is taught by math teachers. The relevance of this category relies

on the fact that the decision is related to the representation of the unit of measurement (money), so it touches directly on the representation of a financial and mathematical concept.

Example:

Lucru în perechi

Găsiţi cât mai multe posibilităţi de exprimare a fiecărei sume, folosind bancnotele date.

a) 685 lei;

b) 490 euro.

Pe fiecare bancnotă din ţara noastră este imaginea unei personalităţi din România. Scrieţi într-un tabel: valoarea bancnotei, numele personalităţii, domeniul de activitate, între ce ani a trăit. Voi ce personalitate aţi propune pentru o nouă bancnotă? De ce?

Translation:

Find out more ways to express each amount using the given banknotes.

[...]

On every banknote in our country is the image of a Romanian personality. Write in a table: the value of the banknote, the name of the personality, the field of activity of the personality, the period of the life of the personality. What personality do you propose for a new banknote? Why?

The decision to be made is simple: finding a personality to put on a bill. However, this decision is situated into a sociocultural context, because the decision is based on socio-cultural knowledge, not on mathematics or finance.

Complex Decision

These activities were of high cognitive demand and usually required students to come with their own ways of solving, as well as making financial decisions that are related to real life situations. Often times, the mathematics also proved to be complex, either mixing different structures, or using different notations and language (algebraic equations, tables, graphs, etc.)

Example:

7 Reprodu tabelul de mai jos şi barează preţul care consideri că este nepotrivit:

creion	1 leu	85 lei	2 lei
stilou	10 bani	32 lei	40 lei
adidaşi	4 lei	85 lei	150 lei
trening	90 lei	120 lei	11 lei
balon	1 leu	3 lei	100 lei
prăjitură	50 bani	4 lei	5 lei
pâine	78 lei	2 lei	1 leu

Translation:
Reproduce the table below and list the price you think is inappropriate: (pencil, pen, sneakers, training, balloon, cake, bread).

ACTIVITATE ÎN ECHIPĂ • • • JOC „De-a magazinul"

Confecţionaţi bancnote de hârtie, pe care notaţi valorile de mai sus şi preţuri pentru diferite obiecte. Oferiţi „vânzătorului" bancnote cu valoare totală mai mare decât preţul obiectului „cumpărat", pentru a primi rest.

Translation:
Teamwork ("On-store")
Make banknotes using paper, which you note the above values and prices for different items. Give the "seller" banknotes with a total value greater than the price of the "bought" object, to get the rest.

The decision to be made is complex, because students are asked to know the value of some familiar items, or they have offered an unlimited choices of items to buy. The mathematics involved in those examples might be considered as simple or complex, depending of the justifications brought up by the students. For instance, in the first example, some students might explain their answer by the mathematical range of the price of the item. Another simple mathematics justification might be because the students know by experience the price of the items. A more complex justification might be because the students assess the value of the items by comparing the price with another item or by the value of the amount of money involved.

5 Discussion

The tasks we analysed, also seen as learning situations, supported the competency development defined in the new mathematics curriculum: *Realization of some value exchanges by using the non-conventional and standard representations in problems-games of the type revenue-expenditure, with the numbers in the limits 0-31/0-100/0-1000; The use of certain standardized instruments and units of measurement, in concrete situations, including to validate certain transformations; Operations with standardized measurement units, using or without transformations.* However, the competencies developed put more the accent on using money in different contexts than making decision in a financial context.

As we said, the new Romanian curriculum emphasizes problem-solving and thus the textbooks presented tasks as problems to solve. Those problems are related to concrete and familiar situations for students: "Problems are exercises of relationship between real life situations and the mathematical models studied in school and they are occasions to reflect on complex relations between reality and mathematics. Students' capacity to issue judgments on the problems' reality reflects a positive attitude towards a realistic mathematical modeling" (Căprioară 2015). However, problem solving does not necessarily involve decision making. As shown in our results, some of the tasks related to FL have only a potential to make a decision (or, in some cases, no potential at all). Not all tasks related to FL demanded students to make a decision.

Among the tasks that require some decision making, our results show that the tasks under the domain "Measurement" emphasize more the mathematics and less the financial education aspects. The financial concepts and notions, mainly about money, are utilized mostly as contexts to apply arithmetical concepts (addition, subtraction, division, multiplication, comparison, decomposition of integer numbers). In addition to that, the notion of measurement is mainly taken from its mathematical meaning, the representational (Savard et al. submitted). A representational measurement is a comparison to a unit that translates attributes to objects and their relationships to numbers (Hand 2016). Our results show that the tasks having a potential decision and those having a simple decision reflect a representational measurement. On the other hand, the complex decision tasks can be seen as pragmatic measurement, where many attributes of an object are taking in consideration to provide a value to something and how measurement will be performed (Hand 2016). For instance, in the first complex decision task presented, students have to find which price is inappropriate. If a student reasoning is about estimating the price of the object in regard to another one (a pencil is cheaper than sneakers), then a pragmatic measurement is involved. On the other hand, if the student reasoning is about looking at the mathematical magnitude of the price (85 is too high comparing to 1 or 2) or based on experience (students already bought a pencil), then the representational measurement is involved. In the second task, students might use a pragmatic measurement reasoning when according a value to an object to sell and buy. It is interesting to think that the value given to an item might be different from the seller or the buyer point of view. Those complex tasks are not only richer in regard to the making-decision opportunity, they are also richer in regard to financial education. It is disappointing to say that in the textbooks analysed, the complex decision tasks are quite rare.

6 Conclusion

Decision making is an essential component in for students in their first school years. This competency can be developed in all disciplines taught in the school, among which mathematics seems to be one of the most important. The mathematical domain of "measurement", which is the closest to the real life, offers to the students situations to develop their decision-making behaviours, and includes financial situations. In this context, the role of the teacher is key to emphasize the decision-making potential of learning situations. Thus, it is relevant to question the preparation of this role: Are teacher preparation programs aim to support teachers in regard to teach Financial Literacy? If yes, are they focusing on making financial decision using a pragmatic measurement reasoning? In fact, more research is needed in regard of teacher preparation and students learning on Financial Education.

References

Arthur, C.: Consumers or critical citizens? Financial literacy education and freedom. Critical Educ. **3**(6) (2012). Retrieved from: http://ojs.library.ubc.ca/index.php/criticaled/article/view/182350

Căprioară, D.: Problem Solving - Purpose and Means of Learning Mathematics in School. Procedia Social Behav. J. **191**, 1859–1864 (2015). ISSN: 1877-0428. http://www.sciencedirect.com/science/article/pii/S1877042815025926

Halpern, D.F.: Thought and Knowledge: An Introduction to Critical Thinking, 4th edn. Lawrence Erlbaum Associates, Mahwah (2003)

Hand, D.J.: Measurement: A Very Short Introduction. Oxford Academic: Very Short Introductions Online (2016). ISBN 9780198779568

Ministerul Educației Naționale (2013) Programa şcolară pentru disciplina Matematică și Explorarea mediului, Clasa pregătitoare, clasa I şi clasa a II-a. București. Retrieved from: http://programe.ise.ro/Portals/1/2013_CP_I_II/25_Matematica_explorarea_mediului_CP_II_OMEN.pdf

Ministerul Educației Naționale (2014) Programa şcolară pentru disciplina Matematică, Clasele a III-a–a IV-a. Bucureşti. Retrieved from: http://programe.ise.ro/Portals/1/Curriculum/2014-12/21-Matematica_clasele%20a%20III-a%20-%20a%20IV-a.pdf

OECD: PISA 2015 Assessment and Analytical Framework: Science, Reading, Mathematic and Financial Literacy. PISA, Éditions OCDE, Paris (2016). http://dx.doi.org/10.1787/9789264255425-en

Paul, R., Elder, L.: Critical Thinking. Tools for Taking Charge of you Learning and your Life. Prentice Hall, Upper Saddle River (2001)

Savard, A.: Making decisions about gambling: the influence of risk on children's arguments. Math. Enthusiast **12**(1–3), 226–245 (2015)

Savard, A., Cavalcante, A., Turineck, L.-P., Javaherpour, A.: Epistemological considerations towards the concept of money in school: money as a unit of measurement. Numeracy (Submitted)

Swartz, R.J., Perkins, D.N.: Teaching Thinking: Issues and Approaches. Midwest Publications, Pacific Groves (1990)

ten Dam, G., Volman, M.: Critical thinking as a citizenship competence: teaching strategies. Learn. Instruct. **14**(4), 359–379 (2004)

Necessities and Expectations in Regard to Dobrudja's Rural Youth Access to Higher Education

Rodica Gabriela Enache, Mariana Floricica Călin and Enache Tușa

Abstract In this chapter, we aim to analyze the particularities of the psycho-educational profile of young people from rural communities in the Romania's Dobrudja region, their professional values and interests as well as the factors that determine their access to studies through university education or professional orientation. In order to obtain a relevant analysis of the results, qualitative research was carried out on the basis of a sample represented by 600 pupils aged 16–20 years in the FDI project, funded by the Romanian Ministry of National Education through the Institutional Development Fund under the project "Gaudeamus Igitur - Equal opportunities for education".

Keywords Access to education · Equal opportunities for university education · Vocational counseling

---

R. G. Enache (✉) · M. F. Călin · E. Tușa
Ovidius University of Constanța, Constanța, Romania
e-mail: rodicaenache29@yahoo.com

M. F. Călin
e-mail: fmarianacalin@gmail.com

E. Tușa
e-mail: enachetusa@gmail.com

D. Flaut et al. (eds.), *Decision Making in Social Sciences: Between Traditions and Innovations*, Studies in Systems, Decision and Control 247,
https://doi.org/10.1007/978-3-030-30659-5_8

1 The Actuality of the Conceptual Perspectives on the Need for Education Throughout the Entire Life and Lifelong Learning

1.1 Aspects of Education Throughout Life and Inclusive Education Implemented in Contemporary European Education Policies

"Education/learning throughout life" (Lifelong Learning-LL) is an educational reality of the twentieth century from the need to cope with the challenges of the current, knowledge-based society. Educational policy discourse on lifelong learning has put a number of emphases on three plans whose lifelong learning contributes to the progress: the individual, the democratic and the adaptive one.

The individual plan refers to both acquisitions in the field of cognitive and metacognitive development through the acquisition of knowledge from various fields of knowledge and skills training.

On a democratic level, lifelong learning ensures its programmatic commitment to promote social justice, equality, and the society development through the participation of informed, empowered, motivated and accountable citizens (Paloș et al. 2007, p. 29).

At the adaptive level, it is a programmatic response to the cultural change in the sense of ensuring development at all levels of social organization (individual, organizational, national, global) through education policies and systems that promote the principle of lifelong learning, thus making it possible to adapt to the changing cultural context.

In 1996, the OECD launched lifelong learning for all. By 2006 all Member States of the European Union were required to have a coherent national strategy on lifelong learning, with objectives, the structure of the offer (which, including all, content, quality and relevance of the training, allocated resources and managerial strategies, roles and responsibilities of the different partners.

In 2003, the European Commission published the Eurobarometer on lifelong learning. Almost 90% of Europeans consider lifelong learning an important way of personal development, even if the lack of time and money are the most important barriers. Preferring informal learning contexts, it is encouraging that most have a positive attitude towards lifelong learning, considering that it "can empower them to cope with change, adapt to the demands of the labor market and assume responsibility for their own lives".

At European level, sustained efforts are being made by investing in education, seen as the most important lever for achieving a 'knowledge-based Europe'. To provide high-quality education and training is an integral part of the European Union strategy that promotes social inclusion and cohesion, mobility, active citizenship, individual and professional development, competitiveness and employment. These are found in the documents of European Commission, reinforced in the Council of Europe resolutions, and resumed in a similar form by the UNESCO and OECD stipulations as strategic action areas that support globalization.

The aforementioned documents are based on 3 strategic objectives:

- Improving the quality and efficiency of education and training systems,
- Facilitating access to education for all,
- Opening up education systems to include the widest and most diverse population for which the following action priorities were established (Copenhagen Declaration, November 2002):
  - Strengthening the European dimension in vocational training through closer collaboration aimed at facilitating and promoting transnational mobility, partnerships and initiatives;
  - Ensuring the transparency of diplomas and qualifications by introducing the transferable credit system of the European CV model, by issuing diploma supplements in a European language of attestation of competences acquired, developing recognition of competences and qualifications system but also a validation of non-formal and informal learning;
  - The development of information and counseling services to facilitate insertion in the labor market and access to education but also greater mobility;
  - Quality assurance by developing detailed criteria and principles in training, by sharing models and methods of teaching and assessment.

In the context of the "Lifelong Learning Memorandum" (2000) and the European Commission communication on "Making the European Area of Lifelong Learning a Reality" (2001), it is necessary to provide the acquirement of basic computer skills, knowledge of foreign languages, social and entrepreneurial skills.

For this, it is necessary to implement the six key messages in the "Lifelong Learning Memorandum":

- Raising investments in human resources;
- Training teachers to implement active methods to make learning more attractive;
- Developing the institutional service and communication infrastructure to make the educational offer more flexible, tailored to individual needs and closer to home (this involves the development of counseling services to facilitate access to education);
- Building partnerships and networking between different social actors, civil society, local and regional authorities, private investors, etc.;
- Promoting learning at the workplace, encouraging ways to validate learning outcomes to build articulation between formal, non-formal and informal learning;
- Bringing educational offers closer to home, in the proximity of the individual, developing them to be more attractive, flexible, introducing alternative ways, for example through distance education.

As a strategic significance, permanent education can be defined at the functional-structural-operational level, highlighted by Suchodolski (1972, pp. 312–313):

- its ultimate goal: to maintain and improve the quality of life;

- its universal (it represents the democratization of education) and dynamic (it allows the adaptation of the materials and the means of learning to the new conditions imposed by the development) character
- its managerial sense (permanent education is an organizational principle for all types of education).

Inclusive education implies a permanent process of improvement of the school institution with the purpose of exploiting the existing resources, especially human resources, to support the participation in the education process of all pupils in the community "(HGR nr.1251/2005, based on a book published by UNICEF and MEC in 1999).

At present, the UN Convention on the Rights of Persons with Disabilities adopted by the United Nations General Assembly in 2006 provides a framework for policy-making and modernization of practices, tools and modalities of community support, leading to unhampered participation of people with disabilities in society.

To ensure the implementation of the Convention, the European Commission adopted on 15 November 2010 the European Disability Strategy 2010–2020: a renewed commitment to a barrier-free Europe, which aims to provide full rights for people with disabilities to participate in life social and economic, by removing barriers in eight main areas of action: accessibility, participation, equality, employment, education and training, social protection, health, external action.

At present, the UN Convention on the Rights of Persons with Disabilities adopted by the United Nations General Assembly in 2006 provides a framework for policy-making and modernization of practices, tools, and modalities of community support, leading to unhampered participation of people with disabilities in society and in the community. However, at different levels, other patterns are observed. Similar percentage of women and men aged 25–64 in the EU (23%) had at most a lower level of education (they were graduates of the gymnasium cycle) in 2016. A lower percentage of women (45%) than men (48%) in the EU had at most an average level of education (graduating from high school or post-secondary non-tertiary education). This pattern has been observed in almost all Member States. In terms of higher education, 33% of women in the EU were graduates of this education cycle, compared with 29% of men. There is a majority of women with this level of education in almost all Member States, with the largest differences between women and men in the Baltic States as well as in Finland, Sweden and Slovenia" (http://www.insse.ro/cms/Women_and_man/index.html).

1.2 Aspects of **Lifelong Learning** *and* **Inclusive Education** *Implemented in Romanian Educational Policies*

According to national statistics, the number of adolescents living in Romania as of July 1, 2011 amounted to 2,012,709, representing 9.43% of the overall population. Of these, 939,538 (46.68%) resided in urban areas and 1,073,171 (53.32%) lived in rural areas. Their make-up by gender was 51.22% male and 48.78% female. Most of Romanian adolescents were enrolled in education (87.36%). As for the rest of 12.63%, statistics don't provide much clarity, further research being necessary.

In the study "State of adolescents in Romania" (Abraham et al. 2013): 93.7% of respondents declared they were attending school, with similar proportions by age-groups (94% in 10–13 years, compared to 93.6% in 14–18 years), but increased proportion in girls (95.0% versus 92.3% in boys) and in urban areas (96% compared to 91.6% in rural). Very few respondents declared they were working (1%) at the time of the survey, suggesting that the topic of adolescents who are neither in education, nor on the labour market needs to be further explored.

Nations don't have only the right but also the duty to know and to support each other to achieve peace, progress, and prosperity through education, culture and science and not chaos and dehumanization.

Many countries have invested massively in education to gain a lasting and thorough development for all social levels. They believed that, when well-targeted and supported by all sectors of society, investment in education is fundamental to triggering and sustaining socio-economic and cultural development (Văideanu 1996, p. 60).

2 The Socio-cultural Significance of University Education Development and the Principle of "Equal Chances for Education"

2.1 Elements Regarding the Evolution of the Romanian Higher Education System

Created in the 19th century, the higher education system in Romania borrowed elements of the French system, but also German influences. Universities teach the exact sciences, the humanities, the arts, but at the same time, special schools were created, like Polytechnic Schools, Commercial Academies.

At the beginning of the 20th century, the real characteristics of the system were observed, the particular effort required by the curriculum of the students, to "compensate" the technological weakness that characterized Romania then. At that time, the number of universities and students was lower than in other countries.

After 1948, the Romanian higher education continued developing, but the pace of this development was lower than that of other Central and Western European countries.

In 1989, the Romanian student population represented merely 0.73% of the total population of the country. It was gathered in 10 university centers: Bucharest (more than 30%), Cluj Napoca, Timișoara, Iași, Braşov, Târgu Mureș, Craiova, Galați, Petroșani, Ploiești. Apart from these, there were also institutes preparing sub-engineers in Suceava, Constanța, Baia Mare, Oradea, Sibiu, Reșita, Pitești and Bacău. There was also a student population at evening classes.

After the revolution of December 1989, the trends, circumstances, and evolution indicate the development efforts of the Romanian higher education system to get closer to the situation in Western Europe. In this respect, the following may be stated:

- developing the institutional network and the number of students,
- the emergence of private higher education,
- efforts to achieve decentralization,
- awareness of the need to diversify sources of funding,
- the reconsideration of the content and the links with the economic agents, especially in various segments of the private sector, to achieve consistency between the formations of future specialists with the current and long-term needs of the country's economic and social activity (Pachef 2008, p. 78).

As shown in the "Evolution of Qualifications in the Romanian University Education 1968-2011" paper, the DOCIS Project, the changes of the political regime and the objectives of education in the 1990s, were reflected by the increased amount of state and private universities, yearly students and graduates' numbers and by setting up private higher education (Sora 2011, p. 3).

Romania became a member of the Bologna Process in 1999 by signing the Bologna Declaration, committing itself to include its goals in the priorities of the Romanian higher education. The Bologna process has led to the creation of the European Higher Education Area, which includes 48 Member States, the essential criterion for accepting a candidate being the adherence to the European Cultural Convention.

According to the Bologna Declaration implemented in the Romanian education system starting with the 2005–2006 academic year, the university studies are organized in three cycles:

- Cycle I—3 or 4 years of undergraduate studies for faculties with the medical and architectural background,
- Cycle II—master studies of 1–2 years' time frame,
- Cycle III—three-year doctoral studies.

On March 3rd 2010, the European Commission published the "Europe 2020: A strategy for smart, sustainable and inclusive growth" communication which has as a general objective the transformation of the European Union into a smart, environmentally friendly and inclusive economy, to provide a higher level of workforce and productivity, to ensure economic, social and territorial cohesion.

2.2 Implementing the "Equal Opportunities Principle" in University Education and Vocational Training

In academic and social organizations and institutions, equality of opportunity—from multiple perspectives (gender, ethnic origin, capacity, etc.) represents a fundamental principle of building and substantiating the policies and practices of societies. This principle is also promoted in higher education institutions around the world.

The principle of equal opportunities at all levels of education in all states, social and cultural areas was proposed and supported in the Coleman Report (Coleman et al. 1966).

In the 50 years since its introduction, the principle has been used by international organizations and UN agencies (OECD, UNESCO, UNDP). In 1993, the UN adopted a resolution on equal opportunities for people with disabilities, including education.

The student, whether entering the adolescent or young adult category, is considered to be an autonomous, independent, mature and wise person who assumes the responsibility of his decisions and choices in life. Learning in the university brings to the same group diverse people with different experiences, from various social backgrounds and locations.

Besides these factors, the authors identified four key social features impacting the teaching-learning process:

- The roles played by teenagers and young people in their personal and professional lives and the responsibilities they have to assume are multiple, leading to a different orientation of learning,
- Physical, psychological and social transformations consistent with age, generating specific needs, specific interests and objectives,
- Anxiety and ambivalence experienced by adolescents can inhibit their needs to become autonomous (Paloș et al. 2007, p. 130).

3 Ovidius University of Constanţa—A Multicultural University that Promotes the Principles of Equity and Non-discrimination in Education

Multicultural education focuses on promoting the rights to education of all citizens regardless of their ethnic or religious background and plays an important role in ensuring equity for learning opportunities and mitigating the effects of racism. It is based on the convergence of three practices: group diversity, interactive learning, and the practice of teaching-learning styles in different cultural contexts.

The "Ovidius" University of Constanța is a public, multidisciplinary higher education institution in which student-centered modern curricula are structured in the bachelor, master and doctoral study cycles in accordance with the Bologna Process of Reforming the European Space of Higher Education.

Ovidius University of Constanța promotes the principles of non-discrimination and equity in education and encourages the participation of young people regardless of their ethnic or religious origin and socio-cultural conditions in the educational activity. It also facilitates the access of young people with special needs for academic programmes in both the frequency and distance learning form.

Ovidius University of Constanța (OUC) is a public higher education institution with legal personality, a component part of the autonomous and open state education system in Romania. Since its inception, in 1961, it has worked unceasingly under various names. The current name, "Ovidius" University of Constanta (UOC) dates back to 1990.

In accordance with the Charter of Ovidius University of Constanța, the assumed mission is to develop a competitive and collaborative environment of education and research aimed at creating, maintaining and distributing knowledge to society.

Ovidius University of Constanța respects the Lima Declaration on Academic Freedom and Autonomy of Institutions of Higher Education (1988), adheres to the Magna Charter of European Universities (Bologna, 1988) and the Bologna Declaration on the European Higher Education Area 1999).

In the academic year of 2018–2019, Ovidius University of Constanța has 170 university degree programs, of which 86 are undergraduate, 76 are master and 8 are Ph.D. At the "Ovidius" University of Constanţa, a transparent policy of recruitment and selection of students is promoted, the admission contest for university studies being organized in the legislative parameters established by its own regulations.

Ovidius University of Constanța carries out all activities of the academic process, including monitoring, modernization and updating, in the coordinates established by a series of Regulations that have the role of ensuring compliance with the mandatory normative requirements, the standards and the performance indicators, according to the requirements of an education quality, focused on student needs, compatible with market economy requirements and synchronized with European requirements.

At present, over 13,000 students study at Ovidius University of Constanța, for which university is a productive experience, outlined by a great diversity and flexibility of the offered study programs, a quality of complementary educational services and educational and professional counseling, the offer of domestic and international scholarships, and social inclusion packages. In the context of the economic and social context of the global financial crisis, the dynamics of the school population and the number of high school graduates with Baccalaureate Diploma, the number of students enrolled at Ovidius University of Constanța registered a dynamic reflected in Table 1.

The educational strategy of Ovidius University of Constanța has been programmatically and systematically carried out in the direction of diversification, the study programs having in view: consolidation of the system of university education structured by levels and organized by education; shaping the academic plans according to the educational priorities deriving from the study of the Romanian and European economic trends; generating competitive qualifications in university education

Table 1 Evolution of the number of students between 2009 and 2017

Year	No.													
	Total bachelor + master			Total Ph.D.		Bachelor			Master		Ph.D.			
	FT	CT	ID/IFR	FT	CT	FT	CT	ID/IFR	FT	CT	FT		CT	
											CF	FF/FR	CF	FF/FR
2009–10	5981	10,554	2181	188	585	4642	8439	2181	1339	2115	146	42	4	581
2010–11	6123	10,830	1728	180	438	4871	8922	1728	1252	1908	180	0	2	436
2011–12	6079	9608	1449	140	423	4889	7947	1449	1190	1661	140	0	85	338
2012–13	5990	9008	1290	85	426	4779	7254	1290	1211	1754	85	0	220	206
2013–14	6064	7954	1124	64	347	4822	6581	1124	1242	1373	64	0	229	118
2014–15	6118	6939	1069	82	244	4879	5953	1069	1239	986	82	0	197	47
2015–16	6160	6723	963	98	250	4942	5788	963	1218	935	85	13	160	90
2016–17	6159	6227	881	108	276	4943	5349	881	1216	878	80	28	134	142

at all levels; the development of low-frequency and distance learning; making sustainable partnerships with the socio-economic environment, encouraging the active involvement of employers in the education and training process (http://www.univ-ovidius.ro/).

4 Institutional Projects Focusing on Educational Inclusion and Vocational Counseling Developed by Ovidius University in Constanţa

4.1 POSDRU Project/160/2.1/S/132690 BUILD YOUR CAREER, STEP BY STEP!

Counseling and Vocational Guidance for Students in Central and Southeastern Regions (2015–2016) aimed at promoting measures to correlate the education system and the labor market through counseling and vocational guidance activities in order to prepare for market insertion the work of 20,000 students in the theoretical and vocational high schools, as well as 8th grade pupils belonging to the Central and South-Eastern regions, in order to increase the employment among the graduates of the public education units.

The project carried out by DONIT Braşov Association in collaboration with "Dunărea de Jos" University of Galaţi, University of Medicine and Pharmacy of Târgu Mureş, "Lucian Blaga" University of Sibiu and Ovidius University of Constanţa, Faculty of Psychology and Education Sciences aimed to increase the adaptability young people to labor market requirements and job-specific requirements, as well as educating the ability to have informed, personalized choices about continuing the training path.

4.2 FDI Project "BE SMART, BE A STUDENT" (2016–2017)

The project was designed to promote the idea that education is the only concrete and sustainable way for a person to evolve and contribute to the overall development of society. At the same time, our approach recognizes the idea that rural students can access university studies and can pursue an academic path that will ensure their development in the medium and long term. The project has been designed to be accessible to all students to whom it is presented.

The project "Be SMART, BE A STUDENT!" is being implemented between October and December 2017 by Ovidius University of Constanta and will carry out a campaign to counsel the students from 15 high schools located in the rural area of Constanța county. The visits will be organized in the communes: Cobadin, Cumpăna, Independenţa, Ciobanu, Nicolae Bălcescu, Poarta Albă, Castelu, Mihail Kogălniceanu, Cogealac, Deleni, Adamclisi, Dobromir, Ostrov, Mihai Viteazu, Topraisar. The purpose of the visits is guiding students to higher education and the University.

The campaign will be directed under the project funded by the Institutional Development Fund for the Romanian public universities—FDI 2017. In this respect, teachers accompanied by a vocational counselor and UOC students will go to high schools in the targeted rural areas where will support the "Student Candidate - The Importance of Informed Choice" workshop. High School students will receive informative materials, attend vocational counseling sessions, and discuss with students. At the end of the event, students will have the opportunity to be included in an online group that will facilitate communication between high school students, teachers, and students through which they will receive information on competitions and events organized for students, as well as information on admission to higher education.

Within the project, we aim to demonstrate there is a direct link between the educational act, the evolution of society and the trust of individuals in their own skills and knowledge. The process of education leads to social integration and is the way individuals acquire the norms and values that apply to the society in which they live. It can be said that an individual is only socially integrated when, thanks to the education process, he has acquired the common culture of the integrating society. Social integration determines the individual to have his own place in society.

The "Ovidius" University campaign, implemented in the project "Be SMART, BE A STUDENT!", has made it possible to effectively collaborate with the "target groups" made up of the high school students mentioned above representing the recruitment group in higher education. Until this date (15 November), I have been visiting high schools in the communes: Cobadin, Cumpăna, Nicolae Bălcescu, Mihail Kogălniceanu, Cogealac, Mihai Viteazu, Adamclisi, Deleni, Dobromir, and Ostrov. The results and success indicators of the project "Be SMART, BE A STUDENT!" consist of:

1. Better visibility of Ovidius University of Constanța in the pre-university environment in Constanța County;
2. Direct communication and exchange of ideas/experience between high school students and volunteer students participating in high-school work visits.
3. A total of 650 high school students who received informative materials and will be included in an online group that will facilitate communication between students, teachers, and students.
4. Dissemination through the online group of information on competitions and events organized for pupils, as well as dissemination of information on the 2018 admission timetable.

The project "Be SMART, BE A STUDENT!" continues a tradition of promoting the educational offer of Ovidius University, materialized in various marketing strategies presented through educational caravans designed and developed over the past three years. In 2017, "Ovidius" University of Constanta conceived and developed a project for promotion, information and educational counseling that started with a three-month Educational Caravan (February–March 2017) in 60 high schools from Constanța, Tulcea, Ialomița, Călărași. We have developed an information platform made up of the website, "student guides", university presentation guides, and awards to promote the educational offer. I visited the high schools in Constanța, Mangalia, Cernavodă, Medgidia, Hârșova, Tulcea, Babadag, Măcin, Năvodari, Ovidiu, Călărași, Fetești, Slobozia, Urziceni, Țăndărei, Eforie Techirghiol. In the educational caravans, we organized visits to the high schools located in the rural areas of Constanța and Tulcea.

4.3 "GAUDEAMUS IGITUR - EQUAL OPPORTUNITIES tO EDUCATION" Project (2017–2018)

Ovidius University of Constanta implemented activities under the project "GAUDEAMUS IGITUR - Equal opportunities in education, CNFIS-FDI-2018-0486 code financed from the budget allocated to the Ministry of National Education, from the institutional development fund, domain 1: social inclusion, and increasing access to higher education, linking educational supply to labor market demand (including career counseling and career guidance). The project coordinated by Ș. L. Ing. Ph.D. Rita Avram ran from 08.06.2018 to 07.12.2018 and with a total value of 328330.00 lei, the financing from the institutional development fund being 280,000 lei. The core objective of the project was to optimize the academic strategy, promote the educational portfolio and strategy for students and graduates at Ovidius University of Constanta, based on the principle of increasing social equity with a focus on social inclusion, increasing access to higher education for young people from the rural area and the correlation of the educational offer with the demand of the labor market.

The following specific objectives were pursued:

1. Promotion of the educational portfolio of Ovidius University of Constanta;
2. Improvement of counseling and professional guidance services;
3. Analyzing and matching up the curriculum with the requirements of the socio-economic environment.

The project's activities were correlated with the strategic objectives of Ovidius University of Constanta and focused in three directions:

a. Promoting the educational offer aimed, especially, towards the rural environment,
b. Counseling young people (UOC students and students) to make informed decisions about their professional career to increase self-esteem, personal self-analysis, awareness of their own potential and changing attitudes towards self-education and education,

c. Updating and adapting the curriculum with reference to both content and didactic of disciplines, to link university education to the requirements of society and the business environment.

The personal development and vocational counseling programs for students aimed at the following objectives and activities:

Objectives:

- Students' awareness of their psychological potential analyzed in terms of cognitive, attitude and social aspects,
- Optimizing self-image and increasing self-esteem,
- Set up a personal portfolio for your career,
- Changing attitudes towards oneself and towards the professional future.

Activities:

- Assessing the personality profile, professional interests, values and attitudes (2 h),
- Developing a professional portfolio for career development (2 h),
- Simulation of interviews for professional selection, debates, presentation of personal career portfolio (2 h) (Enache et al. 2018, p. 5).

5 Exploratory Research on Access to Education, Needs and Expectations of Young People Living in the Dobrudja Area

Within the "GAUDEAMUS IGITUR - EQUAL CHANCES IN EDUCATION" Project (2017–2018), there was an exploratory research on access to education, the needs and expectations of young people from rural areas in the Dobrudja area.

The objective of the research was to identify young people's perception of their interests, social values and their academic and professional development needs.

The group of subjects was represented by 600 pupils from the 11th and 12th-grade pre-university education aged between 16 and 20 years old coming from rural high-schools or residing in rural areas in the counties of Constanta and Tulcea, Dobrudja area, Romania.

The distribution according to gender and ethnicity was homogeneous, there were Romanian pupils and pupils from other ethnic groups: Turks, Tatars, Greeks, Lipoveni, Roma (Figs. 1 and 2).

An anonymous questionnaire survey, completed by pupils, with the consent of parents and teachers, was applied as an investigative method.

The questionnaire for identifying young people's needs, interests and intent to pursue university studies was structured into three components:

a. Assessment of students' professional capacities and interests;

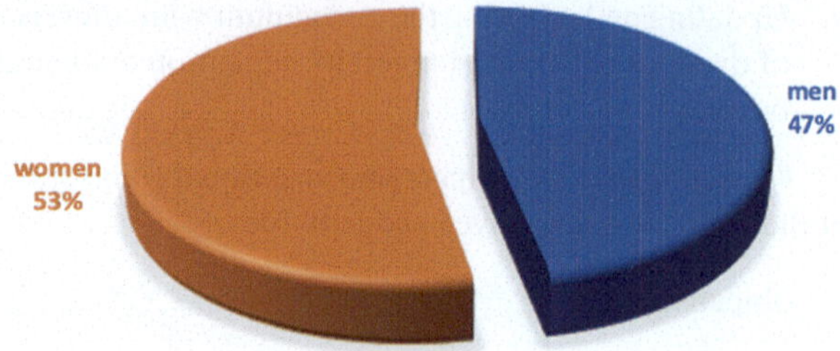

Fig. 1 Distribution of pupils by gender

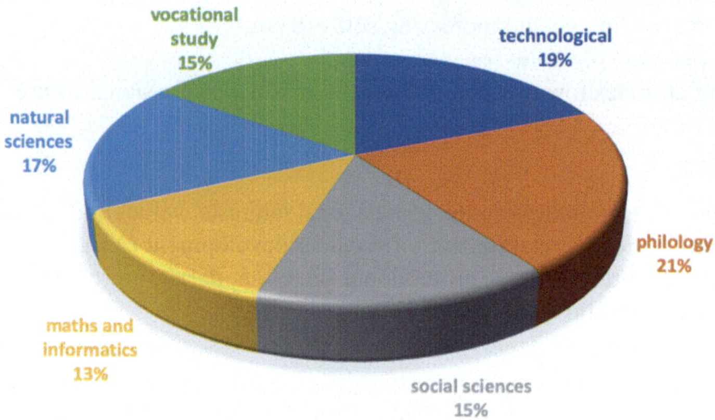

Fig. 2 Student distribution according to the study profile

b. Assessment of intent to undertake university studies and decision-factors;
c. Utility of educational and vocational counseling services.

Description of Research Results

(a) Assessment of Pupils' Capacities and Professional Interests

Assessment of Mathematical, Digital, Practical and Entrepreneurial Competences

Investigated rural students consider they have developed practical and realistic skills that aim at manipulating objects, machines, tools (60% high and 31.17% on average) and also entrepreneurial skills targeting preferences for activities that allow the initiative for self-employment on the basis of professional permits (26.33% in high percentage and 56.50% in average).

As regards the level of digital competence and computer use (only 29.17% claim to have a high percentage of digital competences and 48.33% on average). The obtained results also correlate with the percentages obtained by students with the appreciation of the level of mathematical skills development (only 38.17% claims to have largely mathematical competencies and an average of 42.33%) (Fig. 3).

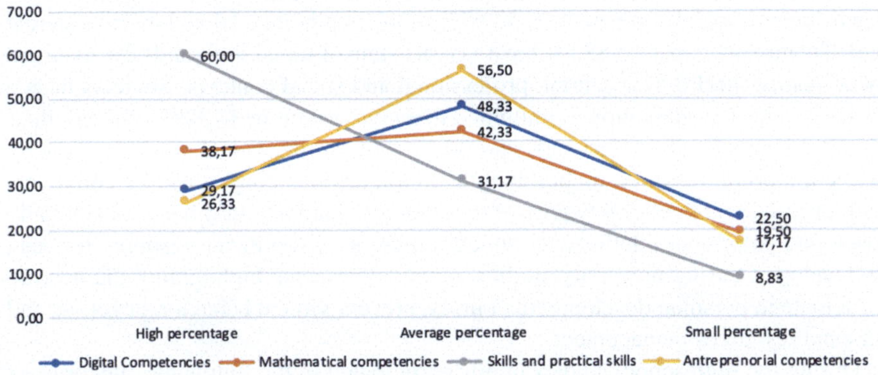

Fig. 3 Evaluation of mathematical, digital, practical and entrepreneurial competences

Assessment of Communication Capacity in Native Language, Foreign Languages, Cognitive Skills and Critical Thinking, Availability of Learning and Academic Intelligence

Students claim they have 69.67% of optimal communication skills in their mother tongue and only 40.83% of communication skills in foreign languages. We observe a high level of 60.17% capacity and availability of learning and a lower percentage of cognitive abilities (analytical capacity, synthesis, algorithmic or heuristic problem-solving strategies).

Students argue that they are interested in optimizing their level of development of academic intelligence (capacity for scientific conceptualization, problem-solving ability and reflexive learning) in 69.67% and critical analysis capacity of 52.5% (Fig. 4).

Assessment of Adaptability, Perseverance, Teamwork, Individual Work, Social Skills, Social Respect and Emotional Intelligence

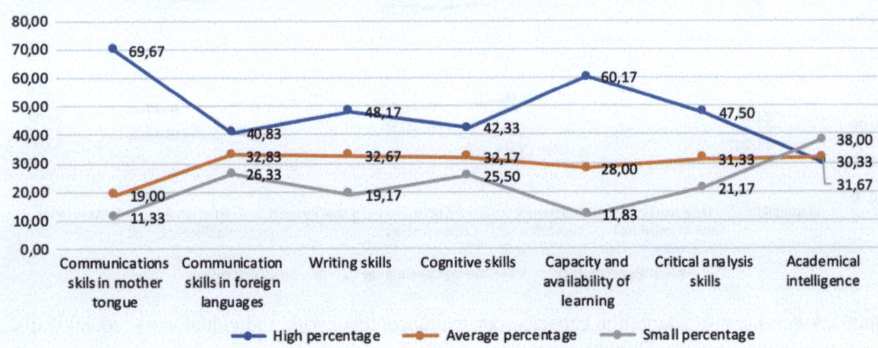

Fig. 4 Evaluation of communication capacity in mother tongue, foreign languages, cognitive and critical thinking skills, learning availability and academic intelligence

Regarding the social competences, 57.67% of the pupils stated that they have a great deal of teamwork capacity, 53.5% have a great extent of social skills and 62% manifest social respect in different school, professional and social contexts. Students have a good capacity for adaptation, highlighted to a great extent by 61.83% of them, their 61% perseverance characteristics, and 60.67% of the individual working capacity can be important resources in building the moral profile being valued by employers in all professional fields. Only 55% of the students claim that they have considerably developed emotional intelligence, which represents a significant resource that can be developed during university studies, many adolescents highlighting the need to participate in personal development courses, prevent violent behavior, negotiate and personal resources management.

During the high school studies pupils participated in the following categories of activities that will ensure their success during university studies: volunteering activities (annual 30.83%, sometimes 39.83%), sports activities (25.5% per year, sometimes 42), professional practice activities (annual 47.83%, sometimes 37.33%), IT courses (23.33% annually, sometimes 32.33%), national projects (annual 22.83%, sometimes 25.17% %), part-time employee (annually 11%, sometimes 32.17%) (Fig. 5).

We can see from the chart that only one-third of the students participated in volunteering activities and sometimes 39.83%. Students were informed that the development of volunteering is definitely a way to follow for the prosocial evolution of the individual as well as society as a whole.

Volunteering is an organized activity within social structures under existing regulations. In most countries volunteering is regulated by law. In Romania a law on this issue was adopted very late, for the first time only in 2001. Law 78/2014, adopted in June 2014, is currently in force.

It is not to be forgotten that international volunteer networks have developed, this type of activity is attracting participation by their global scale, and that the activity

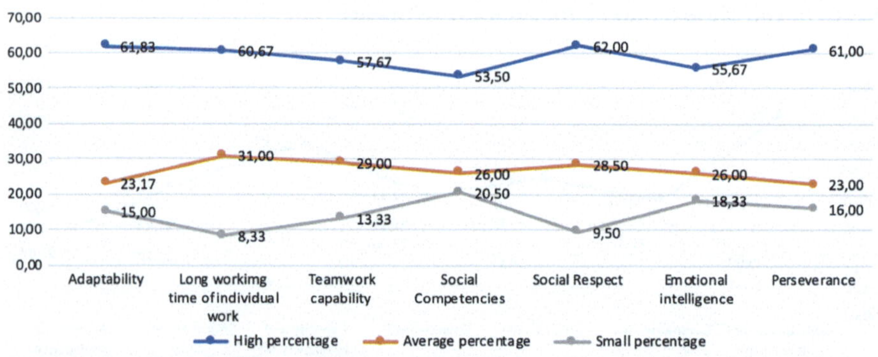

Fig. 5 Assessment of adaptation capacity, perseverance, teamwork, individual work, social skills, social respect and emotional intelligence

is considered a successful social innovation that is increasingly being sustained by the private environment perceives in it a form of assuming civic responsibility.

56.5% of students consider their income will come from their parents during university studies, and half of them are willing to carry out temporary professional activities with a high or less frequent frequency, and 37% consider they should carry out permanent professional activities to support their studies from a financial perspective. After presenting the study opportunities, 55% of them understood they can benefit from the social or study scholarships in academic studies.

(b) **Assessment of Intent to Undertake University Studies and Decision-Factors**

The intention to pursue university studies in Romania was highlighted by 43% of the respondents, 5% claiming they want to study at a university from abroad, and 52% said they either do not want to study for various reasons or they are afraid they will not pass the Baccalaureate Exam.

The decisive factors in the choice of a profile and specialization were 47% of the respondents taking a personal decision to a great extent, taking into account their professional skills and interests, 40% largely followed the recommendations of the family, 31% followed largely the recommendations of friends, 39% were influenced in the decision on distance from home and only 27% were influenced to a very large extent by the amount of tuition (Fig. 6).

(c) **Usefulness of Educational and Vocational Counselling Services**

We identify the need for vocational counseling and personal development at the level of the adolescent group to raise awareness of their bio-psycho-social potential and identifying the professional interests and social values to which they are oriented in choosing their careers. Only 35% of students consider they have benefited to a large extent from vocational counseling and career guidance during high school studies, 27% is claiming they have had some talks with senior teachers about further education opportunities and the benefits of university studies.

Fig. 6 Determinants in career choice

Within the "GAUDEAMUS IGITUR - Equal Opportunities in Education" project, a Guide to Good Practice for Vocational Counseling and Personal Development Activities for Students to Prevent Abandonment of University Studies was developed. Academic and vocational counseling activities attended by students in 11th and 12th high school years aim at discovering their personal resources and forming interpersonal skills needed in the field of professional training and career debut. Other examples of activities to be carried out in school counseling are to develop effective communication, negotiation skills, learning management, time management, conflict management, and entrepreneurial skills development.

Vocational counseling specialists have shown the long-term benefit of career guidance has been the increase in the professional integration rate of young people after a limited period of time since graduation or even in the final year of the faculty.

Deciding on the right career based on the knowledge of vocational identity offers satisfaction, comfort, high professional performance and appreciation from managers and colleagues, and last but not least, building an optimistic perspective on life.

6 Conclusions

Ciolan and Ciolan (2014) show that learning authenticity is therefore framed at the crossroads of several instances: disciplinary mind, individual biography, social context and learning experiences. Social and professional context takes into account the way learner relates its learning experiences and the new learning to the broader social life and eventually link with managing meaningfully different tasks (professional dimension) as result of learning. The social contexts in which learning occurs and the way you can then use learning for transforming social and professional contexts, for receiving validation on own performance in social work or life settings are also significant attributes of learning authenticity. Learning experiences are referring mainly to those who are backed by intentional designs, made by professional educators with the explicit aim to generate or to facilitate learning.

The inclusive university will not only facilitate access to or career debut on the basis of bachelor studies but will encourage access to masters and doctoral studies, providing not only an occupational identity to young people but also a social identity.

Professional development starts with the choice of the university route, builds with career development as a long-term process and can be restructured on the basis of professional reconversion.

Career development involves a succession of activities and professional positions that a person adopts, as well as associated attitudes, knowledge, and skills developed over time. The need for personal knowledge and development of students relates both to the cognitive and social needs of young people in today's globalized society, but also facilitates the optimal choice of the university route and professional insertion.

References

Abraham, D., Abraham, A., Dalu, A.M., Fierbinteanu, C., Marcovici, O., Mitulescu, S., Plaesu, A., Sufaru, I: State of adolescents in Romania. UNICEF, Romania (2013). https://www.unicef.org/romania/UNICEF_Study_State_of_adolescents_in_Romania.pdf

Ciolan, L., Ciolan, L.E.: Two perspectives, same reality? How authentic is learning for students and for their teachers. Procedia Social Behav. Sci. **142**, 24–28 (2014). https://core.ac.uk/download/pdf/81122563.pdf

Coleman, J.S., et al.: Equality of Educational Opportunity. U.S. Departement of Education, Washington, DC (1966)

Enache, R., Călin, F., Raftu, Gh: Program de dezvoltare personală şi consiliere vocaţională pentru studenţi. Editura StudIS, Iaşi (2018)

Pachef, R.C.: Evaluarea în învăţământul superior. Editura Didactică și Pedagogică, București (2008)

Paloș, R., Sava, S., Ungureanu, D.: Educația adulților. Baze teoretice și repere practice. Editura Polirom, Iași (2007)

Sora, A.F.: Evoluția calificărilor din învățământul universitar românesc, Proiect DOCIS Dezvoltarea unui sistem operațional al calificărilor din învățământul superior din România (2011). http://www.invatamant-superior.ro/wp–content/uploads/2013/08/Raport_istoria_calificarilor.pdf

Suchodolski, B.: L'education parallele//Perspectives, Paris, UNESCO, nr. 2 (1972)

Văideanu, G.: UNESCO-50 Educaţie. Editura Editura Didactică și Pedagogică, București (1996)

References

The Tradition and New Approach to the Mathematical Education of Officers in the Czech Army

Jiří Jánský and Radovan Potůček

Abstract This chapter deals with the problems of teaching mathematics at the Faculty of Military Technology at the University of Defence in Brno, originally the Military Technical Academy in Brno, later renamed the Antonín Zápotocký Military Academy and the Military Academy in Brno. The tradition of teaching mathematics at this high military school has its origins in 1951, when this school was established. In more detail, we discuss the innovation of study programs over the last decade. The ranges and thematic areas of mathematics teaching and the focus of relevant study programs are compared both in Bachelor's and follow-up Master's degree programs as well as in Master's degree programs. At the same time, the two-stage study has recently been gradually replaced by Master's degree programs.

Keywords Teaching mathematics · Study program · Bachelor's and follow-up master's degree program · Master's degree programs · University of defence · Czech army

1 Introduction

Before describing study programs at the Faculty of Military Technology at the University of Defence, their development, changes and innovations, we will briefly discuss the history of the University of Defence and the traditions of teaching mathematics at this university preparing military professionals for service in the Czech army as well as experts for the Czech defence industry.

J. Jánský · R. Potůček (✉)
Department of Mathematics and Physics, Faculty of Military Technology, University of Defence, Kounicova 65, 66210 Brno, Czech Republic
e-mail: Radovan.Potucek@unob.cz

J. Jánský
e-mail: Jiri.Jansky@unob.cz

D. Flaut et al. (eds.), *Decision Making in Social Sciences: Between Traditions and Innovations*, Studies in Systems, Decision and Control 247,
https://doi.org/10.1007/978-3-030-30659-5_9

1.1 History and Tradition of the University of Defence

The Brno Military Academy was founded on August 15, 1951 under the command of the President of the Czechoslovak Republic Klement Gottwald under the name of the Military Technical Academy (MTA). The foundations of the school became the pedagogues, buildings and material equipment of Dr. Edvard Beneš University of Technology in Brno (UT, today's Brno Technical University), which was closed on 2 October 1951.

At the time of establishment, MTA consisted of seven faculties: the Faculty of Artillery, the Faculty of Tanks, the Faculty of Aviation, the Faculty of Engineering, the Faculty of Electrotechnical, the Chemical Faculty and the Railway Faculty. At the beginning of 1972, MTA consisted of five Faculties: the Faculty of Command, the Faculty of Engineering—Ground Forces, the Faculty of Aviation and Air Defence, the Faculty of Distance Studies and the Faculty of Foreign Studies. By the end of the 1980s, mechanical engineering, electrotechnical, construction, geodetic and command-staff specialisations were preserved—despite various reorganizations.

Following the Soviet military academy, MTA was named after the then President Antonín Zápotocký in 1954, so was named as Antonín Zápotocký Military Technical Academy (AZ MTA). After the reorganization of military colleges in 1958, it merged with the Klement Gottwald Military Academy in Prague, and the name was changed to Antonín Zápotocký Military Academy (AZ MA). AZ MA also educated foreign students who have studied at the Foreign Studies Faculty. Note that, as a separate AZ MA faculty, the Foreign Studies Faculty was established in the second half of the 1960s to prepare military students mainly from developing countries (Vietnam, Iraq, Afghanistan, Egypt, Libya, Syria, Cambodia, Laos, Ethiopia, Morocco, India, Indonesia, Mozambique, Nigeria, Sudan, Tanzania, Yemen or Cuba) and also from the countries of the former socialist camp (the Soviet Union, the German Democratic Republic, and Poland). In 1990, after the Velvet revolution, the name of the Communist President Antonín Zápotocký was removed from the name of the school.

In 2001 the government of the Czech Republic accepted documents regarding reform of the Czech Republic's Armed Forces. This reform gave rise to a fundamental transformation of the military education system. This transformation aimed to execute such changes to the military educational system so as to achieve a better cost-effectiveness. One of the goals was to make the military educational system more effective.

The Military School of Ground Forces in Vyškov (established in 1947) and the Military Academy in Brno used to be ranked among the most distinguished educational institutions. Medical personnel were educated and trained by the Military Medical Academy (re-established in 1988) of Jan Evangelista Purkyně in Hradec Králové. Military Department attached to the Faculty of Physical Education and Sport of Charles University in Prague was another special school workplace.

The original three independent military schools were merged on September 1st, 2004 to found one national school—the University of Defence (UO). Two of its faculties—the Faculty of Military Technology and the Faculty of Military Leadership (carried the name of the Faculty of Economics and Management until 2014) were moved to Brno, the third—Faculty of Military Health Sciences was moved to Hradec Králové. UO further consists of NBC Defence Institute and three centres—Centre for Security and Military Strategic Studies, Language Centre, and Physical Training and Sports Centre.

A brief history of the University of Defence is described on the websites (Wikipedia Contributors 2018; Pazdera 2018), a treatise on the history of the Faculty of Military Technology can be found on the website (Korecki 2018).

The University of Defence is responsible for education of military professionals and experts engaged in national security system, defence industry and public administration (Svarcova and Navratil 2016; Hoskova-Mayerova 2016). The available Bachelor's, Master's and Doctoral degree programmes, both in full- and part-time mode, focus on military and national security fields. UO represents the Army of the Czech Republic's defence and security research and development centre. The fields of science fostered at UO primarily relate to defence applied research, to forces and population protection, or the economics or medicine fields applied to military. In the Czech Republic, they are exclusive fields, such as weapons and ammunition, combat and special vehicles, air force, aircraft technology, electronic warfare and radiolocation, cybernetic security and robotics, population protection, fire support control, field surgery, radiobiology or toxicology.

1.2 History and Tradition of the Department of Mathematics

An integral part of the military professionals training is not only the teaching of applied subjects and technical subjects for the given specialization, but also the teaching of subjects based in theory, including mathematics and physics. The tradition of teaching mathematics at UO thus dates back to 1951, when the then MTA was founded, and now represents nearly 68 years of teaching in military, civilian but also foreign students. Let us note that the history of teaching mathematics is a very broad topic. One of the sources where this subject is elaborated in detail is the first international handbook of the history of mathematics education (Karp and Schubring 2014).

As well as the Military Technical Academy, the Department of Mathematics also significantly benefited materially from and was partly staffed by Dr. Edvard Beneš University of Technology in Brno. According to an agreement between the administration of TU and MTA, the Department of Mathematics and Descriptive Geometry of TU partly transferred to MTA, creating the Department of Mathematics and Descriptive Geometry (DMDG) of VTA.

Some other teachers of DMDG were called from the army or moved to DMDG from secondary schools. DMDG was also joined by graduates of mathematics studies at the Faculty of Science of Masaryk University (MU) in Brno, which at that time was called the University of J. E. Purkyně. Some MU graduates also completed their military attendance at DMDG.

Among the most prominent teachers of DMDG are Miroslav Novotný, Ivan Kolář, Jiří Rosický (all prominent mathematicians and professors of MU), Josef Diblík (professor of Brno Technical University) Miloš Zlámal (professor of Brno Technical University and founder of the mathematical theory of the finite element method), Michal Greguš (professor of Comenius University in Bratislava and expert in the field of differential equations) and Vladimír Vetchý (Defence Secretary of the Czech Republic in years 1998–2001).

The years 1958–1970 were strongly influenced by personal and political vetting and examinations of AZ MTA staff. A number of departmental staff either left the evaluation alone or, based on the conclusions of these examinations, faced the fact that they were not in the academy for political reasons. Their departure to the quality of the department, including DMDG, suffered greatly. Subsequently, the number of members of the department was extended by new teachers due to the considerable number of hours of teaching, and the fact that teachers were leaving because of the examinations.

The beginning of the 1970–1990 periods was marked by a hard standardization regime. Initially, party reviews (1969–1970) and then a so-called "Comprehensive Assessment of Civilian Workers" (1971) were carried out.

DMDG provided teaching of mathematics and descriptive geometry also for foreign students from 21 countries. Teaching language was Czech, Russian, English, partly German, Spanish and French. DMDG also provided for the preparation of scientific aspirants for their professional candidate examinations. In recent decades, the Department has been educating PhD students in doctoral study programs. The department also devotes many hours to individual and collective consultations with students.

In the 1960s and 1970s, DMDG also participated in the teaching of mathematics and descriptive geometry abroad, primarily at the Military Technical College (MTC) in Cairo and the MTC in Baghdad. The foreign activities of the members of DMDG have always been appreciated very positively.

Individual care for students at DMDG led to the constitution of student mathematical study circles, whose activities began in 1965. From 1970 to 2012, the Department organized a mathematical rehearsal of AZ MA for gifted students every semester. Its goal was to extend the knowledge of mathematics to students and to gain an interest in mathematics. The final of the mathematical rehearsal took place before the start of the exam period. The best researchers were rewarded by the Rector of AZ MA and faculty deans. Based on the results obtained, the head of the department could accept successful results as a semester test in mathematics with excellent evaluation, or its written part also with excellent evaluation.

As a result of the departure of a number of teachers, the department was weakened both professionally and pedagogically. Due to severe reduction of the department, a number of young teachers came to DMDG in the 1970s and 1980s. In these years, DMDG had approximately 25 teachers. On the basis of the Law on Out-of-Court Rehabilitation, in 1990, several teachers returned to DMDG so that DMDG had at that time more than 30 teachers. As a result of the reorganization of the Czech army and the associated reduction of students, the number of teachers in the department gradually decreased.

The Department of Mathematics and Physics (DMP) was established on September 1, 2005 by merging DMDG and the Department of Physics due to organizational changes in accordance with new conception of UO. DMP now consists of Mathematics Group with 8 teachers and Physics Group with 7 teachers.

An important and integral part of the work of the DMDG members and later of DMP was, in addition to pedagogical work, scientific work. The professional and scientific cooperation of DMDG with AZ MA departments, with military facilities, but also with some other Czechoslovakian institutes, especially in solving research tasks, has been very successful.

The scientific focus of DMDG began to be profiled in the mid-1950s. There were two groups formed in DMDG, which deviated from the study of both differential geometry and ordinary differential equations. Both of these groups achieved significant results and published a number of articles and research reports, and their activities were very well evaluated at the other Brno universities and contributed to the reputation of the Brno Geometric School and the Brno School of Differential Equations, headed by Academician Otakar Borůvka.

The department cooperates with a number of scientific institutions, e.g. The Institute of Mathematics and Institute of Nuclear Physics at the Czech Academy of Science, Institute of Nuclear Research in Řež, Technical University Dresden, Kurchatov Institute in Moscow, Masaryk University in Brno, VUT Brno, Žilina University and Vienna University. The members of DMP participate in the delivery of grants, research projects, projects of domestic and foreign authorities (Ministry of Defence, Ministry of Industry and Trade, Grant Agency of CR, Grant Agency of the CR, EU expert committee, the Austrian Institute for Eastern and South-eastern Europe). DMP staff have also taken part in the challenges relating to computer aided teaching of mathematics and physics.

In recent years, the Mathematics Group has been focused on mathematical application in engineering sciences, numerical methods, differential equations, mathematical analysis, discrete mathematics and theory of graphs, algebraic structures theory, half-linear difference equations and variational methods, e-learning and computer-aided. The Physics Group has been focused on the neutron and photon spectrometry, radiation transport calculations, dosimetry of reactor vessels, protection against radiation, the relationship of mechanical properties and structure of metallic and polymeric materials, light scattering systems of thin layers with rough boundaries generally correlated, coherent reflectance and transmittance of thin films, diffuse reflectance of rough surfaces, study of surface quality of thin layers, plasma physics, creating materials for computer-assisted teaching of physics, e-learning and blended learning.

Every two years, from 1999 to 2011, DMP organized a Conference on the teaching of mathematics and physics at the technical universities. Since 2014, DMP is the co-organizer of an annual international Conference MITAV (Mathematics, Information Technology and Applied Science).

2 Military Study Programs of the University of Defence and the Faculty of Military Technology

The University of Defence offers education in accredited courses in Bachelor's, Master's and doctoral degree programmes focused on military, military management, economy, engineering and health and medicine. Besides, it is a lifelong learning center carrying out the non-accredited ultimate defense qualification courses intended for senior officers. The school is open to civilian and international students, too. The University of Defence belongs to the national university education system providing its graduates with education comparable to what they could learn at a civilian university in the Czech Republic.

The university regularly updates provided study programmes to respond to dynamically changing needs of practice and growing standards of advanced European universities. For evaluation and control of study process the university uses credit system based on the European Credit Transfer and Accumulation System (ECTS) which enables mobility of students in frame of European educational programmes and allows easy and transparent recognition of passed study segments within European educational area. The University of Defence is a member of the European Federation of National Engineering Associations (FEANI) whose members that graduate from engineering degree programmes may apply for the European Engineer (EUR ING) title recognized throughout the entire European Union.

The University of Defence organizes higher education in the following forms:

1. Three-year Bachelor's degree program, awarded with the bachelor title (Bc,),
2. Five-year Master degree program of Military Technology with the awarded title of engineer (Ing.),
3. Two-year follow-on Master degree program with the awarded title of engineer (Ing.),
4. Five-year Master degree program of military pharmacy with the awarded title of master (Mgr.), and military dentistry with the title doctor of dentistry (MDDr.),
5. Six-year Master degree program of military general medicine with the awarded title of doctor of medicine (MUDr.),
6. Three- or four-year doctor degree program with the awarded title of doctor of philosophy (Ph.D.).

The Faculty of Military Technology (FMT) guarantees studies in the following accredited study programmes:

1. the Bachelor's study program Military Technology,
2. the follow up Master's study program Military Technology,
3. the five-year Master degree program Military Technology,
4. the Doctoral study program Military Technology.

The main objective of the Bachelor's study program is to prepare academic military experts for the means of various engineering-managerial positions in the Czech Army at the level under independent unit. The study program displays the needs and requirements of the Ministry of Defence and other particular branches of the Czech Army as well as Bachelor's studies general standards. Students must pass final state examination and present their thesis.

The main objective of the Master's study program is to educate academic military specialists on behalf of the various engineering-managerial positions within the Czech Army at the level under independent unit and over. The study program is based on the needs and requirements of Ministry of Defence and other particular branches of the Czech Army as well as the Master's study general standards. Studies are concentrated on the military engineering qualification of the Czech military professionals in the same branches of study as the Bachelor's study program. The Master's study is finished by final state examination and thesis presentation (Cvachovec et al. 2011; Rosicka and Hoskova-Mayerova 2014).

The main objective of the doctoral study program is to educate specialists with the ability to create and use scientific methods and work at the world standard level in their particular branch of study. The graduates of this study program gain the theoretical background, enabling them to be involved in the research projects related to their branch and successfully carry out the tasks linked with their academic preparation. The graduates of this study program are intended to work for the faculty, the university or other universities as well as research institutes, or man the highest technical positions in the military forces or defense industry.

The Bachelor's study at UO was started after 2000 following the adoption of the Bologna Declaration. The Bologna Process is a series of ministerial meetings and agreements between European countries to ensure comparability in the standards and quality of higher-education qualifications. The process has created the European Higher Education Area under the Lisbon Recognition Convention. It is named after the University of Bologna, where the Bologna declaration was signed by education ministers from 29 European countries, including the Czech Republic, in 1999.

In addition to the Bachelor's and follow-up Master's degree programs, since 2014, the 5-year Master's program has been accredited. Over the years, interest in the 5-year Master's degree program has been increasing steadily. One of the reasons for the growing interest in 5-year Master studies in students is the possibility to be profiled from the first year in one of the two basic directions of study—mechanical or electrical. From 2018, only the five-year master degree programme of Military Technology has been opened for military students.

Due to the rise of cyber-attacks and threats, FVT will begin teaching in a newly accredited 5-year master's program Cybernetic Security in 2019. This program is designed for both military and civilian students.

3 Former Study Programmes in Mechanical and Electrical Specializations from the Perspective of Teaching Mathematics and Hourly Grants

During the first 50 years of DMDG, the hour grant in mechanical study programmes ranged up to 480 h in mathematics courses (the basic study of mathematics lasting 4 semesters was completed by the examination every semester. Around 2000 the hour grant was 410 h), the hour grant in electrical engineering study programmes was up to 570 h (all 4 semesters were completed by the examination, around 2000 the hour grant was 400 h), construction and geodesy study programmes were granted up to 400 h (4 semesters completed with the examination), command-staff branches were granted up to 250 h mathematics (3 semesters were completed by the examination).

In 2000, in the main study programmes—mechanical and electrical—represented the teaching of mathematics volume of 480 and 410 h. The teaching of mathematics in the electrical study program lasted 6 semesters and in the machine study program lasted 4 semesters. The number of hours of lectures and exercises, including laboratory exercises, is clearly stated in Table 1.

In the particular semesters, the following hours of individual subjects of mathematics were taught in electrical study program:

1S: Linear algebra and analytical geometry (56 h), Real functions of one variable (20 h), Differential calculus of one variable (44 h)
2S: Integral calculus of one variable (36 h), Differential calculus of several variables (36 h), Ordinary differential equations (36 h), Infinite and power series (12 h)
3S: Multiple integrals (20 h), Vector analysis, line and surface integrals (40 h)
4S: Fourier series (24 h), Integral, discrete Fourier and ζ transformation (36 h)
7S: Complex analysis, integral calculus and Laurent series (30 h), Probability (30 h)
8S: Mathematical statistics (24 h), Numerical methods of algebra and analysis (36 h)

In the individual semesters, the following hours of subjects of mathematics were taught in mechanical study program:

1S: Linear algebra and analytical geometry (46 h), Real functions and differential calculus of one variable (66 h), Integral calculus of one variable I (28 h),

Table 1 The number of hours of lectures and exercises (in the form L + E) in mechanical study program (MSP) and electrical study program (ESP) in particular semesters (from 1S to 8S) in 2000

Study program	1S	2S	3S	4S	6S	7S	8S	Total hours
MSP	74 + 76	60 + 60	44 + 46	×	30 + 30	×	×	420
ESP	60 + 60	60 + 60	30 + 30	30 + 30	×	30 + 30	30 + 30	480

Table 2 The number of hours of lectures and exercises and lectures with exercises in electrical study program (ESP) in full-time and combined form of study in 2005

Study program	1S	2S	7S	8S	Total hours
ESP – full-time form	60 + 60	30 + 30	60 + 60	20 + 20	340
ESP- combined form	64	36	56	24	180

2S: Integral calculus of one variable II (12 h), Differential calculus of several variables (24 h), Ordinary differential equations (32 h), Multiple integrals (16 h), Vector analysis and line integrals (36 h)
3S: Infinite, power a Fourier series (22 h), Vector space and linear map (24 h), Probability and mathematical statistics (44 h)
6S: According specialization were taught two optional programs:

A. Numerical methods (16 h), Partial differential equations (24 h), Stochastic and Markov processes (20 h)
B. Complex analysis, integral calculus and Laurent series (36 h), Integral transformations (18 h), □ transformation (6 h)

In 2005, in the Bachelor's and follow-up Master's degree study programmes—mechanical and electrical—represented the teaching of mathematics volume of 340 h. The teaching of mathematics in the electrical study program lasted 2 + 2 semesters. The number of hours of lectures and exercises, including laboratory exercises, was almost the same in mechanical and electrical study programmes. The numbers of hours of teaching mathematics in full-time and combined form of study in the electrical study program are compared in Table 2.

In the particular semesters (the first and second semester in the Bachelor's study and the seventh and eight semesters in the follow-up Master's study) the following hours of individual subjects of mathematics (Mathematics 1, II, III, and IV) were taught in electrical study program—full-time form of study:

1S: Linear algebra and analytical geometry (44 h), Real functions and differential calculus of one variable (60 h), Integral calculus of one variable I (16 h)
2S: Integral calculus of one variable II (16 h), Differential calculus of several variables (28 h), Ordinary differential equations I (16 h)
7S: Ordinary differential equations II (20 h), Infinite and power series (20 h), Multiple integrals (20 h), Vector analysis and line integrals (24 h), Partial differential equations (16 h)
8S: Probability (24 h), Mathematical statistics (16 h).

Finally, let us note that information on older study programs was drawn by the authors from written documents, and that detailed information on degree programs accredited to the University of Defence since 2004 can be found on the Intranet (Intranet of University of Defence 2019). Interesting information about new trends in education of mathematics around the world with attention given to research, theory, practice and policy can be found, for example, in the in the papers (Even and Ball

2009; Hošková and Račková 2007; Abdulwahed 2012; Hoskova 2007; Hoskova-Mayerova and Rosicka 2015).

4 Current Study Programmes in Mechanical and Electrical Specializations from the Perspective of Teaching Mathematics and Hourly Grants

Since the academic year 2014/2015, in addition to already 10 years being of progress two-stage mechanical and electrical study programmes, i.e. Bachelor's and follow-up Master's studies, began teaching in the 5-year Master study program Military Technology.

The new trends in mathematics teaching brought with them the need to increase the share of teaching and laboratory exercises in computer classrooms. Two fully equipped computer classrooms have been established at the department, which are used mainly for teaching numerical methods and teaching mathematical statistics. At DMP and FMT is licensed mathematical software Maple® and previously used mathematical software Mathcad® is now replaced by software Matlab®.

The study of mathematics in the Master study program Military Technology (MT) lasts 4 semesters and the study of mathematics in the two-stage mechanical and electrical study programmes (TSM and TSE) lasts also 2 + 2 semesters. The number of hours of lectures and exercises, including computer aided exercises, is stated in Table 3 (the number of hours and their distribution to semesters in TSM specialization is similar):

In the particular semesters, the following hours of individual subjects of mathematics were taught in the two-stage electrical study program accredited in the years 2004 and 2006:

1S: Linear algebra and analytical geometry (24 h), Real functions and differential calculus of one variable (56 h), Integral calculus of one variable (40 h)

Table 3 The number of hours of lectures and exercises (in the form L + E) in the study program Cybernetic Security in particular semesters (from 1S to 4S) from 2019 together with the number of hours in the study program Military Technology

Study program	1S	2S	3S	4S	7S	8S	Total hours
TSE	60 + 60	60 + 60	×	×	40 + 40	20 + 20‖10 + 10	380
MT	18 + 18‖12 + 12	18 + 18‖12 + 12	24 + 24	12 + 12	×	×	200

2S: Analytical geometry in space (16 h), Differential calculus of several variables (28 h), Ordinary differential equations (28 h), Infinite series (8 h), Probability and mathematical statistics (40 h)
7S: Multiple integrals (18 h), Vector analysis, line integral (30 h), Complex analysis (12 h), Integral transformations (20 h)
8S: In this semester 2 subjects are taught—Mathematics IV (**M**) and Applications of Mathematics and Physics (**A**):

M: Numerical methods of algebra and analysis (20 h), Partial differential equations (20 h)
A: Stochastic processes (20 h)

In the particular semesters, the following hours of individual subjects of mathematics were taught in the in the Master study program Military Technology accredited in 2014:

1S: In this semester 2 subjects are taught—Mathematics I (**MI**) and Linear algebra (**L**):

MI: Real functions and differential calculus of one variable (18 h), Integral calculus of one variable (18 h)
L: Matrix calculus and its applications (12 h), Vector calculus and analytical geometry (12 h)

2S: In this semester 2 subjects are taught—Mathematics II (**MII**) and Probability and mathematical statistics (**P**):

MII: Differential calculus of several variables (18 h), Ordinary differential equations (18 h)
P: Probability and random variables and vectors (16 h), Mathematical statistics, statistical hypothesis testing (8 h)

3S: Infinite and Fourier series (16 h), Multiple integrals (16 h), Vector analysis, line and surface integral (16 h)
4S: Numerical methods of algebra and analysis (24 h).

5 New Study Program Cybernetic Security from the Perspective of Teaching Mathematics and Hourly Grants

Since the academic year 2019/2020, due to cyber threats, the new full-time study program Cybernetic Security has been accredited. Mathematical subjects that will be taught in this Master's program are only partially coinciding with the subjects taught in the Master's degree program Military Technology. Due to the follow-up of other subjects—especially computer science, programming and cryptography—emphasis

Table 4 The number of hours of lectures and exercises (in the form L + E) in the study program Cybernetic Security in particular semesters (from 1S to 4S) from 2019 together with the number of hours in the study program Military Technology

Study program	1S	2S	3S	4S	Total hours
CS	44 + 46	44 + 46	44 + 46	44 + 46|30 + 30	420
MT	18 + 18|12 + 12	18 + 18|12 + 12	24 + 24	12 + 12	200

is placed much more on algebra. However, algebra and its subjects have not yet been taught at the Faculty of Military Technology, so teaching in the study program Cybernetic Security will put greater demands on teachers of DMP. Greater demands will also be placed on students who have not met or only marginally at secondary schools with the majority of algebra subjects.

The study of mathematics in the study program Cybernetic Security (CS) will last 4 semesters. The number of hours of lectures and exercises, including computer aided exercises, is stated in the following table, which also includes, for comparison, hourly grants for the Military Technology program (Table 4).

In these particular semesters, the following hours of individual subjects of mathematics (Mathematics I, II, III, IV and Graph Theory) will be taught in electrical study program:

1S: Propositions, sets, relations, matrix calculus (44 h), Number theory (22 h), Algebraic structures (24 h)
2S: Vector spaces, linear map (34 h), Applications of matrix calculus (20 h), Algebraic structures (36 h)
3S: Propositional and predicate logic, finite-state machines (56 h), Fundamentals of differential and integral calculus (26 h), Sequences and infinite series (8 h)
4S: In this semester 2 subjects are taught—Mathematics IV (**M**) and Graph Theory (**G**):

- M: Probability (8 h), Random variables and vectors (26 h), Statistics, statistical hypothesis testing (28 h), Regression analysis (28 h)
- G: Graphs and graph algorithms (34 h), Flows in networks, colouring of graphs (14 h), Trees, network analysis (12 h).

6 Conclusion

This chapter briefly discusses the nearly 70-year history of the University of Defence and its integral parts—the Department of Mathematics, now the Department of Mathematics and Physics. The history of the Department of Mathematics, its development and distinguished personalities from its establishment in 1951 to the present, is discussed in detail.

The long tradition of teaching mathematics at the University of Defence ranges from its beginnings concerning the traditional education system to its direction towards the current technological world. The teaching of mathematics has responded and responds not only to the development in society but also to the current challenges in the form of new technologies and computerization of the modern society (Hoskova and Rackova 2010; Rackova and Hoskova-Mayerova 2013).

After 2000, a new trend in teaching at universities has become a two-stage study, which also passed the University of Defence. In recent years, however, there is also a need to move to traditional 5-year Master study programs that more respect professional-oriented and often exclusive education in special fields of military technology. Master's study programs also respond to the interest of the applicants to be profiled in the focus of their interests, i.e. either in mechanical or electrical engineering focus in the study program Military Technology or in the program Cybernetic Security.

The chapter also compares the number of hours teaching mathematics in individual study programs over the last 20 years. In detail is listed in detail not only the number of hours in particular semesters of teaching, but also the number of hours teaching individual thematic units.

The basis of mathematics teaching remains traditional, such as calculus, linear algebra, and probability theory. However, traditional teaching has not only been expanded over the last few years by teaching computer classes, but also extended to many new thematic units such as algebraic structures, number theory and graph theory required in the newly accredited study program Cybernetic Security. Thus, mathematics teaching does not lag behind new trends, but responds flexibly to new directions of learning and new social challenges.

Acknowledgements The work presented in this chapter has been supported by the project "Development of Basic and Applied Research developed in the long term by the Departments of Theoretical and Applied Bases FMT" (Project code: "VYZKUMFVT (DZRO K-217)") supported by the Ministry of Defence of the Czech Republic.

References

Abdulwahed, M., Jaworski, B., Crawford, A.R.: Innovative approaches to teaching mathematics in higher education: a review and critique. Nordic Stud. Math. Educ. **17**(2), 49–68 (2012). Available online at: https://dspace.lboro.ac.uk/dspace-jspui/handle/2134/11988

Cvachovec, F., Hošková-Mayerová, Š., Kohout, J., Komárek, M., Kuben, J., Račková, P., Vižďa, F., Zubík, K.: The results of the current project "Inovation of study program–Military Technology at the department of mathematics and physics. In: Proceedings: Distance Learning, Simulation and Communication 2011, May 10-11, Brno, Czech Republic, CATE 2011, pp. 54–59 (Community-Army-Technology-Environment) (2011). ISBN 978-80-7231-695-3

Even, R., Ball, D.L.: The Professional Education and Development of Teachers of Mathematics (The 15th ICMI Study), 269p. Springer Science & Business Media, Boston (2009). ISBN 978-0-387-09600-1

Hoskova, S., Rackova, P.: Problematic examples at the entrance exams of mathematics at the University of Defence. Procedia Social Behav Sci **9**(2010), 348–352 (2010). https://doi.org/10.1016/j.sbspro.2010.12.162
Hošková, Š., Račková, P.: Experience of mathematics blended—learning at University of Defence, Adult Learnings of Mathematics 14, pp. 177–194. In: Proceedings of Contributions, Ireland, Paperback book (2007). ISBN: 0-9552717-3-8, ISBN: 0-9552717-5-4 Format CDROM
Hoskova, S.: The role of a teacher in the educational process—teaching math at combined study. In: Conference Distance Learning, Simulation and Communication—DLSC, Brno, 6 pp (CATE 2007), 6p (2007). ISBN: 978-807231-240-5
Hoskova-Mayerova, S., Rosicka, Z.: E-learning pros and cons: active learning culture? Procedia Social Behav Sci **191**, 958–962 (2015). https://doi.org/10.1016/j.sbspro.2015.04.702
Hoskova-Mayerova, S.: Education and training in crisis management. In: The European Proceedings of Social & Behavioural Sciences EpSBS, vol. XVI, pp. 849–856 (2016). https://doi.org/10.15405/epsbs.2016.11.87
Intranet of University of Defence (2019) Accredited Study Programs. [Online 15 Feb 2019]. Available online at: https://apl.unob.cz/Akreditace/Programy.aspx
Karp, A., Schubring, G.: Handbook on the History of Mathematics Education, 634p. Springer Science & Business Media, New York (2014). ISBN 978-1-4614-9154-5
Korecki, Z.: History of Faculty of Military Technology (2018). [Online 15 Feb 2019]. Available online at: https://www.unob.cz/en/fmt/faculty/Pages/History-of-Faculty.aspx
Pazdera, P.: University of Defence History (2018). [Online 15 Feb 2019]. Available online at: https://www.unob.cz/en/university/Pages/History.aspx
Rackova, P., Hoskova-Mayerova, S.: Investigation of Problems leading to global extrema. In: CD: ICERI2013 Proceedings, 4745-4750, 6th International Conference of Education, Research and Innovation, Sevila, Spain, 18–20 Nov 2013. ISBN: 978-84-616-3847-5, ISSN: 2340-1095
Rosicka, Z., Hoskova-Mayerova, S.: Efficient methods at combined study programs. Procedia Social Behav Sci **131**, 135–139 (2014). https://doi.org/10.1016/j.sbspro.2014.04.092
Svarcova, I., Hoskova-Mayerova S., Navratil, J.: Crisis management and education in health. In: The European Proceedings of Social & Behavioural Sciences EpSBS, vol. XVI, pp. 255–261 (2016). https://doi.org/10.15405/epsbs.2016.11.26
Wikipedia Contributors: Univerzita obrany, The Free Encyclopedia (2018). [Online 15 Feb 2019]. Available online at: https://cs.wikipedia.org/wiki/Univerzita_obrany

The Development of Education in Emergency Management

Irena Tušer

Abstract Requirements for the training of security specialists, which also takes in the field of emergency management and protection of the public, are changing in response to the changing needs of society. The Czech Republic, acting through its competent ministries, has developed a Concept of Education in emergency Management which was approved by the National Security Council in 2001. Over time, however, these provisions have lost some of their relevance given the needs and rapidly changing demands of society and the security situation. The analysis that has been carried out has confirmed a lack of interconnection and uniformity within educational activities. Therefore, the plan of non-legislative governmental tasks adopted by the Czech Republic for the first half of 2017 aimed to optimise the rules for the training of specialists to bring them into line with current trends and needs. The aim of the updated provision was, in addition to the training of emergency management specialists, to enhance the co-ordination of educational processes, to provide consequent regular information support to trainers, and to provide the opportunity of acquiring lecturing skills and teaching experience. Subsequently, some concerns were voiced that target groups might not be interested in this training, and that the budget for this sector would be reduced. At present, however, adequate systematic branch management is currently lacking in this area in terms of both scale and continuity. The present chapter outlines the gradual development of the education system in emergency management and offers a comparison of educational concepts, emphasising their positive and negative aspects.

Keywords Emergency management · Concept of education · Emergency · Professional education · Protection of the public · Security · System

---

I. Tušer (✉)
Department of Safety and Law, AMBIS College, Prague, Czech Republic
e-mail: irena.tuser@ambis.cz

D. Flaut et al. (eds.), *Decision Making in Social Sciences: Between Traditions and Innovations*, Studies in Systems, Decision and Control 247,
https://doi.org/10.1007/978-3-030-30659-5_10

1 Introduction

A characteristic feature of today's environment is the fact that conflicts taking place beyond the borders of Europe may also threaten our security. This increases the importance of a comprehensive approach in the area of protection and defence combining military and civilian means. Such threats can be avoided or their negative effects alleviated with a combination of legal, diplomatic and economic means. The corresponding competencies in the sphere of preparedness and the capability of prompt response to sudden threats are aspects that also have an influence on state security.

The Czech Republic is developing a system of crisis management and protection of the population within the framework of systematic and continual prevention of emergency events and crisis situations. The aim of this system of crisis management is to reinforce the specialist background of bodies of the public administration, corporate bodies and natural persons engaged in enterprise in the sphere of the planning, preparation, coordination and standardisation of procedures within the framework of their preparation for crisis situations.

The priority in the sphere of crisis management is the synchronisation of tools and measures in crisis and defence planning with a view to the optimal use of means from public budgets, effective planning and taking advantage of the capabilities of bodies of the public administration, the manpower and means of the armed forces and security units, and private sector sources. The government of the Czech Republic is ensuring co-operation between crisis management bodies and the private sector in the area of avoidance of crisis situations and augmenting the robustness of the Czech Republic against their impact, and is adopting certain adaptation measures (The Ministry of Foreign Affairs of the Czech Republic 2015).

The system of education for competent persons in the area of crisis management in the Czech Republic was stipulated in 1999 with the adoption of a "Strategy for Education in Crisis Management". An initial detailed "Concept of Education in Crisis Management", which was approved by ruling no. 211 of the State Security Council of 25 September 2001, was drawn up within the context of the given document for the purpose of initiating the realisation of the educational process.

The gradual realisation of reform to the public administration and the associated changes to the legislation that regulated the preparation and education of state employees and other employees at administrative authorities and officials at local authorities inspired the updating of the given concept in 2004. The updated "Concept of Education in Crisis Management" was approved by ruling no. 14 of the State Security Council on 16 November 2004 and replaced the previous concept of 2001.

This concept, however, also became increasingly irrelevant as it was not in accordance with the needs and demands of a dynamic society. For this reason, the plan for the non-legislative tasks of the government of the Czech Republic for the first half of 2017 set the goal of optimising the rules covering education for specialists so as to correspond to current trends and needs. A system of education for obtaining or extending expert competence and expert preparedness was, for this reason, introduced by the new document "A Concept of Education in the Protection of the Population and Crisis Management" (hereafter merely "Concept"). The currently valid Concept was approved by ruling no. 508 of the government of the Czech Republic on 10 July 2017 and replaced the original concept of 2004. From the viewpoint of education, the Concept also takes in the sphere of protection of the population, the integrated emergency system, critical infrastructure, domestic security and public order, and economic measures for crisis situations and protection of the economy (The Ministry of the Interior—The General Directorate of the Fire Rescue Service of the Czech Republic 2017).

The educational strategy drawn up does not, however, guarantee systematic and continual leadership from the ministries, financing of the costs expended on educating target groups, the all-round competence of lecturers or the interconnectedness of the education of all participating stakeholders. The positive aspects that can be named are the introduced system of reinforcing qualifications, the anchoring of education in the legislation, the development of education with the help of new technologies, and the high level of erudition among lecturers.

2 The Educational Concept in Emergency Management—Tradition and Innovation

An effective aspect of the education for security units, armed forces and employees of local authorities are the functional systems already configured for the attainment and improvement of the qualifications of these people which are regulated by legal standards.

The basic legal standards linked to education in the sphere of crisis management at state and local authority bodies and corporate bodies include the laws passed in 2000—*The Law on Crisis Management*, which legislated for the issue of critical infrastructure and defined the range of requirements for the expert competence of liaison security employees of critical infrastructure entities, and the *Law on Economic Measures for Crisis Situations*, which obliged the Administration of State Material Reserves to ensure, in co-operation with other central administrative bodies, further professional education for employees of administrative authorities and suppliers of mobilisation supplies in the area of economic measures for crisis situations. Other key legal standards include, for example, the *Law on the Integrated Emergency System* and the *Law on Ensuring the Defence of the Czech Republic*, to name just two.

The Analysis of the State of Education of Crisis Management Specialists drawn up in 2016 pointed to the absence of a normative legal act that would stipulate the obligation of elected functionaries (mayors of municipalities and municipalities with extended jurisdiction, regional presidents, city mayors) to undergo education in the area of protection of the population and crisis management (The Ministry of the Interior—The General Directorate of the Fire Rescue Service of the Czech Republic

2016). The given fact has been reflected negatively in the low participation in training organised for mayors in spite of the fact that this target group plays a fundamental role in the resolution of emergency events and crisis situations and has responsibilities defined by the law. (Hoskova-Mayerova 2015).

Education in the area of crisis management and protection of the population is run on a methodical basis by the Ministry of the Interior. Other central administrative authorities also contribute to associated activities according to their competences and jurisdiction. An interdepartmental working group has been established at the level of the Committee for Civil Emergency Planning for the resolution of specialist educational issues in the area of protection of the population and crisis management. This working group is comprised of representatives of the Ministry of the Interior—the General Directorate of the Fire Rescue Service of the Czech Republic, the Department of Security Policy and the Prevention of Criminality at the Ministry of the Interior, the Ministry of Defence, the Ministry of Health, the Ministry of Education, Youth and Sport, the Administration of State Material Reserves and the Police Presidium of the Czech Republic.

Education in the area of crisis management and protection of the population takes place at the level of training and courses at the training facilities of the Ministry of the Interior, at the level of higher vocational education and at the level of university education. The participating stakeholders contributing to education in the area of crisis management and protection of the population are teachers at secondary schools and universities, staff engaged in the given issue, volunteers and accredited lecturers.

Teachers at the individual school levels, however, often lack expert competence in the area in question. Training to obtain the necessary specialisation has been approved for this group within the jurisdiction of the National Institute for Further Education. Another possibility, apart from individual study, is offered by courses held under the auspices of the Fire Rescue Service of the Czech Republic. A certain shortcoming, however, is the fact that no specialist accreditation is awarded following completion of the course. The participating volunteers and other staff, on the other hand, lack the necessary presentational and teaching dexterity. One strong point of this system of education is the high level of expertise and qualifications among the accredited lecturers from the ranks of units of the integrated emergency system, central administrative authorities and regional authorities who regularly conduct the education of target groups according to their area of expertise.

Forms of education for the target groups are realised most often in the form of the organisation of training or in the form of one-off lectures on demand. Another significant form of education is the training of units of the integrated emergency system or crisis management bodies in which the target groups participate. This education is conducted with a certain regularity. Both basic information taken from the legal regulations and information for specialists engaged in the resolution of specific security problems or areas is presented. A negative aspect of this education is the inadequate use of modern technologies (e-learning, audio-video/multimedia screenings) and projects within the framework of the development of security education financed from European structural funds or by the state or regions (Hoskova-Mayerova 2016; Hoskova-Mayerova et al. 2017).

Two-level education, in which the Ministry of the Interior, in co-operation with other ministries and central administrative authorities affected, performed the education of mayors of municipalities with extended jurisdiction and mayors of statutory cities in the form of one-day training, is now being abandoned. Experience gained from this training has demonstrated that the named target group prefers educational topics targeted at practical needs and their specific region over general information "from the centre". Training in 2019 will, for this reason, be organised merely at the level of fire rescue units in the given regions in co-operation with the pertinent regional authorities, the local authorities of municipalities with extended jurisdiction, units of the integrated emergency system and relevant not-for-profit organisations, thereby supporting a practical form of education (Svarcova et al. 2016).

The area of expertise or content of educational programmes is divided into individual "modules" for which study texts have been drawn up. Ten modules have been prepared to date in the area of security. A general requirement, however, is that the prepared modules are regularly updated and that rules are stipulated for their further continual and systematic updating. The response to current needs is not entirely optimal at the present time, i.e. specialisation courses held with a view to newly emerging risks identified in society are not created on a flexible basis. Future expansion of the range and number of specialist courses for all target groups is also associated with this (Resolution of the Government of the Czech Republic 2015).

The system of financial and material provision for education in place operates in accordance with the principles given in the current Concept. The fact that no rules have been set that would enable monitoring of its purposefulness and the amount of financial means allocated by the individual bodies of the state administration and local authorities would seem to be a shortcoming of the system.

The existing educational facilities and classrooms are not equipped with new means for teaching and training, and new training areas, trainers, simulators and associated training equipment for units of the integrated emergency system have not been built. This can also be seen as a negative in relation to newly emerging risks and the anticipated profiling of educational facilities. It is, for the given reasons, essential to the development of education of the target groups that adequate conditions are created for the application of multiple-source financing, e.g. from EU structural funds.

3 Conclusion and Perspectives

The acquisition of specialist knowledge of crisis management and protection of the population is a key factor in the creation of an effective crisis management structure at all levels of the public administration. This is conditional to the creation of an educational system that will ensure the mutual interconnection of individual educational elements and forms and, at the same time, retain continuity with the system of preparation of all participating stakeholders (Ingrassia et al. 2014).

The system of preparation and education of specialists does not, at the present time, take into consideration the threats and implications of the increasing number of current emergency events and crisis situations. The principal cause is the societal economic situation, the result of which is a falling number of experts/staff in the given area. There is also the matter of the disproportionate incorporation of the issue into legal standards and, not least, the global view of the population of the Czech Republic underestimating this range of problems.

For the future development of competent education, it is necessary to deepen co-operation with universities and update educational materials on a regular basis. These materials must be expanded to take in the practical experience of the experts, thereby standardising the level of expertise among university graduates in the given field.

It is recommended that bodies of the state administration define the requirements for research activity, and that the results obtained from the final works of students (degree works and dissertations) are used in practice to the benefit of society (State Security Council Resolution no. 32 2007).

More effective co-operation between universities and research units and the application sphere and an endeavour to realise joint projects that will help further positive developments are essential to the further direction of specialisation.

One of today's trends, and not merely in education, is to develop international co-operation with the sharing of the experience and knowledge of experts and to respond flexibly to current needs and adapt to the environment of newly identified risks. Systematic preparation for these specialists is one of the priorities facing crisis management in the Czech Republic and the European Union.

References

Hoskova-Mayerova, S.: Geospatial data reliability, their use in crisis situations. In: Knowledge Based Organization International Conference, vol 21(3), pp. 694–698. De Gruyter open. https://doi.org/10.1515/kbo-2015-0117 (2015)

Hoskova-Mayerova, S.: Education and training in crisis management. Eur. Proc. Soc. Behav. Sci. (EpSBS) **XVI**, 849–856 (2016). ISSN 2357-1330

Hoskova-Mayerova, S., Hubacek, M., Bekesiene, S., Bures, M.: Vehicle movement modelling possibilities for defense and crisis management. In: Epin, M., Briš, R. (eds.) Safety and Reliability. Theory and Applications, pp. 3035–3039. Taylor & Francis Group, London (2017). ISBN 978-1-138-62937-0

Ingrassia, P.L., Foletti, M., Djalali, A., Scarone, P., et al.: Education and training initiatives for crisis management in the European Union: a web-based analysis of available programs. Prehospital Disaster Med. **29**(2), 115–126 (2014). https://doi.org/10.1017/S1049023X14000235

Ministerstvo vnitra—Generální ředitelství Hasičského záchranného sboru České republiky: Analýza současného stavu vzdělávání odborníků v oblasti krizového řízení. (The Ministry of the Interior—The General Directorate of the Fire Rescue Service of the Czech Republic, 2016. Analysis of the Current State of Training of Experts in Crisis Management). www.hzscr.cz/soubor/analyza-vzdelavani-v-oblasti-kr-po-vcnp-pdf.aspx (2016)

Ministerstvo vnitra—Generální ředitelství Hasičského záchranného sboru České republiky: Schváleno Usnesením vlády České republiky č.508 ze dne 10. července 2017. Koncepce vzdělávání v oblasti ochrany obyvatelstva a krizového řízení. (The Ministry of the Interior—The General Directorate of the Fire Rescue Service of the Czech Republic, 2017. A Concept of Education in the Protection of the Population and Crisis Management). www.hzscr.cz/soubor/koncepce-vzdelavani-oobakr-pdf.aspx (2017)

Ministerstvo zahraničních věcí České republiky: Bezpečnostní strategie České republiky. Ministerstvo zahraničních věcí České republiky, Praha (2015). ISBN 978-80-7441-005-5. (The Ministry of Foreign Affairs of the Czech Republic (2015). Security Strategy of the Czech Republic). https://www.vlada.cz/assets/ppov/brs/dokumenty/bezpecnostni-strategie-2015.pdf (2015)

Svarcova, I., Hoskova-Mayerova S., Navratil, J.: Crisis management and education in health. Eur. Proc. Soc. Behav. Sci. (EpSBS) **XVI**, 255–261. https://doi.org/10.15405/epsbs.2016.11.26 (2016)

Usnesení Bezpečnostní rady státu: Společné minimum pro potřeby vzdělávání odborníků v oblasti bezpečnosti. Usnesení Bezpečnostní rady státu č. 32, ze dne 3. července 2007. (State Security Council Resolution no. 32, 2007. Common Minimum for Training Security Experts.) http://krizport.firebrno.cz/dokumenty/spolecne-minimum-pro-potreby-vzdelavani-odborniku-v-oblasti (2007)

Usnesení vlády České republiky: Zpráva o stavu ochrany obyvatelstva v České republice 2015. Schváleno Usnesením vlády České republiky č. 977, ze dne 2. prosince 2015. (Resolution of the Government of the Czech Republic (2015). Report on the State of Population Protection in the Czech Republic 2015). http://krizport.firebrno.cz/aktualni-situace/all-zprava-o-stavu-ochrany-obyvatelstva-v-ceske-republice (2015)

Waste and Waste Management Awareness Among Teachers: A Phenomenographic Approach

Engin Baysen and Fatma Baysen

Abstract Waste has become a critical problem throughout the world, and it needs to be managed appropriately. The success of waste management primarily depends on education. Therefore, the current study aimed to reveal waste and waste management (WM) awareness of teaching staff. Primary school teachers ($n = 19$) participated in this study, which adopted phenomenography as a research method. There is a variation in awareness among the teachers regarding waste and WM, including conceptions, misconceptions, and the relations among them. The teachers' level of awareness is captured in three categories: low awareness, awareness, and high awareness. Waste awareness includes the waste definition and waste examples, while awareness of WM includes the WM definition, WM importance, WM methods, and WM responsibility. A model of waste management awareness is proposed that features the complexity of WM awareness. Implications and recommendations for education are presented.

Keywords Awareness · Conceptions · Phenomenography · Sustainability · Teachers · Waste management

1 Introduction

The increase in the human population and urbanization and the changes in people's lifestyles have increased the amount of waste in the world (World Bank 2012). Waste has exceeded what nature can handle, and as a consequence, the world's reserves are consumed faster than they are produced. Worse still, humans produce waste that is hard for nature to process (Diaz Silvarrey and Phan 2016). Additionally, humans deposit waste illegally, adding to the waste management problem (Triassi et al. 2015). The problem has reached such dramatic proportions that McDougall, White, Franke,

E. Baysen (✉) · F. Baysen
Ataturk Education Faculty, Near East University, Near East Boulevard Nicosia, North Cyprus, 10, 99138 Mersin, Turkey
e-mail: engin.baysen@neu.edu.tr

F. Baysen
e-mail: fatma.baysen@neu.edu.tr

D. Flaut et al. (eds.), *Decision Making in Social Sciences: Between Traditions and Innovations*, Studies in Systems, Decision and Control 247,
https://doi.org/10.1007/978-3-030-30659-5_11

and Hindle (McDougall et al. 2001) somberly stated that the environment is not an infinite waste sink.

Consistent with the increase in waste problems, waste management research has increased throughout the years. While, according to Web of Science, only three research publications were published in 1975 on waste management, this number had increased to 4568 in 2018. Researchers authoring prior studies have generally considered issues such as awareness, safety, attitudes, strategies, knowledge, and policy development (Baxter et al. 2016; Khan et al. 2016; Mamady 2016; Zaman 2016; Thyberg et al. 2015; Littledyke 2008; Ernst 2007; Vallejo Andrada et al. 2017; Urban and Hoskova-Mayerova 2017). For example, research efforts have focused on the construction sector, which produces enormous amounts of waste while raising buildings (Najafy 2014; Mahrum 2009; Yorucu and Keles 2007), and suggested ways to diminish the industry's waste and develop policies.

A considerable amount of research has also dealt with waste and waste management (WM) issues from an educational perspective (Hasan 2012; Martinez-Borreguero et al. 2019). Authors have tried to find the knowledge (Makisa 2016) and awareness (Cici et al. 2005) levels of preservice teachers, practicing teachers, and students. The research has generated conflicting results. For example, while Karatekin and Meray (2015) found that preservice teachers are aware that waste causes problems, Karatekin (2013) and Cici et al. (2005) found that preservice teachers have insufficient knowledge and awareness. Thus, we can conclude that there is no consensus among the results: some results are favorable, while others are unfavorable. Inconsistency in the results among different studies paints a blurred picture that, in turn, causes us to offer cliché statements about enhancing waste management issues.

Moreover, research has shown similar inconsistencies in groups of peoples' understanding of different aspects of a phenomenon. For example, students may lack knowledge regarding waste, but they know of scientific waste reduction methods (Malandrakis 2008). Children may define waste concretely and make generalizations, but the same children lack knowledge regarding waste processing products (Glažar et al. 1998). We should inquire into the apparent variations in understanding of a phenomenon's different aspects. Do these results reflect real variations in different aspects of waste and WM understanding? To answer this question, we require specific research. Can a teacher discern any aspect of the waste and WM phenomenon differently from others? If so, can we derive a model to represent the content of waste and WM awareness training that has a high probability of clarifying and synthesizing the findings in an aim to enhance learning/teaching procedures?

The next part of this paper introduces waste, waste impacts, waste management methods, awareness of environmental and waste problems, and the relation between waste management and education.

2 Background

2.1 *Waste Definition*

Waste is an inevitable and natural product of living things. According to Directive 2008/98/E.C. of the European Parliament and of the Council, waste can be defined as any substance or object that its holder throws away or intends to throw away (European Parliament and Council of the European Union 2008). Waste produced by living things can become raw materials for other organisms as soon as it is produced.

2.2 *Waste Impacts*

Waste-related issues have created significant problems around the world (Orlins and Guan 2016; Bhutta et al. 2011). Triassi et al. (2015) reviewed waste-related articles regarding the situation in Italy. They concluded that there is a positive correlation between waste and several diseases, such as cancer. Other than the deadliness of waste-induced diseases, there are many other waste-based problems to consider. Rich et al. (1995) reported that hazards can cause disruptions in human lives, and these disruptions can be categorized as physical, such as experiencing clean water shortage; social, such as quarrelling with neighbors; and emotional, such as an increase in the divorce rate because of the stress imposed on families. Summarizing the increase in waste became a great risk for the environment and human itself (Vallejo Andrada et al. 2017; Baysen et al. 2017).

Impact Area (Local or Global)
The impact area of any environmental problem is another prominent aspect that is receiving attention. Some of the environmental problems affect geographically restricted areas and are thus called local environmental problems. For example, cases of polluted soil, water, and air can be defined as local problems (Ilie and Tincu 2016; Saikkonen 2013). These scenarios are small in scale; they can be solved by local authorities and have short-term impacts on the environment relative to global environmental problems. On the contrary, global environmental problems, such as climate change and air pollution, affect the world (Franklin et al. 2015; Pearson and Dowson 2003). Relative to local environmental problems, they are long-term, and their solutions require significant efforts by nations around the world (Funtowicz and Ravetz 1990). In the particular context of this study, waste problems are local problems tackled by municipalities (Eriksson et al. 2017), but they can also be classified as global environmental problems (World Bank 2012; Rootes 1999).

2.3 Waste Management

Solutions to waste problems include waste management (WM), which constitutes prominent approaches, methods, and strategies to manage the waste and, in turn, solve waste-related problems. Pohjola and Pongrácz (2002) described WM as a system composed of physical things, human activities, and relationships between them. Three options that contribute to the minimization of waste problems that are accepted as more sustainable than other methods are Reduction, Reuse, and Recycling. Reduction involves reducing waste by consuming less (Priefer et al. 2016). This choice is the most beneficiary and sustainable method to address waste problems, undoubtedly contributing to economic issues and environmentally friendly solutions, but it is not easy to implement (Zorpas et al. 2017; Tilbury 1999) because it requires strong volition. Reuse refers to the use of an item more than once (Couvidat et al. 2016). An item can be used for the same or a different function. Finally, Recycling refers to breaking down the used item to its raw materials and using these materials again to produce a new item (Jiang and Pickering 2016). The three WM approaches may not be actualized in real-world situations. Barriers exist for sustainable WM behaviors (Marshall and Farahbakhsh 2013; Zachariou and Kadji-Beltran 2009). For economic reasons, local authorities may deposit waste material at dumpsites and eliminate this waste through incineration while ironically neglecting a possible source of income, such as selling methane gas, that can be realized by waste management methods (Lewis 1991). The Landfill is another solution that costs less than other management methods, except for reduction, but landfills cause significant environmental problems, such as methane emission (Krook et al. 2012).

WM Issues in North Cyprus

Similar environmental, mainly waste-related problems are faced in the north of Cyprus because of population growth and urbanization, with construction waste at the top of the list (Najafy 2014; Mahrum 2009; Yorucu and Keles 2007; National Environmental Protection Agency 2017). One other waste problem found on the west coast of North Cyprus is the copper mine abandoned by Cyprus Mining Corporation (CMC), known as the CMC problem (Gündüz et al. 2016).

Municipalities of north Cyprus (28 in total) have WM responsibilities that are governed by the National Environmental Protection Agency (Alkan 2010). Rehabilitation and improvement studies were carried out with the support of the EU. Examples of such projects include wastewater treatment plants in Morphou, rehabilitation of the Dikmen dumpsite, and construction of a new solid waste landfill at Gungor (Aid Programme for the Turkish Cypriot Community 2013). By November 2017, the National Environmental Protection Agency had permitted four private companies to recycle (collect plastic and paper). Additionally, waste metals are collected and exported (National Environmental Protection Agency 2017). There are only four NGOs that deal with environmental issues (TurkishCyprus.com 2017), and of these, only the Association of Lefke Environmental Protection is directly related to waste-related issues through the CMC problem (Baycu et al. 2015).

2.4 *Awareness*

Awareness, which can be defined as perception or knowledge of an event, is accepted as the first step playing a vital role in probable behavioral change and successful education, particularly those related to environmental issues (Hasan 2012). Within the current study, the term "awareness" constitutes conceptions, misconceptions (Baysen and Baysen 2013), and the relations among them. Awareness is one of the most researched topics in environmental studies (Noorhosseini et al. 2017; Ramayah et al. 2012; Sudamadi et al. 2001). Similarly, environmental educators have consistently emphasized awareness regarding environmental issues because awareness is the preliminary step to environmental teaching (Jean-Sigur et al. 2016). Research has produced varying results. For example, research has revealed a high awareness of environmental problems in general (Wong and Wan 2009) but non-satisfactory awareness (Borthakur and Govind 2017; Macias and Piniarski 2016; Pullishery et al. 2016) and moderate awareness (Desa et al. 2010) of waste and waste management.

Developing Awareness of Environmental Issues
Awareness of environmental issues is vital, as the lack of environmental awareness creates barriers to waste minimization (Olgyaiova et al. 2018). Different factors can contribute to making people aware of environmental issues. For example, Palmer et al. (1998) and Palmer et al. (1999) found that being outdoors or in nature increases peoples' awareness in different sample populations from Europe, Canada, and Australia. Palmer et al. (1999) also stated the importance of not only teachers but also the effect of family and other adults for developing awareness.

Researchers have tried to synthesize and represent environmental issues; for example, the Theory of Planned Behavior by Ajzen (2005) considers human behaviors. Regarding environmental awareness, Partanen-Hertell et al. (1999) formed a model to depict environmental awareness development. According to them, environmental awareness develops in four stages, depending on motivation, knowledge, and skills. The first stage is based on concerns about health threats; the second and third stages differ from each other, with the synergy between the acts, motivation, knowledge, and skills; more developed in the latter. Finally, the last stage reaches the level of professional skills and everyday life choices. According to this model, any person or society is in one of the stages of the model.

2.5 *Education and Waste Management*

Educating people about waste and WM is the most crucial action for resolving WM problems (Fortner 1991) because education is the most long-lasting, sustainable, and natural method for increasing awareness. Consistent education regarding sustainable development has been praised in the literature (Negev et al. 2010). Research has demonstrated the important role of teachers in the promotion of environmentally friendly communities (Westing 1993), particularly in regard to WM. First, teachers are critical participants in decision-making processes regarding WM as community members (Morrissey and Browne 2004; Bekesiene and Hoskova-Mayerova 2018); second, they are the primary environmental information source (Malandrakis 2008) and models for their students (Littledyke 2008).

Remembering that humans create waste problems, it is natural to address the human factor for solving waste problems, i.e., to decelerate the producer. Continuous school-based environmental education and sustainability programs have increased (Cutter-Mackenzie 2010; Ferreira et al. 2006; Kyburz-Graber 1999) and enhance waste management awareness (Grodzinska-Juczka et al. 2003).

2.6 Phenomenography

Qualitative research approaches include in-depth studies that aim to analyze how individuals think. One type of qualitative research methodology is phenomenography, which was established by a group of researchers in the late 1960s in Sweden. Phenomenographers use qualitative methods to discover individual variations in the understanding or perception of a phenomenon, mostly by conducting interviews that last 30–60 min (Trigwell 2006) or until saturation (Marton and Booth 1997). Variations arise because individuals are aware of or discern only part of a phenomenon (Marton and Booth 1997).

Phenomenography research includes certain assumptions. The first assumption is that variations occur in the understanding of any phenomenon. Phenomenographers group these variations into a limited and small number of categories that describe these variations. One other assumption refers to the understanding of a small number (10–30) of participants (Trigwell 2006). Phenomenographers define the categories hierarchically, focus on the participants' *collective awareness* (Marton and Booth 1997), and do not report each subject's conception. Phenomenographers add descriptive categories to create a comprehensive form that is called the outcome space. Furthermore, phenomenographers do not generalize their results to a population having common characteristics; it is not even their intention to do so (Marton and Booth 1997). Thus, the goal of this study is to use phenomenography to discover teachers' awareness of waste and WM to enhance learning/teaching environments (Hoskova-Mayerova and Rosicka 2012).

3 Method

3.1 Participants

Turkish primary school teachers ($n = 19$) of schools in the North of Cyprus participated in this study. Of the sample, twelve teachers were female, and seven were male; their work experience ranged between 1 and 20 years, and all participants had graduated from a four-year teacher training program. Thus, purposiveness and variation were considered during sampling, as is recommended in phenomenography (Reed 2006). During the study period, all of the teachers were working in primary schools. Environmental issues in Turkish programs are included in the Life Sciences curriculum, which starts in the first grade of primary schools. All of the teachers said that they had experience teaching subjects that address environmental issues.

3.2 Data Collection

The data were collected by conducting face-to-face interviews (Marton 1986) at the participants' schools and in empty classrooms of this researcher's university, which was familiar to the interviewees. The interviews lasted 25 min on average (ranging between 15 and 40 min), and the questions were open-ended. Probing questions were introduced to obtain any detailed and clear data. The interviews were recorded with the participants' permission. The researcher stopped the interviews when sensing saturation; that is, the data being collected started to repeat itself. Then, the recordings were transcribed verbatim and prepared for analysis. As an outstanding feature of phenomenography, the collected data are seen as a whole.

3.3 Data Analysis

An iterative style of reading that is generally used for phenomenography was conducted. The participants' answers were considered a data source that could reveal differences and similarities. Determining differences and similarities is a method that reveals the variations in the participants' responses. Traces of descriptive categories began to form during the interviews, and the researcher accepted this formation as natural and inevitable. However, the descriptive categories of this study were only identified through an iterative style of reading. The categories were modified, as they are accepted as a starting point for each reading session, and then new categories were applied to all of the remaining data. The reading sessions were repeated until the final results were determined (Carlsson et al. 2001).

Coding and Categorization

During the phenomenography analysis conducted throughout the current study, systematic coding and a hierarchical method of categorization were followed. The analysis revealed three categories. The "low awareness" category is defined by having less knowledge of waste problems while lacking reasoning or in-depth knowledge or having misconception/s and lacking any understanding of impact areas. The "awareness" category for the waste problem refers to having some extent of correct knowledge but only understanding its local effects. Finally, "high awareness" refers to having more knowledge than that of individuals in the low awareness and awareness categories, as well as understanding the worldwide effects and the consequences of local actions and emphasizing the issue. The researcher did not expect participants to state global impacts by giving concrete examples, such as methane emission.

4 Results

The results were grouped into two main sections, namely, Waste and WM (Table 1). Variations existed among the teachers' Waste and WM awareness, except for the definition of waste. The differences were based on conceptions, misconceptions, and relations among the conceptions and misconceptions. The teachers' awareness was described using the three categories of Low Awareness, Awareness, and High Awareness. The teachers' awareness varied by section, subsection, and sub-subsections.

4.1 Waste

Two subsections were established to explain teachers' awareness regarding waste, namely, the Waste definition and Waste examples.

4.1.1 Waste Definition

There was no variation in the definition of waste among the participants. All of the teachers stated that the concept of waste included items that were used and no longer wanted. For example, one participant stated, "We use things and throw them away. Because they are not useful anymore. We have to get rid of them."

4.1.2 Waste Examples

Two categories of this concept were captured, specifically, Low Awareness and Awareness.

1. Low Awareness

Individuals in this category provided examples that included only plastic, metal, paper, and glass, and the teachers provided insufficient waste examples. In other words, they displayed misconceptions regarding waste examples. The following exchange is an example.

> Participant: "Glass, metal, paper, and plastic are waste."
>
> Researcher: "Can you give examples other than these?"

Table 1 Outcome space of teachers' awareness of waste and WM

Main section	Section	Subsection	Sub-subsection	Category	Description
Waste	Waste definition	–	–	High awareness (no variation)	Items used and no longer wanted
	Waste example	–	–	Low awareness	Only paper, plastic, metal, and glass examples
				Awareness	Examples that included all types of waste
WM	WM definition	–		Low awareness	This is the first time they discussed managing waste
			–	Awareness	Stated impact of WM for local environments
				High awareness	Defined WM and its global impacts
	WM importance	–	–	Low awareness	Stated that WM was important, without emphasizing the issue or providing any reasoning
				Awareness	Importance with reasoning
				High awareness	Importance of WM, including global impacts and emphasis on WM

(continued)

Table 1 (continued)

Main section	Section	Subsection	Sub-subsection	Category	Description
	WM methods	Incineration		Low awareness	Foul-smelling and created a blemish on the landscape
			–	Awareness	Causes fatal disease
		Landfills	–	Low awareness	Included definition of the method but not its consequences
				Awareness	Included the definition of the method and its consequences
		Reuse	Reuse at school	Low awareness	Sincere reuse actions
				Awareness	Important reuse actions
			Reuse by authorities	Low awareness	No detailed reuse explanation
				Awareness	Awareness regarding the reuse process
				High awareness	Awareness of the reuse process and its global impacts
		Recycle	–	Low awareness	No detailed explanation regarding the process
				Awareness	Awareness of the recycling process to a certain extent

(continued)

Table 1 (continued)

Main section	Section	Subsection	Sub-subsection	Category	Description
	WM responsibility	WM teaching responsibility	–	Low awareness	Awareness that they were responsible for training students to be environmentally friendly but did not include strategies, methods, etc.
				Awareness	Aware of their responsibility, understood the results of their actions and knew how to be successful but not on a global scale
		WM reuse responsibility	Reuse at school	Low awareness	Reuse was a personal action, not a professional action
			Reuse by authorities	Awareness	Reuse was professional work conducted by authorities
		WM recycling responsibility	Recycling at school	Low awareness	Recycling was limited to waste collection by the teachers
			Recycling by authorities	Awareness	Recycling was a professional activity that was conducted by authorities

Participant: "I said, paper, metal, glass, and plastic…" "…That is all."

Researcher: "Okay."

2. Awareness

Examples in this category included not only plastic, metal, paper, and glass but also other types of waste, but global issues were not mentioned. For example, one teacher stated that it can be paper, metal, glass, plastic, concrete, and nuclear waste. As I said, "waste is everything" that we use and throw away.

4.2 WM

Four subsections were established to explain teachers' awareness concerning WM, namely, the WM definition, WM importance, WM methods, and WM responsibility.

4.2.1 WM Definition

The WM definition is captured in the three categories of Low Awareness, Awareness, and High awareness.

1. Low awareness

Teachers in this category demonstrated little awareness concerning WM. One teacher stated, "I obtained a master's in management…" and "I can quickly develop my understanding of WM, to manage the waste" (no knowledge was added).

2. Awareness

Teachers in this category not only defined WM but also included the impacts of WM on nearby surroundings. One teacher stated that WM means "a decrease in the waste amount in our surroundings…. So, our surroundings become clean."

3. High awareness

For the High awareness category, teachers defined WM and its impacts on a global scale: "WM sustains global effects… If we manage our waste, this will save the whole world."

4.2.2 WM Importance

WM importance emerges in all three categories of Low Awareness, Awareness, and High awareness.

1. Low awareness

For this category, teachers believed that WM was necessary, but they did not provide any reasoning and did not emphasize the issue. One teacher stated, "WM is important…. We have to manage our waste."

2. Awareness

Teachers in this category included reasons for the importance of WM. One teacher stated, "waste is very harmful to the environment, so we have to manage waste materials. Otherwise, waste will damage the soil, the sea, and every living creature."

3. High awareness

Teachers in this category included their reasoning and emphasized the importance of WM. One teacher emphasized the importance of WM by using a Latin phrase: "WM is an issue sine qua non for all the world."

4.2.3 WM Methods

Four subsections were established to explain the teachers' awareness of WM methods. These subsections include Incineration, Landfills, Reuse, and Recycle.

- Incineration. We captured the teachers' perspectives regarding incineration in the two categories of Low Awareness and Awareness.

1. Low awareness

The teachers in the Low Awareness category considered incineration to be a foul-smelling method that blemished the landscape. They stated the unsightly appearance only and neglected its consequences. One teacher stated, "people are unhappy about incineration. Dumpsites smell bad. Whenever they light a fire in the dumpsite, smoke comes out of the dumps and covers the sky."

"Although it is far away from our houses, we do not even want to see piles of waste."

2. Awareness

In the Awareness category, the teachers indicated that the incineration method causes fatal disease but did not demonstrate an awareness of its global effects.

"I am sure that it is hazardous for our health, causing cancer."

- Landfills. The teachers' awareness regarding landfills emerges in the two categories of Low Awareness and Awareness.

1. Low awareness

The teachers in the Low awareness category defined the method but did not discuss its impacts. One teacher stated that they sometimes dig big pits and bury waste in them.

2. Awareness

In the Awareness category, the teachers not only defined but also discussed the impacts of landfills, although not on a global scale. The teachers did not consider landfills a successful method to manage waste. One teacher stated, "they dig big pits. However, this is not good WM because the waste is still there…. It continues to affect our surroundings."

- Reuse. Two sub-subsections of the Reuse subsection were established. These sub-subsections are "Reuse at school" and "Reuse by authorities". Reuse at school is captured in the two categories of Low Awareness and Awareness.

– Reuse at school

1. Low awareness. Teachers in this category stated their actions with sincerity. Some teachers stated that reuse is an art issue: "We use paper rolls to build a Christmas tree."
2. Awareness. In this category, each reuse action was deemed important and a contribution to WM in general. One teacher stated, "Teachers are important. Because we are not only a model for our students, but we contribute to the reuse action as well."

– Reuse by authorities. The teachers' perspectives regarding Reuse by authorities emerge in the three categories of Low Awareness, Awareness, and High awareness.

1. Low awareness.
   In this category, the explanation was not detailed: "I know that they reuse glass bottles."
2. Awareness. In the Awareness category, the teachers' explanations regarding reuse by authorities were detailed and indicated that the authorities used this process for WM. One teacher stated, "they collect the bottles of, let say carbonated beverages' and clean them in the factories and give them those who are going to use them again."
3. High awareness. In the High awareness category, the teachers indicated that they were aware of the reuse process and its global impacts. They collect fruit juice bottles, and after careful cleaning, they use them again. This process is paramount to diminishing the use of raw materials throughout the world.

- Recycle. The teachers indicated that only glass, plastic, paper, and metal were used for recycling, which is consistent with their examples of waste. Recycling in schools and by authorities was represented by the categories of Low Awareness and Awareness.

1. Low awareness. Teachers in this category indicated waste collection awareness of recycling but did not provide a detailed explanation. The teachers mentioned that they collected recyclable materials. For example, they collected plastic lids for a charity that supported disabled individuals. The teachers discussed only the individuals at school who collected recycling materials. Teachers thought that

the authorities were unsuccessful. One teacher stated, "they collect waste. I do not know [if] maybe they did recycle or probably they were unable to do so and sent them to the dump site."

2. Awareness. The Awareness category included an awareness of recycling to a certain extent. One teacher stated, "I went to a factory and saw them chopping plastic things into smaller fragments…and then they melted them too…".

4.2.4 WM Responsibility

Three subsections of the Responsibility section were established to explain the teachers' awareness of WM responsibility, namely, WM teaching responsibility, WM reuse responsibility, and WM recycling responsibility.

- WM teaching responsibility. WM teaching responsibility was captured in the two categories of Low Awareness and Awareness.

1. Low awareness. The teachers were aware that they were responsible for training students to be environmentally friendly. One teacher stated, "we, as teachers, should teach students starting from early childhood and change and maintain some behaviors for our surroundings."
2. Awareness. The teachers were aware of their responsibility, understood the results of their actions, and knew how to be successful and how success impacts this issue locally but not globally: Teachers will succeed if the school is clean and tidy. One teacher stated, "If all wastes are all in the bins and not all around, I can say that we are successful. This act will ensure the protection of our surroundings."

- WM reuse responsibility. The teachers' comments regarding WM reuse responsibility was captured in the two categories of Low Awareness and Awareness.

1. Low awareness. The teachers indicated that they collected waste to reuse at their schools with their students but not in their own homes. The teachers stated that reuse was an insignificant procedure, and not a professional responsibility. One teacher stated (with humility), "Yes, we collect things to reuse in our art class."
2. Awareness. In this category, the teachers indicated that not only they but also the authorities gathered materials for reuse. However, this time, reuse was considered to be a more serious activity, and the teachers indicated that they believed in its effects. The teachers indicated that the authorities were responsible for collecting items such as glass bottles that were cleaned and reused as glass bottles again. For example, one teacher stated, "glass bottles are washed and reused; this is an excellent method that saves the excessive use of raw materials."

- WM recycling responsibility. The teachers' perspectives regarding WM recycling responsibility were captured in the two categories of Low Awareness and Awareness.

1. Low awareness. The teachers indicated that they collected waste for recycling, but their responsibilities were limited to collecting only metals, glass, plastic, and paper. The teachers reported that recycling was conducted in their schools with the students but not in their own homes.
   Interviewee: "Collecting waste for recycling is necessary at such places like at school."
   Researcher: "What about at home?"
   Interviewee: "Not much."
   The teachers also stated that recycling was insignificant and an activity that was more appropriate for professionals.
2. Awareness. The teachers placed the primary responsibility of recycling on the authorities. For example, one teacher stated, "the most important stakeholder for WM is the authorities... They should implement and support activities to raise awareness of the public and promote attitudes toward the environment. I do not see them doing anything for WM; they do not even provide enough waste bins."

5 Discussion

Revealing the variation in the awareness of waste and WM is essential for further teaching/learning and is discussed in the following sections, in which the findings of the present study are discussed and contrasted with the literature findings. The researcher's intention is not to generalize the results but to reveal new perspectives regarding the issue.

5.1 *Waste Definition and Examples*

The teachers sufficiently defined waste, but they provided examples that indicated significant misconceptions. Therefore, the concept of waste was found to be lacking, and this deficiency in knowledge could create problems for teaching. For example, the misconception that there are no other types of waste apart from metals, glass, plastic, and paper limits people to collecting only these materials for recycling. They did not mention other types, such as organic waste, the copper waste at CMC, and construction waste, which appears to be the most important waste problem in the North Cyprus context (National Environmental Protection Agency 2017). Thus, we must be open to new waste definitions that will be more sustainable by addressing these misconceptions about waste. Pongrácz (2009) stated a similar requirement for a dynamic waste definition.

5.2 WM Definition and Importance

Teachers indicated high awareness of the WM definition and its importance. The teachers were emotionally concerned about WM and promoted positive student attitudes regarding the environment. Accordingly, the teachers emphasized WM (Gould et al. 2016) and its consequences. This result is consistent with that found by Olgyaiova et al. (2018).

5.3 WM Methods

The teachers did not compare WM methods; they stated the disadvantages of incineration and landfills separately and superficially. They were expected to provide informed discussions (Australian Government Productivity Commission 2006). Furthermore, the teachers did not recommend or create other original management plans or demonstrate awareness of the less salient WM methods, such as food loss and waste (Lipinski et al. 2013). Regarding reuse and responsibility, teachers had a high awareness of the practices by authorities but a low awareness of the efforts needed at school. Thus, the mindset is that others are more responsible for these issues than the teachers themselves. Teachers lacked awareness regarding recycling. With the exception of one teacher, they did not mention recycling companies. We can conclude that they are not aware of the recycling companies found in North Cyprus.

Interestingly, none of the teachers mentioned Reduction as a WM method, but this result is not interpreted as indicative of the teachers lacking intentions to consume less or lacking care about consuming less. However, they did not relate the practice of reducing consumption to WM and did not perceive Reduction as a WM method; this demonstrates a remarkable lack of knowledge and is regarded as a misconception concerning WM concepts that should be addressed.

5.4 WM Responsibility

The teachers stated that teaching WM, reuse, and recycling responsibilities depend on not only the teachers themselves but also households and authorities. This is consistent with the findings by Palmer et al. (1999). Therefore, the teachers indicated that they were not the only teachers of WM issues and that families may be more effective teachers, and they emphasized that other people and institutions (authorities/government) had primary responsibility. Additionally, their contributions were insignificant, which is consistent with Partanen-Hertell et al. (1999), Negev et al. (2010), and Pongrácz et al. (2004) but inconsistent with the community-driven idea (Quigely et al. 2015). Additionally, in their research with European citizens, Triguero et al. (2016) found that those having less environmental awareness tended to rely on government responsibility, while those with high awareness accepted more responsibility themselves. Seeing others as more responsible can be attributed to a preference to dissociate from the process, hopelessness regarding the success of their teaching, or denigration of the waste management processes carried out in the school context.

5.5 Impact Area

The current research shows that teachers can perceive an environmental problem as local or global. This result can depend on the impact area of the problem and how the problem is discerned (Olander and Nyberg 2014). The result showing that the waste problem is accepted as a local problem is easily predictable. On the other hand, the concept of waste as a global problem must be explained. The reasoning for considering the waste problem as global is thought to stem from its presence on all continents of the world where humans live and the tenet that each local waste problem unites to form a global problem. Moreover, the significant role of waste in the production of methane gas, which contributes to the global environmental problem of ozone layer depletion, and the attention given to the increase in urbanization and WM-related problems are features that enable the discrimination between waste problems that are local issues and those that are global environmental problems. Thus, the present study supports the belief that the waste problem can be accepted as not only a local issue but also a global environmental problem (World Bank 2012; Rootes 1999).

5.6 WM Model

A theoretical WM model focusing on awareness neglecting factors such as motivation and skills, but reflect the complexity of this issue, as is asserted by the present study and shown in Fig. 1. In the schema, awareness boxes represent the quality of awareness of any particular WM aspect from low to high. For example, while an individual has a low awareness regarding one aspect, s/he may have a high awareness of another aspect. An individual may have a high awareness of the WM definition but lack knowledge of recycling and thus be grouped into the low awareness group for the recycling aspect. The greater the complexity and number of different properly interrelated concepts constituting a particular box and the remaining boxes, the higher the quality of the model. The model also comprises the relations, represented by lines, among conceptions, misconceptions (affecting learning negatively), and awareness of the different aspects of WM.

To explain it more clearly, for example, if Box C represents awareness of waste examples, it can include a limited number of waste examples shown by one oval, with a misconception that there are no other types of waste examples apart from the few above-mentioned materials (plastic, glass, etc.). Thus, Box C represents low

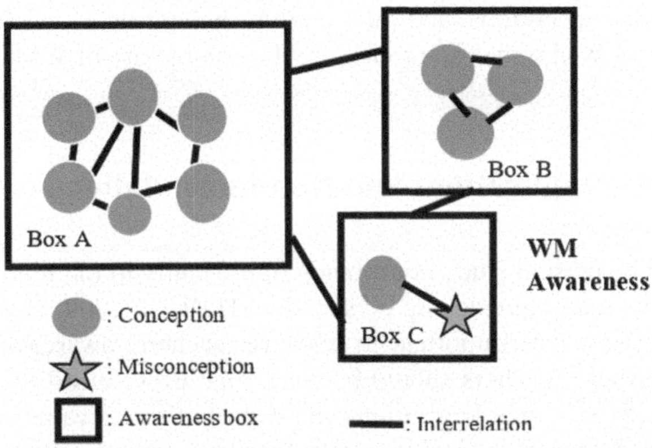

Fig. 1 WM awareness model

awareness (it does not include relationships among qualitative conceptions and, additionally, it includes a misconception). Box B can represent, for example, awareness of incineration as a WM method. Conceptions constitute this box, and there are relationships between them, but there are few conceptions and few interrelations. Teachers holding views consistent with this box are aware that incineration causes foul smells, creates a blemish on the landscape, and causes fatal disease—three conceptions represented by three ovals. Teachers could, for example, relate the foul smell to disease. Finally, Box A, which represents high awareness, is constituted by more conceptions that have suitable relationships among them. Thus, Box A can represent the WM definition. The model represents the relations among the awareness of different aspects. For example, there is a relation between awareness Box C and awareness Box A. Awareness Box C interferes with awareness Box A. Although high awareness of the WM definition is revealed, if we search for in-depth related knowledge regarding waste examples, we can find that there is a lack of waste examples; thus, the schema constructed is not based on proper awareness. In other words, teachers can define WM accurately, indicate its importance, and talk about waste-related global effects, but they cannot, for example, state the same effects considering nuclear wastes.

6 Conclusions

In summary, there is variation in primary school teachers' awareness of waste and WM, except for the waste definition, where participants offered almost the same explanations. The variation shows that these primary school teachers differ in their discernment of waste and WM aspects, including misconceptions. The categories depend on hierarchical relations defined as low awareness, awareness, and high awareness. Awareness of waste is based on the waste definition and waste examples, while awareness of WM includes the WM definition, WM importance, WM methods, and WM responsibility. The WM awareness model, which constitutes conceptions, misconceptions, and interrelations among them, is asserted to represent the waste and WM issues and to feature the complexity of WM awareness.

7 Implications and Recommendations

The present study contributes significantly to the literature of WM awareness from the learning/teaching perspective (Hošková 2009; Hoskova-Mayerova and Rosicka 2015). It is important to improve teachers' awareness of waste- and WM-related issues. Teachers should be aware that every effort that contributes to environmentally sustainable solutions and contributions is valuable. Otherwise, any doubt may result in unsustainable action. Teachers' sincere actions may be rewarded to motivate their awareness. If people are not knowledgeable of their actions, they will not be proud of themselves and will neither repeat nor insist on environmentally friendly actions. This idea agrees with the proposal by Olgyaiova et al. (2018). According to these researchers, following Maslow's model will benefit environmental protection because it states the importance of personal needs and motivation. Educators such as teachers should enhance waste and WM awareness to increase knowledge and remove misconceptions about these issues. They should improve the state of responsibility of teacher candidates; teachers should accept more responsibility for WM issues. The lack of awareness and the misconceptions held by teachers prevent them from creating environmentally friendly students and future waste managers. Additionally, it leaves teachers to rely on clichéd remarks and limited considerations of environmental decision-making instances in their lives.

In the North Cyprus context, the lack of NGOs dealing with environmental issues, the lack of teachers' awareness regarding such organizations and especially waste issues, and the lack of cooperative studies among teachers and both governmental and NGO entities of North Cyprus show an absence of support for fostering waste awareness. The interrelation among such organizations and school employees, particularly primary school teachers, should consciously be strengthened through activities.

The study reveals that there is a variation regarding the awareness of aspects of WM. Thus, one cannot classify learners as simply belonging to one of the categories/stages, as in the above-mentioned case in Partanen-Hertell et al. (1999); the variation must be considered while teaching.

Accepting waste problems as a local problem, global problem, or both may lead to different consequences for education. For example, each idea can result in individuals who step up to the plate to varying degrees. Further research to determine whether and how environmental problems are discerned as local, global, or both can be beneficiary for environmental education.

References

Aid Programme for the Turkish Cypriot Community: https://ec.europa.eu/info/sites/info/files/annex_c_2013_7969_tcc_2013_programme_en.pdf. Accessed on 15 Feb 2018

Ajzen, I.: Laws of human behaviour: symmetry, compatibility, and attitude behavior correspondence. In: Beauducel, A., Biehl, B., Bosniak, M., Conrad, W., Schönberger, G., Wagener, D. (eds.) Multivariate Research Strategies, pp. 3–19. Shaker Verlag GmbH, Herzogenrath, Germany (2005). https://www.czp.cuni.cz/czp/images/stories/Vystupy/Seminare/2010%20Theory%20of%20Planned%20Behavior/Ajzen_2005ii.pdf. Accessed on 15 Feb 2018

Alkan, I.: Kuzey Kıbrıs Türk Cumhuriyeti' nde atık yönetimi [Waste Management in Turkish Republic of Northern Cyprus]; Paper Presented 12, İTU Endüstriyel Kirlenme Kontrolü Sempozyumu, 2010 İstanbul. http://www.academia.edu/25949683/Kuzey_K%C4%B1br%C4%B1s_T%C3%BCrk_Cumhuriyetinde_At%C4%B1k_Y%C3%B6netimi. Accessed on 8 Apr 2018

Australian Government Productivity Commission: Waste Management: Productivity Commission Inquiry Report (No. 38); Australian Government Productivity Commission, Canberra, Australia (2006). https://www.pc.gov.au/__data/assets/pdf_file/0014/21614/waste.pdf. Accessed on 11 June 2017

Baxter, J., Ho, Y., Rollins, Y., Maclaren, V.: Attitudes toward waste to energy facilities and impacts on diversion in Ontario, Canada. Waste Manag. **50**, 75–85 (2016). https://doi.org/10.1016/j.wasman.2016.02.017

Baycu, G., Tolunay, D., Ozden, H., Csatari, I., Karadag, S., Agba, T., Rognes, S.E.: An abandoned copper mining site in Cyprus and assessment of metal concentrations in plants and soil. Int. J. Phytoremed. **17**, 622–631 (2015). https://doi.org/10.1080/15226514.2014.922929

Baysen, E., Baysen, F.: Turkish prospective kindergarten teachers' conceptions concerning some selected atmospheric events. Int. J. Elem. Educ. **2**, 32–37 (2013). https://doi.org/10.11648/j.ijeedu.20130205.11

Baysen, E., Hošková-Mayerová, Š., Çakmak, N., Baysen, F.: Misconceptions regarding providing citations: to neglect means to take risk for future scientific research. In: Mathematical-Statistical Models and Qualitative Theories for Economic and Social Sciences. Studies in Systems, Decision and Control, vol. 104, pp. 177–186. Springer (2017). ISBN 978-3-319-54818-0. https://doi.org/10.1007/978-3-319-54819-7_12

Bekesiene, S., Hoskova-Mayerova, S.: Decision tree—based classification model for identification of effective leadership indicators in the Lithuania army forces. J. Math. Fund. Sci. **50**(2), 121–141 (2018)

Bhutta, M., Khurrum, S., Omar, A., Yang, X.: Electronic waste: a growing concern in today's environment. Econ. Res. Int. **2011**, 1–8 (2011). https://doi.org/10.1155/2011/474230

Borthakur, A., Govind, M.: Emerging trends in consumers' e-waste disposal behaviour and awareness: a worldwide overview with special focus on India. Res. Conserv. Recycl. **117**, 102–113 (2017). https://doi.org/10.1016/j.resconrec.2016.11.011

Carlsson, M., Fülüp, M., Marton, F.: Peeling the onion: student teachers' conceptions of literary understanding. Scand. J. Educ. Res. **45**, 5–18 (2001). https://doi.org/10.1080/00313830020023366

Cici, M., Şahin, N., Görgen, İ., Deniz, S.: Öğretmen adaylarının katı atık kirliliği bağlamında çevresel farkındalık ve bilgi düzeyleri. Eğitim Bilimleri ve Uygulama **4**, 37–50 (2005)

Couvidat, J., Benzaazoua, M., Chatain, V., Bouamrane, A., Bouzahzah, H.: Feasibility of the reuse of total and processed contaminated marine sediments as fine aggregates in cemented mortars. Construct. Build. Mater. **112**, 892–902 (2016). https://doi.org/10.1016/j.conbuildmat.2016.02.186

Cutter-Mackenzie, A.: Australian waste wise schools program: Its past, present, and future. J. Environ. Educ. **41**, 165–178 (2010). https://doi.org/10.1080/00958960903347471

Desa, A., Kadir, N.B.A., Yusoof, F.: A study on the knowledge, attitudes, awareness status and behavior concerning solid waste management. Proc. Soc. Behav. Sci. **18**, 643–648 (2010). https://doi.org/10.1016/j.sbspro.2011.05.095

Diaz Silvarrey, L.S., Phan, A.N.: Kinetic study of municipal plastic waste. Int. J. Hydrog. Energy **41**, 16352–16364 (2016)
Eriksson, M., Osowski, C.P., Malefors, C., Björkman, J., Eriksson, E.: Quantification of food waste in public catering services—a case study from a Swedish municipality. Waste Manag. **61**, 415–422 (2017). https://doi.org/10.1016/j.wasman.2017.01.035
Ernst, J.: Factors associated with K–12 teachers' use of environment-based education. J. Environ. Educ. **38**, 15–32 (2007). https://doi.org/10.3200/JOEE.38.3.15-32
European Parliament and Council of the European Union: Directive 2008/98/EC of the European Parliament and of the Council of 19 November 2008 on Waste and Repealing Certain Directives (Text with EEA Relevance); European Parliament and Council of the European Union, EU (2008) http://eur-lex.europa.eu/legal-content/EN/ALL/?uri=celex:32008L0098. Accessed on 10 Mar 2017
Ferreira, J.A.; Ryan, L.; Tilbury, D.: Whole-School Approaches to Sustainability: A Review of Models for Professional Development in Pre-Service Teacher Education; Australian Government Department of the Environment and Heritage and the Australian Research Institute in Education for Sustainability (ARIES): Canberra, Australia (2006). http://aries.mq.edu.au/projects/preservice/files/TeacherEduDec06.pdf. Accessed on 11 Jan 2017
Fortner, R.W.: Understanding # 2: there must be a-way (to deal with solid waste). Sci. Act. **28**, 16–18 (1991). https://doi.org/10.1080/00368121.1991.9958087
Franklin, B.A., Brook, R., Arden Pope, C.: Air pollution and cardiovascular disease. Curr. Probl. Cardiol. **40**, 207–238 (2015). https://doi.org/10.1016/j.cpcardiol.2015.01.003
Funtowicz, S.O., Ravetz, J.R.: A new problem-solving strategy for global environmental issues. Nat. Forum **70**, 38–41 (1990)
Glažar, S.A., Vrtačnik, M., Bačnik, A.: Primary school children's understanding of municipal waste processing. Environ. Educ. Res. **4**, 299–308 (1998). https://doi.org/10.1080/1350462980040305
Gould, R.K., Ardoin, N.M., Biggar, M., Cravens, A.E., Wojcik, D.: Environmental bahaviour's dirty secret: the prevalence of waste management in discussions of environmental concern. Environ. Manag. **58**, 268–282 (2016). https://doi.org/10.1007/s00267-016-0710-6
Grodzinska-Juczka, M., Bartosiewicz, A., Twardowska, A., Ballantyne, R.: Evaluating the impact of a school waste education programme upon students' parents' and teachers' environmental knowledge, attitudes and behaviour. Int. Res. Geogr. Environ. Educ. **12**, 106–122 (2003). https://doi.org/10.1080/10382040308667521
Gündüz, S., Erbulut, C., Öznacar, B., Baştaş, M.: Determination of consciousness and awareness of the public in Lefka about the Cyprus mining corporation (CMC). Eur. J. Math. Sci. Technol. Educ. **12**, 783–792 (2016). https://doi.org/10.12973/eurasia.2016.1256a
Hasan, S.E.: Public awareness is key to successful waste management. J. Environ. Sci. Health **39**, 483–492 (2012)
Hošková, Š.: Experience with blended (distance) learning study materials. In.: Conference Distance Learning, Simulation and Communication (DLSC), pp. 70–77. CATE, Brno (2009). ISBN: 978-80-7231-638-0
Hoskova-Mayerova, S., Rosicka, Z.: Programmed learning. Proc. Soc. Behav. Sci. **31**, 782–787 (2012). https://doi.org/10.1016/j.sbspro.2011.12.141
Hoskova-Mayerova, S., Rosicka, Z.: E-learning pros and cons: active learning culture? Proc. Soc. Behav. Sci. **191**, 958–962 (2015). https://doi.org/10.1016/j.sbspro.2015.04.702
Ilie, E.M., Tincu, C.: The vision of urban inhabitants on local environmental problems. Valahian J. Econ. Stud. **7**, 23–30 (2016)
Jean-Sigur, R., Bell, D., Kim, Y.: Building global awareness in early childhood teacher preparation programs. Child. Educ. **92**, 3–9 (2016). https://doi.org/10.1080/00094056.2016.1134235
Jiang, G., Pickering, S.J.: Recycling supercapacitors based on shredding and mild thermal treatment. Waste Manag. **48**, 465–470 (2016). https://doi.org/10.1016/j.wasman.2015.10.027
Karatekin, K.: Meray. Attitudes of preservice social studies teachers towards solid waste and recycle. Bayburt Univ. Eğitim Fakültesi Dergisi **10**, 297–314 (2015)

Karatekin, K.: Social studies pre-service teachers' awareness of solid waste and recycling. In: Proceedings of the 5th World Conference on Educational Sciences, Rome, Italy, 5–8 Feb 2013

Khan, D., Kumar, A., Samadder, S.R.: Impact of socioeconomic status on municipal solid waste generation rate. Waste Manag **49**, 15–25 (2016). https://doi.org/10.1016/j.wasman.2016.01.019

Krook, J., Svensson, N., Eklund, M.: Landfill mining: a critical review of two decades of research. Waste Manag. **32**, 513–520 (2012). https://doi.org/10.1016/j.wasman.2011.10.015

Kyburz-Graber, R.: Environmental education as critical education: how teachers and students handle the challenge. Camb. J. Educ. **29**, 415–432 (1999). https://doi.org/10.1080/0305764990290310

Lewis, B.: Manager characteristics in departments supplying a typical urban service: waste collection. Publ. Adm. Q. **15**, 210–223 (1991)

Lipinski, B., Hanson, C., Lomax, J., Kitinoja, L., Waite, R., Searchinger, T.: Reducing Food Loss and Waste: Installment 2 of Creating a Sustainable Food Future; Working Paper; World Resources Institute, Washington, DC, USA (2013). http://pdf.wri.org/reducing_food_loss_and_waste.pdf. Accessed on 11 June 2017

Littledyke, M.: Science education for environmental awareness: approaches to integrating cognitive and affective domains. Environ. Educ. Res. **14**, 1–17 (2008). https://doi.org/10.1080/13504620701843301

Macias, A., Piniarski, W.: Municipal solid waste management problems on a local scale: a case study from rural Poland. Pol. J. Environ. Stud. **25**, 1623–1632 (2016). https://doi.org/10.15244/pjoes/62111

Mahrum, L.: Solid Waste Management Problems in North Cyprus and Its Possible Solutions (Unpublished Master Thesis). The University of Manchester, Manchester, UK (2009)

Makisa, K.: Assessment of integrated environmental management in public and private school in the Copperbelt Province of Zambia. J. Educ. Pract. **7**, 102–109 (2016)

Malandrakis, G.N.: Children's understandings related to hazardous household items and waste. Environ. Educ. Res. **14**, 579–601 (2008). https://doi.org/10.1080/13504620802465046

Mamady, K.: Factors influencing attitude, safety behavior, and knowledge regarding household waste management in Guinea: a cross-sectional study. J. Environ. Publ. Health **2016**, 1–9 (2016). https://doi.org/10.1155/2016/9305768

Marshall, R.E., Farahbakhsh, K.: Systems approaches to integrated solid waste management in developing countries. Waste Manag. **33**, 988–1003 (2013). https://doi.org/10.1016/j.wasman.2012.12.023

Martinez-Borreguero, G., Maestre-Jimenez, J., Mateos-Nuenz, M., Naranjo Correa, F.L.: Knowledge analysis of the prospective secondary school teacher on a key concept in sustainability: Waste. Sustainability **11**, 1–28 (2019). https://doi.org/10.3390/su11041173

Marton, F.: Phenomenography: a research approach to investigating different understandings of reality. J. Thought **21**, 28–49 (1986)

Marton, F., Booth, S.: Learning and Awareness. Lawrence Erlbaum Associates, Mahwah, NJ, USA (1997)

McDougall, F.R.; White, P.R.; Franke, M.; Hindle, P.: Integrated Solid Waste Management: A life Cycle Inventory, 2nd edn. Blackwell Publishing Company, Oxford, UK (2001). ISBN: 978-0-632-05889-1

Morrissey, A.J., Browne, J.: Waste management models and their application to sustainable waste management. Waste Manag. **24**, 297–308 (2004). https://doi.org/10.1016/j.wasman.2003.09.005

Najafy, M.: The Evaluation of Construction Waste Management in Northern Cyprus (Unpublished Master Thesis). Eastern Mediterranean University, Famagusta, North Cyprus (2014)

National Environmental Protection Agency: http://www.cevrekorumadairesi.org/. Accessed on 10 Mar 2017

Negev, M., Garb, Y., Biller, R., Sagy, G., Tal, A.: Environmental problems, causes, and solutions: an open question. J. Environ. Educ. **41**, 101–115 (2010). https://doi.org/10.1080/00958960903295258

Noorhosseini, S.A., Allahyari, M.S., Damalas, C.A., Mogaddam, S.S.: Public environmental awareness of water pollution from urban growth: the case of Zarjub and Gohharud rivers in Rasht. Iran. Sci. Total Environ. **599–600**, 2019–2025 (2017). https://doi.org/10.1016/j.scitotenv.2017.05.128
Olander, M.H., Nyberg, E.: Learning study by variation theory: exemplified by children learning to halve and double whole numbers. J. Res. Child. Educ. **28**, 238–260 (2014). https://doi.org/10.1080/02568543.2014.884030
Olgyaiova, K., Pongrácz, E., Mikkola, T., Skapa, R., Keiski, R.L.: Attitudes toward waste minimization in Finland and Czech. Republic: Barriers and Drivers; Paper Presented Proceedings of the RESOPT Closing Seminar 'Waste Minimization and Utilization in Oulu region: Drivers and Constraints'. http://citeseerx.ist.psu.edu/viewdoc/download?doi=10.1.1.584.446&rep=rep1&type=pdf. Accessed on 15 Feb 2018
Orlins, S., Guan, D.: China's toxic informal e-waste recycling: local approaches to a global environmental problem. J. Clean. Prod. **114**, 71–80 (2016). https://doi.org/10.1016/j.jclepro.2015.05.090
Palmer, J.A., Suggate, J., Bajd, B., Tsaliki, E.: Significant influences on the development of adults' environmental awareness in the UK, Slovenia and Greece. Environ. Educ. Res. **4**, 429–444 (1998)
Palmer, J.A., Suggate, J., Robottom, I., Hart, P.: Significant life experiences and formative influences on the development of adults' environmental awareness in the UK, Australia and Canada. Environ. Educ. Res. **5**, 181–200 (1999)
Partanen-Hertell, M., Harju-Autti, P., Kreft-Burman, K., Pemberton, D.: Raising Environmental Awareness in the Baltic Sea Area; The Finnish Environment Institute: Helsinki, Finland (1999). https://helda.helsinki.fi/bitstream/handle/10138/40477/FE_327.pdf?sequence=29. Accessed on 13 Apr 2017
Pearson, R.G., Dowson, T.P.: Predicting the impacts of climate change on the distribution of species: are bioclimate envelope models useful? Glob. Ecol. Biogeogr. **12**, 361–371 (2003)
Pohjola, V.J., Pongrácz, E.: An approach to the formal theory of waste management. Res. Conserv. Recycl. **35**, 17–29 (2002)
Pongrácz, E.: Through waste prevention towards corporate sustainability: analysis of the concept of waste and a review of attitudes towards waste prevention. Sustain. Dev. **17**, 92–101 (2009). https://doi.org/10.1002/sd.402
Pongrácz, E.; Phillips, P.S.; Keiski, R.L.: Evolving the theory of waste management—implications to waste minimization. Paper Presented in Proceeding Waste Minimization and Resources Use Optimization Conference. University of Oulu, Oulu, Finland (2004). file:///C:/Users/User/Downloads/Evolving_the_Theory_of_Waste_Management. Accessed on 11 June 2017
Priefer, C., Jörissen, J., Brautigam, K.R.: Food waste production in Europe—a cause-driven approach to identify the most relevant leverage points for action. Resour. Conserv. Recycl. **109**, 155–165 (2016). https://doi.org/10.1016/j.resconrec.2016.03.004
Pullishery, F., Panchmal, G.S., Siddique, S., Abraham, A.: Awareness, knowledge and practices on bio-medical waste management among health care professionals in Mangalore—a cross sectional study. Int. Arch. Integr. Med. **3**, 29–35 (2016)
Quigely, C.F., Dogbey, J., Che, S.M., Hallo, J.: Shared understandings: environmental perspectives of Kenyan community members and teachers. Environm. Educ. Res. **21**, 1079–1104 (2015). https://doi.org/10.1080/13504622.2014.966660
Ramayah, T., Lee, J.W.C., Lim, S.: Sustaining the environment through recycling: an empirical study. J. Environ. Manag. **102**, 141–147 (2012)
Reed, B.I.: Phenomenography as a way to research the understanding by students of technical concepts. In: Nucleo de Pesquisa em Technologia da Arquitetura eUrbanismo (NUTAU): Technological Innovation and Sustainability; NUTAU: Sao Paulo, Brazil (2006). http://www.mecheng.uct.ac.za/usr/mecheng/staff/academic/brandon/Reed2006.pdf. Accessed on 11 Jan 2017
Rich, R.C., Edelstein, M., Hallman, W.K., Wandersman, A.H.: Citizen participation and empowerment: the case of local environmental hazards. Am. J. Commun. Psychol. **23**, 657–676 (1995)

Rootes, C.: Acting globally, thinking locally? Prospects for a global environmental movement. Environ. Polit. **8**, 290–310 (1999)

Saikkonen, P.: From unknown problem to an acknowledged local environmental problem. A case of polluted soil in the city of Helsinki. Local Environ. **18**, 888–903 (2013)

Sudamadi, S., Suzuki, S., Kawada, T., Soemantari, S., Tugaswati, A.: A survey of perception, knowledge, awareness, and attitude in regard to environmental problems in a sample of two different social groups in jakarta, Indonesia. Environ. Dev. Sustain. **3**, 169–183 (2001)

Thyberg, K.L., Tonjes, D.J., Gurevitch, J.: Quantification of food waste disposal in the United States: a meta-analysis. Environ. Sci. Technol. **49**, 13946–13953 (2015). https://doi.org/10.1021/acs.est.5b03880

Tilbury, D.: Sustaining curriculum policy and development in environmental education: a European action research project. Eur. Environ. **9**, 24–34 (1999). https://doi.org/10.1002/(SICI)1099-0976(199901/02)9:1%3c24:AID-EET181%3e3.0.CO;2-O

Triassi, M., Alfano, R., Illario, M., Nardone, A., Caparole, O., Montuori, P.: Environmental pollution from illegal waste disposal and health effects: a review on the "triangle of death". Int. J. Environ. Res. Publ. Health **12**, 1216–1236 (2015). 10.1.1.696.7045&rep=rep1&type=pdf

Triguero, A., Alvarez-Aledo, C., Cuerva, M.C.: Factors influencing willingness to accept different waste management policies: empirical evidence from the European Union. J. Clean. Prod. **138**, 38–46 (2016). https://doi.org/10.1016/j.jclepro.2016.05.119

Trigwell, K.: Phenomenography: an approach to research into geography education. J. Geogr. High. Educ. **30**, 367–372 (2006). https://doi.org/10.1080/03098260600717489

TurkishCyprus.com: www.turkishcyprus.com/political-structure-ngos.html. Accessed on 11 Jan 2017

Urban, R., Hoskova-Mayerova, S.: Threat life cycle and its dynamics. Deturope **9**(2), 93–109 (2017)

Vallejo Andrada, A., Hošková-Mayerová, Š., Krahulec, J., Sarasola Sanchez-Serrano, J.L.: Risks associated with reality: how society views the current wave of migration; one common problem—two different solutions. Stud. Syst. Dec. Control **104**, 283–305 (2017). https://doi.org/10.1007/978-3-319-54819-7_19

Westing, A.H.: Education: the global need for environmental education. Environ. Sci. Policy Sustain. Dev. **35**, 4–45 (1993). https://doi.org/10.1080/00139157.1993.9929988

Wong, T.K.-Y., Wan, P.-S.: Lingering environmental pessimism and the role of government in Hong Kong. Publ. Adm. Dev. **29**, 441–451 (2009). https://doi.org/10.1002/pad.530

World Bank: What a Waste: A Global Review of Solid Waste Management. Urban Development and Local Government Unit, Washington, DC, USA (2012)

Yorucu, V., Keles, R.: The construction boom and environmental protection in Northern Cyprus as a consequence of the Annan Plan. Construct. Manag. Econ. **25**, 77–86 (2007). https://doi.org/10.1080/01446190600902356

Zachariou, A., Kadji-Beltran, C.: Cypriot primary school principals' understanding of education for sustainable development key terms and their opinions about factors affecting its implementation. Environ. Educ. Res. **15**, 315–342 (2009). https://doi.org/10.1080/13504620902862902

Zaman, A.U.: A comprehensive study of the environmental and economic benefits of resource recovery from global waste management systems. J. Clean. Prod. **124**, 41–50 (2016). https://doi.org/10.1016/j.jclepro.2016.02.086

Zorpas, A.A., Voukkali, I., Loizia, P.: Effectiveness of waste prevention program in primary students' schools. Environ. Sci. Pollut. Res. **24**, 14304–14311 (2017). https://doi.org/10.1007/s11356-017-8968-7

Phenomenographic Approach to Teachers' Wait-Time Use: Reasons and Consequences

Engin Baysen, Fatma Baysen and Nermin Çakmak

Abstract Researchers put importance on time management, specifically wait-time use in Questioning and Answering while teaching. The purpose of this study is to inquire wait-time use of primary school teachers for the present. Phenomenography is the research approach adopted for this study. The study showed variation in teachers' experiences utilizing wait-time in teaching. We captured teachers' wait-time use in four categories. The categories are the utilization of, Short wait-time I; Extended wait-time I; Extended wait-time I & II; Sophisticate wait-time I & II. Reasons and consequences of wait-time use also emerged accompanying the categories. The present study asserts that we can come up with the most contemporary educational targets through partially by extended wait-time I & II utilization and greatly by sophisticated wait-time I & II use.

Keywords Communication · Questioning and answering · Phenomenography · Primary school teachers · Time management · Wait-time

1 Introduction

Time is an exhaustible but invaluable resource. Human should govern it apprehensively, but its management is peculiar to its proprietor (Çakmak 2016; Hoskova-Mayerova 2016). Each may ascribe a different meaning to time and use it differently. Time management is the prominent aspect of the teaching-learning (Hildenbrand and

E. Baysen (✉) · F. Baysen
Ataturk Education Faculty, Near East University, Mersin, Turkey
e-mail: engin.baysen@neu.edu.tr

F. Baysen
e-mail: fatma.baysen@neu.edu.tr

N. Çakmak
Department of Information and Document Management, Faculty of Letters, Ataturk University, Erzurum, Turkey
e-mail: nermin.cakmak@atauni.edu.tr

D. Flaut et al. (eds.), *Decision Making in Social Sciences: Between Traditions and Innovations*, Studies in Systems, Decision and Control 247,
https://doi.org/10.1007/978-3-030-30659-5_12

Arndt 2016); teachers try not to waste time on non-productive activities. The activities should be well-planned and followed carefully distributing the time efficiently, maintaining a fair amount of time for each particular learning processes. One crucial process covering class time is questioning and answering (Q/A). Teachers use Q/A for many reasons including evaluation, encourage students to participate in learning, support meaningful learning, create debates, divert the flow of learning. Teachers should also manage the time skilfully they allocate for Q/A. The present study considered how the teachers utilize the time they reserve for Q/A in real classes. In this context, the present study focused on a time management strategy in Q/A during teaching-learning activities, wait-time (Rowe 1972; Colburn 2000; Baysen 2003, p. 178; Hošková-Mayerová and Maturo 2018, 2019).

1.1 Wait-Time Definition

Wait-time is the organization of pauses, during the class communications, managed by the teachers in teaching-learning context, which was foregrounded to the research firstly by Mary Budd Rowe. Rowe defines wait-time as the pause applied after a teacher asks a question and after a student answers this question (Rowe 1972). According to Rowe (1986), there are two different types of wait-time used in classes, Wait-time I and Wait-time II. Wait-time I refer the duration teachers waits after they ask a question and student responses to this question. Wait-time II, on the other hand, is the time the instructor waits before the same student, or another student takes the floor after the first response to the question.

1.2 Wait-Time Literature

Researchers conducted an extensive amount of study regarding the state of wait-time use; Beliefs about wait-time; Students' attitudes to wait-time; Difficulty in implementing wait-time; and Favourable/unfavorable experiences to wait-time use. It seems that there are two sides of approaches to wait-time use in teaching. Those supporting and those not. The present study did not interest which group is more populated, but focused wait-time use variation by teachers, as is the tenant for phenomenography.

1.3 State of Wait-Time Use

Many research carried out to reveal the duration of wait-time utilization in a teaching context. Starting with Rowe, many researchers to date showed that teachers allowed an average of 1 s of wait-time which defined as *short* (Rowe 1972; Tobin 1980; Jegede and Olajide 1995; Baysen and Baysen 2010; Bilaloglu et al. 2017). Nevertheless, short wait-time is not the case for all research carried out. About 90% of fourth and fifth- grade teachers in California said to use wait-time as a strategy in practice (Bowers et al. 2010).

1.4 Beliefs About Wait-Time

Beliefs play a significant role for teachers in deciding to use or not to use any method or strategy to their teaching (Holmqvist 2011). Many types of research revealed beliefs regarding wait-time usage. Teachers' wait-time application depends on class population and student needs (Devine and McGillicuddy 2016). Boeck and Hillenmeyer (1973) found that wait-time depends on the type of teacher questions asked. Wait-times should be longer for those high-level. Another aspect of wait-time usage belief is teacher satisfaction. A significant finding to wait-time is that being aware of the need to use wait-time (Grant and Kline 2010) does not guarantee to be satisfied with the performance (Bergman 2015). A preservice teacher swears that s/he should 'rush' so to reserve time for students' activities (Bergman 2015).

1.5 Students' Attitudes to Wait-Time

Students prefer extended wait-time (Bayerbach 1988; Altiere and Duell 1991; Kirton et al. 2007). On the contrary Samiroden (as cited in Altiere and Duell 1991) found that Biology students believe that wait-time consumes class time. In a recent study, King and Aono talked about situational silence anxiety, caused by using wait-time (King and Aono 2017). They found that using silent time may cause discomfort and anxiety to undergraduate students. However, one other study revealed interestingly thinking of 'silent students' (Ghavamani and Ketabi 2015). According to a quiet student, they do not answer because if they try to do so, that will take a long time and the teacher would go to other faster students, so they eventually forgot about answering. For them, this became a kind of habit.

1.6 Difficulty in Implementing Extended Wait-Time

Rowe (1986) stated that extending wait-time is tough. Similarly, Honea (1982) reported that even trying to utilize wait-time cause's teacher anxiety. Additionally, being nervous cause teachers to speak fast not using wait-time (Bergman 2015). Ebadi and Gheisari (2016) stated that teachers might not be patience, reported as a habit, preventing wait-time use. Contradicting Baysen et al. (2003) found that teachers could easily increase wait-time from 2 to 3 s to an average of 20 s solely by encouragement.

1.7 Favorable/Unfavorable Experiences to Wait-Time Use

Positive experiences are found throughout the research regarding the issue. Barnette et al. (1995) found that teachers participated in a program to improve questioning skills showed significant gains regarding wait time knowledge and utilization. There is a vast number of (experimental) research revealing the success of extended wait-time (Rowe 1972; Tobin 1980; Riley 1986; Jegede and Olajide 1995; Baysen et al. 2003; Hoskova-Mayerova 2014; Ebadi and Gheisari 2016; Cardimona 2018). Mary Budd Rowe categorized two kinds of benefits depending on wait-time usage, three to five seconds. She classified as benefits of students and benefits of teachers. She listed the benefits for students as Increase in student responses; Increases in unsolicited but appropriate responses; Decreases in failures; Increased student confidence; Increased in the incidence of speculative thinking. Adding a transition from teacher-centered to student-centered; Increase in evidence-based responses; Increase in student questions and experiments proposed; Increase in "slow student" contribution. Rowe listed benefits for teachers utilizing wait-time in their classes: Exhibit greater response flexibility; Decrease in the number of questions (leaving its place to student communication); Increase in quality questions asked; Change in student performance expectations.

Some other researchers showed contradicting results. For example, Duell (1995) found that extended wait-time did not cause improvement. In another study, Duell (1994) discovered that university students did not benefit from extended wait-time. Kirton et al. (2007) found that extended wait-time may cause boredom for some students. Ebadi and Gheisari (2016) carried out an in-depth qualitative study which was participated by an English as a Foreign Language teacher brought to light reasons why teachers do not prefer to use wait-times in their courses. The main argument is that time is crucial, and we should use it carefully. The teacher should speak more for better learning, such as repeating words, an essential issue for teaching foreign language learning. Otherwise, students use the time gap for mistakenly used words and mispronunciations.

Summarizing it seems that we live a dilemma concerning wait-time use. One side (mostly theoretical) says that it is crucial and must be utilized while the other side (mostly practical) stops the first because in practice wait-time causes trouble.

1.8 Theories of Communication & Wait-Time

Researchers tried to draw patterns of communication. Those considering aspects of verbal communication but excluding issues of media presented in Table 1.

Some researchers interrelate the wait-time use mainly activating theories. Altiere and Duell related the success of wait-time utilization based on information processing model. This approach asserts that learners need time to process information (Altiere and Duell 1991). Jegede and Olajide (1995) related wait-time usage with sociocultural influences: "authoritarianism, goal structure, societal expectation, and traditional worldview." Collecting the notions, the goals of the dense program and others' (shareholders of education) expectations (teach as much knowledge as possible) restricts teachers' time. Teachers behave in a way not permitting students to use precious class-time wondering around the target knowledge, and thus it is wise to reserve the way the teachers themselves are taught, which admittedly is the right way!

Table 1 Theories/models of communication

Theory/model	Assertion	References
Cognitive dissonance theory	Avoid opposing ideas	Matz and Wood (2005)
Communication accommodation theory	Adjust the style of speaking for the audience	Buller and Aune (1992)
Coordinated management of meaning	People create meaning together in a conversation	Brenders (1987)
Groupthink phenomenon	Group members think similarly	Kameda and Sugimori (1993)
Relational dialectics	Wishing to have both contradicting things	Randolph and Holtzman (2010)
Rhetorical theory	To use available means to persuade audience	Bruss (2009)
Social exchange theory	Satisfaction of both people's self-interest is necessary	Ribarsky (2013)
Social penetration theory	People are in particular social standpoints	Edwards (2014)
Structuration theory	Groups create structures which are group's rules	McGarry (2016)
Symbolic interaction theory	People act based on the meanings they assign to people	Aksan et al. (2009)

1.9 Phenomenography

A phenomenography is a research approach to reveal people's variations of conceptions and experiences regarding any phenomenon. Phenomenography considers complete data gathered from a sample of a particular population as one. Phenomenography does not intend to reveal each person's set of experiences, but a collection of all experiences of the sample participated in the research. It interprets the experiences in distinctly different, small and limited in number and interrelated categories. The relation between the groups are mostly hierarchical (Marton and Booth 1997).

1.10 The Purpose

The vagueness about wait-time use discussed above triggered the present research to inquire wait-time issues in more depth. The present study aimed to reveal and discuss the variation in wait-time use among teachers, reasons to wait-time use, and its consequences. The study tried to answer some particular questions: (1) To what degree and quality do the teachers use wait-time in their classes? (2) Is there any halfway taken, did we make headway in wait-time use? (3) Should we categorize teachers only in two: a. Those not using the wait-time at all and b. Those using the wait-time? (4) How do the teachers use wait-time? (5) What are the consequences of utilizing wait-time? (6) Are the theories employed in explaining communication and the wait-time enough? Alternatively, (7) Should we apply for other ideas or adapt to those already employed?

The current researchers believed that many shareholders for education would enjoy the study. The present study will make teachers aware of their wait-time use. This awareness will help them improve their teaching. This research will also promote researchers and programmers. They will understand the current situation for wait-time use. They can base aspects of future in-service courses for improving teachers' Q/A, particularly wait-time use skills.

2 Method

Researchers chose phenomenography, for the current study, because they predict that teachers will indeed state to experience wait-time in their classes to a different extent. Following sections presented will carry aspects of phenomenography.

2.1 Participants

Twenty Turkish primary school teachers participated in the research voluntarily. We made the purposeful sampling as is the case in most qualitative research. Primary school teachers participated in the study because mingling with students they are the most fruitful source for this study. Heterogeneity in sampling was considered to reach variation in experiences, as is critical for phenomenography. We provided heterogeneity in gender, grades taught, and seniority in the profession. Thirteen of the participants were women, and seven were men. Teachers attending the study have job experiences in between 6 and 35 years. Most of the teachers said to teach grades 1–5 at least once, while others stated to have taught only specific grades, such as grades three, four, and five only.

2.2 Data Collection

Data was collected through one to one and face to face interviews lasting 20–40 min. The data collected was accepted as teachers' real wait-time utilization in class. Researchers tried to eliminate any suspicion through probing questions. The interviews were carried out in the investigator's premises and at participants' schools. The researcher introduced questions regarding wait-time use (Appendix) to interviewees. The researcher required the participants to do their best to reckon the wait time they utilize in questioning- answering sessions. The teachers put in extra effort to remember how much they wait during questioning/answering sessions. During the interview some of the teachers imagine those experiences, counting seconds to reach the correct wait-time they use. The teachers were also encouraged to remember to elucidate different or unusual experiences as well. The dialogs between the researcher and the interviewee were recorded by a sound recorder, transcribed verbatim and prepared for analysis.

2.3 Data Analysis

Researcher carried out phenomenographic data analysis adopted from Marton and Booth (1997). According to the phenomenographic analysis, first, all the material collected forms a pool of meaning. The analysis continues with choosing one or a few aspects of the data and neglecting the others. We focused on the preferred aspect to reveal its variation. We iteratively read the data. Its tenant that no one can be aware of everything simultaneously and with the same reality, we accepted the decision changes through the readings. Thus each reading started with the choices we gave at the previous reading. Reaching the categories of the description which capture different ways of experiencing we finally reach outcome space, constituted by these categories of description (Marton and Booth 1997; Asplund-Carlsson et al. 2001). Marton and Booth (1997, p. 125) stated three criteria for the phenomenographic approach which we apply in the present study. These standards ensure the quality of categories captured. The first one is to assert different types. The types must be separate from each other in a distinctive way but should consider every experience manifested related to the issue under investigation. The second criterion is to construct relations with each category. This relationship is mostly hierarchical; that is one type of the higher rank should constitute those underneath. Lastly, participants' experiences should be grouped in as few as groups as possible. This variation should be logical and critical.

3 Results

We captured teachers' wait-time use experiences in two broad categories: Traditional and Contemporary. We arrived Traditional Approach in two sub-categories, Short Wait-time I and Extended Wait-time I. We came up with contemporary utilization in two categories. These are Extended Wait-time I & II and Sophisticated Wait-time I & II. There is a hierarchic relation in between the main and among the sub-categories. A more particular relation between the categories including concepts with examples is shown in Table 2. The following sections include the categories and accompanying reasons and consequences (R/C) for each category.

3.1 Short Wait-Time I—R/C

In this category, teachers stated to the need to use wait-time during class discussions. They guessed that they wait for less than a second and this much is satisfying. Thus this category defines for being aware of the wait-time duration used and may state its importance but not use it as is proposed. One other important finding for this type is that the teachers are not aware of differences of using distinct wait-times, wait-time I and wait-time II.

Whenever asked teachers said that they could not wait longer, they stated a variety of reasons for not using wait-time. They caused that they must catch up on the program. A teacher openly states this reason:

> I cannot wait for much. Otherwise, I cannot finish subjects.

One other teacher said her anxiety regarding expanding the wait-time:

> If I wait then the class disrupts…Everyone starts to shout, even move, displace to another position or walk in the class.

Table 2 Teachers' wait-time utilization

Traditional				Contemporary			
Short wait-time I		Extended wait-time I		Extended wait-time I & II		Sophisticated wait-time I & II	
Concept	Example	Concept	Example	Concept	Example	Concept	Example
Teacher talk more	No time for student talk/discussion	Teacher talk more	No extra time for student talk/discussion	Students talk more than in traditional	Students React each other's responses	Share-holders share ideas	Shareholders distribute the time evenly, each add to each other
Teacher teaches	Teacher gives lectures	Teacher teaches	Teacher gives lectures	Teacher facilitates learning	Students affect each other's learning	Teacher facilitator, Students learn together	Learn how to learn and from others
Teacher assessment	Teacher responds	Teacher assessment	Teacher responds	Students assess each other	Students react to each other	Share-holders assess each other	Shareholders assess each other in team work
Teacher asks questions /poses problems	Student solve the questions (poses problems) by them-selves	Teacher asks questions/poses problems	Students solve the ques-tions/problems by them-selves	Student ask questions/poses problems as well	Students solve the ques-tions/problems helping each other	Act together to a ques-tion/problem posed by share-holders	Shareholders contribute to a question (problem) solution
Group barriers	Students make fun of each other	Group barriers	Students make fun of each other	Belongingness	Respect each other	Self-actualization	Respect each other and other's ideas
Loneliness	Every student should learn themselves	Loneliness	Every student should learn them-selves	Helpful/peer Learning	Students help each other in learning	Cooperative learning	Class works like an organism

Adding teachers put importance to be fair to students, distributing the time evenly among the students. Considering highly populated classes this action creates a further limitation to time reserved for each student. A teacher stated:

> I should communicate with all the students.

One teacher said how fast she must be in questioning- responding sessions:

> I must serially ask them (in the first five minutes of each class time, to diagnose). Otherwise, some of them will feel neglected.

One teacher stated that they must be in a hurry in Q/A otherwise other students will get bored.

> They do not listen to others, and they become tired of hearing each other.

A teacher stated how difficult it is to activate some of her students to take part in discussions.

> "Some of my students are hopeless"… "They do not talk, but I ask those questions directly to activate them."

Summarizing, in this category teachers stated a lot of different and concrete reasons explaining why they cannot think of utilizing wait-time. Thus students find a small number of opportunity and short durations for responding in class caused by short wait-time.

3.2 Extended Wait-Time I—R/C

This category includes the use of extended wait-time I only. Teachers put importance to wait-time use. They stated to wait for more than a second before choosing a student to answer. A teacher reported:

> I have to wait 3–5 s and sometimes more before choosing a student to respond.

One other teacher said that she must give every student the opportunity to think about the question. Because the issue is complicated and, learners lack interest.

> …Some students lack interest you *have to* wait for them when you pause you see that the number of students to answer increases because the question is hard. Waiting is a good strategy to motivate those slow learners.

Summarizing, in this category teachers reasoned that some questions are hard to answer which deserves wait-time. In other words, teachers think that they should use wait-time. Otherwise, they could not understand whether any student understand what is being taught or not. Additionally, they use wait-time I to increase the number of students to respond, but not for discussions or high-level thinking as is aimed at the future contemporary utilization.

3.3 Extended Wait-Time I & II—R/C

In this category, teachers use extended wait-time I and wait-time II in Q/A sessions. They put importance to each wait-time type use. They stated to wait for more than a second before choosing a student to answer and wait for student responses. A teacher reported:

> After asking a question, I wait for some time for my students to raise their hands and then I choose one of them to answer. This wait-time is about 5–10 s.

Following the response wait-time II, is applied. One teacher, differentiating between the two types, said to use wait-time II:

> I wait for student's answer. I wait quite a lot of time. Sometimes I wait more than a minute. I know a lot about my students, so I predict what is going to happen after giving an opportunity to a specific student. For example, I know that student A is going to answer wrongly to that particular question. I use this strategy (wait-time II) to make others react to that student's answer politely. That student who has responded to the question, on the other hand, learns from his classmates.

3.4 Sophisticated Wait-Time I & II—R/C

This category includes using long wait-time I and wait-time II in Q/A, but this time in a mature fashion. Teachers stated that utilizing proper wait-time, they create such a learning environment that their students respect and listen to each other. Students could skilfully use these periods as one organism to fulfill their needs and for their knowledge construction and improvement. Teachers start a discussion by a specific question. The process follows as the teacher asks a question then waits (Wait-time I) for a while until most students raise their hands. The teacher observes to find out if there is any student intent to respond. If so, the teacher encourages that student to answer. Then the teacher chooses one student to respond to that question. The instructor waits for the first student respond to the question (Wait-time II) for that specific student if s/he would like to change the answer or add something else. Wait-time II is a particular moment when the mature discussion starts. The remaining part of this section follows with examples of scripts to show how teachers use both wait-time I & II in an intricate manner. A teacher stated:

> I feel anxious and eager at this moment. I wonder if they will do knowledgeable, beneficiary, and skillful discussions or not.

The second phase starts when the teacher looks around encouragingly and provocatively for the students to react to the first student's response. One teacher stated how skillful she is about using provocative gestures:

> I look around in such a way that other students believe that the answer is wrong. (using her body language)

The third phase is to step back and start observing the students carefully and in total silence, not interrupting students. One teacher stated this behavior clearly:

> I change my place and stand still letting my students see and listen to each other. I stare at my students all ears…. This action makes them patient to each other. I teach them to be respectful of each other. They also wait for each other; they will not interrupt each other. Wait till someone finishes.

A teacher stated how much her students consider and concentrate on the discussion they carry out among themselves:

> Teacher: At this moment they forget me, sometimes cheering.
>
> Researcher: Does not' that interrupt you or the class?
>
> Teacher: Absolutely not.

Another teacher stated to use this period for assessing students:

> …I collect data about each student while they are discussing… These are the moments my students are explicitly there. I can understand how much learning occurred… Sometimes it is better than to apply tests.

One teacher emphasized these periods:

> This is the stage I feel as victorious because students use all their knowledge and communication skills in a very respectful manner. It is very beneficiary for my students; they learn from me and each other.

The discussion period continues with students' staring at their teacher aspiring either to finish the communication or to intervene it. Teachers finally discuss these terms.

> …when they finish or need me. Students remember me. They let me know. They look at me and invite me to the discussion.

One other teacher stated vain arguments at first but for sure beneficiary in the future:

> At first discussions become useless efforts, but these are predicted situations, in time it leaves its place to good communication sessions.

4 Discussions

The present study revealed that there is a variation of wait-time use, although teachers attending the study are aware of the wait-time and its advantages, not all use it. The number of categories regarding the wait-time use of primary school teachers is four, limited and few as phenomenography recommends. Similarly, we related the categories hierarchically with each other. The inclusion of wait-time in the teaching-learning environment varies in between aspects of teacher directed/traditional and contemporary approaches.

The wait-time use might reach its somewhat highest peak-Sophisticated wait-time which teachers defined in steps. This appropriate wait-time utilization in the classroom keeps off teacher-centered traditional education and enhances the most student-centered environment. It seems that we made progress regarding the issue during all the years passed since the conscious wait-time inclusion to the educational literature. The present study contradicts researchers who are the pessimists of the wait-time use and its future, above mentioned.

4.1 Theoretical Discussions

The present study reached such results to claim that apart from those theories, above mentioned and specifically: authoritarianism; goal structure; societal expectation; traditional worldview, wait-time phenomenon to be based and related to other theories. These are Learner Autonomy, Learned Helplessness, Learning Communities (LC), Communities of Practice (CP), and Philosophy for Children (P4C). In the following discussions, the present study reached features, emphasized as italics, targeted in Learner Autonomy, Learned Helplessness, LC, CP, and P4C through wait-time use in the present study categorized in contemporary utilization (wait-time 1 & 2 and sophisticated wait-time).

4.2 Learner Autonomy

Depending on Learner Autonomy, students and teachers organize and distribute wait-times *evenly* among themselves. There should *not be any person dominant* in the learning context, either teacher or any student. Moreover, learners are expected to *participate actively* in their learning and each other. Teachers in this regard play a role to *start, sustain and to end* the wait-time use processes. However, teacher *dominance and management in the process diminish* as the *learner autonomy increases* and both teachers and students get used to it. Adding learner autonomy in action periods give teachers excellent time and the possibility to *assess their students*, in an intact fashion.

4.3 Learned Helplessness

According to Learned Helplessness (Walling and Martinek 1995; Tollefson 2000; Yates 2009; Verma and Gera 2012; Rosicka and Hoskova-Mayerova 2013), failure is uncontrollable (Canino 1981) that is learner responses do not *affect the outcome* (Maier and Seligman as cited in Furio 1987). Learned Helplessness, discussed at literature review as 'silent students,' has refreshed in the present study as, 'hopeless students.' We must integrate Learned Helplessness theory to those theories to better explain wait-time use in teaching. Desperate students discourage teachers regarding wait-time use. Contrarily using wait-time *encourages those students to participate* in class discussions and activities.

4.4 LC

In LC learners work together *support*ing each other's learning (Bielaczyc and Collins 1999). Students deal with issues working with people from diverse backgrounds and views. Hogan and Corey (2001) stated ways to foster scientific literacy. They forwarded social processes for a composite *class culture*. Procedures of which consider each *student's ideas* as paramount, students' taking part in discussions. Adding *peer reviews* and *debates* are necessary strategies in LC. Lehrer et al. (2008) found that LC approaches supported the development of inquiry. In research meetings for example *questions, comments, proposals to difficulties* faced came from all classmates, *resolution* of disagreements among team members and in between teams are all characteristics of LC. Students' resolving any conflict by *compromise* and students feeling like adults are practical aspects of LC enhanced. Students did not consider teachers as authorities, but they accept them as *participants* in the activities.

4.5 CP

In late 1980s Jean Lave and Etienne Wenger forwarded a model of situated learning including CP stating the importance of engaging in learning. According to CP learning is a social process; any humanly related domain can be the subject. CP's have critical characteristics which constitute CP: Domain, the community and, the practice. Through CP people *share* a *relevant* and *recognized domain*. They *learn from each other* and *help* each other in the learning process; they *exchange* information; they take collective *responsibility* for learning; sometimes a leader in the community can govern the group but the *leader can be anyone* in the group; they *solve problems*; increase in performance can be evaluated (Wenger-Trayner and Wenger-Trayner 2015; Hošková-Mayerová and Rosická 2012; Hoskova-Mayerova and Rosicka 2015).

4.6 P4C

P4C (Lipman 1973) tried to oppose traditional education approaches was developed giving learning responsibilities themselves through activating philosophy in learning. Trickey and Topping (2004) summarized positive outcomes created by P4C which resembles those reached in the present study by wait-time use. The results include, *discuss* relevant concepts; class *act together* in inquiry processes; students *support group activities*. The students *respect* each other; students *accept individual differences*; students are *active learners*; students *listen to each other* and *contribute* to each other's learning; *respecting* and behaving *kind*ly to ideas. The teacher takes the role of a proper *climate organizer*. Adding there is a *decrease in negative verbal interactions* between students and an *increase in self-esteem*; *offer reasons* when expressing opinions.

Thus proper wait-time use or just extending the time teachers wait after they ask a question and after students answers, these questions, can be based on theories of Learner Autonomy, LC, CP, P4C, and Learned Helplessness. Additionally, Wait-time can remarkably enhance the goals of Learner Autonomy, LC, CP, and P4C. Additionally, it can prevent the formation and thus consequences of Learned Helplessness. Concluding, wait-time can create those class environments for those long being targets.

5 Conclusion and Implications

The present study brought out significant findings for teachers, programmers, and researchers. Teachers can find clues on how to utilize wait-time successfully. For program developers, regarding the variety of conceptions instead of accepting teachers as the tabula rasa of wait-time knowledge and experience, it is rational to start teacher training programs with a diagnostic assessment to reveal missing information to complete them, misconceptions to delete, and unfavorable beliefs to change because teachers are aware of the wait-time. Researchers can widen the wait-time utilization in different contexts repeating the present research study, adding to the theoretical approach. They can search for wait-time utilizations in a quantitative sense including aspects of such probable sophisticated wait-time use as well.

6 Limitations

The present study based on teachers' explanations as for their real wait-time utilization in class, above mentioned. Although most phenomenography studies solely depend on interviews and many precautions (Marton and Booth 1997, pp. 129–130) were activated to eliminate any doubt caused by mere participants' explanations, researchers may still accept this as the weak side of the study and enable additional data collection methods for further research. For example, to use audio recorders (Rowe 1972; Baysen et al. 2003) to find out the more accurate wait-times of the participating teachers can increase the trustworthiness.

Appendix

Questions

1. Can you please express your experiences about communication during the class?
2. How much time do you reckon you reserve for oral Q/A during class time?
3. How long do you pause before choosing a student to respond to your question? How do you decide that student?
4. What happens next?
5. What happens do you think if you extend the wait-time after a question?
6. What do you do after a student answers your question?

    a. Do you wait for student replies to your question?

7. How much time do you reckon you wait after a student responds to your question?
8. What happens do you think if you wait for more than you usually wait after a student answers your question?
9. What do you do if one other student would like to discuss former students' response?
10. What happens next? What do you do next?
11. What do you do after a student answers your question?

References

Aksan, N., Kısac, B., Aydin, M., Demirbuken, S.: Symbolic interaction theory. Proc. Soc. Behav. Sci. **1**(1), 902–904 (2009). https://doi.org/10.1016/j.sbspro.2009.01.160

Altiere, M.A., Duell, O.K.: Can teachers predict their students' wait time preferences? J. Res. Sci. Teach. **28**, 455–461 (1991)

Asplund-Carlsson, M., Fülöp, M., Marton, F.: Peeling the onion. Student teachers' conceptions of literary understanding. Scand. J. Educ. Res. **45**, 5–18 (2001). https://doi.org/10.1080/00313830020023366

Barnette, J.J., Orletsky, S., Sattes, B., Walsh, J.: Wait-time: effective and trainable. Paper presented at the Annual meeting of the American Educational Research Association, San Francisco, CA, 1–15 Apr 1995

Bayerbach, B.A.: Comparing researchers', teachers' and students' perspectives on a line of research attempting to implement wait time in classrooms. Paper Presented at the Annual Meeting of the American Educational Research Association, New Orleans, LA, 5–9 Apr 1988

Baysen, E.: Fen eğitiminde yeni gelişmeler ve (1960–1985 dönemi) Türkiye'deki uygulamaları [New improvements in science education and it's (1960–1985 period) applications in Turkey] (Unpublished doctoral dissertation), Gazi University, Ankara (2003)

Baysen, E., Baysen, F.: Prospective teachers' wait-times. Proc. Soc. Behav. Sci. **2**(2), 5172–5176 (2010). https://doi.org/10.1016/j.sbspro.2010.03.841

Baysen, E., Soylu, H., Baysen, F.: Soru sorma ve dinleme süresi [Questioning and listening durations]. Kastamonu Educ. J. **11**(1), 53–58 (2003)

Bergman, D.: Comparing the effects of classroom audio-recording and video-recording on preservice teachers' reflection of practice. Teach. Educ. **50**, 127–144 (2015). https://doi.org/10.1080/08878730.1010054

Bielaczyc, K., Collins, A.: Learning communities in classrooms: a reconceptualization of educational practice. In: Reigeluth, C.M. (ed.) Instructional Design Theories and Models: A New Paradigm of Instructional Theory, pp. 269–292. Lawrence Erlbaum Associates, Mahwah NJ (1999)

Bilaloglu, R.G., Arnas, Y., Yasar, M.: Question types and wait-time during science related activities in Turkish preschools. Teach. Teach. Theory Pract. **23**(2), 211–226 (2017). https://doi.org/10.1080/13540602.2016.1203773

Boeck, M.A., Hillenmeyer, G.P.: Classroom interaction patterns during microteaching: wait-time as an instructional variable. Paper Presented at the Annual Meeting of the American Educational Research Association, New Orleans, Louisiana (1973)

Bowers, E., Fitts, S., Quirk, M., Jung, W.: Effective strategies for developing academic English: professional development and teacher practices. Bilingual Res. J. **33**(1), 95–110 (2010). https://doi.org/10.1080/15235881003733407

Brenders, D.A.: Fallacies in the coordinated management of meaning: a philosophy of language critique of the hierarchical organization of coherent conversation and related theory. Quart. J. Speech **73**(3), 329–348 (1987)

Bruss, K.S.: Improving classroom discussion: a rhetorical approach. J. Gen. Educ. **58**(1), 1–20 (2009)

Buller, D.B., Aune, R.K.: The effects of speech rate similarity on compliance: application of communication accommodation theory. West. J. Commun. **56**(1), 37–53 (1992)

Çakmak, N.: Lisans öğrencilerinin bilgi arama süreçleri ile ilgili kavramları, tutumları ve düşünceleri [Undergraduate students' concepts, attitudes and thoughts regarding information search process] (Unpublished doctoral dissertation). Ankara University, Ankara (2016)

Canino, F.J.: Learned-helplessness theory: Implications for research in learning disabilities. J. Spec. Educ. **15**(4), 471–484 (1981)

Cardimona, K.: Differentiating mathematics instruction for secondary-level English language learners in mainstream classroom. TESOL J. **9**(1), 17–57 (2018)

Colburn, A.: Constructivism: science education's grand unifying theory. Clearing House **74**(1), 9–12 (2000). https://doi.org/10.1080/00098655.2000.11478630

Devine, D., McGillicuddy, D.: Positioning pedagogy-a matter of children's rights. Oxford Rev. Educ. **42**(4), 424–443 (2016). https://doi.org/10.1080/03054985.2016.1197111

Duell, O.K.: Extended wait-time and university student achievement. Am. Educ. J. **32**, 397–414 (1994)

Duell, O.K.: Asking questions in college classrooms: Is there a need to wait? J. Adolesc. Adult Lit. **39**(3), 251–253 (1995)

Ebadi, S., Gheisari, N.: The role of consciousness-raising through critical reflection in teachers' professional development: a sociocultural perspective. Cogent Educ. **3**(1), 1–14 (2016). https://doi.org/10.1080/2331186X.2016.1147990

Edwards, G.: Standpoint theory, realism and the search for objectivity in the sociology of education. Br. J. Sociol. Educ. **35**(2), 167–184 (2014). https://doi.org/10.1080/01425692.2012.747588

Furio, B.J.: The relationship between instructor behaviors and student perceptions of control in the classroom. Retrieved from https://files.eric.ed.gov/fulltext/ED291124.pdf (1987)

Ghavamani, M., Ketabi, S.: Voices from the voiceless: iranian EFL students' attitudes toward English. Innov. Lang. Learn. Teach. **9**(2), 102–114 (2015). https://doi.org/10.1080/17501229.2013.849708

Grant, T.J., Kline, K.: The impact of video-based lesson analysis on teachers' thinking and practice. Teach. Dev. **14**(1), 69–83 (2010). https://doi.org/10.1080/13664531003696600

Hogan, K., Corey, C.: Viewing classrooms as cultural contexts for fostering scientific literacy. Anthropol. Educ. Quart. **32**(2), 214–243 (2001). https://doi.org/10.1525/aeq.2001.32.2.214

Holmqvist, M.: Teachers' learning in a learning study. Instr. Sci. **39**, 497–511 (2011). https://doi.org/10.1007/s11251-010-9138-1
Hoskova-Mayerova, S.: The effect of language preparation on communication skills and growth of students' self-confidence. Proc. Soc. Behav. Sci. **114**, 644–648 (2014). https://doi.org/10.1016/j.sbspro.2013.12.761
Hoskova-Mayerova, S.: Education and training in crisis management. Eur. Proc. Soc. Behav. Sci. (EpSBS) **XVI**, 849–856 (2016). https://doi.org/10.15405/epsbs.2016.11.87
Hošková-Mayerová, Š., Maturo, A.: On some applications of fuzzy sets and algebraic hyperstructures for the management of teaching and relationships in schools. Models and Theories in Social Systems. Studies in System, Decision and Control, vol. 179, pp. 343–353. Springer, Cham, Switzerland (2018). https://doi.org/10.1007/978-3-030-00084-4_19
Hošková-Mayerová, Š., Maturo, A.: On some applications of algebraic hyperstructures for the management of teaching and relationships in schools. Ital. J. Pure Appl. Math. **41**, 584–592 (2019)
Hošková-Mayerová, Š., Rosická, Z.: Programmed learning. Proc. Soc. Behav. Sci. **31**, 782–787 (2012). https://doi.org/10.1016/j.sbspro.2011.12.141
Hoskova-Mayerova, S., Rosicka, Z.: E-learning pros and cons: active learning culture? Proc. Soc. Behav. Sci. **191**, 958–962 (2015). https://doi.org/10.1016/j.sbspro.2015.04.702
Hildenbrand, S.M., Arndt, K.: Student teachers' management practices in elementary classrooms: a qualitative study. Teach. Dev. **20**(2), 147–161 (2016). https://doi.org/10.1080/13664530.2016.1143869
Honea, M.J.: Wait time as an instructional variable: an influence on teacher and student. Clearing House **56**, 167–170 (1982)
Jegede, O.J., Olajide, J.O.: Wait-time, classroom discourse, and the influence of sociocultural factors in science teaching. Sci. Educ. **79**, 233–249 (1995)
Kameda, T., Sugimori, S.: Psychological entrapment in group decision making: an assigned decision rule and a groupthink phenomenon. J. Pers. Soc. Psychol. **65**(2), 282–292 (1993)
King, J., Aono, A.: Talk, silence and anxiety during one-to-one tutorials: a cross-cultural comparative study of Japan and UK undergraduate' tolerance of silence. Asia Pac. Educ. Rev. **18**, 489–499 (2017)
Kirton, A., Hallam, S., Peffers, J., Robertson, P., Stobart, G.: Revolution, evolution or a trojan horse? Piloting assessment for learning in some scottish primary schools. Br. Edu. Res. J. **33**(4), 605–627 (2007). https://doi.org/10.1080/01411920701434136
Lehrer, R., Schauble, L., Lucas, D.: Supporting development of the epistemology of inquiry. Cogn. Dev. **23**(4), 512–529 (2008). https://doi.org/10.1016/j.cogdev.2008.09.001
Lipman, M.: Philosophy for Children. Retrieved from http://files.eric.ed.gov/fulltext/ED103296.pdf (1973)
Marton, F., Booth, S.: Learning and Awareness. Lawrence Erlbaum Associates, Publishers, New Jersey (1997)
Matz, D.C., Wood, W.: Cognitive dissonance in groups: the consequences of disagreement. J. Pers. Soc. Psychol. **88**(1), 22–37 (2005). https://doi.org/10.1037/0022-3514.88.1.22
McGarry, O.: Knowing 'how to go on': structuration theory as an analytical prism in studies of intercultural engagement. J. Ethnic Migrat. Stud. **42**(12), 2067–2085 (2016). https://doi.org/10.1080/1369183X.2016.1148593
Randolph, T.H., Holtzman, M.: The role of heritage camps in identity development among korean transnational adoptees: a relational dialectics approach. Adopt. Quart. **13**(2), 75–99 (2010). https://doi.org/10.1080/10926755.2010.481038
Ribarsky, E.: Choose your own adventure: examining social exchange theory and relational choices. Commun. Teach. **27**(1), 29–32 (2013). https://doi.org/10.1080/17404622.2012.737924
Riley II, J.P.: The effect of teachers' wait-time and knowledge comprehension questioning on science achievement. J. Res. Sci. Teach. **23**(4), 335–342 (1986)
Rosicka, Z., Hoskova-Mayerova, S.: Motivation to study and work with talented students. Proc. Proc. Soc. Behav. Sci. **114**, 234–238 (2013). https://doi.org/10.1016/j.sbspro.2013.12.691

Rowe, M.B.: Wait-time and rewards as instructional variables: their influence on language, logic and fate control. Natl. Assoc. Res. Sci. Teach., Chicago Illinois (1972)

Rowe, M.B.: Wait-times: slowing down may be a way of speeding up. J. Teach. Educ. **37**(1), 43–50 (1986)

Samiroden, W.D.: The effects of higher cognitive level question wait-time ranges by biology student teachers on student achievement and perception of teacher affectiveness (Unpublished doctoral thesis). Oregon State University, Corvallis (1983). Retrieved from file:///C:/Users/User/Downloads/SamirodenWalterD1983.pdf

Tobin, K.G.: The effect of an extended teacher wait-time on science achievement. J. Res. Sci. Teach. **17**(5), 469–475 (1980)

Tollefson, N.: Classroom applications of cognitive theories of motivation. Educ. Psychol. Rev. **12**(1), 63–83 (2000)

Trickey, S., Topping, K.J.: Philosophy for children': a systematic review. Res. Pap. Educ. **19**(3), 365–380 (2004). https://doi.org/10.1080/0267152042000248016

Verma, S., Gera, M.: Learned helplessness in adolescents. Int. J. Sci. Res. **3**(7), 930–932 (2012)

Walling, M.D., Martinek, T.J.: Learned helplessness: a case study of a middle school student. J. Teach. Phys. Educ. **14**, 454–466 (1995)

Wenger-Trayner, E., Wenger-Trayner, B.: Communities of practice: a brief introduction—V. Retrieved from http://wenger-trayner.com/introduction-to-communities-of-practice/ (2015)

Yates, S.: Teacher identification of student learned helplessness in mathematics. Math. Educ. Res. J. **21**(3), 86–106 (2009)

Some Aspects of Decision-Making Processes in Social Contexts

The Family: Culture Traditional Keeper and Carrier. Study from the Belint Village

Gabriela-Felicia Georgevici

Abstract The family is a special community type of social will, which puts into motion all social frames and events. It is considered the traditional keeper and carrier of culture. The study realized desired to obtain a general image of the manner in which traditions are kept and held in Belint village, from the Banat region, Romania, starting from the hypothesis: if the family from Belint keeps the old traditions and customs of this place, by maintaining a correct manner of their manifest. Through the interview guide the capture of the manner in which the Belint family keeps its ancient traditions and customs, was attempted, even if the new Romanian realities were more obvious. Regardless of all social transformation, the Banat village, obviously affected by all these, remains the keeper of ancient traditions and customs.

Keywords Family · Traditions and customs · Belint village · Community · Cultural traditions

1 Introduction

The village culture has complex functions „cultural events beyond their great diversity, have the quality of functioning as processes of social valuing for its village activities, for the important moments of the village life—the birth, growth and socializing processes, marriage, death etc." (Bădescu 1981, p. 77).

Traditions, culture, history represent the base of every community. One should be able to respect its foundations and to build on it. Romanians, after the 1989 Revolution, discovered democracy and started to forget traditions in an attempt to become more occidental, more European. Thus the interest to analyse this shift has gained more support. In the past, authors realized analyses of the Romanian culture, the Romanian traditions; at present these analyses need to be updated to show the real evolution of this phenomenon.

G.-F. Georgevici (✉)
"Eftimie Murgu" University Resita, 1-4 Train Vuia Square, Resita, Caras-Severin, Romania
e-mail: georgevicigabriela@yahoo.com

D. Flaut et al. (eds.), *Decision Making in Social Sciences: Between Traditions and Innovations*, Studies in Systems, Decision and Control 247,
https://doi.org/10.1007/978-3-030-30659-5_13

Thus, the purpose of the study presented in this study is to firstly realize an analysis of the village culture, traditions, though the family environment, as seen by the current community and also make a comparison to what ancient tradition and customs meant to the Romanian people.

2 The Belint Village Cultural History

The Belint village is proud owner of rich cultural traditions, even if the community centre was established during the latest years. Its cultural traditions are known through: the mixed chorus of Chizatau, built in 1857, the Belint chorus which dates from 1882 and the Belint theatre team, starting with 1896.

The first village community centre was the private home of Damaschin Cărăbaş, which was later bought in 1938 by the church and used by the community for the organization of cultural activities; and in 1959 a new community centre was built, which could hold entertainment activities. The stage of the community received artists, composers, local music bands and artistic collectives from the Banat area and from all over the country; a good example in this case is the Lugoj Ion Vidu chorus, the State Opera and Philharmonic of Timisoara, the Banat Doina traditional music orchestra from Caransebes, the State Theatre from Resita etc. Thus it may be stated that the Belint inhabitants have a great knowledge of the art world (Dănilă and Gheju 1970, pp. 227–228).

In the past, the great attractions of the Belint village were represented by its theatre, its chorus and the traditional dance group and music band. The amateurs' theatre activity (1896) was connected to that of the chorus and of the traditional dance group, which contributed to the community artistic education. In 1882, the Belint chorus is mentioned for the first time, even if it was established years earlier. While realizing the Belint chorus chronology Dănilă S. stated that "the chorus movement gained its traditions, being the pride of our parents and ancestors" (Dănilă and Gheju 1970, p. 236). The traditional dances group was first mentioned in an article written for the "Drapel" newspaper, in 1905, when the "the national games Căluşerul and Bătuta took place with the help of 15 hardy chanters" (Dănilă and Gheju 1970, p. 242). The activity of the dance groups continued, valuing the area specific folklore.

Nowadays, cultural institutions from the Belint village are represented by the community centres from Belint, Babsa, Gruni and Chizatau and the Belint village library.

The Belint village is also proud of its "Belint Hora" Ensemble, which was gathered for the first time in November 2006. The "Belint Hora" Dance Ensemble, composed for graduates and pupils for the local schools, developed a complete activity, the effort of these young and talented dancers being rewarded by the results obtained and by applauses received for each of their presentments.

Since 2007 the ensemble continued in participating in the "Vetre străbune" (Ancestor homes) Folklore Festival and in the Folklore competition for traditions, songs, customs and ancestor games "Lada cu zestre a neamului" (The nation dowry chest),

Fig. 1 Members of the Belint Hora Dance Ensemble. *Source* http://www.tion.ro/belintul-incinge-hora-pe-coasta-de-azur/945454

organized by the Culture and Art Centre of the Timis county council, where it obtained the first places.

During the month of August 2007 the ensemble participate in the la Gyula Folklore Festival—Hungary and in September 2009 in the Paris Banat Holiday. During 2009 and 2016 they have been invited to participate in international folklore festivals held in Romania, in Deva, Herculane, Lugoj and Timisoara. In 2010 they were invited by foreign countries as cultural ambassadors in folklore festivals held in Gudurita–Serbia and in Velico–Tarnovo–Bulgaria.

The beauty of their traditional costumes can be observed in the following Fig. 1.

In the same order of ideas, the Belint Hora Dance Ensemble received the "Eugen Sismaneanu" award for the performance held at the Paris Romanian Embassy and at the Saint Leu la Foret music school. This award was offered by the president of the Romanian-French Cultural Association, Michele Vuillermoz, as a result of their artistic performance and of the ensemble contribution in keeping the Banat culture and traditions.

In October 2010 the first edition of the "Belint traditions" international festival took place, which reunited on the stage set in front of the community centre, more than 300 participants, from our country and from Serbia and Bulgaria. The young people sang, danced and played out weeding customs, steppe dances and other traditions specific to the area they represented, to the public enjoyment. The tradition was kept through the "Belint traditions" international festival, which is now at its 6th edition, for the community pleasure with folklore programs.

Other participations that can be mentioned are the Basalt Days from Sanovita, the rural holiday from Grindul Petrii–Sacosul Mare, The Banat Sundays, organized by the Village Museum in Timisoara, the Garnic International Festival from the Caras-Severin County, the Faget days and the International festival of all Romanians held in

Valea lui Liman (Liman Valley). They also represented choreographies in folkloric TV shows for TVR Timisoara, Europe Nova Television Timisoara and Europe Nova Lugoj, TVR 2 Bucharest and Favorite television.

The activity of the ensemble is coordinated by the educator Ioan Popovici and choreographies are realized Adrian Zîmbran. The "Belint Hora" Dance Ensemble is sustained by the Local Council and by the Belint Village Hall.

3 Family Cultural-Religious Events

The family is considered culture traditional keeper and carrier, "The sum of all knowledge, attitudes and common behaviours models, that they have in common and are transmitted to a certain society" (Boudon 2005, p. 557). According to R.K. Merton the family represents "the most important transmission belt of cultural rules from one generation to another" (Merton and Mendras 1965, p. 190).

According to monographic researches, the family is considered a fragment of social organization, this is the reason why it hasn't been isolated from the village, but integrated into it "in the ensemble of social conditions and relations where it manifests its functionality" (Stănoiu and Voinea 1983, p. 51).

According to T. Herseni, the family represents the basic and universal unit of the society, where all essential processes take place, meant to satisfy human needs.

Dimitrie Gusti stated that, the family specific resides in the complete integration of individuals by destroying individual will and by being submissive to collective will (customs and common interests). The family is the society natural and fundamental group, which insures the transmission of a biological heritage from one generation to another. In the theoretical context of the Bucharest Sociologic School, the family was defined as "an autonomous ensemble of spiritual, economic, judicial and political exercises, conditioned cosmically, psychologically and historically and integrated to a series of social relations and processes. Opposite to other social units, the family is characterized by a strong life intimacy, its specific social functions being represented by procreation, the individuals' growth and socialization" (Costa-Foru 1945, pp. 30–31).

The family represents a special community type of social will that puts into motion social frames and events, "it represents the individual's reference place, its proper universe and a special communication, cooperation, inter-relating, development and social orientation place, where desirable behaviour models are valued, based on socially accepted rules and values" (Bălan 2004, p. 146).

According to D. Gusti, social life cannot be known but by its events, either economic, spiritual (with a constitutive character) and through moral-judicial and political-administrative events (with a ruling character). In addition, social life can be understood only if we also take into consideration the frames that sustain and condition it: cosmological, biological, historical and psychological. Social life can be seen only through the connection existing between frames and events.

Dimitrie Gusti (Gusti and Herseni 1942, p. 314) considers that the family and the household are the most important social units of any village, this being a reason for a detailed study that would also contain the family compound and its relations, "in order to prove the family composition, the relations between husbands, between parents and children, between brothers, how is life lived inside families, how are activities divided between its members etc.", together with the activities realized inside the household "what does the household produce" and the property type "land ownership" etc.

In this context, the model of the Bucharest Sociologic School inspired the creation of the Banat-Crişana Social Institute, not only from a structural point of view, but also methodologically, the people from Banat being aware of the profound knowledge, preceding any economic, social, cultural and political reforming action. For this desiderate to be fulfilled, the Banat-Crişana Social Institute chose as investigation method the monographic research, practices by the Bucharest monographic School, which registered numerous successes in the social research domain, by applying Dimitrie Gusti's monographic method.

Thus, starting with 1933, in parallel to the Bucharest School, lead by D. Gusti, other research activities were also developed by the Banat-Crişana Social Institute, by using the methodological research schema of the Gusti School, but adapted to local research possibilities and needs.

4 Study Methodology and Description

The research of the Belint village was realized by a monographic team of the Banat-Crişana Social Institute, from 15 August to 2 September 1934 (Georgevici 2013, p. 77), and the material gathered was published in the volume "Anchetă monografică în comuna Belinţ" (Monographic research of the Belint village). The volume contains the reports of each medical-social, cultural, economic, judicial, sociological and political sociological branch and the general report, written by prof. Traian Topliceanu, all of these representing the monographic content.

The basic investigation method used was the enquiry, a standard method, by applying the questionnaire instrument, according to the Bucharest School methodology. The investigators have filled in information with the help of the interview method and that of direct observation (Georgevici 2013, p. 79). The Belint researches have been sustained by the questionnaire, also filled in with data resulted from direct, individual and collective observations.

The content of the monograph realized by the Banat-Crişana Social Institute „Ancheta monografică în comuna Belinţ" (Monographic research of the Belint village), was published in 1838, at the "Tipografia Românească" (The Romanian Printing House) Timisoara, and the construction of this monograph followed the "frame" and "events" schema elaborated by Dimitrie Gusti (Georgevici 2013, p. 79).

The volume also contains the general report on conclusions and recovery solutions proposed by the Banat-Crişana Social Institute written by prof. Traian Topliceanu, together with other 22 reports realized by the five branches and the statistical office and heir proposals for the solving of deficiencies met (Georgevici 2013, pp. 79–80).

The Banat-Crişana Social Institute (I.S.B.C. 1938, pp. 200–213), through its monograph specialists and through professor Traian Topliceanu, studies the Belint folklore.

The cultural branch realized three reports (Georgevici 2013, pp. 82–83):

1. The report on the Belint religious life by M. Bucătură, being represented in Belint by two opposite beliefs, a traditional and a modern one, inherited from a frequent contact with the city. "There are two social processes in Belint:

   - The tradition keeping process.
   - The urbanization process.

   These processes cannot be explained but by two mystical-religious mentalities belonging to those still living strongly under the pressure of ancestor values and traditional forms; and the other belonging to those who rejected traditions and forms, because of a town contact" (I.S.B.C. 1938, p. 190). It also analyzed the family religiosity, that of the "school" and the Church, concluding that "the Orthodox church, not the belief, in its daily practice doesn't satisfy the believers' inner needs, thus creating an institutional decay and a possibility to revive particular religious experiences" (I.S.B.C. 1938, p. 199).
2. The report on the Belint folklore gathered by prof. Tr. Topliceanu, who analyzed traditional poetry, erotic lyrics, birth, marriage and death customs, disenchantments, superstitions and prejudice, which represent the power of any tradition „different aspects of multiple spiritual and material preoccupations are transposed in these creations. This is the reason for not knowing the soul of different peoples, in all its sides, except when the so called folklore science was formulated. The great mass of the people transposes its soul in this folklore, which has become one of the greatest problems of a social cultural nature" (I.S.B.C. 1938, p. 200).
3. The report on the school realized by prof. L. Anţila presented the Belint school history and described the village school institutions, the school for small children, primary school and the secondary household school for girls.

They have gathered and analyzed erotic lyrics, disenchantments and other superstitions and prejudice, thus realizing that the young people from the village don't know any of the old traditional poetry. Sabin V. Drăgoi a published in 1942 „Monografie muzicală a comunei Belinţ" (The Belint musical monograph), which contains 35 carols, 30 doinas, dirges and traditional dances.

Following the example of the monographic teams of the Romanian Social Institute, the Banat-Crişana Social Institute also got involved in the villagers daily life, thus developing an enquiry with a rich cultural activity, built of general interest themes like: health, agricultural knowledge, medical assistance, medical examinations, awards and subventions offered for deserving and to needing households (Georgevici 2013, p. 87).

The folklore, the customs and the traditional art were later also studied by Dănilă and Gheju (1970, pp. 177–203), through a description of customs (disenchantments, winter customs, birth, marriage and death), traditional writings, lyrical poetry (doinas and dirges) and popular art.

This study desired to help obtain a general image of the manner in which traditions were kept and took place, starting from the following hypothesis: is the Belint family keeping ancient traditions and areal customs, by maintaining a concrete application.

The research methodology implied the use of sociologic enquiry, the interview, and the data collection was realized by using the individual, semi-structured interview.

The interview represents "the attempt to obtain information from other persons through questions and answers during a conversation" (Rotaru and Iluţ 1999, p. 61). Stacey recommended the used of interview enquiries when the behaviours studied are difficult to observe "When beliefs and attitudes are researched, written documents not being available in this case" (Stacey 1970, p. 70).

The Bucharest Sociologic School under the management of professor Dimitrie Gusti used the term "Sociologic conversation", meaning the obtaining of information "through the explaining of scientific problems, straight from people's mouth" (Herseni 1940, p. 144).

The methodological option for the use of an interview started from taking into consideration of its advantages (Chelcea 2007, pp. 300–301): flexibility, the insurance of a control on the succession of questions, the collection of spontaneous answers and the insurance of an answer for each question.

Thus the same contained 20 persons that accepted to participate in this study and to which the interview was applied, based on a semi-structured interview guide, which was elaborated on the basis of objectives and hypotheses proposed for this study.

Through the interview guide we have tried to observe if the Belint family is still trying to keep the customs of the place, and if new realities of the Romanian culture are also being accepted.

5 Study Results

After analysing the data obtained from the interview and from the critical analysis of the documents studied, the following could be observed:

- most persons interviewed sustained that their family is a close one and that it kept rituals and ceremonies like: the engagement, the christening, marriage, funerals etc.;
- the meaning of religious holidays in the life of families is a special event, most respondents sustaining that great religious holidays are first of all, a happy opportunity to reunite families;

- all the people questioned sustained that they go to church; an important factors, is that the persons with an age over 60 years go to church at least once a week for messes and Christian rituals, from the other 2 age groups, only 4 persons respect this patters, the other usually participating only during important religious holidays: Easter, Christmas and the Church Patron holiday (The village holiday, Ruga);
- from the customs kept by their forefathers the winter holiday customs were mentioned, the „piţărăilor" carolling held on Christmas Eve, when "children come caroling, its wonderful", the New Year "sorcova" and the "star", a form of traditional theatre. When mentioning the loss of customs, a 71 years respondent stated that "How could one not lose its customs, when our children must leave their village to go to school and get used to life there";
- in the Belint village, the Village Holiday (Ruga) is celebrated for each village from the group at another date, thus the Belint holiday (Ruga) is kept on the second and the third day of Easter, Rand in Babşa the holiday is kept on the Whitsuntide holiday, while for the villages of Chizătău and Gruni it is held on the 8th and the 9th of September, for Saint Mary's Holliday, each year;
- the village holiday (ruga) represents a moment of joy for every inhabitant, because family, relatives and friends celebrate together, being an opportunity to fill their homes with guests:
  - "we are happy to receive guests",
  - "we celebrate together and later we go to the community centre together",
  - "great glory expects us during the Ruga; many guests are expected to come" etc.
- most of the people questioned, characterize their Belint inhabitant status with the word "pride", "Banat is the forehead of our country", "this is our place and our ancestors" and most of the listen and take pride in their Banat folklore creations;
- the fame of the Belint traditional clothing is kept since ancient times due to its beauty and the mastery sewing.

Nowadays, folklore clothing are used only by the representatives of the Belint Hora ensemble, by children during school holidays, and older people put on their traditional costumes during the Ruga or on special holidays.

6 Conclusions

The idea, from which the study started from, was that after the '90 s, Romanians slowly shifted away from their culture, traditions and customs, in order to adapt to the new environment offered by Europe, by the access to new cultures and environments.

Still, this search for the new, for democracy, for development led the Romanian people to the belief that customs and traditions still represent and will represent the basis of our culture, and if lost, history and foundation will be lost.

The conclusion reached after years of adapting to changes, is that culture should be kept and respected, should be exhibited and people should be proud owners of this culture, because it represents our inheritance.

In the case of the village subjected to the study, due to all sociologic transformation, the Belint village has been clearly affected by these, but, it remains the keeper of ancient customs "direct relations between each generation and between each member of the family have facilitated the communication process and the transfer of some cultural dimensions/customs, traditions, norms, roles etc. so that the future children generations could adopt them with no impediments" (Batâr 2003, p. 111).

Urbanization has, unfortunately, had a great effect of the village population; the archaic living style inherited by the young generations is no longer accepted, not even by the older people, over 70 years old. Great religious holidays are kept (Christmas, Easter, Saint Mary etc.) and rituals regarding religious marriage, children's christening and funeral and post-funeral rituals.

Young people participate in family reunions held during great religious holidays, which in the village world possess a special affective feeling. The church represents, at present times, the most prevailing and at hand social frame for the manifest of an allegiance feeling to a family, a group or to the community.

The Belint inhabitants are proudly aware of the fact that they live in this geographic, historic and folkloric region. There are real premises, including younger generations, and the ones to follow, to keep the area cultural value. The civilization process has lead to the disappearance of many ancient traditions of its inhabitants, as Mead stated "our ways of thinking connects us to the past, to our childhood and our adolescence" (Mead 1970, p. 119).

References

Bădescu, I.: Satul contemporan şi evoluţia lui istorică (The contemporary village and its historical evolution). Scientific and Encyclopedic Publishing House, Bucharest (1981)

Bălan, C.C.: Sociologie aplicată. Monografia centrată pe problema (Applied Sociology. Monograph Centered on Problems), Tritonic, Bucharest (2004)

Batâr, D.: Sociologie. Probleme teoretice şi analize ale investigaţiilor de teren (Sociology, Theoretical Problems and Analyses of Field Investigations), 3rd edn. Psihomedia, Sibiu (2003)

Boudon, R.: Tratat de sociologie (Sociology Treaty). Humanitas, Bucharest (2005)

Chelcea, S.: Metodologia cercetării sociologice. Metode cantitative şi calitative. (Methodology of sociological research. Quantitative and qualitative methods). Economic Publishing House, Bucharest (2007)

Costa-Foru, X.: Cercetarea monografică a familiei. Contribuţie metodologică (Family Monographic Research. Methodological Contribution), The Foundation of King Michael, Bucharest (1945)

Dănilă, S., Gheju, C.: Monografia comunei Belinţ (Monograph of the Belint Village) (1970)

Georgevici, G.F.: Monographic Sociology of Dimitrie Gusti. Social Science and Reform. European University Studies, Sociology, XXII/454, Peter Lang GmbH Academic Research, Frankfurt am Main (2013)

Gusti, D., Herseni, T.: Elemente de sociologie (Elements of Sociology). Cartea Românească, Bucharest (1942)

Herseni, T.: Plan pentru convorbirile sociologice (Plan of sociological conversations). In: Gusti, D. (ed.) Îndrumări pentru monografiile sociologice (Guidance for Sociological Monographs). The Romanian Institute of Social Sciences, Bucharest (1940)
Hora Belint Ensemble: http://www.tion.ro/belint/belintul-incinge-hora-pe-coasta-de-azur/945454
Institutul Social Banat–Crişana (The Banat-Crisana Social Institute (I.S.B.C.): Anchetă monografică în comuna Belinţ (Monographic enquiry of the Belint Village) „Tipografia Românească", Timisoara (1938)
Mead, M.: Le Fossé des génération. Denoël-Gonthier, Paris (1970). (1971)
Merton, R.K., Mendras, H.: Elements de théorie et de méthode sociologique. Plon, Paris (1965)
Rotariu, T., Iluţ, P.: Ancheta sociologică şi sondajul de opinie (Sociologic Enquiry and Interviews). Polirom, Iasi (1999)
Stacey, M.: Methods of Social Research. Pergamon Press, Oxford (1970)
Stănoiu, A., Voinea, M.: Sociologia familiei (Family Sociology). Bucharest University Printing House, Bucharest (1983)

The Migration Decision on Economic Level in the Contemporary Social Context

Lavinia Elisabeta Popp, Felicia Andrioni and Ion Petrică

Abstract On national level, as well as internationally, the migration issue occupies an important place on the social policies agenda. The analysis scope aimed at in this endeavour took into consideration the indicators having led to the migration decision, highlighting social change as a necessary condition for the increase of the economic and social potential. The migration decision on microeconomic level is the result of a process by which the potential migrants estimate the benefits of travelling to other countries for a limited time, and takes place when the calculation result is positive, toward the country offering the highest benefits for the individual. The results of the study are focused on the factors and consequences of the migration decision-making process on economic level, which may be correlated to a series of social-economic indicators.

Keywords Migration · Decision · Social change · Macroeconomic · Microeconomic level

1 Introduction

Globalisation is a vast process led by economic, technological and political forces triggering an enhancement of risks and opportunities for individuals and societies or communities, increasing mobility and opportunities to satisfy needs, with implications on the economic and social life components. A positive aspect is the fact that production globalization creates opportunities on the labour market for the economy of migrants' destination countries, especially for those with high unemployment.

L. E. Popp (✉) · I. Petrică
University „Eftimie Murgu" of Reşiţa, Traian Vuia, nr. 1-4, Reşiţa, Romania
e-mail: lavinia_popp@yahoo.com

I. Petrică
e-mail: prof_religie_ion@yahoo.com

F. Andrioni
University of Petrosani, Universităţii Street, 20, Petroşani, Romania
e-mail: felicia_andrioni@yahoo.com

D. Flaut et al. (eds.), *Decision Making in Social Sciences: Between Traditions and Innovations*, Studies in Systems, Decision and Control 247,
https://doi.org/10.1007/978-3-030-30659-5_14

This phenomenon imposed new movements in the international space, and among the most significant are the relatively free movements of people and goods, as well as of capital (Feraru 2011, pp. 53–55).

International migration on macroeconomic level is triggered by the differences between the demand and supply ratio in diverse countries, beside the differences between the levels of salaries and incomes. A labour force flow is created thus, toward the country rich in capital and poor in labour force. These external migration flows become mechanisms of balancing internal deficiencies on the labour market.

The migration decision on microeconomic level is the result of a process by which the potential migrants estimate the benefits of going to other countries for a limited time, and takes place when the estimation is positive, toward the country offering the highest individual benefits (Feraru 2011, p. 14).

Migration is perceived by the neo-classics of the "balanced growth" theoretical model as a form of optimum allotment of production factors to the benefit of migrants'both origin and destination countries (De Haas and Hein 2008, pp. 24–31). The illegal immigration and refugee crisis in Europe in the years 2015–2016 highlighted the specific nature of the problems that this category of immigrants faces so that the phenomenon of external migration should be treated differently (Otovescu and Otovescu 2017, pp. 67–68).

It is proved that migration has the potential to diversify and increase incomes, wellbeing, productiveness and, to a certain extent, to mitigate poverty (Feraru 2011, p. 47).

Taking into consideration the temporary horizon of migration, a distinction is made between permanent and temporary migration. The former implies the definitive change of the permanent domicile county, the latter only a sustained residence in another country where the person has the permanent domicile.

In the formal definitions at the basis of the migration processes, a distinction is made between short-term temporary migration, which means a sustained residence between three months and one year, and long-germ temporary migration, when the residence exceeds one year (Anghel and Horvat 2009, p. 20). As regards the causes of external migration, the salary level in Romania is a priority cause of migration (Otovescu 2016, pp. 228–229).

Migration represents "the travel of a person or a group of persons from one geographic unit to another across an administrate or political frontier, who intend to settle, permanently or temporary, in a place different from the origin place" (IOM 2003, p. 8).

2 Contemporary Trends of Jobs Delocalization from Romania to Foreign Countries

The Romanian labour market in the past 27 years has been strongly influenced by the economic, political and social transformations. Until 2008, after several years of fast economic growth and impressive reduction of poverty, the shock wave of the global economic and financial crisis highlighted the misbalances and vulnerability of the Romanian economy, characterised by a poor economic management and a long list of non-finalised reforms (Montagna 2010).

After the accession of Romania to the European Union, the migration phenomenon has developed in Romania. This has economic and social effects. Romania is one of the most important suppliers of labour force to the European market (Roman and Voicu 2010).

In Romania, the migration impact on labour market is significant, and its causes are largely of economic nature, such as concrete needs of the family, the economic issues challenging Romania because of industry restructuring and bankruptcy of enterprises from the market. Migration has an impact also on the community or family members by the imitation of other migrants' success models from the community, but also on the evolution of unemployment and number of the unemployed persons in Romania (Feraru 2011, pp. 74–75).

In 1990, the collapse of the Romanian economy took place, due to the fact that the number of illegal migrant to other countries increased, although their purpose was not to settle definitively in another country, but to work there (Anghel and Horvat 2009, p. 157).

Thus, in the past years, massive waves of Romanian workers have gone abroad, in search of jobs (Anghel and Horvat 2009, p. 157).

As regards the causes of external migration, the salary level in Romania is a priority cause of migration. In Romania the average net salary is incomparably lower than in other European countries. These salary inequalities generate "differences of status, access to resources, possession thereof, considered to be unjust for the development of one person or of a community" (Ispas 2013, p. 52). The Romanians migration after 1990 becomes a social phenomenon that gets bigger and bigger. In this period, the migration took many ways of manifestation: starting with the ethnic migration (the Rroma ethnics that asked for asylum in different European countries or the Hungarians that migrate in Hungary), continuing with increasing the international mobility of students (especially the IT students) and ending with the abroad migration for work (Androni 2011).

The labour force deficit on national level is declared also in the sectors of healthcare, education, constructions, the number of those who have left the system to migrate being increasingly high. Furthermore, the category of highly qualified persons in Romania is a category attracted by the better labour conditions abroad and by the higher wages.

Romania is considered an emigration country and one of the most important labour force suppliers for the European market. According to the analyses conducted by Roman and Voicu, the migration movements generate crucial economic consequences related to financial aspects and Romanians' life quality, as in the opinion of the aforementioned authors the amounts transferred were extremely large. Other aspects also were underlined about external migration in economic plane, the most important referring to the severe deviations and the multitude of dysfunctions of the labour market; the unemployment rate and its features; mass emigration of

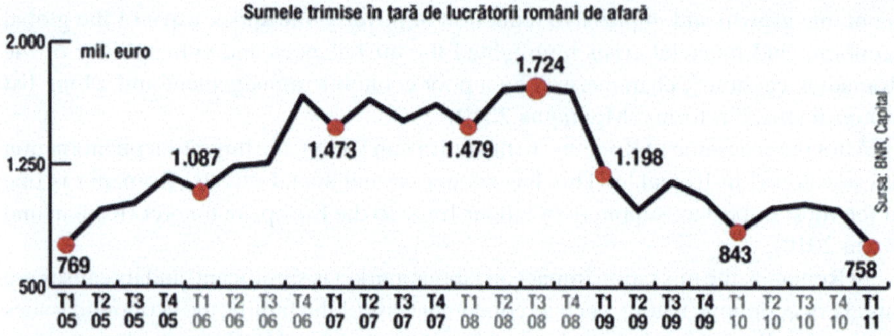

Fig. 1 The amounts sent to the country by the Romanians working abroad. *Source* BNR, Capital processing; Stănculescu et al. (2011, p. 20)

qualified labour force or losses of "brains" capable of creating higher value added, salary distortions and segmentation of the labour force; implications on underground economy ("black" labour"), reduction of local labour potential, and the use of immigrants to replace the shortage of labour force on local level (Roman and Voicu 2010, pp. 50–65).

Additionally, Gabriela Prelipcean considers that a main positive consequence of international migration is represented by the money transfers to the origin country. In the author's opinion, these financial transfers may be endogenous, determined by the altruistic attitude toward the family and the loved ones remained at home, and exogenous, determined by the external environment of financial resources transmission. These money transfers grow the incomes of the country, with repercussions in the plane of local economic development and in the plane of population's living standard (Gabriela Prelipcean, Migration and euphoria or remittances. Implications of current economic crisis, *Politics Sphere Journal*, online access at http://www.sferapoliticii.ro/sfera/137/art02-prelipcean.html). Figure 1 reflects the amounts transmitted by the Romanian workers from abroad to the origin country. Thus it may be remarked that the amount sent to the origin country by the Romanians working abroad in the first quarter of the year (2011) were the value of 758 million euros, compared to 2008 when an increase of remittances was recorded (1724 million euros) which represents a considerable drop caused by the difficult situation on the labour market in the destination countries (Stănculescu et al. 2011, p. 21).

3 The Romanians Beyond Romania

The official statistic surveys conducted by Eurostat show that the weight of the Romania workers in Italy and Spain is around 80% of the total Romanian residents in Europe. Based on the information collected from the CURS opinion poll in 2009, it results that approximately 2,800,000 persons are migrants who left to work in

European states. Moreover, the same CURS opinion poll shows that another 450,000 migrants were currently working in other countries than the European states, raising thus the migrants' volume to more than 3,250,000 persons (jurnalul.ro). This number of immigrants represents the number of those who are temporarily living abroad (Feraru 2011, p. 202).

The opinion polls conducted by CURS in 2008 in Spain show that the persons who left to work abroad who no longer have a household in Romania are about 40% of the total migrants (10% single persons and 30% persons with children who reunited their families abroad, namely they have gradually brought their families with them to the destination countries), and thus the number of persons is 1,200,000, and thus we may estimate that there are around 480,000 households of Romanian migrants abroad. In 2009 the number of migrants remained constant, the arrivals of new migrants being practically annulled by the return to the origin county of some people affected by the economic crisis (Feraru 2011, p. 202).

Certain 2008 poll data show that one third of the Romanian labour force works without labour contract or labour permit, and some even commit crimes. In relation with these aspects, Professor Nicu Gavriluţă presents in his book entitled „Romanians specialists in rape", the drama of Romanians from Italy, of the Romanian community living and working there, claiming that „*not all Romanians may by stigmatised as being thieves, rapists or murderers*" (Gavriluţă 2010, p. 31).

The data provided by the polls also show that a fifth of the migrant Romanians are living illegally in other countries, and thus to the 1,750,000 migrants officially registered in Spain and Italy we may add another 350,000 Romanians, more or less. Most Romanians in Italy come from the Moldova area, which is one of the poorest areas of Romania. In Romania the largest amounts of money were sent from the Rome Province, in a 24% percentage, more precisely 20% of the construction workers, and 11.1% of those working in hotels and restaurants. Certain data show that *remittances* will remain more resilient than other forms of incomes and have become today the most important source of development funding in many developing countries (Roman and Voicu 2010, p. 198).

In 2009, according to BNR, the sum of money transfers from abroad was of 6.6 billion euros, lower than in 2008, when 8.7 billion euros were sent to the country, namely the equivalent of 6.2% of the GDP. The Romanians from abroad transferred to Romania, only in the period 2003-2009, 37 billion euros, and, as a comparison, the direct foreign investments in the same period were of 45.7 billion euros (Roman and Voicu 2010, p. 209).

„*The field details were gathered through a questionnaire and an interview. It is shown that 20%, up to 25% of the Romanians from other countries confessed that they do not intend to return to their own country*" (Otovescu 2017, p. 25). The brains exodus, associated to international migration from underdeveloped countries to more developed countries, has constituted and remains a debate subject when it comes to its consequences on the social and economic development of their origin countries (Anghel and Horvat 2009, p. 215).

As regards this phenomenon, Todaro remarks that „*the irony of the present international migration is that… many of the legal migrants from the poor regions to richer ones are the very persons whom the Third World countries afford to lose the less: the highly educated and qualified inhabitants. As the majority of such migrants leave for ever, this perverse exodus of brains not only represents a loss of valuable human resources, but could prove to become a significant constraint on the future economic progress of the Third World nations*" (Todaro 1996, p. 119).

Furthermore, migration toward a foreign country, for most IT specialists, is a labour decision after graduation and accumulation of labour experience in Romania. The research institutes and IT companies, as well as universities, are open to foreigners and require diplomas and competitive experience standards, motivating also the specialists to be very mobile and to update their labour opportunities (Anghel and Horvat 2009, p. 219).

As regards the capacity of the individual to be re-socialised by labour migration, an important characteristic is the knowledge of a foreign language. Among those who worked aboard, 11% declared in 2004 they speak a foreign language well or very well, and in 2006 the data show a remarkable growth by 19% of those who speak a foreign language. Thus 38% among those who worked abroad declare to speak a foreign language fluently or well, compared to 25% of those who did not work abroad (Anghel and Horvat 2009, p. 277).

4 Strategic Measures, Services and Programmes of Social Intervention Destined to Delocalisation Situations

The fundamental preoccupation for most contemporary sociologists is represented by the study and explanation of the social issues and their causes. Both sociologists and social workers play a central role in the identification of particular situations or conditions such as social issues and also the formulation of public policies and programs for their remedying and improvement (Dan 2007, p. 2).

In their analyses about social issues, sociologists use extremely varied theoretical and methodological frames, called paradigms (Dan 2007, p. 2).

In the sociology dictionary, Zamfir and Vlasceanu (1993) defines the term „paradigm" as being „*A set of concepts, sentences, investigation methods, with a strong normative character, developed in order to guide the research in a certain specific field*". It is also defined as „*The totality of achievements in a certain subject or domain, largely accepted by the respective scientific community, offering the pattern of the research issues and of the solutions explored*" (Zamfir and Vlasceanu 1993).

According to the functionalist standpoint, associated especially with the names of T. Parsons and R.K. Merton, society may be regarded as a system where all segments are somehow interlaced. The change in a certain point of the system may trigger effects throughout the system due to this interdependency, effects which at first glance may seem absolutely unrelated, distinct and individual" (Dan 2007, p. 10).

Another important assertion is that each part of the social system fulfils a function for society, and the pattern resulted from this perspective is that of a system made of interlaced parts, each of them satisfying a certain need of society, contributing to the enhancement of the system's efficiency, or, in another manner, contributes to the maintaining of the system unity by keeping its parts together. In fact, a strong trend in created for a high stability in the system and social change is possible only in certain conditions (Dan 2007, pp. 10–11).

Another idea of the functionalist perspective is that society tends towards consensus as in any society there is a common set of values and generally accepted creeds. Functionalists perceive it as being the fundament of social solidarity, considering that the social solidarity and consensus are important parts of any effective society (Dan 2007, p. 11).

According to Doob (1995) „*the social issue represents an undesirable situation which is considered by an important segment of society as being sufficiently serious to need collective action for reaching significant improvements and enhanced desirability*" (Dan 2007, p. 14).

The manner in which we define a problem as having a social nature and not another one (private, personal, individual), depends on the fulfilment of a relatively high number of theoretical and methodological conditions, as Merton put it (1971). In his opinion a social problem is in fact „*a significant discrepancy between social norms and factual social reality*" (Dan 2007, p. 14).

In the vision of Farley (1992), a social issue represents a situation characterised by the following three elements: it is largely regarded as being undesirable or a source of difficulties, is caused by the action or inaction of people or society, and affects or it supposed to affect a high number of persons (Dan 2007, p. 14).

A more detailed analysis is needed in relation with these three elements, and a certain level of public awareness is expected for a certain situation to be perceived as undesirable. When people do not know anything about this problem, it will not appear as a social problem, and thus it will remain in latent state. Moreover, beside its public awareness, there must be a significant level of awareness that this condition constitutes a problem for society.

In defining certain issues as being social, an important part is played by values, as a value is considered a personal belief about what is good or evil, right or wrong (Dan 2007, p. 14).

Cătălin Zamfir, analysing social issues from the perspective of the social systems functioning, concluded that an important role is detained by „*the activity of confronting different difficulties intervening during their functioning, in other words, the activity of social problems solving. This activity must rely on a theory of social issues which represent "a part of the theory of organisation and operation of social systems", systems which operate as a results of the improvement of their capacity to formulate clearly and in due time efficient solutions for the particular problems they have to overcome*" (Dan 2007, p. 16; Zamfir 1977).

Moreover, the same author considers that "*a particular issue may be of technical, economic or legal nature* etc. *As issue of a social system, it will also be a social issue*". Thus, in general terms, it defines the social issue as being "*a social process, a characteristic, a situation that society or a subsystem thereof considers necessary to change*" (Dan 2007, p. 16; Zamfir 1977).

The main characteristics of a social issue in the vision of Merton and Nisbet (1971), are:

(1) the occurrence of a social issue is indissolubly related to the manifestation of a non-concordance between what people should be and what they are in reality (more precisely between the current social situation and social norms);
(2) a social problem may be, in many cases, the non-anticipated, not desired and indirect consequences of the institutionalised patterns of social behaviour;
(3) different social services have distinct problems depending on the characteristics, values, purposes and distinct interests;
(4) depending on the particularities of social structures the solutions given to these problems are different (the solutions having as purpose and result the social change);
(5) as the component elements of a social structure (which define the system) are in interdependency relation, the actions developed and taken for solving a particular social issue challenging the system shall generate the occurrence of new issues which must be solved and consequently the solutions found will be difficult to concretise in effective policies (Dan 2007, p. 16).

5 Results—Qualitative Analysis of the Perception on Delocalisation Dimensions and on «Circulating Migration» in Caraş-Severin County

5.1 Research Methodology

The general purpose of research aimed at identifying the perception on delocalisation dimensions and the delimitation of the dimension of contemporary «circulatory migration» phenomenon practised by the Romanian population in the current context.

The objective of research was to identify the opinion and attitudes of the subjects related to the necessity of the external emigrational phenomenon, of its economic benefits, of the individual, family or community micro social impact thereof.

The main research method used was the investigation based on an interview intended to collect the data necessary for the case study referring in particular to "circulatory migration" in Caraş-Severin county. When selecting the method of interview-based inquiry we took into account that this method is the most frequently used in the sociologic research (Caplow 1970, p. 203). We also started from the idea that the interview use in scientific research has many purposes, leading to the identification of variables and relation among variables and with the help of interviews we may arrive at formulating interesting and valid hypotheses (Chelcea 2007, p. 106).

The study started from the following hypotheses: (1) If in the near future new jobs are not created in Romania and if salaries do not increase to assure a decent living standard to the population, the phenomenon of circulatory migration will become vaster, increasing the number of persons opting for a job in another country than the origin one, which can offer them a much higher salary; and (2) If the persons choose to periodically migrate to another country for a job offering them a salary much larger than in Romania, their living quality will increase, but beside positive effects certain negative effects may occur.

The study was conducted based on the data collected from the individual interviews applied to the managers of the companies for labour placement abroad, on the one hand, and based on the interviews with the persons who practice circulatory migration, on the other hand, in order to get a higher salary assuring them a better life on individual and family level.

The individual interviews conducted face to face with the managers of three private companies of labour placement abroad in Caraş-Severin county referred to a series of aspects related to the perspective of their current situation, and the phenomenon of „circulatory migration" in Caraș Severin county.

Furthermore, five individual interviews were conducted, face to face, with the persons who practice circulatory migration in order to earn a salary for providing homecare services to senior citizens, working in another European country. The respondents came from different areas of Caraș-Severin county, with varied personal characteristics of age, gender, education, marital and professional status, etc.

The persons selected for the interview were identified based on the snowball method. Several persons were contacted on the phone and from among them five persons were selected with diverse characteristics, who accepted to participate in the interview. The time allotted to the discussion with each person was of approximately 60 min.

5.2 *Analysis and Interpretation of Qualitative Data*

(a) **The analysis dimension regarding the perception about circulatory migration of the respondents who are intermediaries on the labour market from other countries**

The respondents have started their business activity in Romania in the field of labour force placement abroad relatively recently, in 2008, 2010, and 2012. The modality of starting such a business was determined by the needs of the market in this field, as there is a balance between the external labour offer and the availability of internal labour force of Romanian population, more precisely in Caraș-Severin county. Being considered by the respondents as a „profitable business" in the current context, they are content with the profit generated by this segment, as the number of beneficiaries of job searchers abroad has consistently increased, from 20 to 80 persons/month in the initial period of the business, up to 1000 persons/year at present. They also mentioned that despite the fact that the start was timid and difficult, at present this type of business is profitable.

As regards the perception of businessmen about customers, as they claimed, the persons who usually work abroad do not send money home to their families because they avoid the payment of bank commissions, except in emergency situations."The persons bring almost all the money back to the country, the entire salary, personally, as they spend very little abroad" (E.J., 45 years old, Reșița). In the opinion of the businessmen, emigrants do no send money to the home country because they return very fast, at short intervals of time.

According to the assertions of the respondents, the jobs most desired by the consumers of their companies are in Austria and Germany. As for the range of services provided by their firms, the subjects mentioned they provide two-way transport, mediation contract and medical insurance. From their standpoint, there are two main factors determining the person to resort to their services of labour contracts abroad, namely: (a) scarcity of jobs; and (b) very low salary level, insufficient to cover daily costs. In their opinion, the effects of emigration on their customers' families are positive only from the financial perspective (increase of life quality, payment of studies for the children, etc.) but also negative, as the relations deteriorate among the family members, the family is sometimes very affected, as in very many cases this leads even to divorce, and the children are affected the most (they abandon school or worse, end up committing crime).

The respondents declared to be optimistic as regards the prospects for the future, in the sense that their business will definitely be prosperous in the future and they will continue to have the same number of customers or even more, if the situation in the country remains stagnant and nothing changes as regards jobs and wages.

(b) **Analysis dimension referring to the perception about circulatory migration of the persons emigrating to the labour market of other countries**

As regards the verification of hypothesis number 1 „*If in the near future new jobs are not created in Romania, and salaries do not increase to assure a decent living to the population, the circulatory migration will become vaster, increasing the number of persons who will decide to take a job in another country than the origin one, offering them a much higher salary*", this hypothesis was confirmed. In this respect, the interviewed subjects highlighted a series of aspects, such as the fact they started to migrate to other counties as back s 2000, but circulatory migration (short periods of departure and return, of one or two months) has become a practice for them only for several years (since 2007; 2012; 2013; 2014). Among the main reasons which made them leave to work abroad, the subjects listed: the closing of factories, the restructuring of jobs, unemployment, lack of incomes, small pensions, health problems of one of the family members, support of children in college, scarcity of jobs, very low salaries not covering the necessary costs of daily life.

As for the verification of hypothesis number 2 "*If the persons decide to periodically migrate to another country for a job offering them a much higher salary than those in Romania, their life quality will increase, but beside positive effects, some negative impacts may be identified*", this hypothesis was also confirmed.

The interviewed subjects affirmed that, although they would prefer to have a stable job in the home country, they still go to work abroad, out of the need to provide a decent living standard on the family and personal level, the main elements highlighted by the subjects as having led to the decision to work abroad were the lack of a basic qualification because of their poor education and low salaries in the country, which could not cover the decent living costs, generated the decision to go to work abroad. An important element underlined by all the persons interviewed was the desire to remain in the country with the family or the desire to raise their children, things that are possible only if they have a well-paid job.

Among the advantages generated by the practising of circulatory migration, the interviewed persons reminded the fact they are very content financially, having the possibility to live better and to purchase things they could not afford before (apartment, furniture, renovation of the flat, personal effects, payment of the studies for children, clothes, dentist care etc.) but they affirmed at the same time there are also drawbacks: being far from the family and children, the stress and isolation at the place of work (being alone with the person the care for, more inside if this person is confined to bed, the absence of direct interpersonal socialization), the busy programme with the free days (24/7 at the disposal of the person cared for), the longing for home, insufficient time for attending the households in the country, especially for those in the rural environment who no longer take care of the household. They also appreciated the fact that the relation with the spouse have considerably deteriorated.

6 Discussion and Conclusion

There is a general consensus about the fact that delocalisation and circulatory migration may contribute to the improvement of the emigrants' aptitudes as a result of the experience acquired grace to the job abroad (Roman and Voicu 2010, pp. 61–62), but it leads to the damaging of family relations. The implications of the migration phenomenon on the family level are ambivalent, as they belong both to the positive effects sphere and to that of negative impacts. On the one hand, the financial resources provided by emigrants to their family contribute to the growth of their life quality, with implicit positive impacts on family relations. On the other hand, we deal with the family sufferance for the loss of one or several family members during the emigration period abroad, and in this respect the effects on the family are much thornier.

In the opinion of Roman and Voicu (2010, pp. 58–59) the consequences of migration, irrespective of its nature, are sufficient, starting also from the correlation of the demographic factor with the growth or drop rate of the origin country. The migration of adults abroad in search of a job generates a demographic misbalance, registering population drops. Such an implicit effect is the ageing process of the population in the origin country, because of the migration of the young labour force fit for work and implicitly the decrease of the population's birth rate. In the opinion of Androni (2018, p. 133) international migration occurs as a result of the differences between the demand and supply ratio in diverse countries, along with the difference between the levels of salaries and incomes. Thus, a flow of labour force is created toward the country rich in capital and poor in labour force, which becomes a mechanism for balancing the internal deficiencies on the labour market (Andrioni 2011).

From the research conducted, the following important conclusions may be drawn: (a) circulatory migration has favourable implications on the micro- and macroeconomic level; (b) the incomes of the persons working abroad favourably influence the populations' life quality; (c) the incomes of the persons working abroad are influenced by the education level and labour experience; (d) the labour force placement abroad is a profitable business for the active entrepreneurs in the field; (e) despite the fact that the economic level of the individual or family micro system increases, the repercussions of the separation from the family nucleus are serious, generating family relation issues, and the increase of the number of divorces in the origin country; (f) there is a reduction of the local labour force potential in the origin country.

Some of the conclusions mentioned above are supported also by the study *International migration and labour market impacts—Evidence from panel data,* which underlines the fact that "for the origin country, the emigration of part of the labour force has major implications (positive and negative), as it changes the number and structure of the labour force, influences consumption and investments in the source economy, and the conduct of the household members on the labour force, but is also generates negative implications of emigration on the volume of the labour force in the origin countries analysed" (Son and Noja 2012, pp. 48–49).

We may conclude that the migration issue has an important place on the social policies agenda. The indicators having led to the migration decision highlight the social change as a condition necessary for the increase of economic and social potential. International migration occurs as a result of the differences between the demand and supply ratio in diverse countries, along with the difference between the levels of salaries and incomes. Thus, a flow of labour force is created toward the country rich in capital and poor in labour force, which becomes a mechanism for balancing the internal deficiencies on the labour market.

References

Andrioni, F.: The Socio-Economic Effects of Romanian Parents' Emigration on Their Children's Destiny, Annals of the University of Petroşani, Economics, vol. 11, part 4, ISSN 1582-5949, Cod CNCSIS 23, Editura Universitas, Petroşani, pp. 5–12 (2011)

Andrioni, F.: Contemporary Social Work Between Theory and Practice. Academica Greifswald, Germany (2018)

Anghel, R.G., Horvath I.: Sociology of Migration. Romanian Theories and Case Studies (in Original in Romanian). Polirom Editions, Iaşi (2009)

Caplow, T.: L'Enquête Sociologique, Ed., Armand Colin, Paris (1970)

Chelcea, S.: Methodology of Sociological Research (in Original in Romanian). Economic Editions, Bucharest (2007)

Dan, A.: Diagnosis and solution to social issues (Course of Lectures) (in Original in Romanian), Bucharest University Press, Faculty of Sociology and Social Work, Bucharest. https://biblioteca.regielive.ro/cursuri/sociologie/diagnoza-si-solutionarea-problemelor-sociale-24511.html (2007)
De Haas, H.: Migration and Development: A Theoretical Perspective, IMI Working Papers, International Migration Institute (IMI), Paper 9, University of Oxford (2008)
Feraru, P.D.: Migration and Development, Social-Economic Aspects and Trends (in Original in Romanian). Lumen Editions, Iaşi (2011)
Gavriluţă, N.: The Mother of Fools is Always Pregnant (in Original in Romanian). European Institute Editions, Iaşi (2010)
IOM: World Migration 2003—Maps. International Migration Organisation, Geneva (2003)
Ispas, C.: Education—attenuation or enhancement factor of socio-economic inequalities in the Romanian society, in the Annals of the University „Eftimie Murgu" of Resita. Fascicle II. Economic Studies, nr. XX, pp. 52–61 [Online 21 July 2016]. Available online at: http://www.analefseauem.ro/upload/arhiva-revista/2013/Volum_2013.pdf (2013)
Motagna, S.: The Impact of the Crisis on Romanian Labour Market (in Original in Romanian). Available at: http://www.sapereproject.ro/index.php?option=com_content&view=article&id=60%3Aimpactul-crizei-asupra-pietei-muncii-dinromania&catid=23%3Amonografii&Itemid=66 (2010)
Otovescu, A.: Românii din Italia. Monografia unui grup de imigranți, 2nd edn., revised and added, Editura Academiei Române, București (2016)
Otovescu, A.: The monographic research of the romanian diaspora. J. Commun. Positive Pract. **1**, 25–32 (2017)
Otovescu, A., Otovescu, M.C.: The refugee crisis in Europe and addressed solutions. J. Commun. Positive Pract. **2**, 63–73 (2017)
Prelipcean, G.: Migration and Euphoria of Remittances. Implications of Current Economic Crisis (in Original in Romanian) in Politics Sphere Journal, Journal of Political Sciences and Political Theory edited by the Civil Society Foundation no. nr. 137, year XVII, 2009, pp. 10–14 [online 04 June 2016]. Available online at: http://www.sferapoliticii.ro/sfera/137/art02-prelipcean.html (2009)
Roman, M., Voicu, C.: Several social-economic effects of labour force migration on emigration countries. Case of Romania (in original in Romanian). Theoret. Appl. Econom. J. **XVII** 7(548), 50–65 (2010)
Son, L., Noja, G.J.: International migration and labour market impacts—evidence from panel data. Rom. Stat. J. **9**, 40–50 (2012)
Stănculescu, M., Stoiciu, V., Alexe, I., Motoc, L.: The Impact of Economic Crisis on Romanian Labour Force Migration (in Original in Romanian). Friedrich-Ebert-Stiftung, Bucharest (2011)
Todaro, M.P.: Economic Development, 6th edn. Addison-Wesley, London (1996)
Zamfir, C.: Strategies of social development (in original in Romanian), (vol. X of the collection "Theory and Method in Social Sciences"), Political Editions, Bucharest (1977)
Zamfir, C., Vlasceanu, L.: Dictionary of sociology. Bucharest: Babel. 129–187 (1993)

Associations and Groups of the Germans from the Banat Highland 1918–1940

Cristian Rudolf

Abstract The capacity of people to overcome difficulties, their power of continuous adaptation, steady progress and evolution, all come from the fact that they are social beings, they form communities, associating themselves based on different thematic groups. These are formed based on common interests, problems, and future, illustrated by various aspects of personal and public life. Having a clearly defined and explored area of interest, the various associations have a particularly important role in the community social structure. They are meant to satisfy the specific needs of man and woman, in his or her capacity of being creative, political, artful, gregarious or even playful creatures. Bringing together, in a relatively organized setting, more individuals related by ways of thinking, professions and crafts, hobbies, likes and dislikes, groups create inter-personal relationships and contribute to the social human cohesion.

Keywords Social life · The banat highland · German minority

1 Introduction

These links are particularly strong as they are motivated by the winning or keeping of rights and benefits of economic, social or political nature. If we add to the above mentioned, the factor of a minority ethnic group, dominated in the collective mental and behavioral plan by a combative attitude, generated by the desire to preserve their cultural identity, language and religion, then the groups formed based on interpersonal relationships are very important for the social structure of the researched community. The case of the Germans from the Banat Highland during the interwar years is relevant in this regard. Above all, we shall address two issues:

On the one hand, we shall present the associations formed by the Germans from the studied area and from the time interval of interest.

C. Rudolf (✉)
Department of Theology and Social Sciences, Faculty of Social Sciences, "Eftimie Murgu" University of Reşiţa, Piaţa Traian Vuia Nr. 1-4, 320085, Reşiţa, Romania
e-mail: rudi_ec@yahoo.de

D. Flaut et al. (eds.), *Decision Making in Social Sciences: Between Traditions and Innovations*, Studies in Systems, Decision and Control 247,
https://doi.org/10.1007/978-3-030-30659-5_15

On the other hand, we shall describe how the national and international political events affected the social evolution of the ethnic Germans involved.

2 Civil Servants' Associations

In the small social space represented by the village, the priest and the notary played an important role as representatives of the legal and religious authority. Thus, those who were involved in these professions got to have contact with many aspects of public life, for example with land ownership issues, but also private life, with certain aspects of the legacy issues.

The notary's association in Caras was formed on August 29th, 1926 under the chairmanship of Alexander Ruha, succeeded that year by George Berariu. As Vice-presidents of the association we mention Johann Ziepfl from Anina, a representative of the German minority in the county (A.N.D.C.S. Fd. A.N.C., Dos. 1/1926–1938 f. 1). Following the division of the Caras-Severin County, earlier that year, the body of notaries had to form two associations for the two counties based on the members' presence in the territory. Unfortunately, the documentation funds contain very few references related to the notary association from Severin County (A.N.D.C.S. Fd. A.N.C., Dos. 1/1926–1938 f. 3). They reveal how throughout 1926, there were several mentions about the clerk Georg Stricker in Orşova, an ethnic German.

In 1931, we have the first documentary mention that certifies a German notary in the interwar village of Secăşeni, the person in question being Rudolf Socek (A.N.DC.S. Fd. A.N.C., Dos. 1/1926–1938 f. 20). Moldova Veche was the residence of the clerk Sigismund Frauenhoffer and Resita of Johann Zipfl, both of German origin. Konrad Maxim was the notary of Lăpuşnicu Mare, Eduard Servou the Ciclova Romana, Ladislaus Sander of Pojejena, and Franz Müller of Iam, whereas Konrad Merkler of Berliste.

The economic crisis of overproduction had reduced the number of notaries both in the majority population and ethnic minorities. Thus, at the association of notaries in Caras in 1933, from the German community there are mentioned only the notary Konrad Müller in Iam and the notary Ladislaus Sander in Pojejena (A.N.DC.S. Fd. A.N.C., Dos. 1/1926–1938 f. 66).

During 1937, data from the archival documents provides information about the notary Peter Müller of Vrani, about Maxim Martin Konrad of Dognecea, about Sigismund Frauenhoffer of Dalboşeţ, and about Adalbert Wolla of Carasova (A.N.D.C.S. Fd. A.N.C., Dos. 1/1926–1938 f. 131). During the meeting of the notaries' association in Caras held on December 20th, 1937, these four were representing the German minority.

The ascending evolution of the German community from the Banat Highland in economic and social developments can be followed via its elite in these plans of public life manifestation. Thus, in 1938 notaries are mentioned as: Hermann Stich in Brebu Nou, Adalbert Wolla in Caraşova, Maxim Martin Konrad in Dognecea,

Sigismund Frauenhoffer in Dalboşeţ, Michael Santer in Lupac and Eduard Servou in Văliug (A.N.D.C.S. Fd. A.N.C., Dos. 10/1938, f. 49). The notaries Frauenhoffer, Wolla, Konrad and Servou, had been categorized as "set/holder" notaries as of 1926, being employed with a monthly salary of 2450 lei (A.N.D.C.S. Fd. A.N.C., Dos. 9/1937, f. 3.). This quality included a wage increase and certainty to stay a long time in office, sometimes until retirement, in the same boroughs. Once appointed to a village, the notaries concerned were much less likely to be deployed or moved to other locations (A.N.D.C.S. Fd. A.N.C., Dos. 9/1937, f. 8). Such actions could take place only in case of misbehavior, being established as a sanction by an order of the Minister of Interior or Justice.

3 Political and Religious Associations

The temporal authorities have sought to coordinate their youth through organizations that were modeled especially after the interests of the state. Their purpose was twofold: On the one hand it was sought to educate the youth in the spirit of loyalty to the state and monarchy, thus reducing the attraction represented by the political currents of the extreme right and left; on the other hand, the action was related to the introduction of a physical training program designed to improve the performance for boys in the military service and for girls to facilitate their integration in a certain workplace attributable in the case of an of armed conflict outbreak (A.N.D.C.S. Fd. A.N.C., Dos. 9/1937, f. 9).

Any student who comes to school brings with him/her not only his/her intelligence, but a whole complex of emotions, feelings, attitudes, desires, aspirations, needs, interests, etc. that school is conscious of (Zaharia 1981, p. 82). All these are put by the student in the job of training/developing him/her as a rational, creative, independent and self-controlled human being on the situations faced throughout life. It remains to be seen how the school understands to harness the whole human, material potential etc. available, to fulfil its mission to educate and train. This study aims to reveal the students' perception and attitude towards school, based on the consideration that any action aimed at school must consider the realities of it, as perceived by the direct beneficiaries, the students (Ispas 2013a, b, c, p. 79). The human personality is built only through harmonious relationship with the others. The positive interpersonal relationships give the human subject stability, safety, confidence, reinforcing the sense of the belonging to a group (Ispas 2013a, b, c, pp. 85–86).

Ethnic and religious associations were closely monitored by the law enforcement agencies, seeking to enlist young members in various associations founded and owned by the state. The German community was no exception to the rule, its youth being grouped in an organization made at the proposal and under the control of the official structures. This group was called the Association of German Youth in Romania, initiated at the suggestion of the Ministry of Interior under Article 1 of the Act to establishing the Office of Romanian Youth Education, promulgated on May

9th, 1935. It was registered in the scripts of the Ministry of Justice under Act No. 1600/1935 established by the Brasov Court, section I.

Its headquarters were in Brasov and branches around Arad, Anina, Bocşa, Buziaş Caransebes, Deta, Faget, Gătaia, Lugoj, Oraviţa, Orşova, Otelu Rosu, Moldova Noua, Resita, Sânicolau Mare and Timisoara. From the geographical order their balanced ratio stands subsidiary to the Banat. Bearing this in mind, but also the large number of members, albeit voluntarily, we can consider the association in question, as the representative organization for the German interwar youth in the Banat Highland.

The text of the founding document stated that this association had been instituted to promote "patriotic education, moral, professional, physical and religious training for German youth of both sexes aged 7–18 years." Members were divided into two sections, one for boys and one for girls. They wore black and white colors, having the right to wear their own badges and the badge of the organization Straja Tarii (paragraph added to the bylaws in 1938). In official documents addressed to the state authorities, the major language was Romanian, whereas in the internal use it was permitted to use the German. The Office of Romanian Youth Education exercised control over this organization. The goal set was to "strengthen the sense of solidarity both among the urban and rural population and between different provinces of the country" (A.N.D.C.S., Fd. C.P.B.H., Dos. 7/1939, f. 28).

Members could become young men and women who meet the following conditions: they were Romanian citizens of German ethnic origin, clinically healthy, had demonstrated an irreproachable moral behavior in society, respecting the laws and regulations of the Romanian state. Being admitted into the German Youth Association of Romania was based on a written request, accompanied by recommendations from three members of the local government. In case of internal rules breach, guilty members could be excluded up to three months and allowed back only after completing an internship evaluation.

Former members were grouped as follows: those aged between 18 and 21 years were called "Helping Members" and supported the organization through personal examples of good behavior, sharing their experience and involvement in practical activities. Former members who were over 21 years became "Supporting Members" assisting the organization morally and materially and selecting the level of leading cadres (A.N.D.C.S., Fd. C.P.B.H., Dos. 7/1939, f. 29). Among the functions "Supporting Members" there were included: the procurement of hiking donations, the purchase of educational materials and the staffing in specialized vocational training, the provision of items required in the sports fields for physically educating youth, the conducting religious ceremonies, the formation of councils in aiding the boys' and girls' leaders.

Young men and women between 7 and 18 years enrolled in the association were divided based on age: the males aged 7–12 years were called "Young Boys" and those aged between 12 and 18 "Boys". The females aged 7–12 years were called "Young Girls" and those aged between 12 and 18 "Girls". Each member was required to take part on at least four demonstrations of the association, namely: meetings, trips, social evenings at the campfire, community service programs (A.N.D.C.S., Fd. C.P.B.H., Dos. 7/1939, f. 30).

The German Youth Association of Romania uniform consisted of the following parts: for boys: black tie, white shirt with black embroidery on the collar, the badge of the local organization consisting of the initials, black pants, dark boots or tall shoes (depending on the season); the cap and tunic were dark blue or black, decorated with badges organization, whereas for girls: white blouse with black tie, dark blue skirt, boots of matching color (A.N.D.C.S., Fd. C.P.B.H., Dos. 7/1939, f. 31).

This company was coordinated in decision making by the following forums: The Executive Committee, the Leading General Assembly.

The Executive Committee was composed of a president, a vice president (who also served as the Chief of the Boys), the Chief of the Girls, a secretary, and a treasurer. The executive Committee members were elected by the General assembly every 3 years. The Executive Committee held monthly meetings, with the following tasks: to facilitate contact with the Office of Romanian Youth Education, to represent the association in its contacts with public bodies, aimed at the implementation of directives and regulations, to supervise and manage the subsidiaries in the territory, to prepare the general meeting, to co-opt those most capable officers, and to the head of the groups Management Committee (A.N.D.C.S., Fd. C.P.B.H., Dos. 7/1939, f. 32).

The Management Committee was composed of the Executive Committee and of delegates from various branches co-opted by the leadership of the organization. The meetings were biennial, having the duty to inform all delegates of their intentions as well as he Office of Romanian Youth Education and to develop a program of work units. All these forums had the right to convene a General Assembly to decide on important organizational issues or those aimed towards its eventual abolition (A.N.D.C.S., Fd. C.P.B.H., Dos. 7/1939, f. 33).

The General Assembly was formed by those chiefs who ran and headed at least two "gender flocks" composed of members of the association. Several representatives in the General Assembly could delegate a specific member with the right to vote in their names on a certain problem if that person was considered as trustworthy. Among the tasks the General Assembly we note: choosing board members, approving the annual report submitted for decision, appointing three members, approving or challenging the financial statement, annually presented to the Executive Committee, sending invitations 14 days before any meeting to all entitled members, delaying by up to an hour of opening of a meeting unless all participating members were to adopt decisions by a simple majority vote, and dismissing the Executive Committee on a no-confidence vote of two thirds of the entitled members.

Regarding the organization of boys from the he German Youth Association of Romania, there were the following organizational elements: the smallest unit of 4–10 boys called "nest", was headed by a "Nest's Chief", while three nests formed a "flock" group led by a "Flock" s Chief". Flocks of a common urban or rural origin were composing a "group" with the leader called a "Group" s Chief". He called them depending on the heads of flocks and nests for aid and was in turn invested by the Executive Committee. This organizational model proves a striking resemblance to that of the right-wing parties, be it on the native Romanian Legion, on the Spanish Phalanges, on the Italian fascias, or on the storm trooper units from Germany.

The organization was identical girls in the title of the leaders intervening female attributes. Thus, we remember about "Nest's, Flock" s and Group" s Chief ". Both the head of the boys and girls were entitled to more advisers and consultants to help them in current activities. Together with delegates from various counties they formed Central Committee of the Boys and the Girls.

The activities of the association's members were: the patriotic education of young citizens in the spirit of loyalty to the laws of the state; the moral education with the appropriate respect for civil rights, but not for civil liberties, honoring his or her word and assuming obligations undertaken towards others; the professional economic training for raising the homeland and the living standards of fellow citizens; the physical education for development of the body in a smooth and balanced way through intellectual and practical activities; the hygienic information to establish better conditions of housing, food, clothing, leading to the adopting of a healthier lifestyle; and the religious propaganda to infuse every boy and every girl with good quality Christian values.

The meetings began with saying the creed adopted by Straja Tarii, followed by German Youth Association of Romania anthem "Three Colors ..." and salute this organization, "Heil" in German, "Health" in Romanian. In the camps, the state flag and the flag of the organization rose and went down every morning and evening, the ceremony was accompanied by a public prayer.

The descriptive elements in the statutes of the German Youth Association of Romania indicate an organization with deep totalitarian political characteristics. This statement is based on three considerations: firstly, they tended enrolling a larger number of members, only based ethnic reasons, without considering the personal choice of each individual; secondly, they retained the paramilitary organization structure and emphasis on physical and theoretical training program, containing many elements similar to a military training program, thirdly we note the symbolic elements present both in uniform and in badges, flags and salutes of this youth organizations. These features of the political extreme right are not an invention original to the Romanian political space as they could be found also in the Austrian, Belgian, Croatian, Greek, French, German, Italian, Hungarian, Norwegian, Spanish territories. The Romanian decision-making bodies had changed only some components outside of their own specifics, keeping essentially the right-wing social program characteristic from the European continent at the time. Given the ethnic minority concerned, the totalitarian discourse, even changed from a formal point had been accepted without too many objections, especially by young people and adolescents.

Along the course of time there were political associations and other similar organizations sponsored by the religious confessions prevailing in the German minority in the Banat Highland during the interwar decades. They were constantly monitored by the authorities, precisely because they were outside the political control.

Roman Catholic Diocese of Timisoara founded in 1930 was engaged through its subordinated clergy and laity in numerous actions to assist the elderly or suffering. Youth was not ignored, of course. There functioned associations of adolescents and young people led by clerics or lay people closely related to the decision-making structures of the Roman Catholic Church at regional level.

For example, the nun Hildegardis Wulf, coordinated the association "Mädchenkranz" "Laurel girls" which each year organized pilgrimages to Rome and meetings in Herculane. A lot of people in this association were of German ethnic origin. The authorities looked for trips abroad and were interested in taking part along the members of the German community. Meetings held for fundraising were also closely monitored to detect early on any political activity contrary to state interests. Most times the reports had nothing special to mention. However, periodically recommends from the authorities in Herculane and those in Orşova were issued to check Hildegardis Wulf's luggage to find and confiscate any political material considered subversive or seditious. Such a case occurred on August the 13 h 1937. The last action proved void of real results, as the documentary sources of that age stated.

Another association coordinated by the Roman Catholic Church was Roman Catholic Youth Meeting in Banat, founded on May 4th, 1929, with headquarters in Timisoara and branches in Caras- Severin and Timis-Torontal (A.N.D.C.S., Fd. D.P.B.H., Dos. 3/1937, f. 204). In the Banat Highland, this religious organization was represented at Anina, Baile Herculane, Bocsa Montana, Caransebes, Oraviţa, Orşova, Mehadia, Moldova Noua and Resita. Among the founders, we mention: Josef Frank, the first president, Rudolf Hollinger, Michael Stocker, the first secretary, Dr. Franz Schmitz, Josef Trittler, Fritz Albrich and his brother Norbert Albricht, Andreas Willwert, Stefan Fassel, Johann Wendel, Peter Luchs, Franz Marx, Johann Burgos, Andreas Seiler, Josef Krumer, Stefan Gores, Rudolf Beer, Josef Frank, Franz Jakobi, Franz Maurer and Franz Leitnek. Following a spontaneous decision, of that date a coordinating team was chosen unanimously for that association, made up of the following members: chief—Dr. Franz Schmitz, president for assets—Josef Frank, vice president—Friedrich Albrich, secretaries—Michael Stocker and Stefan Fassel, choir—Johann Wandel, sports director—Norbert Albrich, members of the supervisory committee—Andreas Willwerth, Luchs Franz Josef Krumer, Rudolf Hollinger, Peter Marx, Andreas Seiler, Stefan Görres and Johann Bürg (A.D.T., Fd. C.G.E.G., Dos. 1/1929–1944, f. 21).

The legal recognition occurred on October 25th when in the ruling on File No. 4.145/1929, the Timis-Torontal County Court, Section I, ordered the recognition of the Roman Catholic Youth Meeting in Banat as a legal person. Just the name was changed to the Association of the German Roman Catholic Young Men in Banat because the officials felt that this name would have been more effective in avoiding confusion. The Religious Affairs Ministry had already given a favorable opinion with the No. 26.258/July 8th, 1929 and the procedure for legal recognition had thus been completed (A.D.T., Fd. C.G.E.G., Dos. 1/1929–1944, f. 22).

Regarding its statute, the membership organization approved on May 4th, 1929 and subsequently submitted to the law enforcement agencies, provided a description rich in information on the activity and organization of Roman Catholic youth from ethnic Germans in the studied area (A.D.T., Fd. C.G.E.G., Dos. 1/1929–1944, f. 23).

It was specified at the outset that the purpose of the association was to promote the culture, i.e. reading and songs in the countryside, to combat the decay of morals and to improve the quality of life for citizens and to give relevant examples of good behavior in everyday life.

Within the organizational structure there were two categories of members: A and B. Category A was made up of all young people aged 15–21 years old, married or bachelor. Category B consisted of four groups of members: founders, honorary, ordinary and auxiliaries (A.D.T., Fd. C.G.E.G., Dos. 1/1929–1944, f. 24). Members of the class A were exempted from paying any dues. But they had the following duties: to take part in winter evening lectures, participate in major religious and national holidays, take part in the funerals of fellow association members and have an exemplary moral behavior. When receiving membership in the association, each person must sign the following statement:

> I hereby declare on my conscience that my duties membership engaged, which are specified in the present statutes, I quite know, and I will fulfill with all my sincerity and good will. Further I declare that now and, in the future, I will not be part of any company whose trends are against morality, against religion, against the state and against the throne. (A.D.T., Fd. C.G.E.G., Dos. 1/1929–1944, f. 25)

The "Founding members" were those who, from the foundation of the association, acquitted a fee of 200 lei. "Honorary members" were chosen from among those persons of high moral standing and unblemished reputation, and they oversaw the usual activity. "Ordinary members" were those people, married or unmarried, who were over 21 years and who paid the annual fee in January of 21 lei (A.D.T., Fd. C.G.E.G., Dos. 1/1929–1944, f. 26). "Auxiliary members" offered aid society morally or materially. Those members in category B could participate in all activities of the association without paying entrance fees and funeral expenses, in case of death they are supported by members of category A.

Association of the German Roman Catholic Young Men in Banat was conducted by the following officials: the leader or chief, the active president, the vice president, a treasurer, two secretaries, the choir, sports leader and the monitoring committee. The leader was elected by the superior culture of the people who have not violated any civil or religious enactment and showed patriotism and civic spirit, he or she was as active member in category B. The Active President was elected from the unmarried members of the class A, being at least 18 years of age. The mission members to the active Plenipotentiary of the President and vice-president, were elected as the previous one. All other members of the management were established by the same criterion. The People Supervisory Committee were members who had passed the legal age of 18 years and their position was not put to the vote. Their number rose to 18 people, eight were members of category A and 10 of category B (A.N.D.C.S., Fd. C.P.B.H., Dos. 7/1939, f. 122).

Regarding the duties of the association leaders we can mention the following: The leader was tasked to preside over all meetings and celebrations organized by the organization concerned and had the responsibility to inspire youth with the purest moral and religious feelings. He was responsible for the active and passive wealth of the association and was represented by the active President.

It was entitled that in case of vacancy of the leader, the Vice-president would take the specific tasks in turn. The two secretaries were also thought to keep in good order the association documents, namely the minutes of meetings, invitations, summonses, notices for presentation to the Active committee.

The cashier managed associations funds issued receipts of revenue and expenditure but was not allowed to receive or pay higher amounts of 100 lei without prior notification to the Active president or Vice-president. At each meeting the cashier was tasked to show this register of receipts and expenses and the balance of liquidity of the association. Parents were responsible for their children towards the cashier even with their wealth of active and passive, for their work in this regard was registered at the Court of Auditors. In the coffers of the association could not be lodged a sum greater than 1000 lei, money that exceeded this figure was immediately taken to the bank where the cashier had opened an account for storage and saving funds and deposit slips, stored with the cash of the organization's headquarters.

The Supervisory committee meet monthly from October 1st until April 1st in ordinary meetings on April 1st and October 1st in an extraordinary one. In emergencies, the could take place more than one extraordinary meeting. One of the Committee's duties consisted in the validation of new members by secret ballot, in case of a tie, the Presidents vote was decisive. Another task was that of choosing to highlight the financial controllers, the librarian and civil economist. In collaboration with the choir and sports leader, they had specific tasks dictated by their field of activity.

The General assembly held every year in January, had as one of its duties to reconfirm the model approval or round shaped stamp of the association with a cross in the center and the organizations name in German and Romanian. The General assembly could vote to dissolve the company, but this was achievable also by order or requirement of civil and religious authorities or if there were no members of the category A. In the event of dissolution, the company's assets and liabilities would be converted into cash and deposited in a new bank account to be used in case of reestablishment or if another Association for Roman Catholics youth of ethnic German origin should emerge. Financial documents were submitted at the headquarters of the Roman Catholic Diocese of Timişoara.

This information induces the findings of youth associations among the ethnic Germans of the Banat Highland, on two areas: firstly, there is mention of associations based on the state line, on proposal of the relevant ministerial bodies and largely controlled by them; secondly, we remember associations formed spontaneously by Germans, but under the direction and patronage of the Roman Catholic Church in that part of the Banat. The two types of associations (political and religious) were ongoing and as far the information in archival funds discloses, they had a mutual life without conflicts.

4 Sports and Excursions Associations

Education is in constant change. Education is required to prepare the man to face these numerous changes. School reflects social, economic, cultural and political tensions present in any society. It is also required to mitigate the socio-economic inequalities (Ispas 2013a, b, c, p. 52). Socio-economic inequalities translate into differences (of status, of access to resources, of their possession etc.) considered to be unfair for the development of a person or a community. Socio-economic inequalities are deeply rooted in history. We may say they have appeared and developed with humanity (Ispas 2013a, b, c, p. 53).

The Latin proverb: "mens sana in corpore sano", "a healthy mind in a healthy body" was deeply appreciated by the Banat Highland Germans during the interwar decades. They applied the saying in daily practice in two ways: firstly, by creating different sports associations, and secondly, through trips and vacations in the profile organizations. The two types mentioned cannot be treated separately, if only for the fact that the hiking represent a form of exercise, with sporty and organizing holidays and trips and it remains an activity addressable to all socio-professional categories, including amateur athletes or professional ones. Through regular training and participation in competitions, through hiking and vacation leaves made into a proper organized framework, members of the minority have found ways pleasing and useful for leisure, especially for preserving their personal health. In the following we will describe some of them:

"The Tourism Association of Resita", "Reschitzarer Touristenverein", led by the triumvirate consisting of Stefan Haudek, Paul Windhager and Bruno Fulda appeared in 1926 as a swimming club (Ausflugprogram 1924, p. 112). Leaders of the association at the time, persuaded the city's wealthy residents to donate funds for building a swimming pool along the Bârzava Valley. It was inaugurated on June 2nd, 1929. It had a festive aura be known today as the Lido tourist pool. There are still some people in Resita, former officials of the S.T.e.G. who remembered to have booked tickets, when it was opened in 1929. This recreational area was near the hydro power plant Grebla (Brăftălean 1998, p. 86).

Resita's tourist excursions were not limited to this swimming pool, given that next to Anina, there lay the Maialis Glade and the Buhui Valley who were in the terminal lake of the same name. Members of these said tourism organizations realized hikes that targeted the Peaks Ţarcu, Godeanu, Pietrii and Retezat, located at 2000 m altitude, which could easily be achieved during a day's trip, departing from Caransebeş, Oţelu Rosu and Baile Herculane (Brăftălean 1998, p. 87).

In 1938, "The Reschitzarer Touristenverein" headed by Stefan Haudek and Franz Heim had decided to build a cottage on a mountain saddle between peaks Gozna and Semenic, not far from the small glacial lake known as Baia Vulturilor (Reschitzaer 1930, p. 2). Work began in mid-January due to unusually mild weather. To facilitate the transport of materials, the Caras County Prefecture had ordered the paving of the access road from the Canton Prislop to the peaks Gozna and Semenic (Reschitzaer 1930, p. 3).

The German community of the "pemi" villages Garana and Brebu Nou had become involved in the 1930s in seasonal tourism. In 1928, the entrepreneur Josef Weinfurter of Garana opened a hostel in the village and Eugen Loidl one Brebu Nou. Eugen Loidl originally wanted to buy the inn "Markonisch" of Brebu Nou, but he had missed the opportunity due to the lack of funds. For this reason, he acquired a building behind the local church, renovating it for use as a guesthouse. The said location in Garana was equipped with a spacious dining room and one dance hall. It was planned to construct a hotel, project that was not completed until after the end of

World War II. 1934, Martin Köstner, another land owner and Gărâna entrepreneur, opened a hostel for skiers during the coming winter, being helped by Alexander Pap and Paul Eichinger who erected each alternative operating inn, in summer or winter. They were used for passenger transport, buses being purchased after an agreement with the Union of Romanias Car Owners and during winter, the patrons' pensions sending their sleds after visitors up to Slatina Timis.

In 1934 the idea was born to build a motel for Roman Catholic teachers and priests, aided by the Funeral Fund of the Roman Catholic Diocese of Timisoara. The teachers at "Banatia" high- school of Timisoara had registered and contributed to this project. Under the leadership of the school inspector Josef Johann and the teacher Etienne Kiry in Garana, both of German origin, there had been chosen a plot of 1 ha, purchased by the town hall. Construction plans had been drawn up by the architect Franz Leitner in Timisoara. On July 4th, 1937, in the presence of over 1000 guests of the German community in the Banat Highland, the Canon Josef Nischbach hallowed the edifice. The building had 8 rooms on the first floor and 4 on the attic.

The Hiking Association "Reschitzarer Wandervogel-Vereinigung" was founded in 1923 by initiative of the teachers Bruno Windhager and Alexander Tietz. Along with trips there were organized screenings of slides and classical music concerts of opera composers such as: Johann Sebastian Bach, Joseph Haydn, Wolfgang Amadeus Mozart and Richard Wagner. In 1928 the company bought an unfinished building on the outskirts of Resita and its members renovated and converted into office spaces. Wandering through nature, cleaning streams, and song contests seemed to be ways of leisure for the members of the Wandervogel- Vereinigung. The expeditions were organized not only nearby but also on the Danube River, the Black Sea, in Austria and Bavaria, especially during the summer holidays. The association joined in the club "Süddeutsche Vandervögel", "South German Birds Hiking Club" which in turn became in 1932 part of the "National-Sozialistische Selbsthifebewegung", "National Socialist Self- Help Movement". The members of the Wandervogel—Vereinigung were not concerned with politics, although most were enrolled in the group "Sozialistischer Arbeiter Jugend", "Socialist Youth Workers". On its significance, the monograph Karl Ludwig Lupşiasca expressed the following points of view:

> I was, like all young people, of all ages, hungry for events. The association gave them to us. It educated and raised us, us, children of workers and craftsmen. It awakened and nourished in us the love and interest for German culture, strengthened our belonging to the German spirit, without degenerating into chauvinism. Tietz believed a good craftsman is worth as much as an intellectual. What they experienced, marked the members of the association for the rest of their lives. (Lupşiasca 2000, p. 160)

One of the "Wandervogel—Vereinigung" members' frequent destination was Muntele Mic, near Caransebes. In 1933 there was built the cottage "Theresia" by the brothers Edward, Viktor and Oskar Prohaska (Lupşiasca 2000, p. 161). The building could accommodate 35 people. Downstairs was the kitchen and the dining room, while upstairs was the manager's office and the attic alcoves with two beds for guests (Lupşiasca 2000, p. 162). At the initiative of Viktor Prohaska, who had studied in Graz hydraulics, and of the Timisoara Electromechanical Works Director, Cornel

Miklossi, a dam was built not far from the creek Sebesel, near the guesthouse called "Bella Vista", to which had access to members of the "Wandervogel—Vereinigung" (Ghinea 1996–1998, p. 101). It had running water, sewerage, electricity, central heating with steam and telephone. It was composed of the basement, the ground floor, the first floor and the attic. Rooms consisted of the kitchen, the dining room, the living room, 2 bathrooms and 12 bedrooms, with a capacity of 24 beds. The living room was equipped with games, radio, gramophone, records and a library containing 300 volumes donated by "Royal Foundations Publishing House".

The sports activity had not been neglected all this time, Romanian and German workers from Resita founding in 1919 a Section of sport (S.S.M.R.). It was coached by a club of athletes from the company U.D.R. and managed to attract the best players and coaches in the country, in 1929 winning the national championship of the king sport, namely football. Most amateur athletes in Resita frequented the restaurant "Rotter". There resided the Association of cyclists and motorcyclists "Vorwärts", "Forward". Adalbert Rotter was president and Albert Henn vice-president. Motorcycles racing, such as the one of July 5th–6th 1930 were subsidized by the U.D.R. Usually they included two stages on the route: Moldova Noua—Caransebeş, turn and return. Participants were originally from Timisoara, Arad, Orşova, Oradea, Cluj and Bucharest.

Along with organizing motorcycle races, the dispute wrestling matches took place as of November 2nd, 1930 in the National Championship "Ringmeeting" held in Resita. The city had been subsidized by the with commemorative statues on behalf of all contributors. These statues made by workers from the ferrous foundry of the U.D.R. company were of silver and depicted a sports wrestling competitor holding in his left hand a crown of laurel in bronze and in the right a banner of gold that read in Romanian and German "Prize Reşiţa" "Wandpreis der Stadt Reschitza".

Suspicions of extremist political activity prohibited by law had been extended, unfortunately, to the actions undertaken by sports organizations. "The German Association of gymnastics Orşova", "Deutscher Turnverband Orşova" had been under surveillance on August 25th, 1937 as members Czech and Slovak local minority reported to law enforcement agencies, that the meetings in question could carry out anti-state activities, disguised as sports exercises. In conclusion, the Interior Ministry ordered police in Herculane and Orşova to maintain the association with sporting character in a careful and discrete control (A.N.D.C.S., Fd. D.P.B.H., Dos. 3/1937, f. 234).

All Czech and Slovak community members informed the police of Caras and Severin Counties that the young Germans in Sadova Veche, aged between 15 and 21 years had formed a football team called "The Sadova Veche Sport Association", "Sportverein Alt- Sadova", expressing its intention to donate proceeds to the needy residents. The team was coached and led by Armundt Ligl, Johann Covaki and Johan Wolf. The team, numbering 28 persons, had participated in the meeting of June 18th, 1939 of the German youth in Banat, held in Timisoara, at which there had been invited the main leader of the "Hitlerjugend", "Hitler Youth" of Germany, Baldur von Schirach (A.N.D.C.S., Fd. C.P.B.H., Dos. 7/1939, f. 3).

It was however specified in the explanatory note, that the uniform of the team was not playing towards Nazi propaganda or comprising military elements, consisting of black shorts and white shirts with green straps without any pins.

These information lead to the conclusion of an intense sport and tourism activity, carried out by Germans in the Banat Highland in between the world wars. The defining element of it was being tied to character made up of minority associations and their level of organization. On the one hand, we can conclude, backed by bibliographic information and archival funds, that the organizations for sport and tourism had an amateur ethnic issue, being made up of adults and youth with the main concern relating to school education, industrial work, but had little connection with professional sport and tourism. On the other hand, we highlight the degree of organization of trips and sports tournaments, the efficiency of their deployment and good cooperation with the local state authorities and the regional economic firms for successfully managing the necessary resources for the organization and conduct of such events (A.N.D.C.S., Fd. C.P.B.H., Dos. 7/1939, f. 71).

5 Pensions and Mutual Benefit Associations

The Latin classics pointed out that one of the main purposes of community members consisted in providing "consilium et auxilium", meaning "advice and help" for their neighbors in case of need. The often weather- tried Banat Highland imposed a harsh life on its inhabitants, regardless of ethnicity (Ispas 2013a, b, c, p. 80). The land had been a mentor relentlessly struggling life, strengthening them and emboldened them to the vicissitudes of fate, constantly changing, confirming the proverb "harsh craft, rough teacher" (Ispas 2013a, b, c, p. 81). The German minority members assisted each other on many paths: either spontaneously or through specific associations or within specialized institutions and funds, under the patronage of the main industrial firms in the micro-region, namely the U.D.R. The following examples support this claim:

The German farmers' existence in the Semenic Mountains was a difficult one in the first half of the twentieth century. Technical inventions made their place slowly in a conservative society, where animal traction force and the number of laborers accounted for the means considered safe for welfare materials. This detoured prosperity, often even for families with many members and many animals. The land inhabitants known as "pemi", were located at an average altitude of 1000 meters. This allowed them only a reduced cereal crop, the potato remaining the essential food for subsistence. Other branches such as fruit agriculture and animal husbandry were, to a point, more prosperous. Since the nineteenth century there was a custom for each household member to go to work in the exploitation of wood from the surrounding forests or in an industrial center: in Resita, Anina, Oţelu Rosu etc. Additionally, the financial situation was often hampered by bad weather and household accidents with disastrous effects for a whole village community (A.N.D.C.S., Fd. C.P.B.H., Dos. 7/1939, f. 83).

The monographs Karl Ludwig Lupşiasca and Josef Schmidt supported the view that the existence of German villages in Banat Mountains was marked by frequent natural disasters. The events held this view, showing that life for the "pemi "communities was far from being a happy one (A.N.D.C.S., Fd. C.P.B.H., Dos. 7/1939, f. 99). Floods and fires from Sadova Veche and Garana proved to be particularly destructive for them. A major fire took place on November 9th, 1929 when a household barn owned by Franz Köstner, resident of the village Garana lit up. The wind had carried the flames to the other barn and the house of the unfortunate owner, and then to other buildings in the neighborhood, stopping just in front of the brick walls of the Franz Pfaffls household. Through the vigorous intervention of firefighters and volunteers from Garana and Brebu Nou the fire had been stemmed with no casualties. Instead, it had produced major damages 11 homes and 12 outbuildings were consumed by flames. The value of the goods destroyed amounted to 250,000 lei. Until November 9th, 1930 much of the damage was compensated by generous donations from: the local municipality- 41,000 lei, the German community in Resita- 64,000 lei, in Brebu Nou- 12.510 lei, the U.D.R. General Directorate- 10,000 lei, the County German Organization based in Lugoj- 4400 lei and the Văliug Town Hall- 3270 lei. Thus, they gathered in a calendar year 135,180 lei, but the difference of 114,820 lei was covered by the owners who suffered damages. We note that the aid had risen to more than half the total amount, but that it had failed to compensate all damages.

In the context of increasing external pressure and prefiguration of the Second World War, the German minority had been carefully monitored by officials (A.N.D.C.S., Fd. D.P.B.H., Dos. 10/1940, f. 60. Given that it was anticipated that the Third Reich would trigger a military conflagration, associations formed by ethnic Germans in the Banat Mountains were attentively monitored. The action found its justification in the desire to ensure a climate of social calm, but mostly because the Romanian state wanted to prevent acts of sabotage in the most industrialized part of the country. The following information is supporting this view.

On June 26th, 1939 the police station in the town Herculane reported that in Sadova Veche, the ethnic German citizens over the age of 21 had formed a mutual aid committee in case of accidents, natural disasters, war, etc. It was led by Augustin Milota and counted 34 members. Each of them possesses a book that apply stamps worth 2 or 5 lei per month. These stamps were manufactured in the Third Reich and brought monthly from Timisoara by the leader of this organization.

The information notes of July 5th, 1939 sent by the Herculane police to the Interior Ministry, it was noted that a mutual aid committee, recalled in Sadova Veche and Garana had spread in the area, although the number of members of both villages was only 32 people. The organization adopted the name N.A.F.: "Arbeitsfront Nazionale", "National Labor Front", the contribution of each member ranging between 10 and 25 lei per month. Stephan Neff in Resita, a professional photographer, was identified by the police as the leader of this organization.

The history of family social work, as practice activity, unorganized, may be mistaken by human history. The affirming of family social work as organized professional activity, specialized, is considered more recent and can be placed at the level of the 20th century. In Romania, the domain of family social work is far too little exploited, the problem of family social work is being approached more collaterally and little through ideas referring to this subject. This state is in fact easy to observe by analyzing specialty literature in this domain. On a national plan, studies referring to family social work are extremely reduced in number and theme approached. This fact leads us to think that the decisions taken domain of social politics that aim at the family aren't complete in the plan of their scientific foundation. Thus, without any further discussion, the quasi-absence of a solid theoretical-practical support based on family social work is also sensed at the level of those activating in this domain.

The management sent a request Land Timis Police Herculane on 8 July 1939 on the petitions "of German private individuals and the German community in Banat", related to the creation "of homes for children in German communes".

Stated that before approving this request, police stations in the territory of the province had to send detailed reports about the nature of these charities. Herculane enforcement bodies had the task: "to collect in the most unobtrusive information on these homes and that is the true purpose of their establishment" outcome of the investigation will be communicated until July 13.

Information note of July 9, 1939 brought some additions in the discussion shelters for German children. The leaders of county of Timis showed that it was a draft from "Social Sectors of the German community in Timisoara" to open in villages with population of native previously called across the counties of Arad, Timis-Torontal, Caraş and Severin, hospices day "for supervision of children 2–12 years older Germans, while working in the fields and harvest i.e. between July 9 and August 25 this year." Police representatives are claiming land territory to monitor closely these institutions care for minors to impose strict adherence to the timetable.

On January 13, 1940 Police in Herculane sent a report by the management of State Security in Timisoara which stated that members of the German town had produced a collection worth 15,000 lei for the establishment of a "social center German" local "Deutsches Heim". Given the fact that this collection of funds for an institute of social support appear to be generalized to the whole Banat requesting follow carefully this project to detect whether under the guise of charitable activities were held events political subversive. On June 12, 1940 law enforcement agencies in Herculane and Orşova Interior Ministry reported on the mechanism of fundraising elements within the German community:

> Each local organization of Germans gather monetary donations, they are sent Gauleiter of Timisoara, and from there is sent by special courier to the organization's headquarters Germans in Romania, located in the capital. Among the most couriers are chosen little-known law enforcement agencies to be spared supervision.

In collaboration with spontaneous initiatives and help association formed by Germans, the respective members have played a role in some major pension institutions coordinated by employing local companies. Given that U.D.R. CPC was the main industrial enterprise and mining area, many minority contributions submitted and received aid from its pension Institute. Together with the Romanians, the ethnic majority, but also with others, those concerned had integrated in a system of private pension and mutual assistance they had provided the necessary funds for subsistence.

The institution was mentioned above, succeeding the organization "Bruderlade," "brotherly house" appeared on January 9, 1925. Its tasks are to ensure pensions for old age, sickness and accident vocational training for workers, officers and directors UDR and provision of allowances for widows and orphans of workers concerned. Body have a management committee composed of 14 members: 1 president, usually union leader, one vice president called it, 4 members elected by workers, four appointed by the management's U.D.R. and 4 selected from retired workers. Voting is done by direct ballot, equal, secret and freely expressed, and the top finishers were advanced management for validation.

Members involved had to contribute 20% of your first paycheck, including rent allowance, 35% of salary increase for a promotion and 5% of usual monthly income. In the case of a 3-month unpaid leave, employees assigned to the pension scheme had to pay 10% of their last monthly emoluments due. The same was true where a worker was mobilized into the army for six months. In case of abandonment of payments that exceeded half a year, covering the unpaid amounts could be done to return to work. If a member retires after 10 years, he could continue the payments in a manner like the case furlough. Any employee who had contributed at least a decade entitled to a pension from the body cited. This totaled 30% of the last monthly salary. If a subscription for two decades are added 2% each to the final salary of the U.D.R. for each year of the next segment contributions. For every year exceeding the limit established by levies an additional 3% of the remuneration last received. If a worker suffered an occupational accident and have more than 30 years aside, then a decade of contributions is automatically added to the number of years already contributed for calculating pension (Reschitzaer 1930, p. 2).

Widows were entitled to a survivor's pension if you were married at least one year before retirement employee spouse. The duration could be shortened only if the deceased was suffering from a chronic disease or if an accident was professional. Widows, spouses who were over 50 years old or older were older than 15 years they were not entitled to a survivor's pension. Those whose husbands had paid between 10 and 13 years received a pension worth 23% of the last monthly salary, supplemented with 1% if the contribution period ranged between 14 and 40 years. Who remarried lost the right to survivor's pension. Orphans of one parent were getting 8.33% of the last monthly salary father, brothers or sisters if they had, the percentage being 16.66% if they were sole survivors. Orphans of both parents had the same pension rights as widows. Pensions are paid up to the age of 22 years only if the boys did not have a job, and for girls, if they were not married. In the 1930s, against the background of the economic crisis of overproduction, had set the pensions are 28% of the last monthly salary for a fee of 53% a decade of uninterrupted same amount for her contributions between 11 and 35 years (Zaberca 1996, p. 102).

Analyzing the activity of associations of mutual aid and pensions, both those made by the Germans, and those that were active members of this ethnic group, we observe a permanent care to ensure financial means to help contributors to overcome more easily than certain unfortunate events of their lives and continuing concern for financial security interval following termination of professional activity. Lay constant pension funds and mutual German minority community members had hoped to be able to be callable funds in an emergency. They largely succeeded in this, as demonstrated by the information discussed in the chapter on economic life.

6 Professional Guilds and Trade Unions

Banat mountain was, without doubt, one of the most industrialized areas of the interwar Romania. The presence of large numbers of skilled workers or qualification course, the secular tradition in the Mining and Metallurgical abundant underground resources, which contributed to the development of an active life economic center of gravity focused on the main industries. Since the second half of the nineteenth century industrial workers formed union professional organization type, part of the dual monarchy similar organizations. The first part of the twentieth century, these unions have won an important increase of authority in social and economic life of the area by the successful completion of negotiations with employers to improve living and working conditions.

After the Union of 1918, trade unions in mountainous Banat established contacts and cooperation with similar organizations across unified Romania, taking part in economic and social activity. It is interesting to note that some organizational structures of trade unions group "Men of trust", "Vertraungsmänner", founded in the decades before 1918 had continued to operate successfully. Germans were good many members of local unions, with political beliefs or social democratic socialist interpellation fact evident from the following data:

Most of them members of professional bodies of miners and metallurgists, manifested sympathy with the center-left political program. Many were members of the Social Democratic Party. On 25 May 1919 conference held in Timisoara in which Socialists leader Josef Mayer proposed the establishment of an organization for Banat, keeping the previous program and misalignment world conflagration at any social-democratic central to clarify the territorial belonging of Banat. Zonal workers were represented by their German colleague in Resita, Georg Petkovics (Hromadka 1982, p. 91). On the 7 and 8 September in Sibiu took place a meeting of the Social Democratic organizations in new provinces joined the Old Kingdom that participated two representatives of socialists in Banat. It had fallen to agree on a statement of principle by providing a unified socialist organization in Moldova, Muntenia, Oltenia and Dobrogea.

Divergences within the socialist organizations in Banat became evident when the conduct of the congresses II and III of the Socialist International, Geneva and Moscow in February and March 1919 (Hromadka 1982, p. 92). Participants from Resita and Anina, unlike those in Timisoara requesting more powerful alignment with the third International. At its meeting of 20 January 1920, the extreme left wing of the Socialists in Banat succeeded Josef Mayer's removal from the leadership of the regional organization (Hromadka 1982, p. 94). On April 20th, it was decided change of name of the political organization of the Social Democratic Party Socialist Party of Banat. Unlike their colleagues and co-ethnics in Timisoara, the German union of Mountain Banat adopted the thesis of class struggle and the dictatorship of the proletariat welcoming the formation of the former Soviet Republics in Russia, Hungary and Bavaria (Hromadka 1982, p. 95).

On 15 August 1920, held in Cluj-Napoca congress socialist parties and trade unions Banat and Transylvania. Starting a union federation had been carried out without problems, however he could not complete the unification of the socialist parties (Brăftălean 1998, p. 73). September 12 of that year had convened the first meeting of postwar Metalworkers union leader, Johann Mountain Banat under Staudt. Regional power trade unions from Banat, based in Timisoara, participated Socialist leaders and Rudolf Josef Gabriel Weltner. The proposal issued by the latter, the Assembly decided Metalworkers Union accession founded in Cluj-Napoca. Staudt union leader submitted his mandate because it had expired in 1918 and in its place, were two other elected union leaders, namely: Josef Ebert and Peter Bernau (Brăftălean 1998, p. 77).

Due to differences in joining III Socialist International, metallurgists' workers from Resita decided to December 20, 1920 of the separation of socialist organizations in Timisoara. It had not been ruled out, however, from the start collaborating, communicating those trade unionists in Timisoara Resita that if some delegations undertaken for their support of common interest will be given unreserved material:

> The local organization Resita decided that before the establishment of a central country to remain passive. Please support our members who come to you, as the accompanying table, and we pledge to pay your bills. Instead, members properly support local organizations belonging to Central or county organizations, if they move and do not ask us anything, we currently having enough funds. Please activate all the power so that much-desired and much needed country central to life as soon. (Brăftălean 1998, p. 95)

The legislative framework for the conduct socio professional groups had been established by law to organize labor unions of 26 May 1921. Per the trade unions to be confirmed as legal by a court in the county training. Although the law in question allowed only purely professional organizations without political affiliates, though unions had been and remained closely linked to the socialist parties, through their experience of fragmentation and fusion. For example, at 19–21 July of the same year, it took place in Ploiesti parties and socialist congress at which it was decided formation of the Blacksmiths Workers, Metallurgists and Chemists in Romania and the Mineworkers Union (Brăftălean 1998, p. 96).

Many German workers in these industries specific urban centers joined the Banat mountain top called professional associations. German workers elected leadership of these unions of employees we mention Anton Daum, Peter Bernau, Franz Wein and Leopold Geislinger. Socialists in Romania have decided their accession to the Union of Socialist Parties in Europe in Vienna. Resita, was formed on August 6, 1921 the Steel Workers Central Trade Union, led by Peter Bernau. They had followed them as union leaders Peter Rohm, Franz Klein in 1930 and in 1935. In the same year Anina had formed a syndicate led by mining Eftimie Gherman and then Rudolf Hirschvogel and Johann Banks. In 1938 they were succeeded as leader of the communist union activist by Rudolf Moser (Lupşiasca 2006, p. 97).

In 1923 he formed the first trade union in communist Mountain Banat, led by Koloman Müller, International Federation of Trade Unions affiliated to the Communist headquarters in Amsterdam. He was mostly composed of employees of the mining company U.D.R. A syndicate consisting of forestry workers have U.D.R. Communist sympathetic and options was formed in the same year and on January 27, 1924 was composed of a Communist union workers' metallurgist. To alleviate the animosity between workers with socialist convictions and those with Communist sympathies had been convened between 7 and May 9, 1927 Congress trade unions in Romania. The workers from Resita they delegate to the congress Victor Brăftălean and those from Anina, Johann Nemcsok. Congress decided the creation of a unified central union for Romanian Banat region divided into sections for each locality that operated industrial workers (Lupşiasca 2006, p. 98).

Parliamentary elections in June 1920 have applied, the Romanian workers, Hungarians, Serbs, Croats, Czechs, Slovaks and, not least, of the German Banat Mountain, Josef Mayer for a deputy seat and Johann Staudt one senator. Was first elected to Parliament from Caras-Severin county without its mandate have been validated. Six years later, Koloman Müller ran as deputy in Timis-Torontal, Caras Severin and. It had obtained 4.88% of the vote in Timis-Torontal and 12.6% of the vote in Caras Severin and 2.87% in the Parliament but had not entered because the party represented Bloc Workers' and Peasants, had not reached the level of 2% of the vote representation at national level. A year later, in 1927, the situation is repeated: massive votes helped the German workers the same social-democrat candidate had obtained 5.73% of the vote, Timis-Torontal, 9.18% and 4.37% in Caras in Severin not serve as deputy electoral procedure due to the problem of the foregoing (Reschitzaer 1938, p. 1).

On November 10th, 1928 National Peasant Party formed a government with the collaboration of the Social-Democratic Party, German and Hebrew Party. Romanian and German Social Democrats in Caras Severin and they elected as deputies on Eftimie Gherman and Johann Steinbauer and those in Timis-Torontal Josef and Josef Pistiner Proschaszka. Strength Social Democrats had been confirmed to the local elections in February 1929. In his book, Victor Brăftălean appreciate the contribution German workers alongside Romanian, Hungarian, Czech, Slovak, noting that they had contributed to strengthening local socialist organization, "although they and he had to fight against apathy and disinterest part of the proletariat".

Of the 15 members of the local Resita, elected in 1925, seven were social democrats, being chosen from workers Romanian, German, Hungarian, Czech and Slovak. Per a list of trade union presence at the meeting of 20 February 1926 their names were: Peter Bernau, Bela Brodnyanszky, Franz Drescher, Ferdinand Gavron, Peter Rohm, Adalbert Schimek, Johann Steinbauer. They are supplemented by two alternates, namely: Sandor and Johann Franz Stoikovics (Reschitzaer 1938, p. 2). We note that a significant number of councilors Social Democrats: Bernau, Drescher Gavron, Rohm, Steinbauer were of German origin, which indicates the percentage increased their minority in the social category of industrial workers, especially in the organization's socio professional union type (Reschitzaer 1938, p. 3).

Following the local elections of 1929, the situation had improved social democrats, they occupy nine of the 18 stations of Resita Local Council. In 1932, the ratio had become more favorable party members previously mentioned, it owns 12 of the 18 seats of local councilor in Resita. Anina situation was even better than in Resita. Thus, in 1926 he was elected mayor in the village Social Democrat candidate in the person of German union leader Rudolf Hirschvogel. During his leadership brought a rise dramatically Social Democrats who vote in local elections in 1932 had obtained 68% of votes cast (Jurca 1978, p. 82).

Romanization policy of the interwar governments materialize sometimes less moral or legal measures to address minority representatives or trade unions. Thus, during 1938 they were attacked in broad daylight union leaders from Anina and Oraviţa: Rudolf Moser and Eftimie Gherman. This incident was not the only chicanery: if hiring and promotions U.D.R. factories, workers complained that the foremen and minor officials belonging to national minorities, few were promoted to leadership positions as senior officials of that company. Regarding this situation problematic newspaper "Reschitzaer Zeitung" reported the following 22 May 1938.

> Before only some graduates were employed in these enterprises of U.D.R.'s as practitioners with a minimum wage. They had to work for years among the ambition and refine order to gradually advance. When it is known that the main qualification consists of national affiliation, it can be seen on the human innate common sense, as these young people consider protecting not work and continuous improvement as prerequisites for advancement. Instead they behave arrogant, juvenile exuberance succumbs to make propaganda nationalist, but are reluctant to work in factories or mines (Jurca 1993, p. 72).

Also, then another newspaper article cited was a harsh indictment against the company's policy of promoting U.D.R.:

> Every child in Resita knows that factory production is based primarily on a solid middle class, well-educated and with work experience who is recruited from among the local skilled workers and foremen. They have demonstrated their expertise over the years. In contradiction, many of the new employees' fresh engineers and workers, caused the company U.D.R. damage million lei, if we remember the last lawsuits lost by this company. (Baumann 1989, p. 27)

To keep the story fairness, we note that the company often criticized local publications U.D.R. for failing to hire enough workers and Romanian officials compared to the number of people belonging to ethnic minority groups. Thus, an article in the journal "Glasul Bârzavei" appeared in the January 21, 1939 shows the names of newly hired workers above named company: Szilághy, Ungermann, Strobl, Kabal, Bohn, Schweg and Pajsz. The article ends with an unspoken accusation against the management of the firm in question:

> We do not comment policy frameworks, because we know that anything, I write about this, is very well-known director Mr. Bálint. (Baumann 1989, p. 28)

On August 19, 1938 Resita Metal Workers Union called a meeting of its Members, with the following points of discussion topic: collective bargaining contract with U.D.R. site and setting membership fees for sports activities. The negotiating committee were elected German and Romanian workers following: Franz Klein, Peter Bernau, Karl Lindner, Iosif Musteţiu, Ion Popeţi and Michael Gemmel (Baumann

1989, p. 29). Sporting contributions had set at 0.5% of the salary of each worker, which decrease as the contribution of net salary for musical activities charged. Of the total amount collected for sports, physical education department of the union retained 60% and the U.D.R.-'s Sports Club received 40%. That meeting was the last meeting metalworkers' union workers from Resita during the interwar years. A week later trade union were dissolved and replaced with crafts guilds and collective bargaining negotiations were postponed until 14 November of that year (Scurtu 2008, p. 190).

Per the corporate model, supported extensively by a major economic theorist of interwar Romania, Manoilescu guilds were limited to activity in the industrial, agricultural, commercial, technical and cultural retrospective being excluded any social and political manifestation (Schmidt 1938, p. 125). The guilds functioned landing only national, but not international professional contacts with foreign countries to be approved by the Ministry of Labor. In this way, the government hopes diminishing contacts between unionized industrial workers in the country and their fellow majors like abroad (Schmidt 1938, p. 126). These contacts were due invoice existence of strong historic ties, the social-democratic branch, the local unions in certain industries and those of other European countries specializations close, if not identical. The unions of industrial workers in Banat, Crisana, Maramures, Transylvania and Bucovina had multiple ties with their counterparts in Hungary, Czechoslovakia, Austria, Poland, given that before 1918 they had supported parties invoice social-democratic and socialist of the Dual Monarchy both of Cisleithania and Transleithania (Brăftălean 1998, p. 90).

Royal Decree of October 12, 1938 provided for the establishment of guilds of workers, intended to replace professional unions. Metallurgists workers from Resita confronted with this situation on 2 December, when local government was established, which is the name change earlier union leadership. Remained in office: Franz Klein, president, Peter Bernau and Emil Theodore, vice-chairmen, Hans Langhardt secretary general. At December 22, 1938 Musteţiu Joseph and Karl Lindner signed collective agreement for workers. It contained some provision favorable to workers: general salary increases of 6%, payable retroactively from 1 October; build housing for workers by investing U.D.R.'s work and equipping each polling with shower facilities. Even workers who had German and Hungarian fervent communist beliefs, such as: Josef Bayer, Ferdinand Klipitsch, Johann Augustin, and Bido Virág (UTC-'s Secretary for Banat), appearing to agree with this contract, although in autumn 1938 had protested vehemently against the abolition of trade unions.

The management organization "Community German People in Romania" "Deutsche Volksgemeinschaft Rumäniens" used the occasion disappearance of trade unions to negotiate the formation of a polling own in the guilds of workers based on the following considerations: Body raised had been and remained exponent interests of Germans in Banat Mountain stronger than the Social Democrats; if those workers were distinct part of their organization, they could score a greater number and there is a specific section within their crafts guild members could work more easily with employers and fellow Romanian. In general workers' union concerned, and then the guilds had avoided contact with representatives sent by "Deutsche Volksgemeinschaft" to force adherence to this because the organization in question was bourgeois and clerical interests being subordinated to the capitalist employers (Bender and Kehr 1988, p. 21).

10 days after the promulgation of the Decree on the formation of the German Ethnic Group, on December 1, 1940, was founded the organization "German working class in Romania", "Deutsche Arbeiterschaft Rumäniens" (D.A.R.). Meanwhile crafts guilds were disbanded. Branch D.A.R. the mountains of Banat, was led by Julius Baumann, and the Resita Hans Langhardt. Workers' rights were at least partially met, even in the militarization of domestic units and plants, during 1940. In the last collective bargaining agreement ended when the company U.D.R. stated that workers enjoyed the following rights: the working day of 8 h, 12 holidays per year with an additional three religious holidays Roman Catholic, paid leave between 7 and 30 days per year per seniority at work, bonuses wage for overtime, subsidies for the purchase of wood for heating and lighting-based electricity, the recognition of "men of confidence." Romanian and German leaders of workers from Resita who signed this contract we mention Julius Baumann, Elijah Fărcăşescu, Michael Gemmel, Hans Langhartd, Aurel Minea and Iosif Musteţiu.

Also, in the fall of 1940, he had been arrested several German workers and union leaders of the Romanian city on charges that they were members P.C.R. They were freed in the coming years, depending on the evolution of world conflict. Among those detained we mention Anton Breitenhofer, Adolf Drucker, Robert Grabovsky, Fedor Josef, Franz Hromadka, Georg Hromadka, Karl Lindner, Ion Popeţi and Josef Puvak.

One of the main concerns of professional unions in the area between wars is to ensure adequate logistical framework for their activities with social and cultural. This meant finding or building buildings functionally suitable for carrying out certain actions or organizing events. If the early decades of interwar industrial workers' unions had opted for micro-region renting such spaces by the end of the interwar already had enough funds to build offices deemed necessary. The following information that illustrate the facts:

1918 marked the end of the First World War, and a threshold for the reorganization of professional unions in Resita. They divided the 5 categories of trades: the steel workers, builders, carpenters, tailors and shoemakers. The meetings usually took place in the homes of some traders were maintaining appropriate links with socialist and social democratic circles.

For example, the union of the steel workers be combined in space on the ground floor Weiß family in Resita. These measures, however, were only short-lived and did not offer only a temporary solution.

Members of trade unions, regardless of their ethnicity were requesting their leaders for ascension to the premises allowing meetings with many people. These meeting had to be held on the premises that could accommodate a billiard club, a library and the expanded spaces corresponding to the Associations of Hiking, Choirs and Brass Bands of workers from Resita. The new project required significant funding, which is why the workers had formed a joint action to allow members to achieve financial contributions in exchange for a purchasing percent. On December 11th, 1921, 5 trade unions published in the local newspapers a plan for the "House of Workers

Resita Stocks Company" in German "Reciţarer Arbeiterheim Aktiengeselschaft". The founding committee, numbering 45 people put on sale 3750 shares worth 2000 lei each. In terms of compensation payment, 75% of the share was paid on the spot and subsequently the remaining 25% were paid later plus a fee of 15 lei for stamps. At the general meeting of March 23rd, 1924 that the company had already established a capital of 1.5 million, but the share value had dropped to 200 lei. The direction of the company was taken over by the German worker Peter Rohm. He had ordered the purchase and demolition of three buildings in the old center of Resita to create a plot of land large enough to build homes for workers. The main shareholders were steel workers and builders, members of the Social Democratic Party, the Workers' Associations of Music, Sports and Hiking, comprising mostly ethnic Germans.

The building was completed in 1934, providing space for: The Department of Women of the Social Democratic Party, Resita subsidiary, the "Red Falcons" Children Association, the local management of the Social Democratic Party, and the trade union committees for music and sports. The opening ceremony took place in 1936. The main hall was a vast meeting room, equipped with a stage and 888 seats per the concept Social Democrat timesharing: namely, 8 h for work, 8 h for rest and 8 h for the personal use of each worker.

However, unionized workers working in metallurgy and mining focused not only on Resita. Based on a law from 1927, each village had to possess a cultural center, having legal personality. Mayors had the duty to allocate a plot of land for the construction and had to seek financial sponsors for raising the building. The activity was successful, given the fact that in only 11 years (1927–1938) in Caras- Severin there had been built 15 cultural centers. At the insistence of the ethnic Germans from Steierdorf and Sigismund, two neighborhoods of Anina, the Mining Trade Union Secretary, Eftimie Gherman and the Minister for Health and Social Care, Traian Haneş triggered the bureaucratic process for building a community center. Donations from German workers of the two neighborhoods have provided most of the funds, allowing the erection of a building with a large hall for meetings and performances, six offices, a courtyard and garden with a double track bowling alley. More than 2000 workers had made cash donations to raise the cultural center. Work began on August 20th, 1932 and was completed on September 2nd, 1934.

7 Conclusion

The existence of professional unions from the Banat Highland where the Germans held management positions was closely linked to that of social-democratic parties and socialist bills voted into laws. Relying on broad popular support among industrial workers they managed to influence the results of local and parliamentary elections to their advantage. When the place held within the organizational structures of workers in industry, had been taken by crafts guilds, the political activity diminished but did not disappear completely, being replaced by a heightened interest on economic issues, raised by the members concerned. One of the main problems solved by unionized workers in the area, during the interwar years, had been the building of functional premises to ensure adequate logistical bases for various activities as well as different social and cultural events.

Acknowledgements The work presented in this chapter has been supported by the Research Grant: "Digitalizarea documentelor referitoare la studentii evrei de la Universitatea din Cluj in perioada 1919–1945", "The electronic referencing of the Jewish students in Cluj from the period: 1919–1945", financing institution: Turo College, New York, S.U.A., manager: Cercetător Ştiinţific II Dr. Varga Attila Carol.

References

Baumann, J.: Geschichte der Banater Berglanddeutschen Volksgruppe, pp. 27–29. Eckertschriften Verlag, Viena (1989)

Bender, Á., Kehr, F.: Auf und nieder immer wieder, în „Banater Berglanddeutsche", XIV, München-Viena, p. 21 (1988)

Brăftălean, V.: 25 de ani de mişcare muncitorească la Reşiţa: 1903–1928, Reşiţa, Editura InterGraf, pp. 73, 75, 77, 86–87, 90, 95–96 (1998)

Ghinea, D.: Enciclopedia geografică a României, vol. II, Bucureşti, Editura Enciclopedica, p. 101 (1996–1998)

Hromadka, G: Reschitza - eine Hochburg der Sozialdemokratie im europäischen Südosten, în „Arbeiterbewegung und Arbeiterdichtung", Band 5, Stuttgart, Seliger Archiv, pp. 91–92, 94–95 (1982)

Ispas, C.: School through the eyes of the pupils in the fourth and the eighth grade. J. Educ. Sci. **1**(27), An XV, [online 21 Apr 2016]. Available online accessed on http://www.resjournal.uvt.ro/files/Arhiva/c2en2013.pdf (2013a)

Ispas, C.: Education—attenuation or enhancement factor of socio-economic inequalities in the Romanian society, Analele Universitii „Eftimie Murgu" Resita. Fascicola II. Studii economice, nr. XX, pp. 52–61, [online 14 July 2016]. Available online accessed on http://www.analefseauem.ro/upload/arhiva-revista/2013/Volum_2013.pdf (2013b)

Ispas, C.: Family social work in Romania—between present and future, Revista Analele ştiinţifice ale Universităţii „Alexandru Ioan Cuza" din Iaşi (serie nouă). Sociologie şi asistenţă socială, TOM VI/Nr.2, [online 15 May 2016]. Available online accessed on http://anale.fssp.uaic.ro/index.php/asas/article/view/51/42 (2013c)

Jurca, N.: Mişcarea socialistă şi social-democrată din România: 1934–1944: poziţia sa faţă de fascism şi război, p. 82. Editura Litera, Bucureşti (1978)

Jurca, N.: Social-democraţia în România: (1918–1944), p. 72. Hermann Verlag, Sibiu (1993)

Lupşiasca, K.L.: Dem Emporbringen und Aufblühen dieser Bergwerke. Eine Geschichte des Banater Berglands in der Zeitspanne 1855–1920, Bucureşti, pp. 160–162. ADZ-Verlag (2000)

Lupşiasca, K.L.: Höhepunkt ihrer Geschichte. Eine Geschichte des Banater Berglands in der Zeitspanne 1920–1948, Reşiţa, Editura Banatul Montan, pp. 97–98 (2006)

Schmidt, J.: Die Deutschböhmen im Banate, Timişoara, pp. 125–126. Deutsche Buchdruckerei Verlag (1938)

Scurtu, I.: Civilizaţia românească interbelică: 1918–1940, Bucureşti, Editura Fundaţiei "România de Mâine", p. 190 (2008)

Zaberca, V.M.: Momente din istoria economiei naţionale româneşti, Reşiţa, Editura "Eftimie Murgu", p. 102 (1996)

Zaharia, G.: Ausflugprogram des Toursistenvereins die Naturfreunde für das Jahr 1924, Reşiţa, Buchdruckerei Anton Leeb, p. 112 (1924)

Zaharia, G.: Archivium Dioecesanum Timişoarensis, Fond: Corespondenţa Grupului Etnic German, Dosar 1/1929–1944, ff. 21–26
Zaharia, G.: Arhivele Naţionale: Romanian National Archives: Caraş-Severin county branch, Fond: Asociaţia Notarilor Caraş, Dosare: 1/1926–1938, ff. 1, 3, 20, 66, 131; 9/1937, ff. 3, 8-9; 10/1938, f. 49
Zaharia, G.: Arhivele Naţionale: Romanian National Archives: Caraş-Severin county branch. Fond: Comisariatul de Poliţie Băile Herculane, Dosar 7/1939, ff. 28–33, 99, 122
Zaharia, G. Arhivele Naţionale: Romanian National Archives: Caraş-Severin county branch, Fond: Detaşamentul de Poliţie Băile Herculane, Dosare: 3/1937, ff. 204, 234; 7/1939, ff. 3, 71, 83; 10/1940, f. 6
Zaharia, G.: Reschitzaer Zeitung, Reşiţa, XLIII: nr. 38–20 septembrie 1930, p. 2; XLIV: nr. 44-1 noiembrie 1930, pp. 2–3; LII: nr. 17–24 aprilie 1938, p. 3, nr. 19–22 mai 1938, pp. 1–2
Zaharia, G.: Politica de apărare naţională a României în contextul european interbelic: 1919–1939, Bucureşti, Editura Militară, pp. 82, 85–86 (1981)

Asceticism for Society—*Integrating Body, Soul and Society. In Searching of the 'Real Self' that Actually Is 'Clothed with Christ'*

Nichifor Tanase

Abstract The greatest challenge which is facing Christian civilization and culture today is the following: The atrophy of spiritual senses. So, we can ask ourselves, What are the most effective steps that we can take to overcome this challenge The answer lies in the early Christian ascetics (patterns from antiquity can be a response to modernity). In this chapter I explore how Christian asceticism can be made relevant to a modern culture in which the idea of "ascetic holy man" has lost much of its power. Regarding the model of holy man, many scholars continue to assume that a distinction must be made between an ascetic and a monk, as every monk is an ascetic, but not every ascetic is a monk. Peter Hatlie says that "*Although spiritual authority and "the holy" remain fertile topics for discussion among early Christian and late antique scholars, it receives considerably less attention from Byzantinists working in the generations to follow*" (Hatlie in Portraits of Spiritual Authority. Religious Power in Early Christianity, Byzantium and the Christian Orient. Brill, Leiden, 1999, p. 195). In the context of the perceived disintegration of the secular world, monks showed how ascetic renunciation of the world could provide a new style of civic leadership. Susan Ashbrook Harvey manages to capture the relationship between ascetism and society in this way: "*During the fourth century monasticism flowered across the Christian realm, and with it a critical role for the ascetic – the holy man or woman – to play in society. By their discipline and their conscious imitation of biblical models, especially from the Gospels, the ascetics enacted the image of Christ. To the public this was more than imitation: in the image of Christ, the holy one could do what Christ had done. The ascetics could intercede for divine mercy, and they could be instruments of divine grace in this world; they were a channel between humanity and God that worked in both directions. The ascetic was the point at which the human and the holy met. Often seen as an attempt to leave the worldly for the spiritual, asceticism in fact carried heavy responsibilities in relation to the larger Christian society*" (Ashbrook Harvey in Asceticism and Society in Crisis: John of Ephesus and The Lives of the Eastern Saints. University of California Press, Berkeley, 1990, pp. 20–21). In the desert St. Anthony the Great redefined

N. Tanase (✉)
"Eftimie Murgu" University of Resita, 1-4 Train Vuia Square, Resita, Caras-Severin, Romania
e-mail: pr.nichifor_tanase@yahoo.com

D. Flaut et al. (eds.), *Decision Making in Social Sciences: Between Traditions and Innovations*, Studies in Systems, Decision and Control 247,
https://doi.org/10.1007/978-3-030-30659-5_16

the ascetic as one who fights with the Adversary face-to-face, in the desolate and un-Christianized wilderness. Antony made "the desert a city", sanctifying a place where God had not been present. And he did more: he brought that strength back into Christian society. By the sixth century, the ascetic's role in society had both expanded and become an orderly part of how society functioned. "*We should seek holiness, not clothing, food and drink*", says St. Neilos the Ascetic, because "*possessions arouse feelings of jealousy against their owners, cut off their owners from men better than themselves, divide families, and make friends hate one another [...]. Why do we abandon hope in God and rely on the strength of our own arm, ascribing the gifts of God's providence to the work of our hands? Job considered that his greatest sin was to raise his hand to his mouth and kiss it* (Job 31:27)" (St. Neilos 14, in *Philokalia* 1, 1979, p. 208). Now when bodily concerns predominate, "everything in man is asleep: the intellect, the soul and the senses" (St. Neilos 16, in *Philokalia* 1, 1979, p. 210), and this indicates the state of one whose reason is closely absorbed in physical things. John the Baptist lived in the wilderness and the population of entire towns came out to him; the miraculous life of this humble desert-dweller is acclaimed until this day, and his memory is greatly revered by all. For "*the renown of holiness is eternal, and its intrinsic virtues proclaim its value*" (St. Neilos 20–21, in *Philokalia* 1, 1979, p. 214). But false teachers are blind to such examples, and arrogantly tell men what to do. For in their foolishness they have extinguished the light of contemplation. So their contemplative understanding is immediately destroyed.

Keywords Real self · Holy man · Aesthetics of asceticism · Theosis

1 Introduction: The "Real Self" and a "Theology of Shame". Asceticism for a 'Sinless' Society and the Neuroaesthetics in the Age of Complexity

Shame, not guilt is the focal point at which atonement must aim today (Pattison 2000, pp. 186–274) and Christianity has the power to heal it through the inclusive ministry of Christ. An emergent issue for practical theology is that, if "*shame is no pervasive, so universally evident, then surely we require a theology to understand and address it*" (Stockitt 2012, p. 7). The mission of Jesus into a shamed-filled world will form the substance of Stockitt's book: "*The action of Jesus and his words to women, alluding Isaiah's declaration* (62, 11), *announced that healing for the people of God, who for so long had suffered the shame of exile, was indeed possible and that such healing would take place in the body for those have the faith to reach out and touch. As Elisabeth Moltmann-Wendel puts it, «liberation takes place in the body»* (*I Am My Body, p. xi*). *Jesus declares to her that she is now healed from her illness and that she can «go in peace», a phrase that would have evoked the Hebrew «šālôm», meaning «wholeness». She leaves the scene not only in her body but re-integrated into society. The crucial theological significance of this event is that the encounter between their bodies that healing was communicated*" (Stockitt 2012, pp. 94–95;

Moltmann-Wendel 1995, pp. 35–37). Starting from this "biography of the sinless self" (Stockitt 2012, p. 79), the next step is to go farther to the assertion: "Sin isn't relevant anymore". Alan Mann takes seriously this oft-heard assertion and instead goes in search of the real plight at the heart of contemporary Western society. He says that "*even the idea of free choice to sin has been brought into question with the advent of the modern genetics... because nobody has a free choice to do anything*" (Mann 2005, p. 22). Thinkers like Scot McKnight, Joel Green and Mark Baker are calling the church to rethink the message of the cross of Christ in a multicultural, multilingual, postmodern, post-Christian society (McKnight 2007, pp. 15–24; Green and Baker 2000, pp. 199–220). Mann summarizes: "*What residue of sin that may remain in us, what responsibility and obligation we have failed to shake off onto meaningless metanarratives, oppressive institutions, and impersonal systems, can at last be put down to the fact that ultimately we are the victim of our own genetic make-up*" (Mann 2005, p. 22). While the post-industrialized self is able to push away the sins of moral delict, this self has generated a chronic, internalized disease, typically labeled 'shame'. In this new world, individuals still seek intimacy with other humans, but no longer out of a desire to know another; instead, "*the self is primarily looking to satisfy its own need*". There is a deep hunger for self-realization, and a latent understanding that relating to others is a necessary step to that self-realization. However, in the end, it is always about project "I". But the persons become cynical about relationships or social interaction and, says Mann, "*thus a vicious cycle ensues: we long for intimacy, to have a deep sense of connectedness with ourselves and with 'Others'. However, the project of self-realization ultimately pushes the 'Other' away. Alone, the project of self-realization collapses in on itself, laying the seedbed for chronic shame to grow... With such fears the chronically shamed person hides behind masks, never truly connecting with others, never satisfying his or her need for intimate, mutual, undistorted relating - and so always falling short of what it means to be a human being, created in the image of the relational, Trinitarian God revealed in the Bible*" (Mann 2005, p. 28). We live in a 'sinless' society because without the 'Other' there are no subjects to have wronged. Sin is first and foremost a relational concept. But even more, the postmodern self is an isolated self, people have lost any sense of "others". Furthermore, where any concept of 'sin' exists it does so within the context of being 'sinned against' or guilty before God, rather than being a sinner. In other words, the 20th century therapeutic culture has successfully convinced people that "it's not your fault" and "We're all victims". Since the people of this culture live in a 'sinless' society, any message beginning with concepts of sin and guilt ('the penalty of that guilt is death' or 'Christ died in your place') will fall on deaf ears. A chronically shamed person fears being exposed and tries to live an unexposed life, fearful that others will see the true self rather than the ideal. The result is a distortion in every relationship with others, with God, and even with themselves. If we begin to discuss atonement, at its most basic level, the story of Jesus and the cross, can we ask ourselves how we ever reveal our "real self"? Jesus was completely true to himself and the cross was the proof of that, because, paradoxically, Jesus' pursuit of death in actuality becomes the pursuit of life. The postmodern fears that, in living for the 'Other', we die to ourselves. At the cross, Jesus subverts this rationale by showing us that living for the ‘other’ (even unto death) leads us away from a place of isolation, alienation and meaningless insufficiency (Piper and Taylor 2007, pp. 103–148). “Postmodernism is an elusive term—even for its advocates! But if we can say anything for certain about postmodernity, it is that the concept of accessible, knowable, objective truth is antithetical to standard, postmodern epistemology” (Piper and Taylor 2007, pp. 51–52; Groothuis 2000, pp. 17–32; Spong 2002, pp. 219–233). Judas easily becomes the recognizable figure to the post-modern self, seeking value and worth in the absence of true relationship. The post-industrialized self reads the story of Judas and recognizes someone traumatized by the *disease* of chronic shame: “*The chronically shamed fear exposing the reality that the way they narrate themselves to others is not their real self. They are insecure in their relating, constantly aware of the need to cover the self from the ‘Other’ for fear of being found socially unacceptable. The shamed person lives lives in permanent state of hiding, even when interacting with others. Only ever seeking to story their ideal-self, he or she never wants their real-self to be found*” (Mann 2005, p. 41). But, the cross is not a place of judgment for the inadequacies and insufficiencies of human relations. Indeed, it is a place of acceptance, of embracing the human condition. Therefore, atonement is the presence of the ‘Other’ without condemnation. A redeemed person is a person who has allowed himself to be exposed before the ‘Other’, namely Jesus, and therefore, to be made coherent in Christ. Mann talks about the “restorying of the self” (Mann 2005, p. 151) through ritual (particularly Communion or Holy Eucharist) but, unfortunately, for him, in this Jesus is simply an inspiring character, not a Saviour. Significant for our study, Mann emphasizes also the consumerist character of the postmodern assertion that ‘sin is dead’ and we live in a ‘sinless’ society: “*Sin has been caricatured, a tool for advertisers to suggest that a product is good and pleasurable. Far from being something to be avoided, sin is now part and parcel of the human desire that drives consumerist mentalities*” (Mann 2005, p. 31). Mann’s main premise is that the word “sin” has become meaningless, semantically diluted, in the Western culture. He is not alone in his views. “Sin doesn’t really exist as a serious idea in modern life”, wrote the journalist Bryan Appleyard.

Further, we will refer to the current neuroaesthetics in this age of complexity. Thus, according to Bryan Appleyard, already exists the foundation of a bridge connecting the mind and the brain: “*with the ever-increasing power and precision of the scanners, it seems there will be no limit to our ability to know the workings of the human mind. Results of brain scans are being studied for insights inti economics, politics, advertising, marketing, sociology, anthropology, religion and art*” (Appleyard 2011, p. 4). If it seems that we have found a way of solving the hard problem, “the mind is not explicable in purely physical terms” and “seems to create the universe every day, lies beyond materialism”. So, says Appleyard, “*the problem is that the human brain may be the most complex thing so far discovered in the physical universe and, as any good neuroscientist will admit, it is far from clear what we are seeing when we look at fMRI scans*” (Appleyard 2011, p. 5).

In this brief analysis I noticed some dangers in the concept of Christian anthropology: the dispelling of inwardness from the modern myth of the self (Robinson 2011, pp. 77–135), absence of mind (Minsky 2007, pp. 298–345), a reinventing of the Sacred (Kauffman 2008, pp. 31–43: "The nonreducibility of biology to physics", pp. 230–245: "Living into Mystery" and pp. 281–287: "God and reinventing the sacred"), artificial intelligence and the future of the human mind (McKibben 2006, pp. 40–80), when humans transcend biology (Kurzweil 2005, pp. 56–71, 167–193, 299–319 and, in particular, pp. 382–390: "Who Am I? What Am I? The Singularity as Transcendence") and how we become posthuman (Hayles 1999, pp. 25–49, 283–290).

2 Ethics as Ascetics in "After Virtue" Age. Perfectible Society for Perfectible Man –Aesthetic, Ethical and Religious Modes of Life

Agreeing with Kierkegaard who contrasted ethical and aesthetic ways of life (Kierkegaard 1998, pp. 501–663), for MacIntyre the dialectic of the virtue is a movement "from the virtues to virtue and after virtue" (MacIntyre 2007, pp. 226–243). He says that "*when Kierkegaard contrasted the aesthetic, ethical and religious modes of life in* Enten-Eller (Kierkegaard 1971, pp. 361–458)*, he argued that the aesthetic life is one in which a human life is dissolved into a series of separate present moments, in which the unity of a human life disappears from view*" (MacIntyre 2007, pp. 241–242). The existence of the individual [*das einzelmenschliche Dasein*] is for Kierkegaard the scene [*Schauplatz*] of ontology (Adorno 1995, pp. 158, 121). Thus, within the central culture of liberal or bureaucratic individualism new conceptions of the virtues emerge and the concept of a virtue is itself transformed. Every moral philosophy, says MacIntyre, has some particular sociology as its counterpart in the culture of bureaucratic individualism and "*conceptions of the virtues become marginal and the tradition of the virtues remains central only in the lives of social groups whose existence is on the margins of the central culture*" (MacIntyre 2007, p. 225). The Enlightenment project that MacIntyre decries in *After Virtue*, as a tradition with practices and virtues set within a community, represents a stage in MacIntyre's return to Christianity. Jonathan R. Wilson says that the weakness in MacIntyre's account in *After Virtue* is the following: "*although he advocates a recovery of the moral tradition, no specific moral tradition is present. He argues for a conception of practices but advocates no specific practices. He argues for virtues but no particular virtues*" (Wilson 1997, p. 56). MacIntyre shows that our moral fragmentation largely results from the loss of the conception of the *telos* of human life. But, he moves toward a more Christian and theological conception of 'the good life' by drawing on Thomas Aquinas's assertion that the human *telos* is "*that state of perfect happiness which is the contemplation of God in the beatific vision*" (Wilson 1997, pp. 57–59). The purpose of God's work is that "all of us come to the unity of the faith and of the knowledge of the Son of God, to maturity, to the measure of the full stature of Christ" (Eph. 4:13). The human *telos* is revealed in the gospel of Jesus Christ. Regarding the issue of "The End of Teleological Ethics", MacIntyre teaches us to think teleologically, to identify the human *telos* and order our lives toward it (Slote 2010, pp. 35–38; Davidson 1994, pp. 63–80).

According to some recent studies about religious consciousness, like the one of Jasper (2009, pp. 63–86), a particularly ascetic form persists into modernism, despite the loss of an openly theological context. Harpham calls this "*a Modernism unembarassed about its asceticism despite its repudiation of theology*" (Harpham 1987, p. 199). Henry Michael Gott upholds that the ascetic paradigm can be described as a "*religious sentiment – conceived as an aesthetic urge*" (Gott 2013, p. 137). He exhibits a fluctuation between metaphors of '*control*' (the discipline of the monastic life) and the metaphors of '*abandon*' (the mystical drowning). Consistent with these ascetic paradigms, Gott understands the "*complete knowledge as a form of loss*" (Gott 2013, pp. 54 and 138). He concludes that "*throughout modernism runs a sense that, abstracted from a specifically religious context, ascetic experience could become both more universal and more personal*" (Gott 2013, p. 162).

Over the last seventy years, theorists in social sciences and humanities, says Richard Valantasis, "*have explored asceticism as a vital component of sociology, social history, and hermeneutics*" (Valantasis 1998, p. 544, 2008, pp. 3 and 212–233, for here, see the chapter: "ascetic subjectivities in the Letter to the Galatians"). An inside look to the problem of the self in moral space or to this "moral topography" related to the many conflicts of modernity, we find at Charles Taylor and especially in his "Sources of The Self. The Making of the Modern Identity" (Taylor 2001, pp. 25–52, 111–115, and conclusions pp. 495–521). He distinguishes between the 'theorists', those who understand asceticism as a large and pervasive cultural system, and the 'historians', those who view asceticism as specific religious practices (social withdrawal, restriction of food, regulation of sexuality, and the formation of religious community). Valantasis presents the ascetical theories of the three primary ascetical theorists of this century (Max Weber, Michel Foucault, and Geoffrey Harpham) and, in order to develop a theory of asceticism aligned to what, he calls "*the social function of asceticism*". Weber (2001, pp. 102–124) treats asceticism as part of sociological theory and he develops the theory of '*innerworldly asceticism*'. Michel Foucault explores the place of asceticism in the context of ethical formation. He distinguishes four aspects of what he calles "the relationship to oneself": (1) the ethical; (2) the mode of subjection (the mode to relate to moral obligations, such as revelation or divine law); (3) asceticism, or self-forming activity (the changes to oneself in order to become an ethical subject); and (4) the *telos,* or goal (result of ethical formation). Although Foucault identifies asceticism as one aspect of this process of ethical formation, he also views asceticism as the heart of the entire process of formation: "*No technique, no professional skill can be acquired without exercise; neither can one learn the art of living, the techne tou biou, without an askesis which must be taken as a training of oneself by oneself*" (Foucault 1990a, b, p. 36). There is no specific moral action that does not call for the forming of oneself as an ethical subject and "*no forming of the ethical subject without 'modes of subjectivation' and 'ascetics' or 'practices of the self' that supports them*" (Foucault 1990a, b, p. 28). Geoffrey Harpham develops a theory of asceticism in relation to contemporary structuralists, poststructuralists, and postmodern theories of literary criticism (M. Bakhtin, J. Derrida, M. Foucault). He widens the arena of ascetical studies by exploring the relationship of asceticism and culture and develops the theory of asceticism as "*the 'cultural' element in culture*" (Harpham 1987, see especially: pp. xii about the 'ascetical basis of culture', p. 196 about the 'self-transformation of culture' and pp. 24–28, 36–41 for 'ascetic structure or *image* of self'). Harpham argues that asceticism is related to culture because asceticism is that which 'enables communication in a culture'. Like Foucault, he emphasizes "*the ethical nature of culture itself, arguing that there is an inherent level of self-denial necessary for a person to live within a culture*" (Valantasis 2008, p. 546). Valantasis' theory of asceticism follows Harpham's orientation. Valantasis asserts that the basis of ascetical activity is the cultural foundations: "*At the center of ascetical activity is a self who, through behavioral changes, seeks to become a different person, a new self; to become a different person in new relationships; and to become a different person in a new society that forms a new culture. As this new self emerges (in relationship to itself, to others, to society, to the world) it masters the behaviors that enable it at once to deconstruct the old self and to construct the new. Asceticism, then, constructs both the old and the reformed self and the cultures in which these selves function*" (Valantasis 2008, p. 547). Asceticism links individual to culture. So, asceticism "*inaugurate an alternative culture, to enable different social relations, and to create a new identity*" (Valantasis 2008, p. 548). Its goal of creating new persons through patterning of behavior involves the performance of certain acts. The ascetic becomes a different person molded to live in a different culture (repeated prayer, consistent withdrawal from society, continuous silence, repeated physical acts of fasting, sleep deprivation, and manual labor). These acts become natural activities for the monk, perceived within the ascetic culture. Thus, the goal of ascetical life is the transformed life, an 'alternative culture' and a 'new subjectivity'. For Valantasis asceticism performs four major social functions. First, asceticism enables the person to function within the re-envisioned or re-created world through ritual, new social relations, different articulations of self and body; second, the ascetic culture relies upon narrative, biography, demonic and angelic psychology, as well as systems of theological anthropology and soteriology; third, asceticism requires intensive perceptual transformation at the most basic perceptual level of the senses, i.e. an ascetic retraining of the senses and perceptions based upon the theological culture; fourth, asceticism functions as a prism through which the light of other domains of knowledge are refracted into a new cultural environment, which gives to the old knowledge a new interpretation (Valantasis 2008, pp. 550–551).

The relationship between asceticism and society is linked to two other important components: world conditions and searching for a perfectible society. Thus, according to Alfred Braunthal "*there is doubtless a close connection between world conditions and these new trends in Christian thinking*". Therefore, the relative peace and prosperity that characterized the Western world in the second half of the nineteenth century drew to a close (enabling the transition from liberal theology to existential theology). World War I brought with it a feeling of crisis and anxiety. So, "*Salvation was no longer expected, as liberal theology had assumed, from the teachings of a superior human being the role that liberal theologians had given to Jesus*" (Braunthal 1979, p. 221). But then humanists based their faith in the perfectibility of human society on two doctrines: man's natural rights and the law of progress. Secular humanism grew out of the Jewish-Christian eschatological vision of the kingdom of God as an age of unending peace and welfare, justice and love. This vision was secularized in the form of a perfect society, achieved not through divine intervention, but through the faith in the existence of laws ensuring progress toward the perfection of society. So, the ultimate goal of secular humanism, the perfection of society through human efforts, presupposes "*not the grace of God, but rather the full responsibility of man for his own thoughts and deeds*" (Braunthal 1979, p. 275). Braunthal makes the connection between asceticism and society in terms of the individual man searching for his own salvation and happiness. For him "*the quest for perfectible man had yet to become the quest for a perfectible society*" (Braunthal 1979, p. 277).

3 God's Presence and the Holy Man ("*Self-made Man-«Objectivity Personified»*"). the Movement from the *Disjunction Between Social and Divine* to the Urban Asceticism ("*Integration Between Society and the Holy*")

The word 'asceticism' reaches today the original meaning which *askeo* has in old greek: "*to work materials, to form by art*" (like in Homer). The word *askesis* had come to mean by Plato's time: "to practice, exercise and train" so as to make oneself fine or beautiful in a specific way. Paul, speaking before the Roman governor, says: "I do my best to keep (*asko, ekein*) a clear conscience at all times before God and man" (*Acts* 24: 16). In the current use of the word 'asceticism' designates the practice to attain "the fine self, the virtuous or excellent person". Still, says J. Giles Milhaven, the word "asceticism" has kept, more or less, its common meaning. "*One part of what it usually means is a consistent, rigorous renunciation of certain pleasure. I say certain pleasure because ascetics have their own pleasures, most obviously the pleasures of self-mastery, God's presence, the pleasure of serving God, the pleasure of anticipating union with God, or the pleasure at repentance of a sinner*" (Milhaven 1998, p. 376). So, there is an attempt to redirect asceticism toward the Christian tradition, from which it was split. To those 'liberal' views of those which have attempted to make sense of Christianity within the rational framework of modernity, we have to remind them that in the early church, like Frances Young expounds, there was always a "*tendency to criticize philosophy*" for its production of many conflicting opinions (its *haeresēs*). The true theologian has to go through a process of moral and intellectual stripping, an *askēsis* of the mind, which in humility acknowledges that the subject-matter of theology transcends the capacities of created human reason: "*The critical process had to be absolutely radical, and to reach any knowledge of*

God, however dim, this asceticism of mind had to be accompanied by strict asceticism of the body" (Young 2013, pp. 38, 71–72). To Young ascesis is related to the *philanthrōpia* of Christ: "remarkable are the occasions when Jesus is named as exemplary: he appears as an example of brotherly love, of *philanthrōpia*, asceticism, prayer, watching, chastity and virginity, humility, poverty and contempt for money. The citations indicate that his incarnation – his readiness to become human, humbling himself and taking the form of a slave, and being born of a virgin – is just as exemplary as his fasting and conquest of temptation; and the demonstration of his *philanthrōpia* is by appeal to the fact that he created humankind for good deeds" (Young 2013, pp. 14–15).

The goal of all asceticism is perfect union with God and eremitism is seeking to embody and exercise supernatural power, as Peter Brown says (Brown 1993, p. 12). Brown emphasizes the social role of the eremite; but the real power of the eremite was held in reserve. As Brown observes, "heavenly" power was defined "*quite simply as power that was not to be used*" (Brown 1971, pp. 80–101, 1982, 1993, p. 93). The power of transcendence was kept pure and unworn by use. According to Brown, the rise of the holy man in the fourth century compensated for the increasing silence of the pagan oracles. But, says Harpham, "*the Holy Men and the power they bore imposed rigors that were utterly alien to paganism, which had fostered a sense of easy access to the divine, of a comfortable interlocking of the natural and the supernatural*" (Harpham 1987, p. 22). A Holy Man such as Anthony spent his life engaging, as Brown says, in a "solemn ritual of dissociation-of becoming the total stranger". Going to live in the desert the eremite rejected human culture "not merely by leaving it but by existing at the extremities of human capacity" (Brown 1971, pp. 91–92). Therefore, as Harpham highlights "*the 'new mood' of Christian asceticism presumed a total disjunction between the social and the divine (Luke 14:26). According to the doctrine of the soul, each human being contained an essence, the 'grace of God within', or the 'Christ in me'* (Harpham 1987, p. 23)." So, the desert was an ideal site for ascesis, to reinvent himself. This was why the eremite was a "*self-made man-«objectivity personified»* (Brown 1971, p. 93)" (even beyond his own subjectivity). In this sense, Gregory of Nyssa regards asceticism as a repetition of Christ's original "taking-form", the act of becoming "*a beauty in accord with the character of the Archetype*", by the way in which Christ is an "image of the invisible God". Gregory says that "*every person is the painter of his own life, and choice is the craftsman of the work, and the virtues are the paints for executing the image*", and we must, then, "*prepare the pure colors of the virtues, mixing them with each other according to some artistic formula for the imitation of beauty, so that we become an image of the Prototype through activity as a kind of imitation, as did Paul, who became an 'imitation of Christ', through his life of virtue*" (Gregory of Nyssa, "On Perfection", in Deferrari et al. 1967, pp. 110–111). Thus, through its conversion the essential "perfect" self is called simultaneously to imitation and to an original condition. Instead, as Harpham points out, pagan ideology of self-formation sends to "on individual uniqueness and cultural accomplishment, cold comforts in the long run". Therefore, he says, asceticism is both dynamic and static "constantly progressing, but never arrives": "*while the pagan self was well formed, it manifestly lacked transcendence, and belonged to the phenomenal and contingent world, a world that Platonism systematically devalued. In place of the self-sufficient self formed on the model of the work of art, Christianity proposed an identification of self with a text of transcendence, an experience that shattered all self-sufficiency and opened up the self both to other selves who had made this identification, and to idealization and closure. In the theology of conversion the self can be synchronized with a transcendent origin; it is invited to accede to the design that is already intrinsic to it*" (Harpham 1987, pp. 41–42). He suggests that eremites (as transcription of the spoken Word) and cenobites (as a copy of a manuscript) stand in relation to each other as speech and writing: "In the dialectic between eremite and cenobite we can see, for example, a conflation of the very oldest oppositions of the culture of writing. The eremite *looks* at his empty world and illegible temptations; the cenobite *reads* the narrative of this regard. The eremite *signifies* through his ascesis; the cenobite *imitates* that signification" (Harpham 1987, p. 43).

In counterpoint with Harpham's understanding of a "*total disjunction between the social and the divine*", Ashbrook Harvey argues for an "*integration between society and the holy*" (Ashbrook Harvey 1990, p. 20). And because the holy could operate anywhere, she emphasizes asceticism practiced within an urban setting. So, to her, the ascetic has to become responsible for society as in John of Ephesus' "urban asceticism".

Hagiography is about a theology of activity, in accordance with fluctuations in the values of society. In addition to an end to martyrdom, the fourth century brought "*a shift in the Syrian Orient from Christianity as an ascetic religion to Christianity as a religion with asceticism as a possible vocation*" (Ashbrook Harvey 1990, p. 10). The shift is marked in the writings of Aphrahat the Persian and Ephrem Syrus. During the fourth century monasticism flourished across the Christian realm, and with it "*a critical role for the ascetic – the holy man or woman – to play in society*" (Ashbrook Harvey 1990, p. 13). By their discipline and their conscious imitation of biblical models, especially from the Gospels, the ascetics enacted the image of Christ. To the public this was more than imitation: "*The ascetic was the point at which the human and the holy met*" (Brown 1983, pp. 1–25, 2008). Often seen as an attempt to leave the worldly for the spiritual, "asceticism in fact carried heavy responsibilities in relation to the larger Christian society". Thus, Antony made "the desert a city", sanctifying a place where God had not been present. And he did more: he brought that strength back into Christian society. Indeed, as Moses, Elijah, John the Baptist and as Christ himself, Antony faced the wilderness as prelude to a career that involved much public ministry (Guillaumont 1979, pp. 67–88). Therefore, the fifth century brought the full articulation of Syrian asceticism and established its place in relation to Christian society (a key figure was Simeon the Stylite, c. 386–459). The Man of God had just this task as his work: "*to reveal the presence of the holy in the midst of human* life" (Ashbrook Harvey 1990, p. 20). The Man of God's life would seem to undermine an entire picture here, because "*he does not withdraw from the world: he goes to it*", as Harvey highlights: "*He enters the harsh reality of the destitute in a major urban center, 'For I am one of them'. He lives among men and women, unmarked by clothing, company, or conduct. Without even a name, he has no identity as a holy man*". This he did by the power of his presence alone, sanctifying the world itself and causing good works to be done by those around him, "*not miracles, but actions of concrete import in human society and possible for any person to perform*". So, the man of God forged an "integration between society and the holy, for the holy could operate anywhere" and emphasises the paradigm of asceticism as "both an external expression and an internal reality" (Ashbrook Harvey 1990, pp. 20–21). The ascetic still inhabited a realm outside the temporal world but also had become settled in the midst of its society, "the ascetic had become responsible for society" (Ashbrook Harvey 1990, pp. 52–53). We notice, here, a shift of ascetic practice from the hermit of late antiquity, as separate from the space of civilized society, to the Syrian asceticism "*practiced within an urban setting or in close contact with village populations*" (as in the case of John of Ephesus' urban asceticism), for his usual ideal is that of an "*asceticism ministering to a crisis-ridden society*" (Ashbrook Harvey 1990, pp. 65, 70). The human and divine were dichotomized arenas. However, differences still persist being noteworthy, as such: "*The vitality in John of Ephesus' urban asceticism, the distant clarity of Cyril's monasteries, and the stillness of John Moschus' desert, seem to speak of altogether different worlds*" (Ashbrook Harvey 1990, p. 135).

4 *Askesis* Encounters *Theosis*—Integrating Body, Soul and Society (Towards a Model for Societal Relationships). The 'Social' Understanding of the Trinity and Asceticism as a 'Social Discipline'

Arisen from ethical pluralisms, practical soteriology (salvation as *praxis*) represents another approach that claims that "*doctrinal affirmations were second order reflections emerging out of Christian practices*". This form of 'applied theology' is, also, concerned with belief before praxis. Jürgen Moltmann articulated a particularly 'social' understanding of the Trinity (Moltmann 1981) and, after this, a number of subsequent 'social' and 'ecclesial' models of the Trinity have emerged. Some studies have also appeared that have argued along similar lines for thinking about the Trinity as a model for the ways that societies and churches might operate in more egalitarian and just ways (Boff 2005; Fiddes 2000; Gunton 1997; Lacugna 1991; Bacon 2009; Volf 1998). Those works have suggested that the doctrine of the Trinity provides a useful and positive model for ecclesial practice and societal relationships. Orthodox theologian Kallistos Ware explains, "*salvation is social and communal more particularly because of our faith in the Holy Trinity*" (Ware 1996, p. 68). Zizioulas, likewise, understands this process of becoming God as a process of becoming God as Trinity, to grow into the likeness of God as a relationship of persons (*others*) in loving communion with one another (Zizioulas 2006, pp. 165–166). 'Otherness', which affirm 'uniqueness' and 'irreplaceableness', is a quality of all human persons and provides an important corrective to omogenizing tendencies. Therefore, the contribution of Orthodox thinkers to current debates in religious epistemology, regarding the relationship between virtues and relevant epistemic goods, concerns the relationship between *praktike* and *theoria*. To Aristotle Papanikolaou worship and sacraments aim at *theosis*; *theosis* entails the acquisition of the virtues; and the virtues are cultivated within the lives of the faithful by their repeated activity together as a sacramental community. He says, that "*human flourishing is understood as the progressive realization of theosis – a godly mode of being cultivated through liturgy and askesis*" (Hamalis and Papanikolaou 2013, pp. 271–280).

"*Without asceticism*, says Ware, *none of us is authentically human*" (Ware 1995, p. 13). And it is precisely here that *askesis* encounters *theosis*. The early Egyptian and Palestinian desert fathers underline rigorous discipline and, in this regard, they prefer to speak more about askesis, rather than about theosis. While the end of *askesis* may be the vision of God or *theosis*, the way of *askesis* is none other than the "*daily life of self-knowledge or integrity, carved out of the ordinary experience of everyday life perceived in the extraordinary light of the eternal kingdom*". Asceticism is a way of intimacy and tenderness, a way of integrating body, soul and society. In this respect, "*asceticism is essentially a social discipline*" (Le Bras 1964, 21–26). Unfortunately, however, says John Chryssavgis, centuries of misunderstanding and abuse have tainted the concept of asceticism, "*identifying it either with individualism and escapism or else with idealism and angelism. Both tendencies verge on the point of dis-incarnation, promulgating enmity towards the world*" (Chryssavgis 2009, pp. 160–162).

A. N. Williams emphasizes the "refusal to separate theology and contemplation from ascesis and morality" (Williams 2007, p. 208; Jennings 2010, pp. 35–112). For her, the necessary and systematic relation of mind and will indicates a further relationship: that of contemplation to ethics, of theology to ascesis. Like *theoria* in Greek (sic!), she says, *contemplatio* implies both a sensuous activity (physical seeing) and an intellectual one ('seeing' the truth of an idea, for example). This is linked to a moral and ascetical theology insistent upon the mind's governance of the body and the will, and its fulfillment in contemplation: "*Asceticism by itself is accorded little value, however; as Antony also notes, some have afflicted their body, but lack discernment. Similarly, Agathon likens the human person to a tree, with asceticism as the leaves, and vigilance, the fruit. However desirable, indeed necessary, bodily discipline may be, it is subordinated to the mind, to mental acts of judgement, but also to obedience*" (Williams 2007, pp. 142, 178, 215).

Asceticism has had a central place within Christianity (Vaage and Wimbush 1999; Rader 1982, pp. 80–87; Krawiec 2008, pp. 764–785), not only in monastic communities and as part of the practice of the laity, but also in theology and anthropology. Gavin Flood highlights the way in which "*the concern with asceticism has been accompanied by an ambivalent attitude towards the body, which on the one hand has been evaluated in positive terms as part of creation and on the other negatively as the locus of the passions, the locus of sin that keeps us away from God. This negative attitude towards the body has been associated with a dualism inherited from the Greek world that associates matter and body with impurity and constraint*" (Flood 2004, p. 145). But, while the body is the locus of the passions, it is also formed in the likeness of God and it is this divine nature of the body, its true nature, which has to be realised in asceticism. The ultimate goal of Christian asceticism has been the reconstitution of the pre-fall state through withdrawal (*anachoresis*) and self-mastery (*enkrateia*). Asceticism in positive terms is the refashioning of the self rather than mere renunciation, developing contemplation (*theoria*), practice (*praxis*) and purity of heart (*hesychia*). Deification is the goal of the ascetic life because God and man (*anthropos*) are in a reciprocal relationship, as Maximus says in his *Difficulties* (*Ambigua*): "*For God and man are paradigms one of another, that as much as God is humanised to man through love for mankind, so much is man able to be deified to God through love, and that as much as man is caught up in God to what is known in his mind, so much does man manifest God, who is invisible by nature, through the virtues*" (Maximus 1996, p. 101). The construction of the ascetic body, therefore, looks back to an origin, the pre-fall body of the *logikoi*, and forwards to a telos, the eschaton. The goal of becoming God-like, of crossing beyond sin, is achieved, says Flood, when tradition and scripture are brought to life in subjectivity, because "*asceticism is not simply the control of the passions but the imitation (mimesis) of God and the cultivation of a life of beauty as a reflection of God's being: an imitation or cultivation which is a recapitulation of tradition in interiority and memory. But to interiorize tradition and to imitate Christ, whose life contained both beauty and pain, is to undergo intentional suffering, as suffering and beauty are integral aspects to the Christian ascetic path*" (Flood 2004, pp. 164, 154). The hypostatic union in Christ provides the structure that allows for human theosis, which involves a unity of mind (*nous*), reason (*logos*) and sense (*aisthesis*).

5 The *Aesthetics* of Asceticism (Freedom, Beauty and Joy). Spiritual Beauty—A Dazzling Beauty of Radiant, Light-Giving Personality. Ascetic Eucharists—Cultivation of a Religious Self as a Striving After the Real Presence

We are not allowed to experience Christ without the presence of Beauty. Aesthetics (the beauty of the Church) and ascetics (the spiritual disciplines of the Church) emerge from the one faith. To emphasize freedom/beauty found in asceticism, Ware quotes Nicolas Berdyaev and Paul Florensky and by this he tries to shows us that "*a certain measure of ascetic self-denial is thus a necessary element in all that we undertake, whether in athletics or in politics, in scholarly research or in prayer*" (Ware 1998, p. 3). Ware speaks in defense of asceticism by examining two aspects: *anachoresis* and *enkrateia*. The rationale of ascetic *anachoresis* is summed up by St. Seraphim of Sarov: "*Acquire the spirit of peace, and then thousands around you will be saved*" (Cavarnos 1980, pp. 93–122). To emphasize freedom and beauty found in asceticism, Ware quotes Berdyaev and Florensky: "*Freedom, beauty, joy: that is what asceticism meant to Berdyaev, Florensky, and the Syrian monks. But most people in our present-day world have a radically different perception of what asceticism implies: to them it signifies not freedom but submission to irksome rules; not beauty but harsh rigor; not joy but gloomy austerity*" (Ware 1998, p. 4). Thus, the ascetic serves society by transforming himself through prayer, and by virtue of his own self-transfiguration he also transfigures the world around him. Every Christian is an ascetic and without asceticism none of us is authentically human (Granger and Pahman 2016, pp. 215–227). Fr. Pavel Florensky has written: "*Asceticism produces not a good but a beautiful personality; the characteristic peculiarity of great saints is not the goodness of heart which is common among carnal and even very sinful men, but spiritual beauty, the dazzling beauty of radiant, light-giving personality, unattainable by carnal men weighed down by the flesh*" (Florovsky 1974, pp. 125–126). To Dylan Pahman in Orthodox asceticism Christians behold the beauty of God, become beautiful themselves: "*in Ortodox asceticism, we become beautiful and are able to see the beauty of all things in the light of God, who is the Beauty that alone transcends all things and from whom all that is beautiful derives its glory*" (Granger and Pahman 2016, pp. 215, 226–227). To what extent is learning an art not just a metaphor for asceticism but itself an ascetic endeavor? He concludes that art and asceticism ought not to be separated as has too often been the case in modernity (Pahman 2013, pp. 183–188, 2016, pp. 489–504).

Human beings *are* their bodies, because "*the body is not an object like others because one cannot distance oneself from one's own body so that it can give itself as a totality. I myself am that body*" (Wyschogrod 1990, pp. 16–17). But, as Jean-Pierre Vernant argues, for ancient Greeks the human body was conceptualized as "*a dim version of the dazzling bodies of the gods*" (Vernant 1989, pp. 18–47). For them the human body is not fully a body. In fact, in the Christian asceticism of late antiquity the body was also perceived to be problematic, not because it was a body, but because it was not a body of plenitude. Brown suggests that the ascetic view of the human body oscillated between two modes of visual perception of the body: *dimness* or particularity and *dazzle* or grandeur. As he has explained, Adam's body "*had been unimaginably different from our own, it had been a faithful mirror of a soul which, itself, mirrored the utterly undivided, untouched simplicity of God.... It was like the diaphanous radiance of a still midday sky*". Further, as Brown has observed, "*Through the Incarnation of Christ, the Highest God had reached down to make even the body capable of transformation*". Metaphors of light, as evocations of the true body and the lightning flash irradiating on the face of a Desert Father, mark the turning point for what we find to be a 'language' that makes possible the description of ascetic self-identity. So, if asceticism means to see the body as signifier, the body of plenitude signified an existence that would defy the constraints of time and space. Hence, says Patricia Cox Miller, "*ascetic Christians typically used metaphors of light to convey what this exemplary self would be like. The face of Abba Pambo shone like lightning, for example, and Abba Sisoes' face shone like the sun. In the glare of such brilliance, theologians like Gregory of Nyssa caught a glimpse of the dazzling plenitude of Adam, exemplar of an original humanity once lost but retrieved by Christ as the sign of human destiny*" (Cox Miller 1998, pp. 281–283; *Pambo* 12 and *Sisoes* 14, in Ward 1975, pp. 197, 214–215).

The characteristic asceticism of the Cynics was not merely a rejection of luxury but a rejection of the symbolic centre of the society. Thus, "*Pagan asceticism is a public and even civic practice. Christian asceticism, by stark contrast, concentrates exclusively on the self*" (Harpham 1987, pp. 27–28). For the earliest Christian ascetics "*it is the activity of eating, rather than food itself, that assumes symbolic significance for the creation and cultivation of a religious self*" (Miles 1995, p. 550). Just as Jewish dietary law could be a means of self-definition and opposition in the face of persecution, Andrew McGowan argues that "*Christian uses of food could express self-definition and separation of the group from the wider society*" (McGowan 1999, pp. 87–88). He began with a critique of the approaches to eucharistic meals which have been characteristic of most liturgical scholarship until quite recently (Grimm 1996, pp. 13–32). Veronika Grimm pictures Christianity as a sort of dietary psychopathology, emerging from the integrated eating of an all-too-monolithic Judaism. As Geoffrey Harpham has observed, "*While asceticism recognizes that desire stands between human life and perfection, it also understands that desire is the only means of achieving perfection, and that the movement toward ideality is necessarily a movement of desire*" (Harpham 1987, p. 45). St. Ignatius of Antioch sees the Eucharist as the means of God's intelligent self-expression in creation. To choose to eat only vegetables, not meat, and to drink only water, not wine, is natural asceticism; but it is unnatural to make intentionally our food and drink repulsive, as was done by Isaac the Priest, who after the Eucharist emptied the ashes from the censer over his food, and by Joseph of Panepho, who added sea water to the river water that he drank (*Apophthegmata*, Isaac the Priest 6, Eulogius the Priest 1; Ware 1998, pp. 9–10).

A fortiori, in the Orthodox view, the reading of a patristic or ascetic text becomes a striving after the Real Presence, Christ: "*Where we read truly, where the experience is to be that of meaning, we do so as if the text … incarnates (the notion is grounded in the sacramental) a real presence of significant being. This real presence, as in an icon, as in the enacted metaphor of the sacramental bread and wine, is, finally, irreducible to any other formal articulation, to any analytic deconstruction or paraphrase*" (Steiner 1971, p. 127). Mental concentration (προσοχή, περίνοια) together with the indispensable faculty of piety help the reader to traverse the density of phonetic and semantic particulars until he arrives at the meaning of the Presence which the religious text incarnates. Thus, asceticism also means piety, as a respect for the Word, insomuch as Petropoulos pointed out: "*Εσβεια may be likened to a light illuminating the pious reader's ascent through the cloud of opaqueness (γνόφος) which surrounds the Unseeable (θεώρητον, θατον) and Unknowable (κατάληπτον), i.e., the divine determinant of meaning*" (Petropoulos 2017, p. 133). St. Gregory of Nyssa, for example, is unequivocal in his condemnation of secular knowledge per se: "Secular education, in very truth, is infertile, always in labor, and never giving life to its offspring" (St. Gregory of Nyssa, *De Vita Moysis*, II, 36; Chrysostomos and Patapios 2017, p. 56). The Church Fathers considered Christianity to be the "science of sciences" and the unique and singular path to human restoration through the therapeutic application of its precepts and the mysteries of the Church. So, "their admonitions about secular knowledge falsely elevated to something that it is not; i.e., to a stature that exceeds its proper place in facilitating the human ascent to knowledge beyond the mind to the heart, or beyond reason to noetic enlightenment" (Chrysostomos and Patapios 2017, p. 57).

Yannaras demonstrates the usefulness of asceticism for theology and philosophy, specifically in ethics and anthropology: "*Every voluntary mortification of the ego-centricity which is "contrary to nature" is a dynamic destruction of death and a triumph for the life of the person. The culmination comes when man shows complete trust by handing over his body, the last bastion of death, into the hands of God, into the embrace of the "earth of the Lord" and into the fullness of the communion of the saints*" (Yannaras 1984, p. 116). Once again, the goal of this self-negation is nevertheless positive: it is part of the ascetic process by which a person apprehends and attains his/her true self-identity (Valantasis also connects asceticism and identity formation, as we saw above). For Florensky, asceticism carries with it a new perspective on life and death: "*The ascetic saints of the Church, are alive for the living and dead for the dead*". To him, "*ascetical practice helps the creature overcome its egoism and empty identity, and in this fashion enables it truly to center its life in the objective, created order, which enjoys an importance in itself as an independent creation of the same Godhead*" (Slesinski 1984, p. 169).

6 Conclusion

Mircea Eliade has spoken of the "*irruption of the sacred*" (Eliade 1960, p. 15) within life and of the complex ways in which myths and dreams are rooted in the manifestations of the divine within. Most often, asceticism is located under assertions like this. "*Such searchers who say that in life's crises they look within themselves for answers rather than to an outside power like the Christian God*" (Wells 2007, p. 27). An outside God, such as we find in biblical faith, is comprehensible because he is self-defined in his revelation; the inside god is not. But, spirituality, in the contemporary sense, spawns no such debate because it makes no truth claims and seeks no universal significance. It lives out its life within the confines of private experience, "*Truth is private, not public; it is for the indi-vidual, not for the universe*" (Wells 2007, p. 28). The result is that this kind of spirituality is inevitably experimental and even libertarian or this spirituality is "gnosti-cism". We are also seeing the convergence between spirituality and a resurgent paganism through the *autonomous self.* This is the self that is not subject to outside authority. "*Many gnostics, then, would have agreed in principle with Ludwig Feuerbach, the nineteenth-century psychologist, that «theology is really anthropology»* (*the term derives, of course, from anthropos, and means «study of humanity»*). *For gnostics, exploring the psyche became explicitly what it is for many people today implicitly – a religious quest*" (Pagels 1989, p. 123; see, especially, the chapter VI: "Gnosis: Self-knowledge as knowledge of God", pp. 119–140).

Some who seek their own interior direction, like the radical gnostics, reject religious institutions as a hindrance to their progress. Others, like the Valentinians, willingly participate in them, although they regard the church more as an instrument of their own self-discovery than as the necessary "ark of salvation". People stumble, suffer, and make mistakes not because of sin but because of ignorance. Therefore, Elaine Pagels adds that "*the gnostic movement shared certain affinities with contemporary methods of exploring the self through psychotherapeutic techniques. Both gnosticism and psychotherapy value, above all, knowledge – the self-knowledge which is insight*" (Pagels 1989, p. 124). Pagels' Gnosis or "*Self-knowledge as knowledge of God*" is related to Bloom's "*Self-reliance as mere gnosticism*" (Bloom 1997, pp. 13–34). This pursuit of the knowledge of the self (that we find, also, in the postmodern expression of asceticism) rests upon an emptiness of human experience which is the outcome of alienation from God. God is at once deep within the self and also estranged, infinitely far off, beyond our cosmos (Taylor 2001). But all this modern research has also an influence on theologians who study early Christianity and late antiquity. So it is difficult to determine if there is a direct and uncritically influence on the latter or rather it is about of an apologetical process of inculturation of recent modern research in their approaches. Many of those pseudo-*morphosis* of the asceticism we have listed above. Holy Fathers use the image of the holy man *'clothed with Christ'*, an image very common in Syriac ascetic contexts yet not limited to such ascetic contexts but rather applicable to every Christian (Brock 1981, pp. 11–38). As much as Christ put on human nature like a garment in the Incarnation, so human beings put on Christ in Baptism. Every Christian therefore is 'clothed with Christ'. Beyond this basic theological fact, however, it seems that by this emphasis Christian's closeness to Christ stresses that the Christian ascet was 'living with Him and in Him' and was 'drawn near Christ in spirit' (Horn 2006, pp. 143–145). Christian's ascesis was inspired early on by ascetic features perceivable in the lives of members of his community. Also, a virtuous life was not only founded in God, but also corresponded to the Gospel and was a truly evangelical life. A life in imitation of Christ and the Gospel brought the ascetic in much closer contact with God. Lay people acknowledged this when they regarded ascetics as a different kind of being by calling them 'angels'. They gave to the ascetic a considerably greater authority, that of being transformed into Christ's likeness.

References

Adorno Theodor, W.: Kierkegaard. Construction de l'esthétique. Payot, Paris (1995)

Appleyard, B. (2011), The Brain is Wider Than the Sky: Why Simple Solutions Don't Work in a Complex World. Weidenfeld & Nicolson, London

Ashbrook Harvey, S.: Asceticism and Society in Crisis: John of Ephesus and The Lives of the Eastern Saints. University of California Press, Berkeley (1990)

Bacon, H.: What's Right with the Trinity? Conversations in Feminist Theology. Ashgate, Aldershot (2009)

Bloom, H.: Omens of the Millennium: The Gnosis of Angels, Dreams, and Resurrection. Riverhead Books, New York (1997)
Boff, L.: Trinity and Society. Wipf and Stock, Eugene (2005)
Braunthal, A.: Salvation and the Perfect Society: The Eternal Quest. University of Massachusetts Press (1979)
Brown, P.: The rise of the holy man in late antiquity. J. Rom. Stud. **61**, 80–101 (1971)
Brock, S: Clothing metaphors as a means of theological expression in syriac tradition. In: M. Schmidt (ed.) Typus, Symbol, Allegorie bei den östlichen Vätern und ihren Parallelen im Mittelalter. Eichstatter Beitrage, vol. 4, p. 11–38. Regensburg (1981)
Brown, P.: The Cult of the Saints: Its Rise and Function in Latin Christianity. University of Chicago Press (1982)
Brown, P.: The saint as exemplar in late antiquity. Representations **1**(2), 1–25 (1983)
Brown, P.: Making of Late Antiquity. Harvard University Press (1993)
Brown, P.: The Body and Society: Men, Women and Sexual Renunciation in Early Christianity. Columbia University Press, New York (2008)
Cavarnos, C.: St. Seraphim of Sarov. Institute for Byzantine and Modern Greek Studies, Belmont, MA (1980)
Chrysostomos, A., & Patapios, A.: The greek fathers and secular knowledge. In: Chrysostomos, A. (ed.), The Sculptor and His Stone. Selected Readings on Hellenistic and Christian Learning and Thought in the Early Greek Fathers, pp. 53–66. James Clarke and Co Ltd. (2017)
Chryssavgis, J.: The spiritual way. In: Cunningham, M.B., Theokritoff, E. (eds.) The Cambridge Companion to Orthodox Christian Theology, pp. 160–162. Cambridge University Press (2009)
Cox Miller, P.: Dreaming the body: an aesthetics of asceticism. In: Wimbush, V.L., Valantasis, R. (eds.) Asceticism. Oxford University Press (1998)
Davidson, A.A.: Ethics as ascetics: foucault, the history of ethics, and ancient thought. In: Goldstein, J. (ed.) Foucault and the Writing of History, pp. 63–80. Blackwell, Oxford (1994)
Eliade, M.: Myths, Dreams, and Mysteries: The Encounter Between Contemporary Faiths and Archaic Realities (trans. Mairet, P.). Harper, New York (1960)
Fiddes, P.: Participating in God: A Pastoral Doctrine of the Trinity. Darton, Longman & Todd, London (2000)
Flood G.: The Ascetic Self. Subjectivity, Memory and Tradition. Cambridge University Press (2004)
Florenski, P.: The Pillar and Ground of the Truth (trans. Jakim, B.). Princeton University Press, Princeton, NJ (1997)
Florovsky, G.: Christianity and civilization. In: Christianity and Culture, The Collected Words of Georges Florovsky, vol. 2, pp. 125–126. Nordland, Belmont, MA (1974)
Foucault, M.: The History of Sexuality, Vol. 1: An Introduction (trans. Hurley, R.). Random House, Inc. Vintage Books, New York (1990a)
Foucault, M.: The History of Sexuality, Vol. 2: The Use of Pleasure (trans. Hurle, R.). Random House, Inc. Vintage Books, New York (1990b)
Gott, H.M.: Ascetic modernism in the work of T.S. Eliot and Gustave Flaubert. Taylor & Francis: London (2013)
Granger, S., Pahman, D.: Art and the Art of Arts: On Aestheticism and Asceticism. In: McGuckin, J.A. (ed.) Orthodoxy and the Sacred Arts. Sophia Studies in Orthodox Theology, vol. 10, pp. 215–227. Theotokos Press, New York (2016)
Green, J., Mark, B.: Recovering the Scandal of the Cross: Atonement in New Testament & Contemporary Contexts. IVP Academic (2000)
Gregory of Nyssa: On perfection. In: Ascetical Works, vol. 58 (trans. Callahan, V.W.). In: Deferrari, R.J., et al. (eds.) The Fathers of the Church, pp. 95–122. Catholic University of America Press, Washington, DC (1967)
Grimm, V.: From Feasting to Fasting, the Evolution of a Sin: Attutudes to Food in Late Antiquity. Routledge, London (1996)
Groothuis, D.: Truth Decay: Defending Christianity against the Challenge of Postmodernism. Inter-Varsity Press, Downers Grove, IL (2000)

Guillaumont, A.: Aux origines du monachisme chrétien: pour une phénoménologie du monachisme, Spiritualité orientale 30. Abbaye de Bellefontaine (1979)
Gunton, C.E.: The Promise of Trinitarian Theology, 2nd edn. T&T Clark, London (1997)
Hamalis, P.T., Papanikolaou, A.: Toward a godly mode of being: virtue as embodied deification. Stud. Christ. Ethics **26**(3), 271–280 (2013)
Harpham, G.G.: The Ascetic Imperative in Culture and Criticism. University of Chicago Press, Chicago, IL (1987)
Hatlie, P.: Spiritual authority and monasticism in constantinople during the dark age (650–800). In: Drijvers, J.W., Watt, J.W. (eds.) Portraits of Spiritual Authority. Religious Power in Early Christianity, Byzantium and the Christian Orient. Brill, Leiden (1999)
Hayles, N.K.: How We Become Posthuman: Virtual Bodies in Cybernetics, Literature and Informatics. University of Chicago Press, Chicago (1999)
Horn, C.B.: Asceticism and Christological Controversy in Fifth-Century Palestine The Career of Peter the Iberian. Oxford University Press (2006)
Jasper, D.: The Sacred Body: Asceticism in Religion, Literature, Art and Culture. Baylor University Press, Waco, TX (2009)
Jennings, N.G.: Theology as Ascetic Act: Disciplining Christian Discourse, Peter Lang (2010)
Kauffman, S.: Reinventing the Sacred: A New View of Science, Reason and Religion, Basic Book, New York (2008)
Kierkegaard, S.: Either/Or, Vols. I and II (trans. Swenson, D., Swenson, L.M., rev. and foreword by Johnson, H.). Princeton University Press, Princeton (1971)
Kierkegaard, S.: Stages on Life's Way, Writings, vol 11 (trans. Hong, H.V., Hong, E.H.). Princeton University Press, Princeton (1998)
Krawiec, R.: Asceticism. In: Harvey, S.A., Hunter, D.G. (eds.) The Oxford Handbook of Early Christian Studies, pp. 764–785. Oxford University Press, Oxford (2008)
Kurzweil, R.: The Singularity in Near: When Humans Transcend Biology. Viking, New York (2005)
Lacugna, C.M.: God for Us: The Trinity and Christian Life. HarperCollins, New York, NY (1991)
Le Bras, G.: Place de l'ascéticisme dans la sociologie des religions. Arch. Sci. Soc. Relig. **18**(18), 21–26 (1964)
MacIntyre, A.: After Virtue: A Study in Moral Theory. University of Notre Dame Press, Notre Dame, Indiana (2007)
Mann, A.: Atonement for a 'sinless' Society: Engaging with an Emerging Culture. Paternoster, London (2005)
Maximus: Ambiguorum 10. In: Louth, A. (ed.) Maximus the Confessor. Routledge, London, New York (1996)
McGowan, A.: Ascetic Eucharists. Food and Drink in Early Christian Ritual Meals, pp. 87–88. Clarendon Press, Oxford (1999)
McKibben, B.: The End of Nature. Random House, New York (2006)
McKnight, S.: A Community Called Atonement. Abingdon Press, Nashville (2007)
Miles, M.: Religion and food: the case of eating disorders. J. Am. Acad. Relig. **63**(3), 549–564 (1995)
Milhaven, J.G.: Asceticism and the moral good: a tale of two pleasures. In: Wimbush, V.L., Valantasis, R. (eds.) Asceticism. Oxford University Press (1998)
Minsky, M.: The Emotion Machine: Commonsense Thinking, Artificial Intelligence and the Future of the Human Mind. Simon & Schuster, New York (2007)
Moltmann, J.: The Trinity and the Kingdom of God: The Doctrine of God. SCM Press, London (1981)
Moltmann-Wendel, E.: I Am My Body: A Theology of Embodiment. Bloomsbury Academic, London (1995)
Pagels, E.: The Gnostic Gospels, Vintage Books, a Division of Random House, Inc., New York (1989)

Pahman, D.: What makes a society? an orthodox perspective on asceticism, marriage, the family, and society. In: Love, Marriage and Family in the Eastern Orthodox Tradition. Sophia Studies in Orthodox Theology, vol. 7, pp. 179–193. Theotokos Press, New York, NY (2013)

Pahman, D.: Alive from the dead: asceticism between athens and jerusalem, ancient and modern, east and west. St. Vladimir's Theol. Quart. **60**(4), 489–504 (2016)

Pattison, S.: Shame: Theory, Therapy, Theology. Cambridge University Press, Cambridge (2000)

Petropoulos, J.C.B.: In defense of piety: respect for words and respect for the word. In: Chrysostomos, A. (ed.) The Sculptor and His Stone. Selected Readings on Hellenistic and Christian Learning and Thought in the Early Greek Fathers. James Clarke and Co Ltd. (2017)

Piper, J., Taylor, J. (eds.): The Supremacy of Christ in a Postmodern World. Crossway Books, Wheaton, Illinois (2007)

Rader, R.: Christian pre-monastic forms of asceticism. In: Skudlarek, W. (ed.) The Continuing Quest for God, Monastic Spirituality in Tradition and Transition, pp. 80–87. Liturgical Press, Collegeville, MN (1982)

Robinson, M.: Absence of Mind: The Dispelling of Inwardness from the Modern Myth of the Self. Yale University Press, New Haven (2011)

Slesinski, R.: Pavel Florensky: A Metaphysics of Love. SVS Press, Crestwood, NY (1984)

Slote, M.: Essays on the History of Ethics. Oxford University Press (2010)

Spong, J.S.: A New Christianity for a New World: Why Traditional Faith Is Dying and How a New Faith Is Being Born. HarperSanFrancisco, San Franciso (2002)

St. Neilos the Ascetic: Ascetic discourse. In: Philokalia, vol. 1. Faber & Faber, London, Boston (1979)

Steiner, G.: In Bluebeard's Castle: Some Notes towards the Re-definition of Culture. Faber, London (1971)

Stockitt, R.: Restoring the Shamed: Towards a Theology of Shame. Wipf and Stock, Eugene, OR (2012)

Taylor, C.: Sources of The Self. The Making of the Modern Identity. Harvard University Press, Cambridge, MA (2001)

Vaage, L.E., Wimbush, V.L.: Asceticism and the New Testament. Routledge, New York, London (1999)

Valantasis, R.: A Theory of the social function of asceticism. In: Wimbush, V.L., Valantasis, R. (eds.) Asceticism. Oxford University Press (1998)

Valantasis, R.: Making of the Self: Ancient and Modern Asceticism. Wipf and Stock, Cascade Books, Eugene, OR (2008)

Vernant, J.-P.: Dim body, dazzling body. In: Feher, M. (ed.) Fragments for a History of the Human Body, vol. 1, pp. 18–47. Zone, New York (1989)

Volf, M.: After Our Likeness: The Church as the Image of the Trinity. Eerdmans, Grand Rapids, MI (1998)

Ward, B.: Sayings of the Desert Fathers. Cistercian Publications, Kalamazoo (1975)

Ware, K.: The way of the ascetics: negative or affirmative? In: Wimbush, V., Valantasis, R. (eds.) Oxford University Press, Asceticism (1995)

Ware, K.: How are we Saved? The Understanding of Salvation in the Orthodox Tradition. Light and Life Publishing, Minneapolis (1996)

Ware, K.: The way of the ascetics: negative or affirmative? In: Wimbush, V.L., Valantasis, R. (eds.) Asceticism. Oxford University Press (1998)

Weber, M.: The Protestant Ethic and the Spirit of Capitalism (trans. Parsons, T., with an introduction by Giddens, A.). Routledge, London, New York (2001)

Wells, D.: The supremacy of Christ in a postmodern world. In: Piper, J., Taylor. J. (eds.) The Supremacy of Christ in a Postmodern World, pp. 21–49. Crosswa Books, Wheaton, Illinois (2007)

Williams, A.N.: The Divine Sense. The Intellect in Patristic Theology. Cambridge University Press (2007)

Wilson, J.R.: Living Faithfully in a Fragmented World. Lessons for the Church from MacIntyre's After Virtue. Trinity Press International, Harrisburg, PA (1997)
Wyschogrod, E.: Saints and Postmodernism: Revisioning Moral Philosophy. University of Chicago Press, Chicago (1990)
Yannaras, C.: The Freedom of Morality, St. Vladimir's Seminary Press (1984)
Young, F.: God's Presence: A Contemporary Recapitulation of Early Christianity. Cambridge University Press (2013)
Zizioulas, J.D.: Communion & Otherness. T&T Clark, London (2006)

The Rural Population of Muntenia in the 1930s: A Brief Analysis Based on Statistical Data

Daniel Flaut and Cristina Flaut

Abstract This chapter features a brief analysis of the structure and evolution of the rural population of Muntenia in the 1930s, as they emerge from statistical data. The period selected, albeit marked by social issues and crises preceding the beginning of the Second World War, was one in which the population was able to face up to all the challenges entailed by holding on to traditions as well as by accepting and harmoniously assimilating elements of novelty.

Keywords Rural population · Structure of population · Birth rates · Death rates · Natural increase

1 Introduction

In 1930, Romania had a surface area of 295,049 km^2 and was divided into a variety of administrative units—9 provinces (Oltenia, Muntenia, Dobruja, Moldavia, Bessarabia, Bukovina, Transylvania, Banat, Crişana-Maramureş), 71 counties, 322 subdivisions, 172 towns and cities, and 15,201 villages (Georgescu 1937, p. 68; Manuilă and Georgescu 1938, p. 134). On the 29th of December 1930, Romania had 18,057,028 inhabitants, with 79.8% of the total population living in rural areas (Anuarul statistic al României 1940, pp. 44–46).

The 1930s were characterised by profound social crises, preceding the beginning of the Second World War. Such issues notwithstanding, the countryside represented a world in which traditions were preserved and parents acted as role models for their children. Most people carried on living in their place of birth, families were large and tightly knit and the number of divorces was much lower than in urban areas.

D. Flaut (✉)
Faculty of History and Political Science, Ovidius University of Constanţa, Constanţa, Romania
e-mail: daniel_flaut@yahoo.com

C. Flaut
Faculty of Mathematics and Computer Science, Ovidius University of Constanţa, Constanţa, Romania
e-mail: cflaut@univ-ovidius.ro

D. Flaut et al. (eds.), *Decision Making in Social Sciences: Between Traditions and Innovations*, Studies in Systems, Decision and Control 247,
https://doi.org/10.1007/978-3-030-30659-5_17

Nevertheless, elements of novelty were not rejected but emerged naturally and were assimilated organically, slowly and irreversibly into the traditional way of life. Literacy was one such example, with parents becoming increasingly eager to send their children to school. Even though the number of literate people was much lower in rural than in urban areas, more than three quarters of the children between the ages of 8 and 12 in rural Muntenia could read, with dramatically declining percentages in the case of adults.

A study of the structure and evolution of the rural population can provide considerable insights into village life and people's attempts to find a balance between preserving traditions and embracing novelty, irrespective of the shape taken by the latter. For the purposes of this case study, we have selected Muntenia, the Romanian province with the largest number of villages at the time.

2 The Structure of Muntenia's Rural Population in 1930

At the time of the 29 December 1930 census, Muntenia had a surface area of 52,505 km^2, which accounted for 17.8% of the country's total surface. The province was divided into 12 counties (Argeş, Brăila, Buzău, Dâmboviţa, Ialomiţa, Ilfov, Muscel, Olt, Prahova, Râmnicu-Sărat, Teleorman, Vlaşca) and 56 subdivisions (Georgescu 1937, p. 68; Manuilă and Georgescu 1938, p. 135).

In 1930 Muntenia 2,927,242 people, representing 72.7% of a total population of 4,029,008, lived in villages. The highest percentage of rural dwellers was recorded in Olt County (93.9%) and the lowest in Ilfov County (35%). Muntenia was the Romanian province with the largest number of rural settlements (3357). Villages were generally small in terms of surface area, as was also the case in the other provinces of the Old Kingdom. In terms of rural population density, Muntenia came third, with 55.8 inhabitants per km^2, behind Bukovina (59.9) and Bessarabia (56.1). The average number of inhabitants per village, 872, was lower than the country average of 948. According to the data collected in the 1930 census, rural Muntenia totalled 650,415 households and 637,451 buildings (Anuarul statistic al României 1940, pp. 44–46; Georgescu 1937, p. 68; Manuilă and Georgescu 1938, p. 135). The average number of inhabitants was 4.5 per household and 4.6 per building (Manuilă and Georgescu 1938, p. 135).

The number of men was below that of women in rural Muntenia, with the 1930 census recording 1,427,352 men and 1,499,890 women. Men accounted for 48.8% of the province's total rural population. Across Muntenia's 12 counties, the highest percentage of male rural dwellers was recorded in Brăila (50.4%) and the lowest in Argeş (47.3%) (Anuarul statistic al României 1940, p. 46).

The population of rural Muntenia featured an overwhelming percentage of children and young people. According to the 1930 census, people under 19 accounted for 50.5% of the province's total rural population. The percentages were progressively lower in the case of adults, with people over 65 representing only 3.3% of the total population of rural Muntenia. 187 of the 95,483 villagers belonging to this category were aged 100 or older (Anuarul statistic al României 1940, p. 56).

As regards civil status in 1930 Muntenia, the data featured in Table 1 reveal that the percentage of married people was higher in rural than in urban areas (58.6% compared to 49.6%). The percentage of unmarried people was correspondingly higher in towns and cities (39%) than in the countryside (31%). As far as widows and widowers were concerned, there was no significance difference between the two environments. The same table also highlights the fact that the percentage of divorcees in Muntenia was more than 4 times lower in rural than urban areas, indicating that village family life was much more stable.

The percentages of married people and divorcees in rural Muntenia were lower than the ones recorded in the Romanian countryside as a whole, 61.1% and 0.5% respectively. The percentage of widowed villagers in this province was identical to the national countryside average (10%) (Georgescu 1937, p. 69).

The 1930 census revealed that the vast majority of widowed inhabitants was made up of women, both in urban and in rural areas. The percentage of widows was higher in towns and cities (84.6%) than in villages (77%). In 1930 150,971 (79.1%) of the 190,842 widowed villagers of Muntenia were female, and only 39,871 (20.9%) were male (Anuarul statistic al României 1940, pp. 51–52). As most women married older men, it was common for the husband to die first. Given their advanced age and the frequent existence of children, widows found themselves unable to remarry. Men found it harder to stay single and remarried faster (Millea 1933, p. 20).

The data gathered in the 1930 census highlight the fact that 99.3% of Muntenia's villagers were Eastern Orthodox, 0.2% Roman Catholics, 0.2% Adventists, etc. 97.6% of the population distributed among the province's 3357 villages was Romanian. The rest comprised Hungarians (0.1%), gypsies (1.9%) and other ethnicities (0.1%). The mother tongue of 98.9% of Muntenia's villagers was Romanian, with 0.1% speaking Hungarian, 0.6% Romani, etc. (Anuarul statistic al României 1940, pp. 36–38).

On the 29th of December 1930 the percentage of literate villagers in Romania (51.3%) was much lower than in the case of urban dwellers (77.3%). In rural Muntenia they amounted to only 48.8%, compared to 78.4% in urban areas. Moreover, literacy levels were considerably higher among men than women, both in urban and in rural areas. In the countryside the gap between the two sexes was much wider (64.9% compared to 38.7%). Only 68.5% of the male and 30.4% of the female villagers of Muntenia could read and write, indicating an obvious cultural disparity between men and woman (Manuilă 1936, pp. 938–940).

The overwhelming majority of literate villagers in Muntenia (94.3%) had only attended primary school. Only 3.1% of the province's literate villagers had benefited from secondary education, fewer than was the case in towns and cities, with 9.4%. The percentage of professionally trained people was extremely low in rural (1.8%) compared to urban (9.4%) areas. Only 0.1% of the literate women in rural Muntenia had benefited from higher education, compared to 2.5% in towns and cities (Table 2).

The percentage of literate Romanians in both urban and rural areas tended to be inversely proportional to age. In 1930 75.1% of the children aged 8–12 were literate,

Table 1 Civil status in 1930 Muntenia

Province and type of area	Inhabitants aged 13 and older		Unmarried		Married		Widowed		Divorced		Undisclosed	
	Thousands of inhabitants	%	Thousands of inhabitants	%	Thousands of inhabitants	%	Thousands of inhabitants	%	Thousands of inhabitants	%	Thousands of inhabitants	%
Muntenia	2786.6	100	932.9	33.5	1554.7	55.8	273.3	9.8	16.8	0.6	8.7	0.3
Rural	1918.6	100	594.5	31.0	1124.3	58.6	190.8	10.0	5.7	0.3	2.9	0.1
Urban	868.0	100	338.3	39.0	430.3	49.6	82.4	9.5	11.0	1.3	5.7	0.6

Source Anuarul statistic al României (1940, pp. 50–52)

Table 2 The percentage distribution of literate people in 1930 Muntenia according to their level of instruction

Province	Sex	Urban						Rural					
		Out-of-school	Primary	Secondary	Vocational	University	Other additional schools	Out-of-school	Primary	Secondary	Vocational	University	Other additional schools
Muntenia	Total	1.9	66.5	16.2	9.4	3.7	2.3	0.5	94.3	3.1	1.8	0.2	0.1
	Male	1.5	66.8	13.8	9.7	4.6	3.6	0.5	94.0	3.1	2.0	0.3	0.1
	Female	2.4	66.1	19.3	9.0	2.5	0.7	0.5	94.8	3.2	1.4	0.1	0.0

Source Anuarul statistic al României (1937, p. 40)

with a gradual decrease in relationship with age, reaching only 12.2% in the case of people over 65. Women accounted for a considerable percentage of illiterate people, both in towns and cities and in villages. In rural Muntenia 32.3% of the girls aged 8–12 and 54.7% of those aged 13-19 were illiterate. Female illiteracy in the countryside reached 81.8% in the case of those aged 20–64 and only 2.5% of village women over 65 could read and write (Table 3).

As far as birthplace is concerned, an analysis of the 1930 census figures reveals a considerable disparity between Romanian towns and cities and villages. In rural areas, the overwhelming majority of the inhabitants counted in the census were still living in their place of birth (83%). On the other hand, almost half of urban dwellers (49.6%) had been born in other places than the ones in which they were living at the time of the census. In rural Muntenia 2,428,175 of the total 2,927,832 inhabitants, that is 83%, were still living in their place of birth. The percentage of men was higher than that of women (85.5% compared to 80.5%) (Anuarul statistic al României 1940, p. 100; Measnicov 1942, pp. 394–395). The province's remaining rural dwellers had moved in from other areas. The 1930 census revealed that 14.9% of the women and 9.1% of the men living in rural Muntenia had been born in a different village (of the same county) than the one in which they were living at the time of the census. A higher percentage of women than men is to be noted in this respect, due to the fact that when young people from nearby or distant villages got married it was usually the woman who changed her domicile and moved into her new husband's home. Only 3.2% of the women and 3.1% of the men had migrated to rural Muntenia from the province's other counties. The percentage of inhabitants coming in from other provinces or from abroad was extremely low (Fig. 1).

In terms of occupation, the population of Romania could be divided into two large categories: active and passive, or dependant, population. The active population comprised those inhabitants who had an actual profession, whereas the second category consisted of women, children, the disabled and the elderly, who did not work and were supported by others (Georgescu 1937, p. 71; Manuilă and Georgescu 1938, p. 154).

The active population of Muntenia was 57%, a lower percentage than the national one of 58.4%. The active population of rural areas reached higher values (59.5%) than the province's towns and cities, whose active population only amounted to 50.6% (Anuarul statistic al României 1937, p. 46).

In 1930 Muntenia 1,746,000 people (or 76%) out of a total number of 2,298,600 inhabitants who had an actual profession lived in the countryside. An extremely high percentage of the province's population, amounting to 71% of the total active population, worked in agriculture. In the countryside, the overwhelming majority of active inhabitants (89.6%) worked the land. Only 4.4% of the active population of villages worked in the industrial sector. The percentages of active inhabitants accounting for the other professional categories were negligible: 1% in commence-finance, 1% in transport, 2% in public institutions and 0.3% in public health and sport, etc. In the countryside 50.8% of the active population were men and 49.2% were women. Almost all the active females in rural areas, that is 95.3%, worked the land (Table 4).

Table 3 The percentage of illiterate people in 1930 Muntenia, according to sex and age group

Province	Sex	Total				Urban				Rural			
		8–12	13–19	20–64	65 and older	8–12	13–19	20–64	65 and older	8–12	13–19	20–64	65 and older
Muntenia	Total	22.1	32.0	47.8	78.1	11.9	14.2	23.2	53.1	24.9	38.6	59.6	87.8
	Male	16.0	17.4	27.8	58.4	9.9	8.2	12.8	38.8	17.6	20.9	35.3	78.4
	Female	28.4	45.6	66.5	87.5	14.0	20.0	33.4	64.5	32.3	54.7	81.8	97.5

Source Anuarul statistic al României (1937, p. 39)

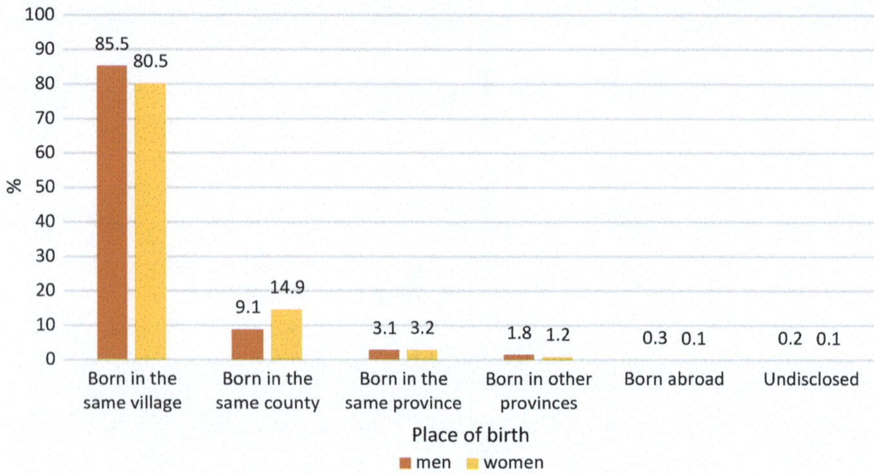

Fig. 1 The rural population of 1930 Muntenia according to their place of birth (expressed in percentages relative to the total population). *Source* Measnicov (1942, p. 395)

More than half of the active rural population of Muntenia (50.5%) consisted of auxiliary family members, working in agriculture alongside the head of the household. A high percentage (33.2%) was represented by independent farmers, working their own land. Salaried employees only accounted for 5.3% of the province's active rural population. There were very few landowners, landlords and pensioners (0.7%), self-employed (0.3%) and apprentices (0.3%), etc. (Anuarul statistic al României 1937, p. 49; Georgescu 1937, p. 72).

3 The Evolution of the Rural Population of Muntenia in the 1930s

3.1 Birth Rates

According to the data collected in the 1930 census, there was a difference between rural and urban birth rates in Romania. The birth rate was higher in the case of the rural (38.1‰) than of the urban population (22.7‰). Throughout the 1930s birth rates decreased, reaching 29.9‰ in rural and 20.9‰ in urban areas in 1939 (Anuarul statistic al României 1933, p. 44, 1940, pp. 151–153).

In 1930 the birth rate in rural Muntenia reached 41.4‰, which was above the country average for rural areas (Anuarul statistic al României 1933, p. 44). In the 1930s even Muntenia's birth rate decreased, but was still higher than the values recorded in rural areas at country level. The highest birth rate (44‰) was recorded in 1932 and the lowest (32‰) in 1939. Among the province's counties, the highest birth

Table 4 The active population of 1930 Muntenia according to professional category, sex and area, expressed in percentages

Province	Sex	Area	Total active population		Agriculture	Industry	Commerce-finance	Transport	Public institutions	Public health, sport	Other categories, undisclosed
			Thousands	%							
Muntenia	Both sexes	Total	2298.6	100.0	71.0	9.6	4.4	2.7	5.8	1.5	5.0
		Urban	552.6	100.0	11.7	26.2	14.9	8.1	18	5.2	15.9
		Rural	1746.0	100.0	89.6	4.4	1.0	1.0	2.0	0.3	1.7
	Male	Total	1273.0	100.0	61.3	14.0	5.6	4.4	8.3	1.6	4.8
		Urban	385.3	100.0	8.6	29.8	15.4	10.0	20.4	4.4	11.4
		Rural	887.7	100.0	84.1	7.1	1.4	2.0	3.0	0.4	2.0
	Female	Total	1025.6	100.0	82.9	4.2	2.9	0.7	2.7	1.2	5.4
		Urban	167.3	100.0	18.9	17.9	13.6	4.0	12.6	6.9	26.1
		Rural	858.3	100.0	95.3	1.6	0.8	0.1	0.7	0.1	1.4

Source Anuarul statistic al României (1937, p. 48)

rate (49.8‰) was recorded in Ialomiţa in 1932 and the lowest (28.6‰) in Muscel in 1939. Over the nine years under discussion, the highest average birth rates were recorded in Ialomiţa (41.9‰), Brăila (40.8‰) and Râmnicu Sărat (40.4‰) counties, and the lowest in Muscel (33.3‰), Teleorman (33.9‰) and Vlaşca (34.4‰) counties (Table 5).

As regards the sex of live born children, statistical data analysis reveals a greater number of boys than girls. Between 1930 and 1938, 2,466,822 out of a total number of 4,794,988 live born children recorded in the Romanian countryside were boys and 2,328,166 were girls. The average number of live born boys per 100 girls over the 9 years under discussion was 105.95. During the same period, 559,462 of the 1,087,934 live born children in rural Muntenia were boys, and 528,472 were girls. The average number of live born boys per 100 girls over the 9 years under discussion was 105.86 (Table 6).

Between 1930 and 1938 most of the total number of live births recorded in rural Muntenia occurred in October (109,223), April (98,695) and March (98,126), the respective children being conceived in January, July and June. The smallest number was recorded in December (56,627), January (75,343) and November (82,516), that is in the case of children conceived in March, April and February. Throughout the nine years, more boys than girls were born between January and November. With the exception of 1935 and 1936, more girls than boys were born in December. The birth rate percentages in decreasing order for each month of that period are as follows (Table 7).

Between 1930 and 1938 most of the 559,462 live born boys were recorded in October (55,991), April (50,981) and March (50,260). Over the same period, most of the 528,472 live born girls were recorded in October (53,232), March (47,866), May (47,724) and April (47,714). The birth rate percentages per sex, in decreasing order for each month of that period, are as follows (Table 8).

Illegitimate children were born less frequently in rural than in urban areas. Between 1934 and 1938, 232,148 (8.88%) of the 2,614,402 children born in the Romanian countryside were illegitimate. Over the same period, 65,341 (17.26%) of the 378,674 children born in Romanian towns and cities were illegitimate. Between 1934 and 1938, 64,291 (10.80%) of the 595,131 children born in rural Muntenia were illegitimate. Their percentage decreased from 11.82% of the total number of live born children in 1934 to 9.61% in 1938 (Table 9).

Very few of the births in rural Romania were assisted by a doctor or a certified midwife. The majority of births were assisted by people with no medical training or received no assistance whatsoever (Measnicov 1937, p. 162). In 1937, for instance, 59 of the 116,873 live born children in rural Muntenia were delivered in hospital under medical supervision, 116,809 at home (of which 666 were assisted by a doctor, 38,218 by a certified midwife, 76,969 by a non-certified midwife, 931 received no assistance whatsoever and 25 were assisted by undisclosed persons) and 5 elsewhere (Anuarul statistic al României 1940, p. 158).

Table 5 Live births in rural Muntenia between 1931 and 1939 (expressed in percentages per 1000 inhabitants for each county)

Counties	1931	1932	1933	1934	1935	1936	1937	1938	1939
Romania (rural areas)	36.1	39.1	34.7	35.0	33.0	33.8	32.9	31.5	29.9
Muntenia	39.4	44.0	37.9	38.4	36.0	37.7	34.8	33.5	32.0
Argeş	37.4	40.3	34.7	35.3	34.7	34.6	33.0	32.7	30.5
Brăila	44.9	48.9	43.1	43.6	39.2	39.8	38.3	34.8	35.0
Buzău	41.3	45.5	41.5	40.6	38.4	38.6	36.7	36.4	33.8
Dâmboviţa	39.7	42.2	35.5	37.5	34.9	35.8	33.6	31.9	31.3
Ialomiţa	42.4	49.8	45.1	43.3	40.3	42.9	39.5	37.5	36.0
Ilfov	39.4	46.2	40.0	40.2	37.8	39.8	34.9	33.5	33.5
Muscel	37.1	38.1	33.7	33.7	32.8	33.6	30.5	32.0	28.6
Olt	38.3	42.3	33.9	37.6	34.6	37.5	34.8	34.0	30.7
Prahova	39.8	42.0	37.0	37.8	36.5	36.6	33.2	33.7	32.2
Râmnicu-Sărat	44.5	49.1	43.5	43.0	38.7	38.3	38.1	35.1	33.5
Teleorman	34.4	40.8	32.9	34.0	31.2	37.9	34.1	31.1	28.7
Vlaşca	36.9	43.0	35.1	34.8	32.5	35.5	32.1	30.4	29.3

Sources Anuarul statistic al României (1935, pp. 44, 64, 84, 1937, pp. 53, 59, 1939, pp. 93, 113, 1940, pp. 145, 151)

Table 6 Live births according to sex in rural Romania and rural Muntenia between 1930 and 1938

Year	Romania (rural areas)			Muntenia (rural areas)		
	Male	Female	Total	Male	Female	Total
1930	280,116	263,212	543,328	62,085	58,094	120,179
1931	272,278	256,869	529,147	60,763	57,843	118,606
1932	300,026	283,491	583,517	69,301	65,826	135,127
1933	269,542	255,052	524,594	61,010	57,881	118,891
1934	276,520	261,879	538,399	63,265	59,482	122,747
1935	264,098	247,708	511,806	60,262	56,365	116,627
1936	274,285	258,170	532,455	64,282	60,186	124,468
1937	269,380	254,621	524,001	59,870	57,003	116,873
1938	260,577	247,164	507,741	58,624	55,792	114,416
	2,466,822	2,328,166	4,794,988	559,462	528,472	1,087,934

Sources Anuarul statistic al României (1933, p. 45, 1935, pp. 50, 70, 90, 1937, p. 64, 1939, pp. 98, 118, 1940, p. 156)

Table 7 Rural Muntenia birth rate percentages per month, in decreasing order for the period between 1930 and 1938 (percentage of live born children)

October	10.04%	May	8.97%	August	8.83%	November	7.58%
April	9.07%	September	8.87%	June	8.47%	January	6.93%
March	9.02%	July	8.84%	February	8.17%	December	5.21%

Sources Anuarul statistic al României (1933, pp. 45–46, 1935, pp. 50–51, 70–71, 90–91, 1937, pp. 64–65, 1939, pp. 98–99, 118–119, 1940, pp. 156–157)

Table 8 Rural Muntenia birth rate percentages per sex and month, in decreasing order for the period between 1930 and 1938 (percentage of live born children)

Boys		Girls	
October	10.01%	October	10.07%
April	9.11%	March	9.06%
March	8.98%	May	9.03%
September	8.93%	April	9.03%
May	8.92%	August	8.87%
July	8.88%	July	8.80%
August	8.78%	September	8.80%
June	8.48%	June	8.47%
February	8.22%	February	8.12%
November	7.51%	November	7.67%
January	7.17%	January	6.67%
December	5.01%	December	5.41%

Sources Anuarul statistic al României (1933, pp. 45–46, 1935, pp. 50–51; 70–71; 90–91, 1937, pp. 64–65, 1939, pp. 98–99, 118–119, 1940, pp. 156–157)

Table 9 Live born children in rural Muntenia compared to rural Romania between 1934 and 1938 according to civil status

Year	Romania (rural areas)						Muntenia (rural areas)					
	Total	%	Legitimate	%	Illegitimate and undisclosed	%	Total	%	Legitimate	%	Illegitimate and undisclosed	%
1934	538,399	100.0	487,618	90.57	50,781	9.43	122,747	100.0	108,241	88.18	14,506	11.82
1935	511,806	100.0	462,670	90.40	49,136	9.60	116,627	100.0	102,879	88.21	13,748	11.78
1936	532,455	100.0	484,386	90.97	48,069	9.03	124,468	100.0	111,016	89.19	13,452	10.81
1937	524,001	100.0	481,182	91.83	42,819	8.17	116,873	100.0	105,280	90.08	11,593	9.92
1938	507,741	100.0	466,398	91.86	41,343	8.14	114,416	100.0	103,424	90.39	10,992	9.61

Sources Anuarul statistic al României (1937, p. 69, 1939, pp. 103, 121, 1940, p. 162)

3.2 Death Rates

In 1930 Romania death rates were higher in rural (20.0‰) than urban (17.0‰) areas. Throughout the 1930s death rates decreased in the countryside and increased in towns and cities. In 1939 rural death rates were lower than urban death rates (18.5‰ compared to 19‰) (Anuarul statistic al României 1933, p. 44, 1940, pp. 151–153).

On the 29th of December 1930 the death rate in rural Muntenia was 21.0‰, higher than the country average for rural areas (Anuarul statistic al României 1933, p. 44). Throughout the 1930s, the death rate in rural Muntenia decreased and was lower than the values recorded in villages across the country. The lowest death rate (16.8‰) was recorded in 1939 and the highest (21.3‰) in 1931. Across the province's counties, the highest death rate (24.2‰) was recorded in Ialomiţa in 1931 and the lowest (14.6‰) in Prahova in 1939. Over the nine years under discussion, the highest average death rates were recorded in Ialomiţa (20.9‰), Ilfov (20.4‰) and Buzău (19.8‰) counties, and the lowest in Muscel (17.8‰) and Prahova (17.2‰) counties (Table 10).

A total number of 2,809,933 people died in the Romanian countryside between 1930 and 1938, comprising 1,442,376 men and 1,367,557 women. The average number of male deaths per 100 female deaths over the 9 years under discussion was 105.47. During the same period, a total number of 559,307 people died in rural Muntenia, comprising 288,174 men and 271,133 women. The average number of male deaths per 100 female deaths over the 9 years under discussion was 106.29 (Table 11).

Between 1930 and 1938 most of the total number of 559,307 deaths recorded in rural Muntenia occurred in March (51,557), April (49,981) and December (49,556) and the fewest in June (38,901), September (42,202) and May (44,637). The death rate percentages, in decreasing order for each month of that period, are as follows (Table 12).

Between 1930 and 1938 most of the 288,174 recorded male deaths occurred in March (26,873), April (25,878) and February (25,596) and the fewest in June (19,806), September (21,619) and May (22,952). Most of the 271,133 female deaths recorded during the same period occurred in March (24,684), April (24,103) and December (24,023) and the fewest in June (19,095), September (20,583) and May (21,685). The death rate percentages per sex, in decreasing order for each month of that period, are as follows (Table 13).

Table 14 highlights the distribution of deaths according to age groups in rural Muntenia between 1934 and 1937. It can be noted that most of the deceased were children. 127,855 (50.8%) out of the total number of 251,389 deceased people were children under the age of 5, followed by people older than 65, who accounted for 18.7% of the total number of deaths in the rural areas of this province.

Child mortality rates were higher in rural than urban Romania. Between 1932 and 1934, average child mortality was 18.1‰ in the countryside and 17.6‰ in towns and cities. During the same period of time, average child mortality in rural Muntenia was 17.6‰, same as the one recorded in the country's urban areas. Child mortality values were not identical across the country's provinces. Between 1932

Table 10 Deaths in rural Muntenia between 1931 and 1939 (expressed in percentages per 1000 inhabitants for each county)

County	1931	1932	1933	1934	1935	1936	1937	1938	1939
Romania (rural areas)	21.4	22.3	19.0	21.3	21.5	19.9	19.4	19.0	18.5
Muntenia	21.3	21.0	18.5	19.9	19.7	19.2	18.1	17.7	16.8
Argeş	20.3	19.6	18.2	19.2	19.2	19.2	18.2	17.1	16.5
Brăila	22.0	23.7	15.9	21.0	17.7	19.0	17.6	16.7	17.4
Buzău	21.7	22.2	18.6	19.8	19.5	20.5	17.9	19.2	19.0
Dâmboviţa	20.1	20.1	19.1	18.9	19.2	18.9	18.9	17.5	14.9
Ialomiţa	24.2	23.0	20.4	23.0	20.0	20.1	19.0	19.3	19.5
Ilfov	23.0	23.1	20.0	20.7	21.1	21.0	18.4	19.1	17.0
Muscel	20.5	18.2	17.8	18.6	18.3	17.1	16.8	17.2	16.1
Olt	20.3	19.9	20.3	19.4	21.8	20.4	18.1	17.3	16.0
Prahova	18.6	18.7	16.5	18.0	17.2	17.5	17.4	16.1	14.6
Râmnicu-Sărat	20.7	22.2	17.1	21.9	18.5	19.4	16.7	17.5	17.7
Teleorman	22.7	20.3	17.9	18.2	21.4	18.6	18.4	17.3	16.5
Vlaşca	21.3	21.0	18.6	20.6	21.8	18.2	18.0	17.8	17.0

Sources Anuarul statistic al României (1935, pp. 44, 64, 84, 1937, pp. 53, 59, 1939, pp. 93, 113, 1940, pp. 145, 151

Table 11 Deaths per sex in rural Romania and rural Muntenia between 1930 and 1938

Year	Romania (rural areas)			Muntenia (rural areas)		
	Male	Female	Total	Male	Female	Total
1930	146,552	138,586	285,138	31,588	29,363	60,951
1931	161,332	153,038	314,370	33,164	30,965	64,129
1932	170,894	161,912	332,806	33,387	31,131	64,518
1933	147,693	139,825	287,518	29,671	28,145	57,816
1934	167,278	159,937	327,215	32,590	30,901	63,491
1935	171,578	162,635	334,213	32,991	30,951	63,942
1936	160,696	152,216	312,912	32,852	30,526	63,378
1937	158,576	149,908	308,484	31,078	29,500	60,578
1938	157,777	149,500	307277	30,853	29,651	60,504
	1,442,376	1,367,557	2,809,933	288,174	271,133	559,307

Sources Anuarul statistic al României (1933, pp. 48–49, 1935, pp. 54–55; 74–75; 94–95, 1937, pp. 66–67, 1939, pp. 100–101, 118–119, 1940, pp. 164–165)

Table 12 Rural Muntenia death rate percentages per month, in decreasing order for the period between 1930 and 1938 (percentage of deceased people)

March	9.22%	February	8.82%	November	8.28%	May	7.98%
April	8.94%	August	8.66%	October	8.27%	September	7.55%
December	8.86%	January	8.39%	July	8.07%	June	6.96%

Sources Anuarul statistic al României (1933, pp. 48–49, 1935, pp. 54–55; 74–75; 94–95, 1937, pp. 66–67, 1939, pp. 100–101, 118–119, 1940, pp. 164–165)

Table 13 Rural Muntenia death rate percentages per sex and month, in decreasing order for the period between 1930 and 1938 (percentage of deceased people)

Males		Females	
March	9.33%	March	9.10%
April	8.98%	April	8.90%
February	8.88%	December	8.86%
December	8.86%	August	8.84%
August	8.51%	February	8.75%
January	8.48%	November	8.33%
October	8.33%	January	8.30%
November	8.24%	October	8.22%
July	8.06%	July	8.07%
May	7.96%	May	8.00%
September	7.50%	September	7.59%
June	6.87%	June	7.04%

Sources Anuarul statistic al României (1933, pp. 48–49, 1935, pp. 54–55, 74–75, 94–95, 1937, pp. 66–67, 1939, pp. 100–101, 118–119, 1940, pp. 164–165

Table 14 Deaths per age group in rural Muntenia between 1934 and 1937

Age group	Number of deaths	%	Age group	Number of deaths	%
Total deaths 1934–1937	251,389	100	25-44 years old	21,024	8.4
Under 1 an	83,779	33.3	45-64 years old	28,828	11.5
1–4 years old	44,076	17.5	65 and older	47,053	18.7
5–14 years old	14,260	5.7	Undisclosed age	51	0.0
15–24 years old	12,318	4.9			

Sources Anuarul statistic al României (1937, p. 73, 1939, pp. 107, 123)

and 1934, average child mortality in rural Muntenia was higher than in rural Oltenia (15.5‰), Transylvania (16.4‰) and Banat (17.4‰), but lower than in rural Moldavia, Maramureş (19.0‰), Bessarabia (19.1‰), Dobruja (20.4‰) and Bukovina (20.5‰) (Measnicov 1937, p. 161).

Between 1934 and 1937, 533 of the 251,389 deaths recorded in rural Muntenia were caused by typhoid and paratyphoid fever, 41 by exanthematous typhus, 2850 by measles, 1299 by scarlet fever, 1076 by whooping cough, 392 by diphtheria, 1336 by influenza, 18,908 by respiratory tuberculosis, 4413 by other types of tuberculosis, 1232 by syphilis, 313 by malaria, 3606 by other infectious and parasitic diseases, 2367 by cancer and other malignant tumours, 173 by non-malignant tumours, 244 by rheumatism or gout, 85 by diabetes, 317 by alcoholism, 3885 by pellagra, 1696 by other common illnesses and chronic poisoning, 355 by locomotor ataxia and general paralysis, 2549 by cerebral haemorrhage, 4712 by other nervous system diseases, 9274 by heart disease, 1265 by other circulatory system diseases, 3769 by bronchitis, 42,262 by pneumonia, 6412 by other respiratory system diseases, 19,125 by diarrhoea, enteritis, under the age of 2, 4527 by diarrhoea, enteritis, above the age of 2, 474 by appendicitis, 1084 by liver and biliary tract disease, 3636 by other digestive system diseases, 6495 by nephritis, 299 by other genitourinary disorders, 1137 by septicaemia and puerperal infections, 601 by other puerperal diseases, 339 by skin or bone conditions, 53,290 by congenital malformations and early childhood illnesses, 34,628 by senility, 562 by suicide, 545 by homicide, 4868 by violent and accidental death, 4415 by unspecified and poorly defined causes (Anuarul statistic al României 1937, pp. 74–75, 1939, pp. 108–109, 124–127).

3.3 *Natural Increase*

On the 29th of December 1930 the Romanian rate of natural increase was over 3 times higher in rural than urban areas (18.1‰ compared to 5.7‰). Throughout the 1930s, the rate of natural increase diminished, reaching 11.4‰ in the countryside and only 1.9‰ in towns and cities in 1939 (Anuarul statistic al României 1933, p. 44, 1940, p. 151–153).

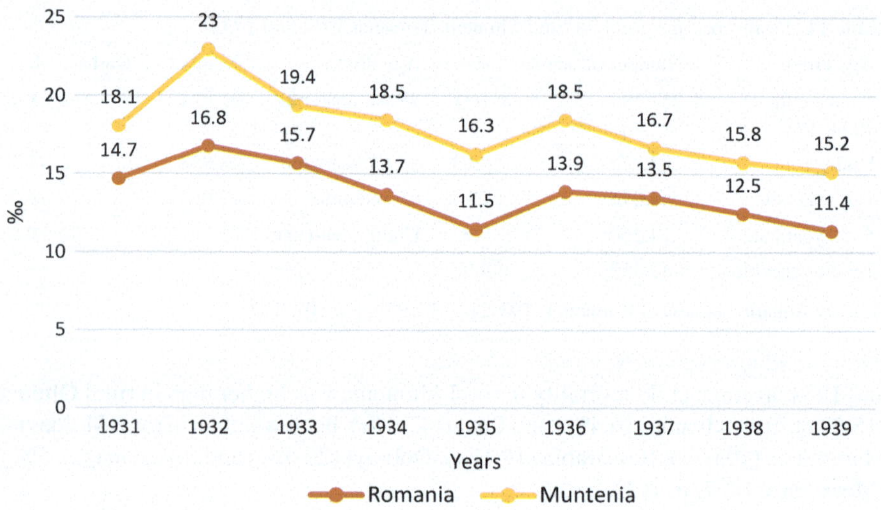

Fig. 2 The evolution of rural Muntenia rates of natural increase compared to rural Romania rates between 1931 and 1939. *Sources* Anuarul statistic al României (1935, pp. 44, 64, 84, 1937, pp. 53, 59, 1939, pp. 93, 113, 1940, pp. 145, 151)

In 1930 the rate of natural increase in rural Muntenia was 20.4‰, higher than the country average for rural areas (Anuarul statistic al României 1933, p. 44). Throughout the 1930s the rate of natural increase also diminished in rural Muntenia, but reached higher values than the country average for rural areas. The highest rate of natural increase (23‰) was recorded in 1932 and the lowest (15.2‰) in 1939 (Fig. 2).

An analysis of the data in Tables 5 and 10 reveals that the highest rate of natural increase across Muntenia's counties (27.2‰) was recorded in Brăila in 1933 and the lowest (9.8‰) in Teleorman in 1935. Over the 9 years under discussion the highest rural population natural increase was recorded in Brăila (21.8‰), Râmnicu Sărat (21.3‰) and Ialomiţa (20.9%) counties, and the lowest in Teleorman (14.9%), Vlaşca (15.03‰) and Muscel (15.5‰) counties.

4 Conclusions

On the 1st of September 1939 the Second World War broke out. The years of war had a negative impact on the population evolution of rural Muntenia. In 1944 the birth rate in rural Muntenia was lower than that of 1939 (27.1‰ compared to 32‰). It was however one of the highest in the country, only surpassed by that of rural Dobruja (31.4‰). Among the province's counties, the highest birth rate was recorded in Ialomiţa (30.9‰) and the lowest in Teleorman (24.8‰). In the course of the same year, the death rates of rural Muntenia averaged 18.5‰, with the highest values recorded in Muscel County (20.2‰) and the lowest in Teleorman County (16.2%).

In 1944 the rate of natural increase in the province's rural areas was 8.6‰, compared to 15.2‰ in 1939. It was however one of the highest in the country, only surpassed by that of rural Dobruja (12.8‰) (Buletinul demografic al României 1948, p. 10).

Four years later, in 1948, death rates in rural Muntenia averaged 16.3‰, lower than the values recorded in the 1930s. Birth rates averaged 25.4‰, a lower percentage even than the one recorded in 1944. The rate of natural increase resulting from these values is 9.1‰, somewhat higher than in 1944, but very far from pre-war figures (Buletinul demografic al României 1950, p. 70).

This chapter made use of statistical data for a brief analysis of the structure and evolution of the rural population of Muntenia in the 1930s, a period marked by profound social issues and crises. Notwithstanding the social turmoil, the data presented also reveal positive aspects: there were numerous people who lived up to and past the age of 100 and divorce rates were very low in rural areas, traditional families being extremely tightly knit. On the other hand infantile mortality values were very high and pneumonia and tuberculosis featured among the chief causes of death. Nowadays, scientific progress has drastically cut the number of such deaths and many of that period's devastating illnesses have either been eradicated or are now considered curable.

Depending on how they are interpreted, statistical data can yield a wide range of information, as well as insights into a certain society at a given moment (Urban and Hoskova-Mayerova 2017). We cannot argue that this kind of data can provide us with information regarding the level of happiness experienced by the people of those times, but we can assume that, irrespective of the situation, there has always been a tendency to hope for a better life.

References

Anuarul statistic al României: 1931–1932, 399pp. Monitorul Oficial şi Imprimeriile Statului. Imprimeria Naţională, Bucureşti (1933)

Anuarul statistic al României: 1934, 494pp. Tipografia Curţii Regale F. Göbl Fii S.A., Bucureşti (1935)

Anuarul statistic al României: 1935 şi 1936, 421pp. Institutul central de statistică, Bucureşti (1937)

Anuarul statistic al României: 1937 şi 1938, 714pp. M.O. Imprimeria Naţională, Bucureşti (1939)

Anuarul Statistic al României 1939 şi 1940, 762pp. M.O. Imprimeria Naţională, Bucureşti (1940)

Buletinul demografic al României, anul XIII, 1944. Institutul central de statistică, Bucureşti (1948)

Buletinul demografic al României, anul XVIII, nr. 1–2, ianuarie-februarie 1949. Institutul central de statistică, Bucureşti (1950)

Georgescu, D.C.: Populaţia satelor româneşti. Sociologie românească, Anul II, nr. 2–3, februarie-martie, pp. 68–79 (1937)

Manuilă, S.: Ştiinţa de carte a populaţiei României. Arhiva pentru ştiinţă şi reformă socială, XIV, pp. 931–959 (1936)

Manuilă, S., Georgescu, M.: Populaţia României. In: Gusti, D. (ed.) Enciclopedia României, vol. I, pp. 133–160. Bucureşti (1938)

Measnicov, I.: Mortalitatea populaţiei rurale ţărăneşti. Sociologie românească, Anul II, nr. 4, aprilie, pp. 158–167 (1937)

Measnicov, I.: Migraţiunile interioare în România. Sociologie românească, Anul IV, nr. 7–12, iulie-decembrie, pp. 394–411 (1942)
Millea, G.: Contribuţii la demografia Municipiului Bucureşti 1831–1930, vol. 1. Populaţia după recensăminte. Natalitatea. Mortalitatea, 88pp. Tipografia Curţii Regale F. Göbl Fii S.A., Bucureşti (1933)
Urban, R., Hoskova-Mayerova, S.: Threat life cycle and its dynamics. Deturope **9**(2), 93–109 (2017)

Vernacular and Urban-Influenced Housing Types in Constanţa County in the 20th Century

Enache Tuşa and Ilie Iulian Mitran

Abstract The evolution of housing in Constanța County was influenced by a multitude of factors from one decade to the other, this created some significant variations in the way homes were designed and built. The changes had to do mostly with the fact that peasants began to massively import aspects of urban lifestyle, along with new design and building techniques. The current study aims at mapping the evolution of rural housing from Constanța County, beginning with the last decade of the 20th century, up to the late 1960s. The study highlights the way in which many rural communities suffered various social changes, that greatly affected the mindset that was previously in place when it came the housing design and household management.

Keywords Constanța · Rural society · Household vernacular · Peasant · House · Urban housing · Village · Hut · Settlement

1 Introduction

The current paper is motivated by the lack of detailed chapter that are focused on the evolution of local housing types in Constanta County, adding to this, there is a lack of resources written in English that could contribute to making this topic more accessible to research that is done by scholars that are not speakers of Romanian. Another important aspect that we need to highlight is the fact that there are very few studies that analyze the actual evolution of housing in the area that we focuses on, Constanta County, in the timeframe that work with, more precisely the period between the 1910s and 1970s. The houses that we include in this study are from central and south-eastern area of the county, each of these areas have a series of particularities when it comes to their topography, proximity to any nearby urban center, ethnic make-up,

E. Tuşa (✉)
Faculty of History and Political Science, Ovidius University, Constanta, Romania
e-mail: enachetusa@gmail.com

I. I. Mitran
Faculty of Sociology and Social Work, University of Bucharest, Bucharest, Romania
e-mail: mitraniulian@europemail.com

D. Flaut et al. (eds.), *Decision Making in Social Sciences: Between Traditions and Innovations*, Studies in Systems, Decision and Control 247,
https://doi.org/10.1007/978-3-030-30659-5_18

as well as some minor aspects that are more tightly connected to the local economy. Darabani is situated near the border with Bulgaria, its current inhabitants are mostly Romanian, the economy is entirely based on agriculture, with some households being also involved with animal husbandry. The topography is not entirely flat, the village's outskirts are dominated by a few mild hills that together outline the shape of the flat valley in which the majority of households are concentrated. Văleni, on the other hand, is also located in a valley, but the hills that are situated on its outskirts are more prominent, thus giving the sensation that they valley is deeper. Romania's coastal region experienced some significant changes in housing, especially in the two decades that followed World War II. The beginning of the 1900s brought a series of novelties when it came to the configuration of peasant houses, the building materials that were put in use, and the amenities that were made available. Poorly built huts dominated many settlements at the end of the 19th century, from this, locals switched to a two-room house, usually constructed from mud brick or from rock masonry, being kept together by a mixture of mud and dung. The transition to a modern housing type was made somewhere at the beginning of the 1930s, from this period, and continuing tot the 1960s. This period was marked by the construction of modern houses which had a layout that clearly separated the living quarters from those that were destined for household activities. The houses that were built in this period usually exhibit bold color schemes, porches sustained by thick columns with arched openings, and various decorative patterns which fill the facades. The current paper attempts to give insight on the way in which housing styles evolved in Constanța County throughout the last century. The study will focus on four three-dimensional models which will recreate a few representative housing types from the 20th century, highlighting both the functionalities and the distinctive cultural footprints embedded their style.

A huge difference that we can notice between the two previously mentioned villages is that Darabani, in comparison to Văleni, is situated nearby Negru Vodă (formerly Cara-Omer), which for decades acted as a regional hub for commerce and craftsmanship. Văleni is quite isolated, far away from any regional town, thus we can see very few examples of urban-type homes there, also some of the basic amenities that are available are lacking, by comparison to Darabani. The paper is not only interested in giving a detailed description of the houses' architecture, and the amenities that were present in the era in which they were built, it is also interested in giving insight in the way in which some of the factors that we previously mentioned influenced the spread of a certain style of housing in certain villages.

The development of style is not seen only through a technical perspective, the cultural context is highly stressed, the author highlighting some general changes that occured for the entire community. The houses that the paper focus on were selected by taking a few criteria into consideration. First of all, how well does the house fit into the general housing pattern present in the village, this obvious for the houses from Darabani, Comana and Văleni. We are clearly talking here just about some general style patterns, mostly connected to the layout and the home's functionality, the house from Rasova is an exception here, due to the fact that one-story private homes are quite uncommon in the region, but their discreet presence show us the development of craftsmanship in certain settlements.

The architectural heritage of Dobrudja did not constitute a subject of great interest for the academic community, most of their efforts that were put into research projects were usually centered around the non-material heritage, mostly folklore, that was indigenous to both Dobrudjan Romanians, but also to the other minorities that inhabited the region. Works about the now well-known *multiculturalism dobrogean* (Dobrudjan multiculturalism) are quite numerous, and they already started to follow a very repetitive pattern, this usually involving stereotypication and self-stereotypication. Dr. Maria Magiru, director of the Museum of Ethnography and Traditional Art from Constanta coordinated a few books, that were the outcome of extensive field research, which documented the vernacular architectural styles of Constanta County. Besides the works that were coordinated by Dr. Magiru, we can also extract some valuable information about the housing styles popular among the Greeks of Izvoarele (Tulcea County), study singed by Narcisa Ştiucă. Her study does not focus on a community from Constanța County, but is describes very well the midset that motivated the spread of multifamily housing in Dobrudja.

2 Description of the Study Area

Romania's south-eastern region, which roughly corresponds with the historic province of Dobrudja, known for its semi-arid climate, mainly characterized by harsh summers dominated by periods of intense draught and the presence of intense winds that blow yearround, went through some dramatic changes when it comes to the way in which homes are designed and built. The shift from the vernacular two-roomed homes, which still greatly present in many rural settlements until the beginning of the 1900s, to the three-room single family homes, made from mud brick with wooden porches that dominated the entire length of the main facade, to urban-inspired homes made entirely of masonry (stone, abode, or brick] with colorful geometric patterns that adorned the facades that were concealed with cement plastering, which was often painted in a dark shade in order to make the decorative elements that were highlighting the outline of the doors and windows stand out. The focus of our study will be Constanta County, one of the forty-two counties of Romania, which incorporates the southern half of the historic province of Dobrudja (Fig. 1a–d).

In order to understand the way in which housing evolved in Constanta County in the 20th century, we need to knowledgeable in regards to some of the physical and social-demographical features that directly impacted housing development patterns within this area. The county is characterized by a predominantly flat terrain which takes the form of a plateau, even so, there are some notable local variations that are hard to ignore. The county's northern half usually has a more homogenous topography, by comparison the southern half, which is usually characterized by the presence of the so-called canarale—narrow valleys that branch across the surface, which is almost entirely covered in less, underneath which we find the plateau's chalky interior. It should be noticed that due to the local topography, the villages

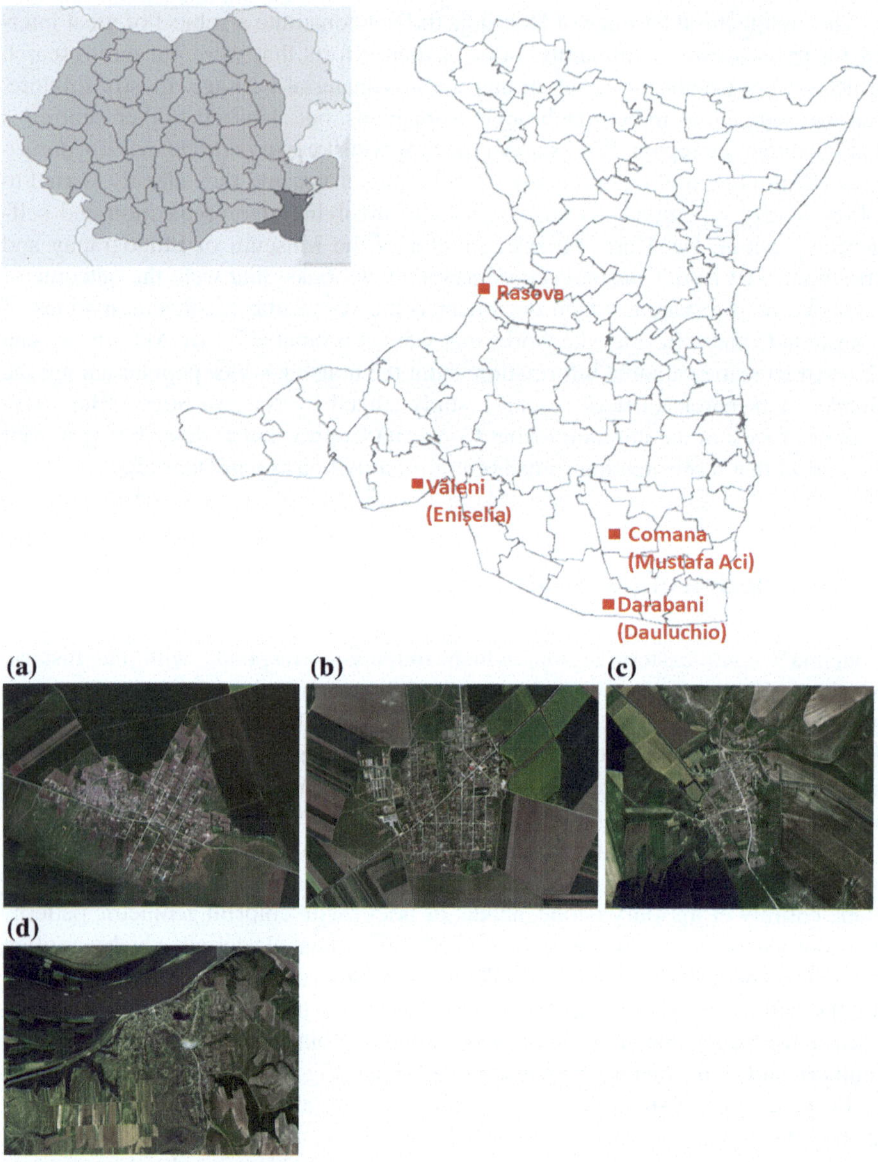

Fig. 1 Map of studies area: **a** Darabani/Dauluchioi; **b** Comana/Mustafa Aci; **c** Văleni/Enişelia; **d** Rasova

from the county's southern half are less when it comes to size and population in comparison to the ones from center and north.

Dobrudja's topography, and the overall physical characteristics of the terrain, greatly influenced the layout of many rural settlements, and in the same time some of the characteristics of some of the vernacular homes. The villages that are situated within canarale, narrow valleys with flat bottoms carved within masses of chack, tend to have a irregular layout, which is directly shaped by the shape and wight of the valley (Magiru 2010b, pp. 2–6). These villages usually have maximum two rows of houses parallel to the main road that extends throughout most of the village. The size of the households is somewhat smaller than in the case of villages that are located on flat pains, the overall layout of these villages is more complex, usually have a way more well-defined downtown area which usually host all of the key-institutional buildings. Interior Dobrudja lacks significant rivers, the southern half of Constanta County is usually dominated by semi-permanent creeks are usually dry for most of the year, receiving a consistent volume of water during the beginning of spring when the snow melts. The northern area is dominated by the Casimcea, one of the few permanent rivers from this region. The Casimcea discharges into Lake Tașaul, situated just north of the city of Navodari. The river is characterized by a very low annual volume of water, during the summers when the region is affected by intense draught usually it totally dries out.

There are also a few other permanent and semi-permanent rivers and creeks that flow through the region, namely—Taița [Tulcea County], Telița [Tulcea County], Tabana (Tulcea County flows through Babadag], Valea Seacă and Cartal (both affluents of the Casimcea], Cabul (Constanța County, creek], Dereaua [Constanța County, creek]. As a direct result of the low volume of water that these rivers in creeks receive year-round, flooding is not a serious issues, especially during the summertime. In the same time, it is important to acknowledge that the low volume of water makes them unreliable when it comes to irrigations and supplying nearby household with drinkable water. The county's south-western sector has the most rugged topography, mainly characterized by irregular network of narrow rocky valleys with a flat bottom. This did not benefited the development of too medium-size and big rural settlements.

Also, this area suffers greatly from a shortage of drinkable water, usually, in the most villages, there are a few wells that supply the whole village. Most wells are not actually owned by a particular family, they serve the entire community, they are usually built along various interior roads that do through the village, or on the main road that connects the village with the neighboring communities. Some of these wells were built during the Ottoman era, their depth can reach 1181–1377 inches, they usually have a rudimentary form and are handled by manpower. In some villages, there would be a well in front of the church, the water that was extracted would usually be used within various religious and folk-based rituals. In this area, villages vary in terms of form, from small linear villages—most households are positioned around the main road [Brebeni]; others that have a well-outlined center, which hosts the most significant institutions and commercial buildings, their outline is more regular; and the last type is represented by villages with a highly irregular outline, they usually are situated in narrow valleys and their settlement pattern usually coincides with bottom's outline [Lespezi]. Most villages that fall within the previous mentioned categories are to be found in county's south-western sector, the area that is usually between the Adamclisi and Băneasa. The villages that lay within this area are usually ranked among the county's poorest communities, with a significant number of households experiencing extreme poverty, some lacking even the most basic amenities.

The village of Văleni, previously known as Valea Rea [lit. Wicked Valley], is assembly one of the poorest settlements from the area, and from the entire county. The village's landscape is dominated by poorly built abode houses, with a highly irregular settlement pattern, mainly characterized by roads constantly twist and turn, the exact boundaries of some households is hard to determine due to the lack of fences, or if they are present they do not surround the entire property. Văleni is, maybe, one of the few settlements from the region that preserved, in part, several characteristics of Dobrudjan villages in the period that immediately followed the Independence War.

Specifically, low-rise buildings, mainly made out of abode, with two rooms, the streets are usually narrow and have an irregular pattern. The lack of sanitation and clear delimitation between the sectors of the households that are destined for raising the livestock and those reserved for human activities is quite obvious. As households evolved, beginning with the early 1900s up to the present-day, some significant changes were made in the way in which the house was positioned one the property by comparison to the stables and other annexes. Since urban influence became more and more powerful within many rural communities, the complementation of the property became more well-defined, each sector being reserved for a specific type of activity (Fig. 2). We can see, in most villages from Constanta County that the property is mainly divided into two areas.

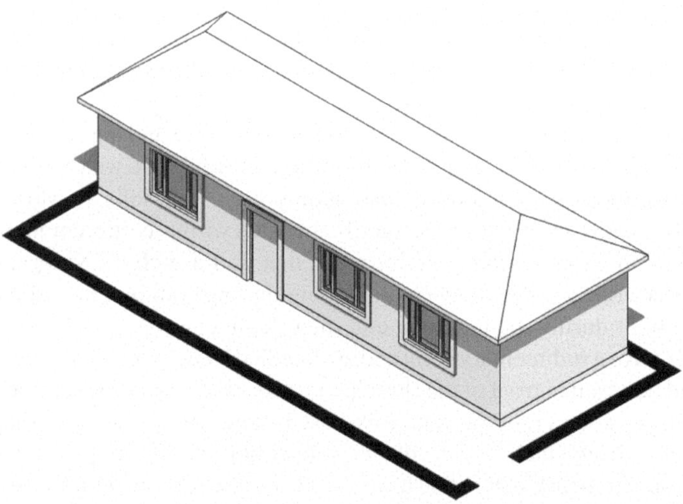

Fig. 2 Urban house on small lot, typical of several Constanta neighborhoods

The first one includes the house and the strip of land that is next to the street opening, this area is usually hosts the mainly house, sometimes we also have here the "chiller" or "bucătăria de vară"—a small building that is entirely detached from the main house and serves as a cooking space during the summertime, and the flower garden that usually stands between the main house and the main gate that opens to the street. The flower garden is good indicator of the degree of influence that urban culture had on rural communities, this being especially the case after WWII, after which many peasants started to assimilate certain behaviors tat were rather indigenous to city dwellers. If we were to put side by side a vernacular household and one that we would usually find nowadays in most of the villages of Constanta County, we immediately notice that the vernacular household usually has a bigger degree of ambiguity when it comes giving a precise destination to the majority of spaces. There would be always a specific section of the yard that would be exclusively reserved for the vegetable garden. Birds [e.g. chickens, ducks, goose] would usually run free, the stables would be near the main house. During the summertime, some of the food would be prepared in an oven situated in the yard, near the main house.

As time passed, the yard started to be divided into quarters, each one of them serving a specific goal, in some cases they were actually divided by wooden fences. The quarter that is stretches along the street front is usually reserved for presentation purposes and for leisure, it is the part of the property that is seen from the street, it is usually very well maintained, from this space usually the guest will walk enter the house where the good rooms are found. The good rooms, or as they are called by the local camerele bune, are usually not used as living quarters, especially if the house has some extra rooms besides them. These windows of these rooms usually face the street, one of them is usually reserved for guests, and the other one is usually used as family museum. Here the most precious belongings that the family has are put on display, the items can vary from china pieces to home made tapestries, or we can also find beautifully ornate pillows that have embroideries or lace borders. This room is also used to store the good clothes, usually the clothing pieces that would be worn at various events (e.g. weddings, baptisms, funerals, Sunday clothing]. This room would also host the wedding picture of owners, it would usually hang above the big tapestry that would on the wall which faces the bed. The picture, in some cases, would be an actual photo taken by a photographer, or it can also be a sketch, as photography used to be quite expensive. As most houses tended to have their main facade facing interior of the property, thus the narrow sides of the house would face the area which is near street fence, the good room would usually have its windows(s) facing the roadside. As the illiteracy level from the countryside started to slowly decline due to the policies implemented by the communists, books became more accessible to the rural population (Gaivoronschi 2009).

As a result, the good room became the place that hosted the family's collection of books, depending on their number we can talk about collection or a small family library. The room could also host various modern gadgets that would be used for entertainment purposes. But, this depended greatly on the actual social and economic status of the family in cause, there was a greater probability that the family that constituted the village's intellectual and economic core would be more determined to host a family library in one of the rooms than a family was solely involved in agricultural activities. We have to take into consideration that in Dobrudja, but not only, spaces reserved for displaying the family heritage, or the wealth that was accumulated by the current owners of the property, were not uncommon before the influence of the urban started to slowly penetrate in most rural settlements (Magiru 2010a, pp. 3–6). We are specifically talking about temporary exhibits, spaces from within the home that were used to display the dowry, for instance, for a temporary period, during the wedding, this a well-known practice among the Tatars.

"The good room" basically gives to the outsider a clear image on the values that are commonly shared among the members of a family, if the family still values hand crafted textiles the room will usually be filled with woven carpets, usually hanged on the walls, or decorative bed and pillow sheets made by the women. On the other hand, one can notice a room where hand-crafted goods comprise a small number out of the total number of items that are put on display. For instance, the woven carpets, that mostly decorate the walls, are replaced by oil paintings, bought from the city or from one of the few Sunday markets that are weekly organized in the region. When it comes to religious objects, things changed a little bit as time went by. Before printing technology was perfected, icons were produced only in the traditional techniques–oil painting. They were produced only by iconographs, artists that were specialized in Byzantine religious painting, which possessed the necessary skills to create icons, or to paint various other religious objects (e.g. crucifers, candle supports]. In some cases, those that painted icons would usually spend a bigger amount of time painting the interiors of churches. Due to the fact that icons were hard to acquire, and quite expensive for the average profits of a peasant family, a family would own mostly one icon. In many cases, the icon would be passed on within a family from one generation to the other, becoming an integral part of the family heritage.

Modern printing techniques were put in use for the production of religious objects, including icons, thus, icons became mass-produced and less costly. Besides this aspect, they became easier to find, as a result, the family would own more than a single icon. Before the spread of paper-print icons, the oil-paint icon would usually sit in the room where the family members would sleep. In some rural homes, there will be just one icon, located in the sleeping quarter, while the church calendar would usually be found in the kitchen, or the spaces that was reserved for cooking. The overall outline of some post-WWII houses reflected some patterns that were already present in many cities, that of basing the overall layout on a single axis. More specifically, houses tended to be long an narrow, most of the room were positioned in a linear pattern, which formed an axis. Mostly, this pattern was rarely broked, and when this was the case, the secondary axis was usually made up of rooms that served as cooking or deposing spaces.

The single-axis layout was popular within cities due to the shortage of building spaces, this being strongly connected to the way in which lands were split by the inheritance. After WWI many veterans received lands, new neighborhoods, and even entire villages, were formed. As time went on, specifically in case of lands from within the cities, the children of the veterans divided the lands among themselves, the next generation did the same thing, and the pattern continued (Ştiucă 2011, pp. 3–8). A major consequence of this ongoing pattern, the properties became long and narrow, leaving no other solution other than building houses that would have no more than 157 inches in wight, while the total wight of the property could go to around 196 inches.

This had negative consequences on the living standards from within many residential neighborhoods that were built on lands granted to veterans in the aftermath of WWI. First of all, the houses are too close to each other, meaning that there is very little privacy. Other aspects can be related to the reduced volume of sunlight that the house received due to its proximities, this can have a series of native consequences both on the building and its inhabitants. The overall natural ventilation of the house is put in danger, and in the same time there is greater risk for fires to spread more rapidly due to the close proximity between the properties.

This is one of the reasons why such properties tend not to perform very well on the local housing market, but there are basically no real solutions that can be applied in order to entirely correct all of the previously mentioned dysfunctions household's layout is usually determined by various factors, we are considering the size, the terrain—especially the topographic particularities, and ethnic background of the owners. Besides the style patterns that are directly influenced by the terrain and the building materials that are available in a certain region, we also need to consider some aspects that are directly linked to the cultural background of the people that inhabit.

The houses of Old Russian Believers, the most homogenous community from Constanta County is concentrated in Ghindărești, just a few km north of Rasova, on the eastern bank of the Danube. The houses of Old Believers from Dobrudja are well-known for their highly distinctive layout, which is centered on a straight axis, starting from the roadside, with the main house, and gradually progressing into the interior with various annexes that form a strong linear pattern. Traditionally, the annexes that are directly linked to the main house are usually used as cooking or storage spaces, while the ones that are last usually serve as shelters for animals. The household that are owned by the descendance of Oltenian colonists that were colonized in Constanta County in the aftermath of the Treaty of Craiova. The treaty had a direct impact on the demographic profile of Southern Dobrudja, also known as Cadrilater.

The Romanians and Aromanians that were colonized in this territory which constituted the interwar counties of Caliacra and Durostor. The Wallachian Romanians that were resettled in the central and north-eastern regions of Constanta County preserved most of their ancestral costumes that were practiced in Oltenia, one of them was concerned with the way in which the household was to be inherited by the sons, The Wallachians that settled in villages they were left behind by the Bulgarians that were resettled in their homeland, preserved the patrilineal character of their culture. More specifically, they preserved the costumes that gave priority the youngest son when it came to inheriting the property. The costume also dictated that the son that inherited the house and the land would be granted with the responsibility of taking care of the parents, after the passed away they would receive full ownership of the property. Usually, during the period in which the ownership agreement was settled, the younger son would start building a new house within the same yard as the house belonging to the parents. In most cases, after the youngest son would get married, the newly-weds and the groom's parents would live in the same house hold. The house that would be built is usually entirely detached from the old one [casă bătrânească], thus the two couples would have shared a common space, represented by the yard, but would also have private spaces. The Greek community from Izvoarele, Tulcea County, would also rely on multi-family households for assuring this way a strong bound between the generations.

The local costume stipulates that the son needs to stay with the patents, usually the young couple would build some additional rooms in which they would concentrate their family life. Thus, the houses from Izvoarele usually have a narrow and highly elongated layout, the new additions would usually be situated more into the interior, while the old house would be positioned near the roadside. Multi-family households are still quite common among many Aromanian families, we can notice that within a single household we can see the members of two, or even three generations, in some cases, that live together. This type of setup brings along with it a whole series of particularities when it comes to the way in which the people belonging to the generations, that were mentioned above, interact and socialize with each other (Magiru 2010a, pp. 7–10). The elderly, in the case of the Aromanians, would periodically, or constantly, take on the duty of teaching the children the most basic elements of human interactions. Later on, as the children would grow, the grandparents would pass on to the costumes, beliefs and rituals that are rooted in their community. This way, traditions are well-preserved within the family and the community, acting as an important factor for the preservation of the community's distinct ethnic identity. Besides the actual costumes, many of the ancestral crafts were also passed on through this interaction. In fact, they acted as a preparation stage for the adult life that would begin after the wedding takes place.

3 Methodology Research

The four homes that are included in this study were selected on the basis on their level of representativity for the studies area, and for the timeframe that is of interest to us. In order to reconstitute in 3D the houses, we used a few different techniques. Street view images that were made available by Google came were extremely helpful, combined with the knowledge that the author had about the villages from on-site exploration, this being the case for Darabani and Comana, and through the study of archive documents (e.g. monographies, images), as it was the case for Rasova and Văleni. The 3D models reproduce both the shape, layout, and the building materials that were used for each house. The author made no changes to the layout, shape, ornaments, or color schemes that were used by the builders. The models that are presented in this research paper were designed with Trimble SketchUp 2017, a 3D software that relies on polygonal modeling. The models we created by Iulian Mitran, they represent 3D representation of already existing houses from Constanța County.

They recreate the geometry of the buildings along with materials, textures and color schemes that were used during the decades in which they were built.

In order to create the models, Trimble SketchUp, a 3D CAD software, mostly used for architectural design, was used. The software's features made possible for us to show all of the architectural details associated with the housing styles that were analyzed. The study is qualitative, as it rather tries to bring forward some particularities that are present in four settlements, without assuming that the development of housing in the studies communities could ever follow a single pattern, taking into account the fact that without the existence of clear building standards, and the lack of construction teams made up of qualified workers, the housing styles evolved differently even from one village to the other, but in the same time, mostly, they followed a general style present in the era. As a result, our study is more interested in presenting some houses that are more representative for the village itself, and not so much for the entire region, or country, when it cames to the way in which homes were general built in that era.

4 Results and Discussion

By carefully selecting the most appropriate 3D modeling techniques, we were able to recreate the four homes that we selected for this study. Recreating them in 3D was essential for this study, as no available image could actually give a very clear view of all the architectural details, and other particularities, associated to each house.

Through the techniques that we used, we managed to present each house in its finest details, ranging from aspects that are related to its general geometry, or others that have to do with textures and colors. Thus, each house is presented from multiple angles, thus making certain particularities to be very easy to spot. The end result consists of all the images that are present within this study, this study demonstrates, once again, the immense benefits that working with 3D technology can bring to the study and preservation of our material heritage. If we were to rely on satellite and street view images, or even traditional photos, we would stumble across a series of major problems:

- first of all, there would be great difficulties in seeing very clearly the general outline of the building, this is due to the rich vegetation present in many yards that can partially cover various sides of the buildings when it is photographed;
- the same thing can be said about other neighboring buildings and annexes that have the potential of covering various sides of the building when it is photographed from various angles;
- the colors can get highly distorted, depending on the among of sunlight that is present in a certain period of the day, this usually takes away from the intensity of the colors;
- depending on the angle from which the photo is take, the proportions can get highly distorted, making some part of the building more prominent then other.

Another big advantage that 3D modeling gives us is the fact that we have the advantage to make modular models, this basically means that we have models that we can show in pieces, we can show them with the surrounding environment, or not, we have the possibility o highlighting or hiding certain components, depending what we are most interested in studying at one point.

4.1 Description of the Vernacular and Urban-Influenced Houses in Constanta County

4.1.1 Urban-Style House from Dauluchioi

The village used to neighbor, to the east, the village of Grăniceru [Canlî-Ciucur], which is today is entirely depopulated, its last inhabitants left the village during the 1980s. To the west the village neighbors Valcelele [Valala], along with Darabani and Graniceru are the three villages that are part of the town of Negru Voda, once a prosperous settlement which based most of its economy on agriculture, but also having a well-represented economic sector which included trade and craftsmanship. Darabani went through a dramatic shift in its demographic make-up in the aftermath of the War of Independence, after which Dobrudja, more particularly what constitutes nowadays the counties of Constanta and Tulcea, were granted to Romania through the Treaty of Berlin. After 1879, Romanian families, mostly originating from villages of what constitutes nowadays the territory of Dolj County, settle into Darabani. In the decades that followed, marriages between the inhabitants of the villages and those from neighboring ones became a common practice (Fig. 3a–d). Another major wave of settlers came to Darabani during the 1960s and 1970s when families from Moldavia came to the region, fleeing the poor living conditions that they had to endure in the native villages. Nowadays, there are no Muslims left in the villages, most of them left during the 1980s and settled in Techirghiol, Tuzla, Mangalia, and Constanta (Magiru 2003, pp. 32–34). There are very few houses built at the beginning of the last century that are still standing, the few remaining were usually built somewhere between the 1920s and the 1930s. Even so, some of the inhabitants recall the way in which their ancestors' homes used to look. A good example is constituted by the houses of our great-grandmother, which was built somewhere in the first decade of the last century. It was one of the few remaining examples of old vernacular architecture from the villages, the term „old" does not refer in our case to something from a very distant historical era, but something that usually was built at the end of the 19th century and managed to be preserved until recently. This is mainly due to the fact that the region was severely affected by Russo-Turkish War of 1877–1878, when a lot of villages where destroyed and a significant part of the local population fled the region, usually seeking refuge in the regions situated more to the south (Ştiucă 2011, p. 5).

The house that we mentioned had a simple layout, made up of two rooms, one of them served as a living space, while the other was more used for storage. The height

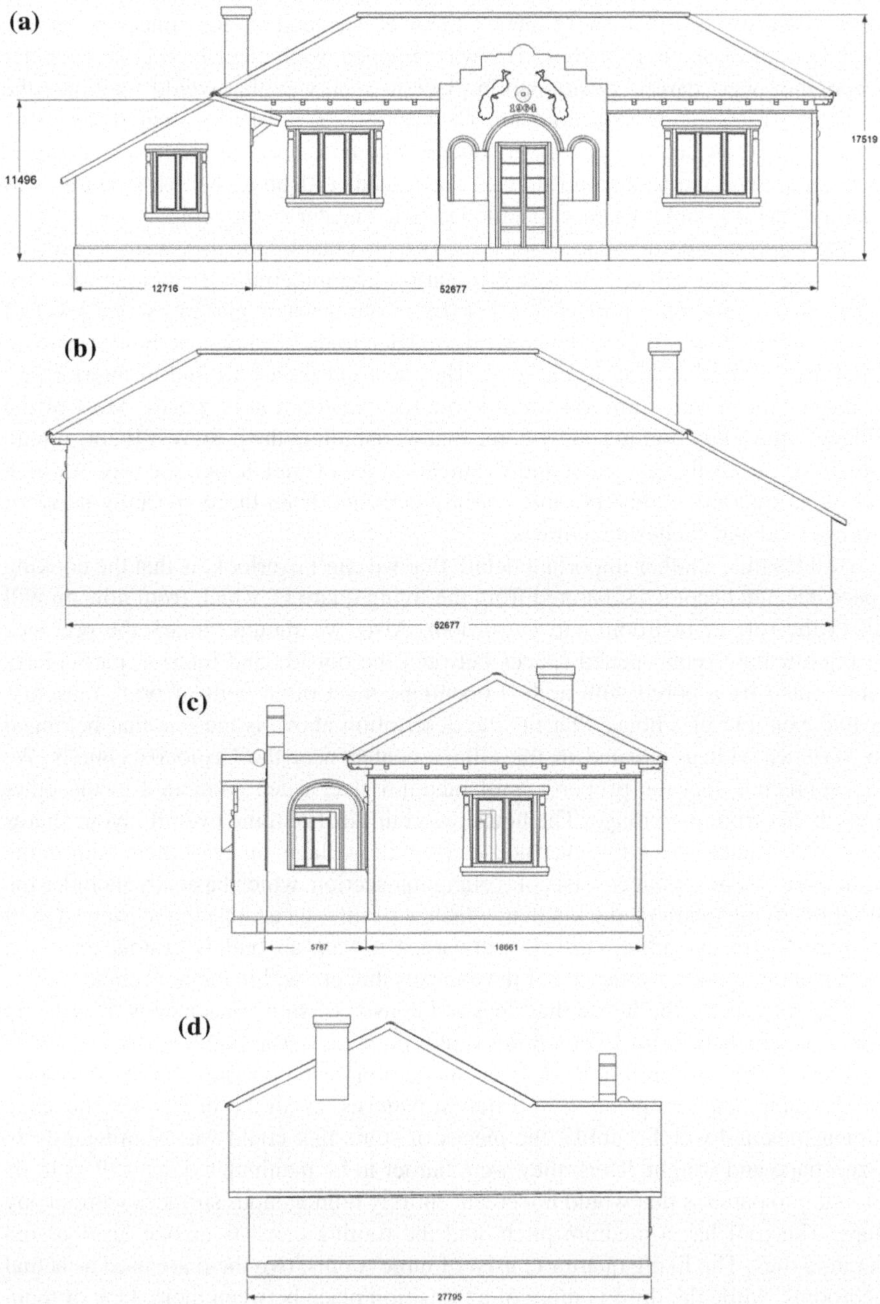

Fig. 3 Urban-style house from Darabani—elevations: **a** main façade; **b** rear façade; **c** right façade; **d** left façade

of the room was a little over 78 inches, the wall was mad of stone masonry, having a total width of about 19 inches. The was chopped, each piece having an irregular shape, they were careful positioned in place in such way that would minimize the existence of large gaps between them. The interior was finished with mud plastering over which a few coats of paint were added. The house had gambrel on both sides, the gambrel was made of wood sidings, positioned vertically. The living quarter was equipped with a cooking space, positioned in the center of the room.

The two rooms could be accessed directly from outside, and they were also linked in the interior through a door. The original roofing material was made up of cobs, put in two layers, on top of each other. Later on, it was replaced by one layer of ceramic tiles. "Coceni" [cobs] were preferred due to their capacity to hold water and humidity from infiltrating in the attic. They were also easy to find in interior part of the region, where stuff was unavailable for construction purposes. Most of the village is presently dominated by home that were built in the post-WWII era, among which we can easily spot some improvements in the overall layout, the way in which the living quarters almost became entirely detached from those specially used for storing food and sheltering animals.

Besides this, another important detail, that we can't overlook, is that the cooking space became becomes separated from the living quarters, which from now on will fill in the role of bedroom and guestroom. Also, we should remark the presence of intermediate semi-opened spaces between the outside and interior, particularly represented by a porch with arched openings, most often built of brick masonry. A fine example of a house that fits the description above is the one that belonged to Veronica Velicu, situated in the village center, near the Orthodox church. We can appreciate that the property is of medium size, when compared to the other households from the village. The house was built in 1964, the overall layout shares a lot of characteristics with other houses from the village, many of them built in the same period. The layout consists of rectangular section, which basically includes the all of the living quarters, the kitchen, which is positioned against the narrow edge of the main house, extending a little bit outward. The walls are built with stone masonry, the same thing goes for the rest of the annexes that are within the household.

The only sector the house that does not consist of stone masonry is the opened porch which shelters the main entrance into the house. The porch is made of brick masonry, brick was preferred for the construction or the porches as they were easily to manipulate and position in different patterns, as all the bricks had the same dimensions and weight, unlike the pieces of stone that could vary significantly in size, shape and weight. Thus, they were harder to be manipulated, as well as to be placed into patterns that would not create entirely homogenous surfaces without any gaps. The roof has a medium pitch, and the roofing consists of one layer of red ceramic tiles. The living quarter consist of three rooms, two them are used as actual bedrooms, while the third is more of a transition place between them. One of them bedrooms actually served its original purpose, that of a space that would host the bed, and would facilitate sleep and rest. The other bedroom would serve a rather different purpose, that of hosting and putting on display some of the family's most prized possessions, this room was, in some respect, a miniature museum that made known to outsiders the family material heritage. Some of the objects that are put on display in this room are items that were inherited by various members of the family, some of which, a consistent portion of them, were actually given as wedding gifts. Other from the items that were inherited, the rest of them were handcrafted by the women of the house.

These items were usually were decorative carpets, that would adorn the room's walls, always one of them would be placed on the side of the wall that is just above the bed. "Fantasy pillows" and pillows that would have a more interesting design, usually an elaborate lace border or various embroideries. The room that would stand between the two bedrooms, locally referred as "sală" (saloon], it would host little furniture, you could also find here various valuable pieces within this room, usually in the form of china put on display in a glass cabinet, or an open wooden cabinet (Fig. 4a–c).

The room was also used for ceremonial purposes, when a family member passed away, the coffin would stay in this room, where the wake would take place, until the day of the funeral. We are talking about an age, and even today, where funeral homes were still not heard of in the countryside, thus the family cared entirely for the dead body, from the moment that death accrued until the body was put to rest in the village cemetery. It is interesting to notice that the villages from the region where Darabani is located usually have village cemetery, not graveyards, meaning that the burial plots are not located in the church yard. The main facade overlooks the interior of the yard that stretches along the main road. The overall volume made up by intersecting three distinct bodies, the one that makes up the main living quarter, the porch, and the kitchen. Unlike the facades that face the interior of the yard, the back facade is entirely blind, and unlike the other ones mentioned lacks and type of ornamentation (Fig. 5a, b).

The ornamentation from the other facades is usually concentrated in three distinct areas, the interior part "soclu" [pedestal], the windows, the doors, and superior part of walls—where we usually have the crown molding. One important difference that is to be noticed between the post-WWII built homes, and the ones built prior, is that the ones built in the 60s and 70s had very well-defined exterior ornamentation, while the older ones lacked any true ornamentation. Or any type of ornamentation would actually be an integral part of a structural item, in some cases these items would be shaped in such a way that they would have a more unique appeal. This was often the case when it came to the ends on the wooden beans that would sustain the part of the roof that extended over the porch. The facades would also receive a slight touch of color by painting the area next to the windows the same color as the woodwork. The exterior ornamentation is in s stark contrast with the color of the walls, which has a purple-grayish tint, while the ornamentation uses yellow, white and medium blue. Blue is used for the platform, windows and doors, while yellow is used for the window ornaments, corners and crown molding (Fig. 6a–e).

White is used for as a transitional color, we can see that is used for the portion of the window where the wood frame and glass are bind together. The front wall of the porch is above the wall, the elevated section has a stair-type pattern, with the middle point being the mostly highly elevated. The upper surface of this wall is dominated by

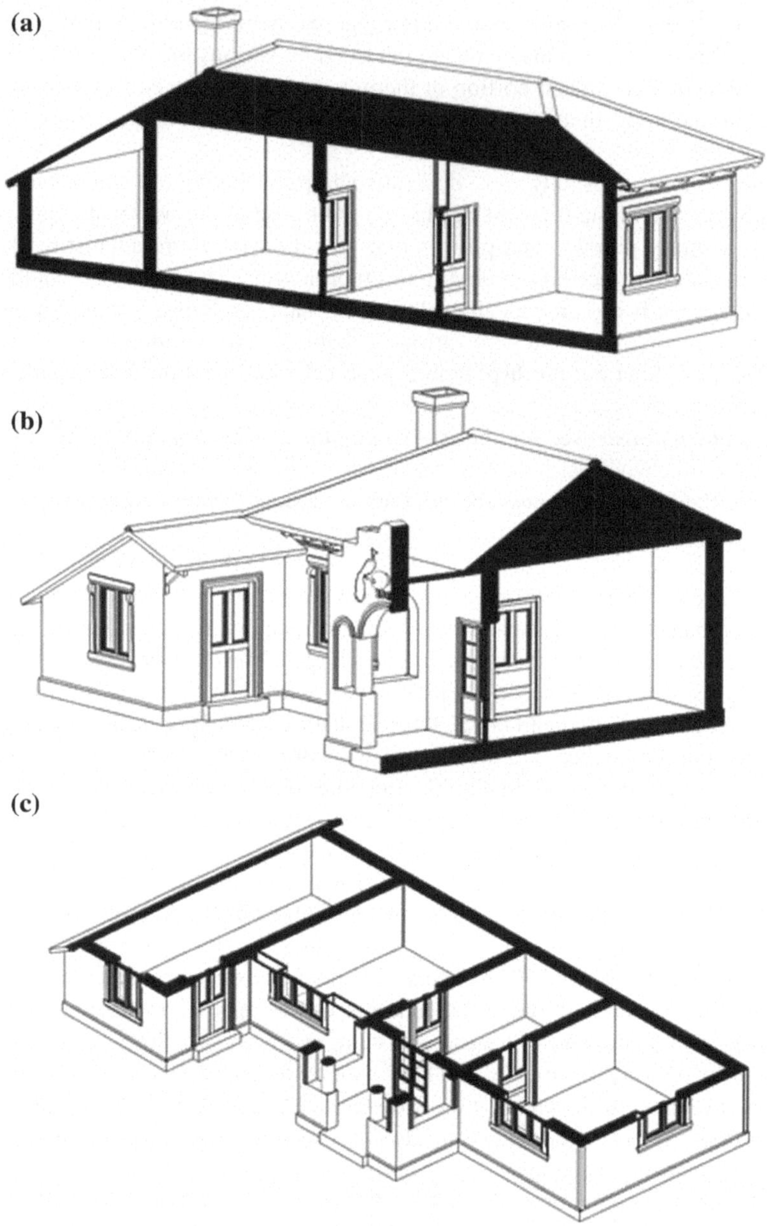

Fig. 4 Urban-style house from Darabani: **a** longitudinal section; **b** transverse section; **c** roof section

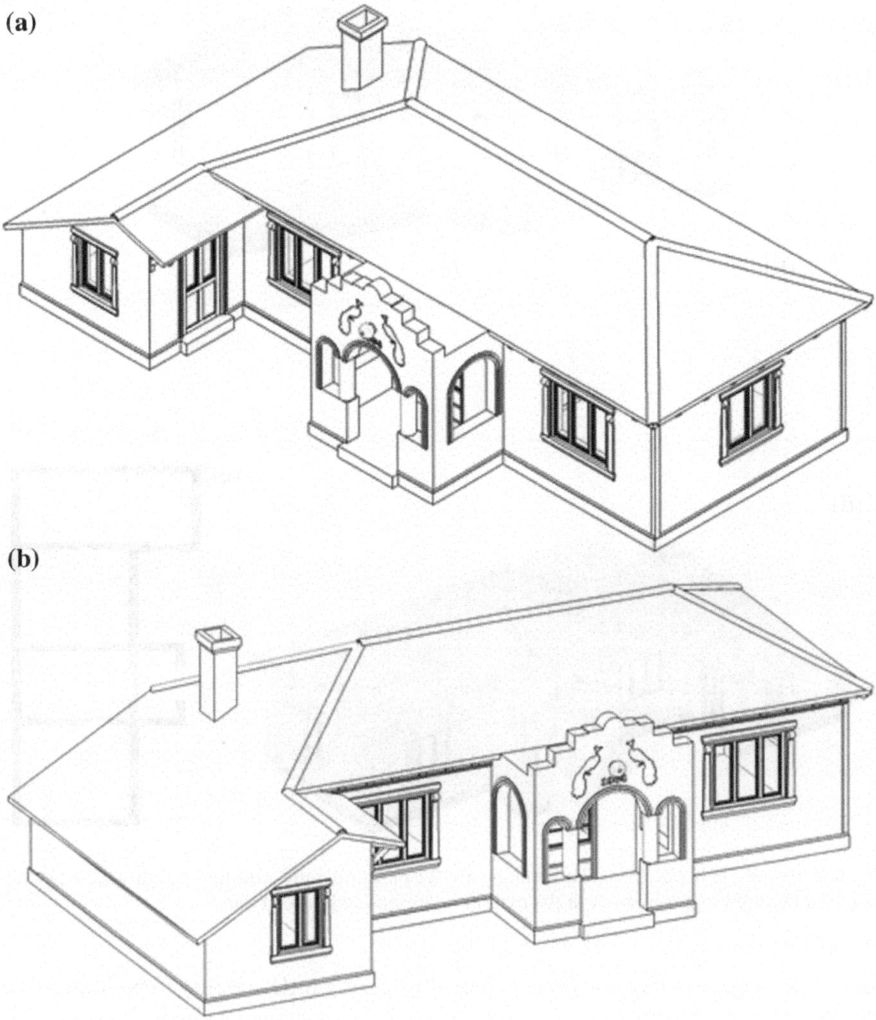

Fig. 5 Urban-style house from Darabani—orthogonal projection: **a** left façade view; **b** right façade view

the presence of two peacocks, underneath the peacocks the year "1962" is displayed, signifying the year in which the house was built. The front wall of the porch, which is also the main entrance into the house, has two thick white columns that sustain part of the weight of the porch's upper frame. The topography of the property is not entirely flat, the portion facing the street has a lower elevation, as we progress to the other limit of the property the elevation gradually increases (Fig. 7a–c).

We also need to consider the composition of the soil, we don't find the same type of soil throughout the entire property, but more two types. The portion on which the house stands is made up of red soil, more precisely it has the property to expand and

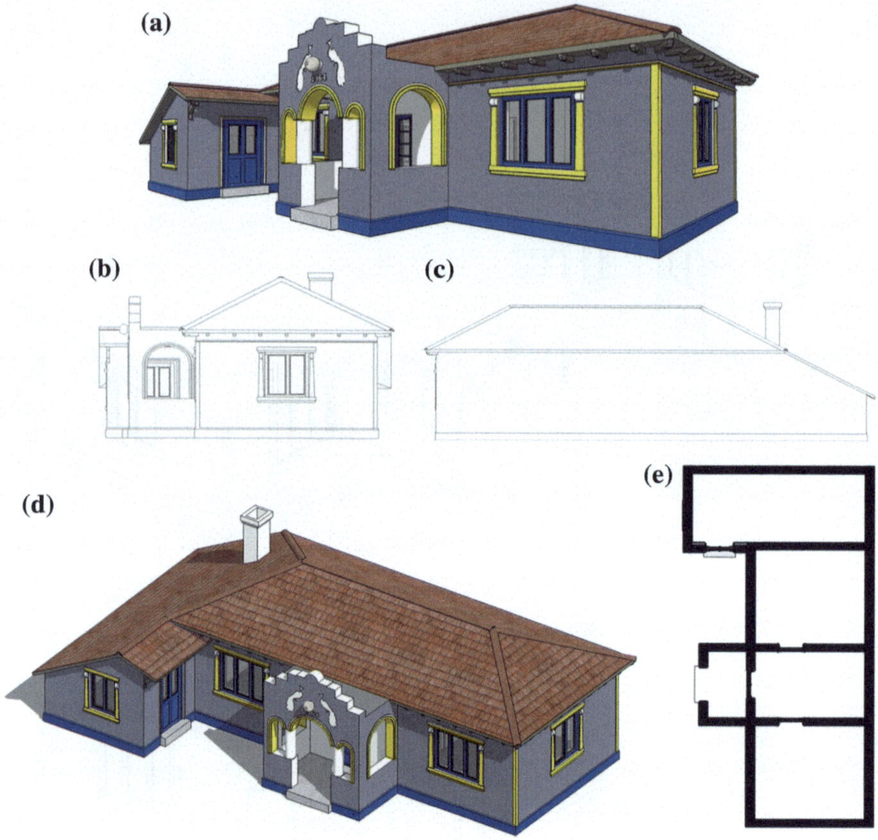

Fig. 6 Urban-style house from Darabani: **a** exterior plastering and paintjob; **b** right façade; **c** rear façade; **d** textured representation of the exterior—air view, **e** house layout

slide when it comes in contact with water, thus giving little stability to buildings that are built on it. This is the reason why the house is experiencing some major structural problems at the moment; the interior is full of cracks and the building looks like it is slowly sliding towards the roadside. The fence that delimitates the property from the space nearby to the roadside is built out of stone and wood, its patter is based on a succession of blotchy pillars, the space between them is close with evenly spaced wooden planks with pointy edges. The fence has a very monumental appearance, the fences that were at the beginning of the 20th century were usually made just of wood planks, or were made of stone but would go 15–19 inches above ground level (Fig. 8).

The heating system is not centralized, in both cases still mostly based on burning fire wood. Not all of the rooms are heated, in the living quarter the bedroom that has a common wall with the kitchen has its own heating system, while the bedroom that faces the roadside, which is referred to as "camera bună" (the good room], does

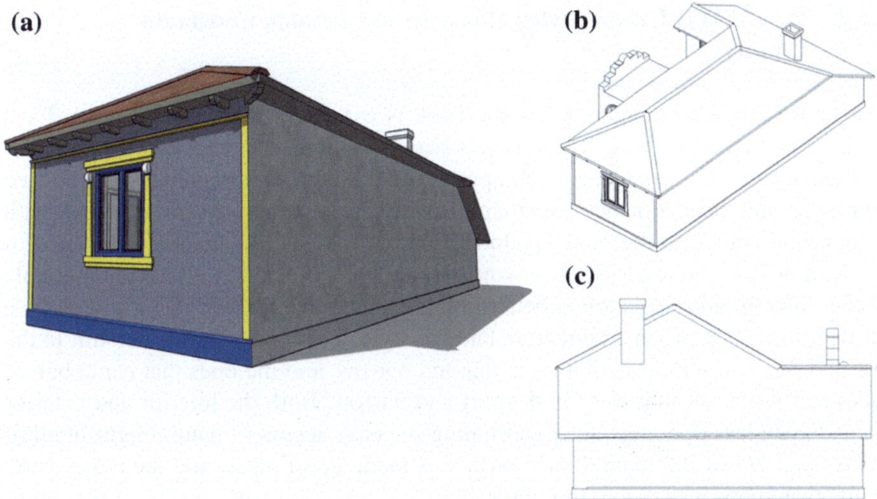

Fig. 7 Urban-style house from Darabani: **a** exterior plastering and paintjob—rear view; **b** back façade—air view; **c** left façade

Fig. 8 Darabani (Dauluchioi) household general layout

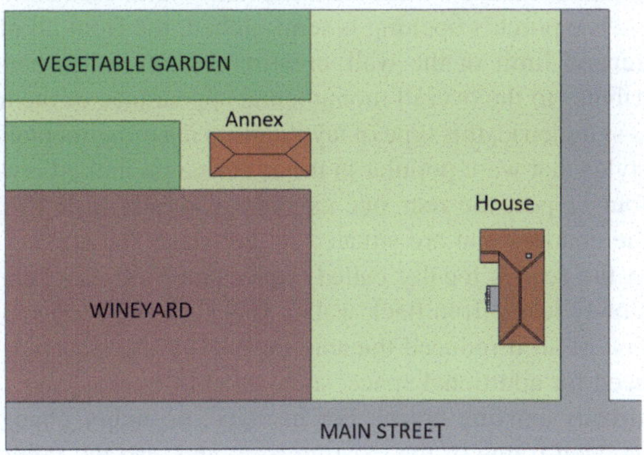

not have its own heating source. This is why this room is also referred to as the "summer bedroom". The interiors are painted in vibrant colors, blues and greens dominate, while the ceilings are left entirely white. Even though the exterior is richly ornamented, the interior is quite rudimentary when it comes to ornamentation, more precisely, patterned plastering and molding are entirely absent, the brightly colored walls and the painted lines that delimitate the limit between most of the wall's surface and the superior ¼ gives the rooms a touch of color and personality.

4.1.2 Post-WWII Urban-Styled House from Comana, Constanta County

The house has a rectangular plan, the back is extended, thus making place for a cooking and storage space. The house is basically divided into spaces, on the basis of their usage. The first one is comprised of the area that encompasses the two bedrooms and the saloon between them. Both bedrooms share a similar layout, with a door that leads into the "sală" [saloon], and with a window that faces the area of the yard next to the roadside. The windows are both wide and tall, having a double frame, thus thermic isolation is better obtained. The size and window framing give of the impact that urban architecture had the style. The homes that were built in the 20s and 30s had windows that were tall and narrow, and the ones that came before them had windows that were both short and narrow. Both the interior and exterior doors have glass openings, thus facilitating the easy access of natural light in all of the rooms. 2/3 of the main door's surface is made up of glass, and the thing wood frame that keeps it in place (Fig. 9a–c).

The central segment of the main facade is dominated by a monumental half built-in porch with thee openings, one of them acts as the door, thus making access to the main house possible. It should be noticed that ½ of the porch's width extends beyond the main facade's wall, the other half within the confines of the main layout.

The porch's opening is semi-arched, the front all of the porch extends above the general limit of the wall, creating a beautifully-shaped front ending, which contributes to the overall monumental appearance of the main facade. The main facade is symmetric, this type of layout gives it a monumental look, reminding of the homes styles that were popular in urban Romania in the between war period. The roof has four slopes, the rear one extends gradually in a 45-degree angle, thus sheltering the quarters that are situated in the back. The style of the extension that is located in the rear is locally called "aplecătoare" (rear extension], due to the fact that is slowly leans from back walls. This style was convenient for a number or reasons, first of all it reduced the amount of building materials used. Secondly, if there was need for additional space, such an addition was easy to build and to connect to the already existing house. For instance, no major changes needed to be done to the wooden frame of the existing roof, and also the volume of roofing material would be drastically diminished.

The back side of the house is partitioned in half, one half serves as a cooking space, while the other is used as a storage room. The storage space is accessed through the side of the house, while the cooking space is accessed through the "sală" (saloon]. The color scheme that dominates the exterior is made up by a dominant shade of cement-grey, which basically represents the exterior plastering, which has a highly textured. The ornamentation follows the same pattern as with the previous house that we presented. For the current house, the color scheme is made up of yellow, for the crown molding, the windows, the porch openings and the front end of the porch. Green is used to accentuate the platform, as well as for the door and window frames. It actually came upon to symbolize wealth within the community (Fig. 10a–c).

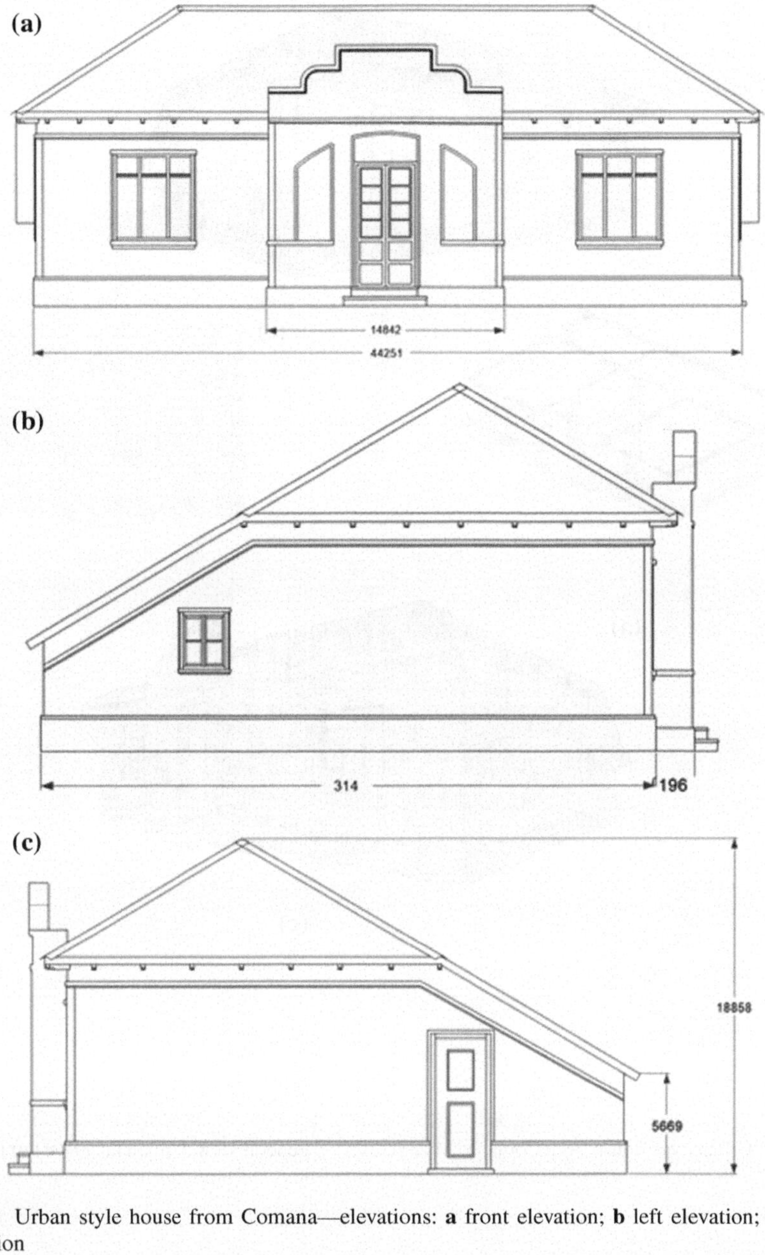

Fig. 9 Urban style house from Comana—elevations: **a** front elevation; **b** left elevation; **c** right elevation

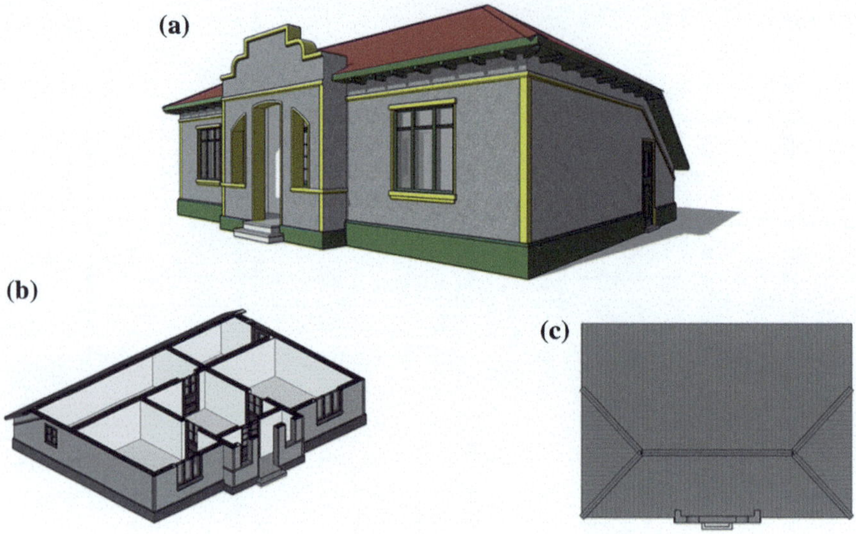

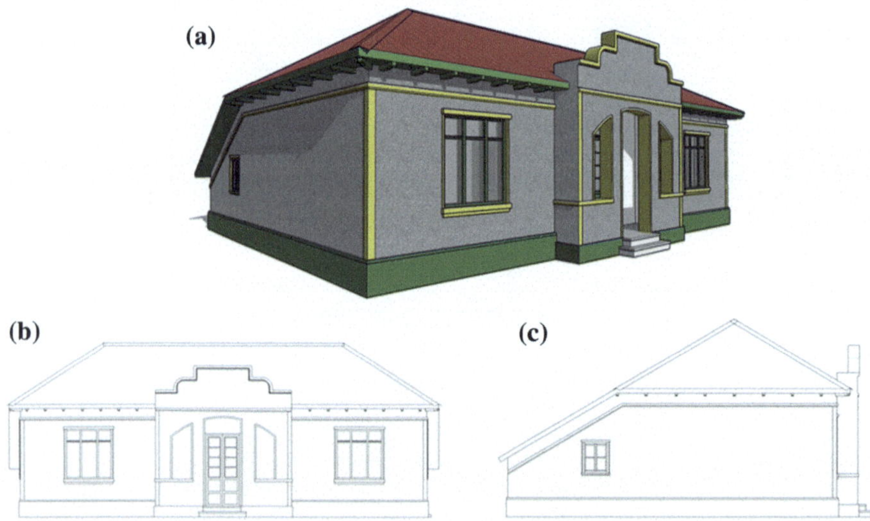

Fig. 10 Urban-type house from Comana (Mustafa Aci): **a** general view, **b** main facade, **c** left facade

4.1.3 One-Story House from Rasova, Constanta County

One-story houses are quite a rare sight in the region, one-story buildings could be more commonly seen in cities, and to a limited extent in urban-type settlements where trade and craftsmanship were well-represented in the local economy. But we must keep in mind the fact that these buildings served primarily a commercial purpose. Houses with one-story are also common with families in which the members were involved in some sort of craftsmanship. The house from Rasova has mainly a rectangular layout, with a porch, a or this case better said the balcony, extends on the full length of the first level. The balcony is actually positioned inward, not suspended above the ground level. As with the house from Valeni, this one also has a cerdac connected to the balcony, but extending above the ground, sustained by two thick wooden pillars (Fig. 11a–c).

The space from underneath the *cerdac forms a nieche, extending inward.* This nieche divided the main facade of the ground level into two parts, both symmetrical, each side has a door and a narrow window with a door on one side integrated into the door frame. Due to the lack of information, we cannot give a very accurate insight on the interior layout, especially when it comes to the ground-floor. In case of the first floor, is easy to tell that it has the same configuration as the layouts of the homes that we already talked about. The windows are quite large, they are actually have double openings, the door that leads to the sala has a door with a wide glass opening, and with two narrow side windows integrated into its frame. The roof has four slopes, and the part that extends over the cerdac only two, the extension ending with a gambrel close by wooden planking. The ground level is linked to the first level through some stairs located in the side of the building, leading up to the *cerdac* area. The roof is covered by one layer of ceramic tiles of rectangular shape, unlike the ones from the Valeni house which had a semicircular shaped and were put into two layers (Fig. 12a–c).

4.1.4 Vernacular House from Valeni, Dobromir, Constanta County

It is highly representative of the styles that was popular in the first three decades of the 20th century, it combines a closed space, built of masonry, with a semi-opened one, built from a wood frame. The large porch dominates the entirety of the main facade, it is supported by a number of wooden pillars. The layout of the porch is shapes as the letter T, the middle part extend beyond the limits of the part of the porch that faces the wall, also the roof extends along with it. This extension is locally referred to as *cerdac*, and it was usually used as a place for leisure during the summer period. Due to the topography of the land, the foundation extends far beyond the ground line. The access into the house is made possible through the side of the porch through some wooden stairs (Fig. 13a–c).

Unlike some other vernacular homes from other ethnographic regions, the niech's opening was not sealed by a wooden door. The part of the house built of masonry is divided into three living quarters, two bedrooms and a *sala*. Each bedroom has two windows, narrow in wight, but relatively tall. The *sala* gets its light through the door's

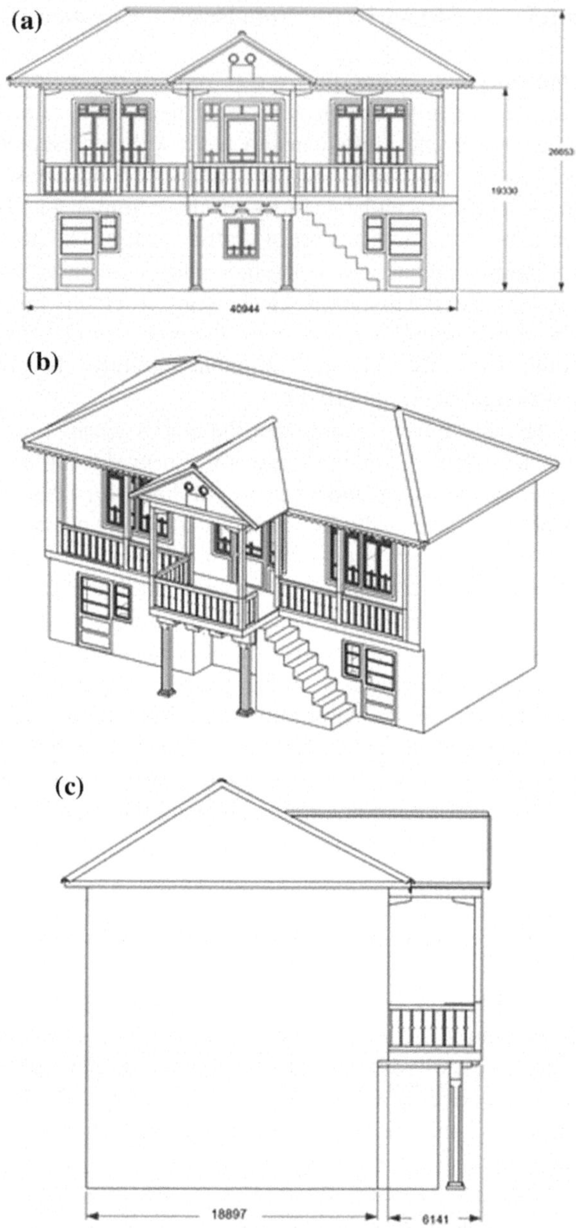

Fig. 11 One story house from Rasova: **a** front elevation; **b** orthographic projection—air view; **c** left elevation

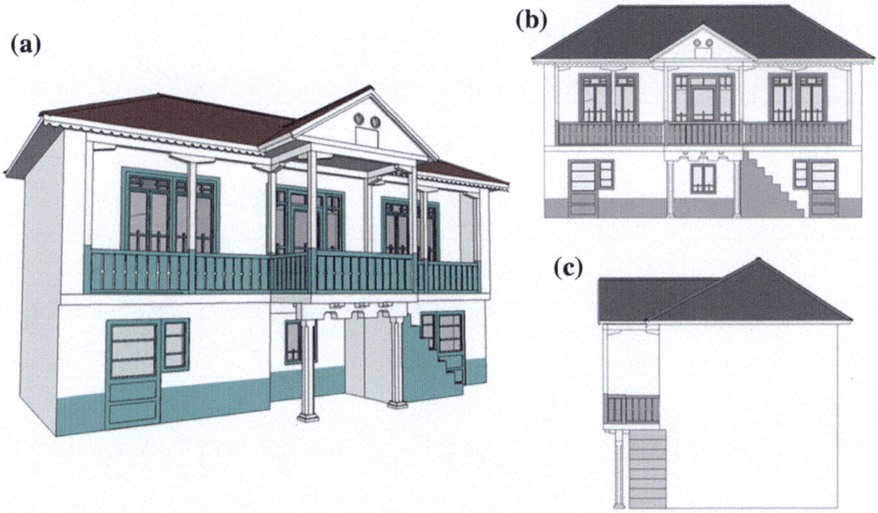

Fig. 12 One-story house from Rasova: **a** general view; **b** main façade; **c** right façade

glass openings, and the narrow windows that side the door in both sides. In some cases, these side windows are integrated into the window's frame, in other cases they are separated from the door by a segment of masonry. The main section of the roof has four sides, covered with ceramic tiles, in two layers, the portion that extends over the *cerdac* is covered with the same roofing material. Both the exterior and interior is covered with mud plastering, which is painted over with white slaked lime. The roof has two front openings that facilitate natural ventilation and a minimum volume of light (Fig. 14a–c).

5 Conclusions

Drawing on the final conclusions, we can state that there are significant differences between the buildings that we included in this study, when it comes to their functions and overall aesthetics. To start off, the pre-WWII houses tended to have a more reduced size, this being mainly due to the fact that they hosted only spaces designated as living quarters. Secondly, the homes built in the era that we mentioned lacked any actual ornaments, both on the interior and exterior. The interior hight of the rooms was way lower, especially if we are talking about homes built in the 1910s, or even sooner. The homes that were built after WWII integrated into their layout rooms designated for some essential household talks, such as cooking and storing goods. The shape of windows gradually changed, shifting from small narrow ones, to ones that were wide and double framed, thus providing more thermic isolation.

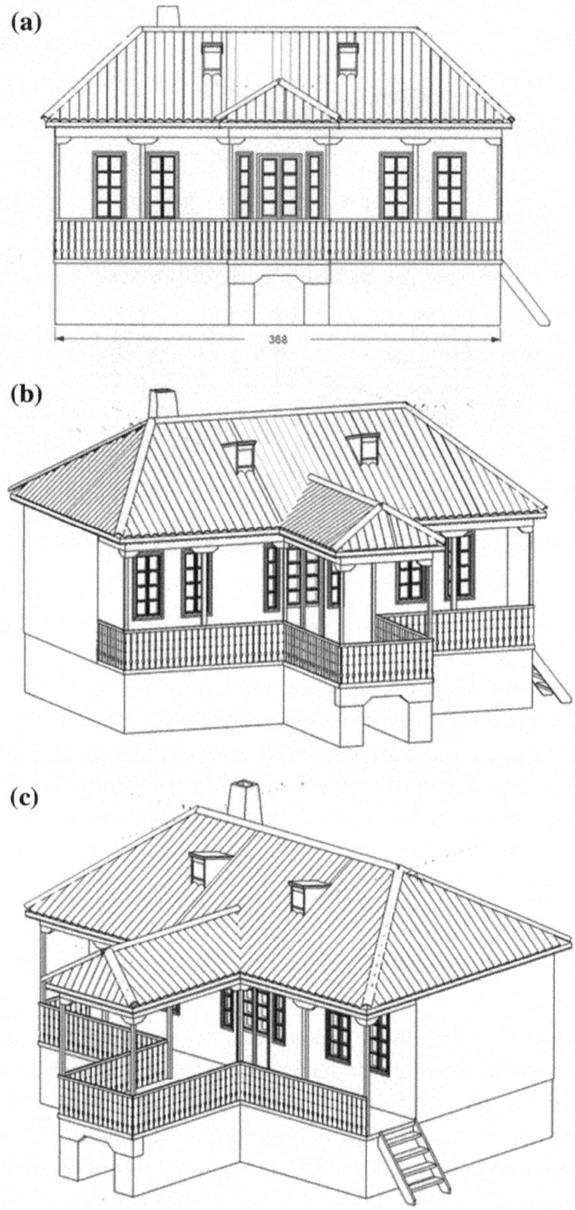

Fig. 13 Vernacular house from Văleni: **a** front façade; **b** orthographic projection—left angle; **c** orthographic projection—right angle

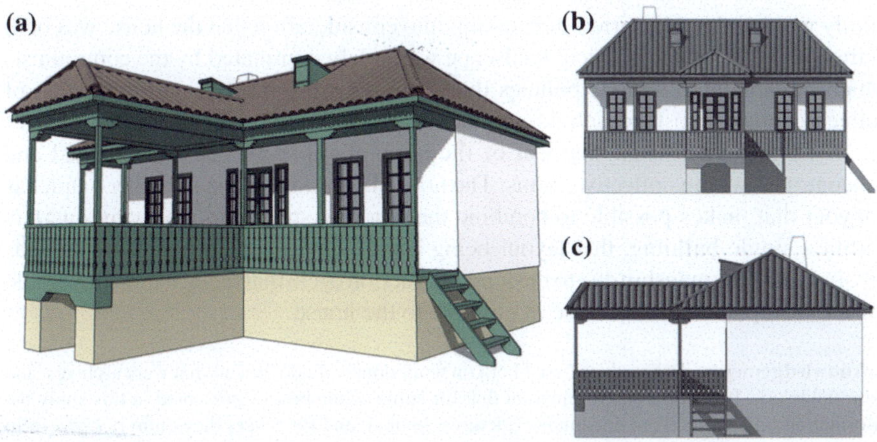

Fig. 14 Vernacular house from Văleni: **a** exterior plastering and paintjob; **b** front facade; **c** right façade

As time progressed, houses were built with intermediate spaces between the interior and exterior, semi-opened spaces usually. They usually took the form of wooden porches, later on they took the form or porches built out of masonry. Post-WWII houses also seen some significant improvements in construction materials that were used, thus the exterior plastering, which previously was made of mud, was replaced by a cement-based plaster, usually having a textured surface. This improved the durability of the exterior walls in time. Also, in the decades that followed WWII synthetic dyes became widely spread, thus it became highly fashionable in the countryside for homes to be painted in vibrant colors. By comparison to the vernacular homes that were painted reddish and earth-based tones.

The appearance of transitional spaces between the exterior and interior meant the apparition of a new type of mindset when it came to the hygiene, and general cleanliness of the house. More precisely, these spaces were mean for acting as a filter for the dirt that would be brought from the outside in. This could also be seen in the fact that the spaces that were used for raising animals were distanced as much as possible from the house. Heavy, or inflatable, materials were no longer used for as roof covering, semi-circular ceramic tiles were replaced with rectangular-shaped ones. To conclude, we can state that the homes that were built in the post-WWII era underwent some serious improvements, both when it comes to their layout, aesthetic, and used materials, but they did not fully transition into a urban format, this mainly due to the fact that the countryside was still struggling with poverty, and people had a very slow pace when it came to accept and adapt new ideas and to integrate them in their everyday lives (Gaivoronschi 2009).

The buildings that were presented within this study share a core-layout, which is represented by the living quarter made up of three rooms positioned on a straight axis. Their particularities are mainly by two categories of factors—terrain and the profession practiced by the occupants. The house from Darabani was owned by a family was involved in agriculture, taking into consideration that the house was built in the 1960s, when the political landscape was solely dominated by the communists which implemented various policies that were meant to implement the process of collectivization, most households gradually simplified their layout as a result of the fact that a significant segment of the agricultural work was externalized and concentrated within collective farms. The house from Rasova, on the other hand, has a layout that makes possible to combine the working space and the living quarters within a single building, this layout being more convenient for a family that was involved in craftsmanship due to the long production cycle that made it more favorable for the workplace to be as close as possible to the house.

Acknowledgements The study is based both on work done with document that were archives, and from fieldwork. It is important to mention that for some of the houses presented in this study the information only came from documents (pictures, books], and for others they entirely came from on-site observations. This is why there are certain inconsistencies when it comes to describing some aspects related to the structural characteristics and interior layout. The models that are present in this study were created in Trimble SketchUp 2017. The author used real-life pictures of the homes, gathered from on-site work, books and various online sources.

Appendix

Old vernacular house, similar in style to the houses that were built at the beginning of the 20th century. It consists of only two living quarters and an additional room used for storing food and various household objects (Fig. A.1).

Urban-type house with two side gambrels built of masonry. The porch, which is centered on the main façade, has opening only on front that wall has openings. The porch gives the façade a monumental appearance, while the rest of the house lack any type of complex ornamentation (Fig. A.2).

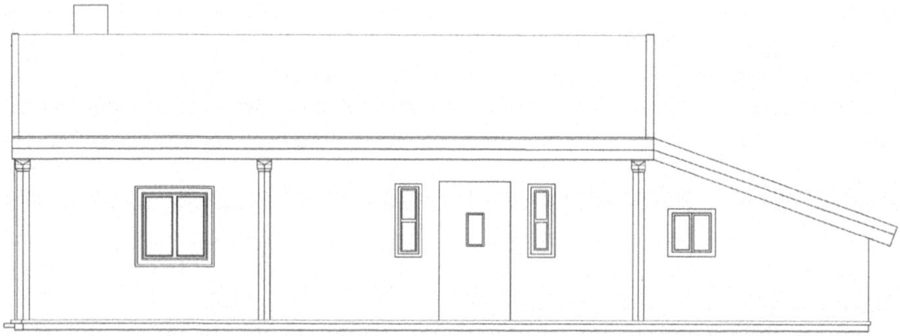

Fig. A.1 Vernacular type: front façade

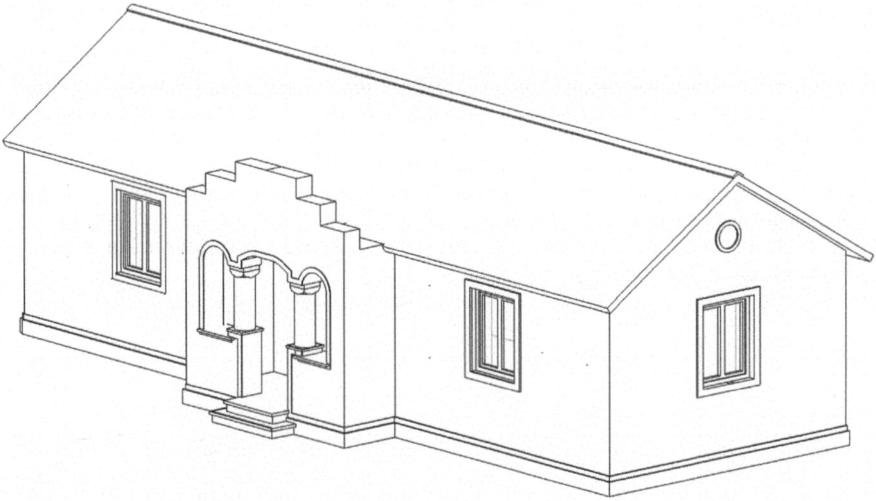

Fig. A.2 Urban-type house: main façade

Urban-type house with three rooms, unlike the other houses that were presented here, it has no intermediate space between the inside and outside. The interior ornamentation is also quite minimum, the overall appearance is dominated by simplicity and symmetry (Fig. A.3).

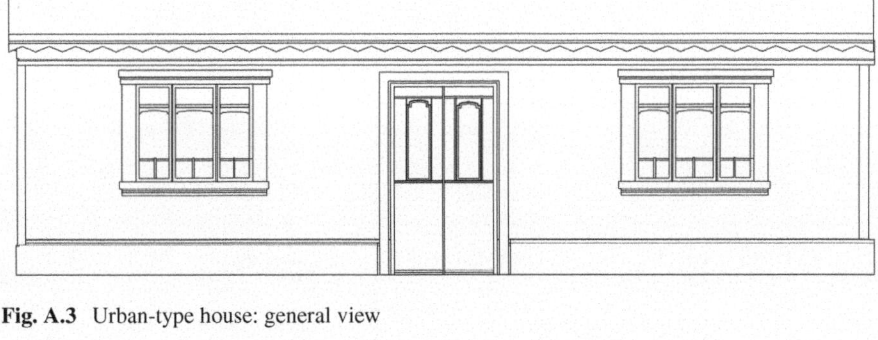

Fig. A.3 Urban-type house: general view

References

Gaivoronschi, V.: Matricile spaţiului tradiţional. Editura Paideia, Bucureşti (2009)
Magiru, M.: Dobrogea studiu etnografic – Românii autohtoni, vol. 1. Muzeul de Artă Populară, Constanţa (2003)
Magiru, M. (coord.): Arhitectura tradiţională din Dobrogea, no. 2 (2010a)
Magiru, M.: Caracteristicile arhitecturii tradiţionale dobrogene de pe Limesul Dunării. Arhitectura tradiţională din Dobrogea, no. 3 (2010b)
Ştiucă, N.: Comunitatea elenă din Izvoarele (2011). http://old.unibuc.ro/prof/stiuca_n_a/docs/res/2011noiGrecii_izvoarele.pdf

The Dimension of Criminality in the Context of the Variation of Migration Floods Trends

Elena-Ana Iancu, Ioana-Raluca Rusu and Nicolaie Iancu

Abstract In this chapter, we want to present some reflections on international migration in a context where there are variations in migration flows in the period 2015–first half of 2018 according to official statistics. To this end, we provided theoretical references concerning the notion of the state border and we explained its significance in international relations. Moreover, we used the method of comparison in order to highlight the differences in values regarding facts which affect fundamental values protected by law, by incriminating them. We have also presented the major migratory routes in a context where dialogue, cooperation and collaboration between the Member States of the European Union and third states are the main ways to identify existing risks, to avoid any possible shocks resulting from illegal migration and to maintain the greatest responsiveness to the size of migration flows.

Keywords Criminality · Migration · Statistics · Security · Risks

1 Introduction

Since ancient times migration and expansion of tribes or people gave rise to conflicts and the proximity and interdependence of human groups have determined the characteristics of the state border. A crossing point but also "a bridge that links in various

E.-A. Iancu (✉)
Faculty of Legal and Administrative Sciences, Agora University of Oradea Municipality, Oradea, Romania
e-mail: anaelena2009@yahoo.com

E.-A. Iancu · I.-R. Rusu
Doctoral School of Law, "Alexandru Ioan Cuza" Police Academy, Bucharest, Romania
e-mail: rusu.ioanaraluca@yahoo.co.uk

N. Iancu
Euroregional Centre for the Prevention and Fight Against Cross-Border Criminality, AGORA Foundation, Oradea, Romania
e-mail: niancu2009@yahoo.com

D. Flaut et al. (eds.), *Decision Making in Social Sciences: Between Traditions and Innovations*, Studies in Systems, Decision and Control 247,
https://doi.org/10.1007/978-3-030-30659-5_19

levels" (Stamate 1997, p. 16), that unites and separates, the state border imposed transformations in the neighboring territories.

The state border divides a state territory from another state territory (Stamate 1997, p. 26), being a complex of territorial elements, as per international law, namely together with „the soil, the subsoil, waters and aerial space above ground and waters" (Geamănu 1965, p. 289).

In retrospect, state borders bear traces of the ancestors, they show us the evolution and involution of space within which people exists with their related identity values. The meaning of state border was different depending on the periods when the idea emerged in the field, starting from the significance of natural borders, military borders, astronomical borders, conventional borders or internal or external borders of the European Union. As a line of separation between states, the border appears in a triple dimension (Scăunaş 2007, pp. 238–239): national, economic and security related.

Currently we emphasize cooperation in the European Union (EU) to reduce the incidence of illegal migration to EU countries and to reduce irregular movements between the European countries, taking into account the meanings ascribed to the state border resulting from the internal regulations of each state (for example, GEO no. 105/2001 on the state border of Romania). Both cooperation on migration and specific police surveillance activities of cross-border flows of people, goods and clandestine movement of illicit goods and illegal immigrants (Iancu 2013, p. 177) are necessary also in order to prevent crimes and to allow the European states to control their own borders, in the same time, by ensuring the security in what concerns external and internal borders.

The migratory movement due to emigration and immigration for different reasons (historical, economic, social, political, etc.), voluntary, has various consequences both in the country of origin and in the country of destination, including in what concerns transnational crime. Where due to war or political problems existing in the country of origin, the neighboring countries face imposed migration (National Geographic Society, Human Migration Guide 2005, pp. 1–2) there will be reported variations in statistics, especially in terms of legal immigrants composition (asylum seekers and refugees), protected also by international law.

The migration routes, from geopolitical point of view, are mainly relatively wide geographic areas which can contain territories representing one or various states.

The areas affected by these routes are closely related to the destabilizing geopolitical factors affecting various third countries neighboring Schengen or European Union countries. Usually these areas are perceived as hostile since the cooperation with the politicians or government members from those countries is either very cumbersome or often absent for various reasons. However, because of gaps in the law of a state and/or an inconsistent approach, sometimes coupled by inconsistency in adopting programs of prevention, investigation and combating transnational crime in Member States of the European Union and/or in Schengen countries can result in situation when, based on a migration, there are carried out actions and/or inactions of some subjects that influence the migration routes and create the premises for certain activities which will fall within cross-border crime and criminality.

The development of coherent penal policies in relation to existing policies at European level will also lead to the application of better practices in the countries under imposed migration, which will have the effect of reducing the negative impact in the countries of transit and/or destination. Although the positive effects of migration are well known, there are political, economic and electoral controversies on the answers which may be made to questions such as: immigrants consume more public services than they pay, they create competition in the labor market, they are selective in choosing occupations, they put pressure on social security systems, they are active subjects of certain crimes in greater numbers.

In early June 2017, the EU Council recommended the implementation and use in surveillance procedures and border control of biometric data (Risk Analysis for 2018, p. 31) in order to identify illegal migrants, thus improving the methods used by authorities with border and national security responsibilities.

Many methods used to estimate international migration are based on mathematical deterministic patterns or on various quantitative techniques, but, in contrast to deterministic methods, the stochastic (probabilistic) tools for analysis and forecasting of migration are rooted in the theory of probability (Iancu et al. 2017, p. 395).

2 Migration Routes in Europe

Nowadays, Europe, with reference to the segment between the border points also known as "green border" or, as appropriate, "blue border", is affected by at least two main routes of illegal migration, namely the eastern route and the southern route. In turn the two routes of illegal migration can be divided as follows:

2.1 *Eastern Route*

(a) Norther and central-eastern route, which generally occurs at the border with Russia, Ukraine and Belarus, reported to which we can take into consideration that Russia is in a permanent diplomatic and economic divergent dialogue with the European Union's states;
(b) South-Eastern route, including the area of the black Sea and the border of Romania cu Ukraine and Republic of Moldova;
(c) Balkan route, having a rather big rate in the following states: Albania, Bosnia and Herzegovina, Macedonia (former Yugoslav Republic of Macedonia), Montenegro and Serbia, including Kosovo region. Some states have recognized Kosovo as independent country while others consider it a region with special status. This route became active as being an unstable area which can become vulnerable both politically and in a context of nationalist ideas with a strong ethnic character.

2.2 Southern or Mediterranean Route

(a) East Mediterranean route, mainly affecting the Greece area;
(b) Central Mediterranean route, especially affecting Italy (Sicily);
(c) West Mediterranean route, affecting Spain.

It may be noted a peculiarity regarding the Balkan route since in these areas illegal migration has been identified as active in both senses of the region (Greece—the Balkans and the Balkans—Greece area). The explanation is simple when referring to foreign fighters transit coming from the residence states or residence towards the conflict countries in Syria and Iraq and their return home.

A similarity for the two routes of illegal migration (East-Mediterranean route and Balkan route) is the origin of most migrants from the Near and Middle East. In what concerns the routes in central and western area of the Mediterranean, the majority of illegal migrants is from African countries (Chart 1).

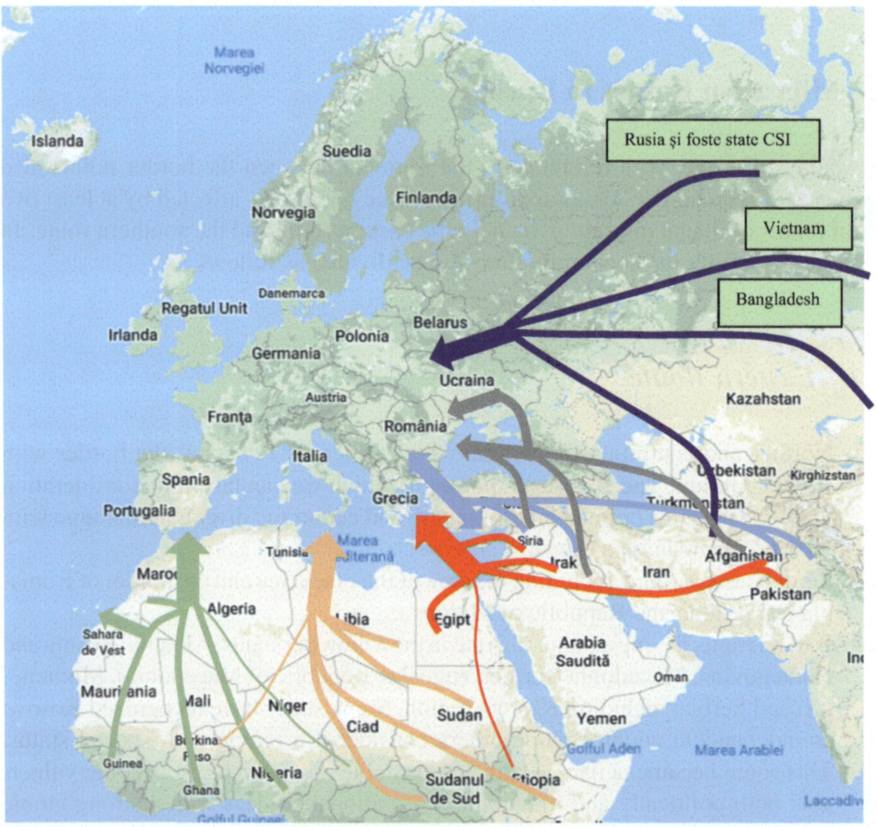

Chart 1 Main illegal migration routes during 2015 and first semester of 2018

Northern and Central-Eastern route

According to the statistical data concerning the citizens, as mentioned in Eastern Partnership, Risk Analysis Network Quarterly, Quarter 4 • October–December 2016, p. 10 and Eastern Partnership, Risk Analysis Network Quarterly, Quarter 4 • October–December 2017, p. 9.

South-Eastern route

According to the data concerning foreign citizens involved in illegal border crossing, as mentioned in the Analysis of Romanian Border Police for 2015, pp. 16–17. *Source* https://www.politiadefrontiera.ro/ro/main/i-analiza-activitatii-politiei-de-frontiera-romane-pe-anul-2015-6678.html (site accessed on January 18th, 2019).

Balkan route

According to the statistical data concerning citizens, as mentioned in Frontex · Western Balkans Annual Risk Analysis 2017, Table 1. Illegal border-crossings between BCPs, p. 47.

East-Mediterranean route

Central-Mediterranean route

West-Mediterranean route

For East-Mediterranean, Central-Mediterranean and West-Mediterranean routes, there have been used the statistical data regarding citizens as mentioned in FRONTEX · Risk Analysis for 2016, pp. 16–20, FRONTEX · Risk Analysis for 2016, p. 16, Frontex · Risk Analysis for 2017, p. 18 and Frontex · Risk Analysis for 2018, p. 18.

2.3 Northern and Central-Eastern Route

With reference to the number of migrants trying to cross the border illegally on this route since 2015, we can notice that there were identified 12,818 people who tried to illegally cross the border in EU countries. Of these, 6663 people are from countries that belonged to the former Soviet Union and the other 6155 people (48%) are mainly citizens of Afghanistan, Syria, Bangladesh, Iraq or Vietnam. Most reported cases of illegal border crossing (between 70 and 80%), involving citizens from Russia, Ukraine, Moldova and other former CIS states (Commonwealth of Independent States—organization of independent states that were part of the Union of Soviet Socialist Republics to which Georgia, Ukraine and the Baltic countries (Estonia, Lithuania and Latvia) did not joined, do not related to illegal migration, being cases of smuggling various goods.

In the analyzed period there were engaged in guidance, recruitment, hosting etc. activities (facilitators) 676 people mostly from Russia, Belarus, Ukraine and other former CIS states, including the Baltic States.

In the period under review there were submitted asylum applications by citizens from third parties in the responsible structures of the European Union Member States and Schengen, 332,505 applications for a form of protection. Out of these requests, 43% (143,695) were submitted by citizens from Iraq, Afghanistan and Syria. Because since February 2014 Ukraine lost the Crimea, there have been found more and more cases of Ukrainian citizens seeking asylum in the European Union member states or in those located in the Schengen area, as appropriate. Most states have registered asylum applications of citizens from Ukraine were Sweden, Netherlands, Germany, Italy and Poland (FRONTEX, Eastern European Borders Quarterly Quarter 1, January–March 2015, p. 12).

With reference to the use of forged documents, we can see that the citizens of Moldova and Ukraine highlight regarding the use of these documents in the border crossing points.

(There have been used and processed the data mentioned in the Statistical Annexes from FRONTEX, Eastern European Borders Quarterly Quarter 4, October–December 2015, pp. 11–14; Eastern Partnership, Risk Analysis Network Quarterly, Quarter 4 • October–December 2016, pp. 10–16; Eastern Partnership, Risk Analysis Network Quarterly, Quarter 4 • October–December 2017, pp. 9–12; Eastern Partnership, Risk Analysis Network Quarterly, Quarter 1 • January–March 2018, pp. 9–12.)

Table 1 Illegal border crossing on Northern and Central-Eastern route

	2015[a]		2016[b]		2017[c]		2018[d]	
	Illegal crossing	Guides	Illegal crossing	Guides	Illegal crossing	Guides	Illegal crossing	Guides
Total persons	5189	187	3832	327	3298	140	499	22
Migrants outside CIS space	2704		1779		1482		190	
Use of forged documents	412		354		722		187	
Asylum requests	254,115		53,357		23,094		2939	

[a]Data according to FRONTEX, Eastern European Borders Quarterly Quarter 1, January–March 2015, pp. 8–9, 18–21; FRONTEX, Eastern European Borders Quarterly Quarter 2, April–June 2015, pp. 8–9, 16–19; FRONTEX, Eastern European Borders Quarterly Quarter 3, July–September 2015, pp. 6–7, 11–14; FRONTEX, Eastern European Borders Quarterly Quarter 4, October–December 2015, pp. 6–7, 11–14

[b]Eastern Partnership, Risk Analysis Network Quarterly, Quarter 1 • January–March 2016, pp. 6–7, 11–14, Eastern Partnership, Risk Analysis Network Quarterly, Quarter 2 • April–June 2016, pp. 6–7, 12–15, Eastern Partnership, Risk Analysis Network Quarterly, Quarter 3 • July–September 2016, pp. 6, 10–13, Eastern Partnership, Risk Analysis Network Quarterly, Quarter 4 • October–December 2016, pp. 6–7, 11–12, 15–16

[c]Eastern Partnership, Risk Analysis Network Quarterly, Quarter 1 • January–March 2017, pp. 6–8, 11–14, Eastern Partnership, Risk Analysis Network Quarterly, Quarter 2 • April–June 2017, pp. 6–8, 11–14, Eastern Partnership, Risk Analysis Network Quarterly, Quarter 4 • October–December 2017, pp. 6–7, 9–12

[d]Data according to Eastern Partnership Risk Analysis Network Quarterly Quarter 1, January–March 2018, pp. 6–7, 9–12

2.4 South-Eastern Route

Although on this migration route there is not recorded a significant migration pressure (Table 2), the measures taken for surveillance and border control must not be minimized. We believe that any measures taken to stop migration on other presented routes may lead to an escalation of migration flows from the East or Africa regions whose transit route is the Black Sea and the Romanian border with Ukraine or Moldova.

Guides, carriers, recruiters, etc. (facilitators) are both Romanian citizens and citizens of third countries who have a legal form of stay in Romania but who usually do not have a steady job. As a proportion, third parties citizens are between 65 and 70% of the facilitators.

Citizens who are originally from Afghanistan, Iraq, Syria and Iran are among the top nationalities acting on the eastern border of Romania, using both the land route from the border with Ukraine or Moldova, and especially the one from the Black Sea, and in order to reach Romanian soil they use Turkish boats.

The low migration pressure on this route is mainly due to proximity to the Balkan route and the east-Mediterranean route, two routes which are more affordable to migrants since the Balkan route offers the attraction of support from Muslim citizens, much closer to the ideology of the majority of migrants, and the east Mediterranean route offers the attraction of direct entry into the Schengen area, with the possibility of moving further out of control in any other country in this area, thus avoiding controls at border crossings to third countries or countries which are only members of the EU.

In Romania, citizens seeking asylum in the period 2015–2017 come usually from the following member countries (according to the General Inspectorate for Immigration/About Us «Reports and Studies» Assessments/"Activity assessment for 2015, pp. 24–25; Activity assessment in 2016, pp. 24–25; Activity assessment in 2017, pp. 10–11, source http://igi.mai.gov.ro/ro/content/analiza-activitatii, site accessed on January 18th, 2019):

Irak—3428 persons (43%)
Syria—2302 persons (29%)
Afghanistan—433 persons (5.5%)
Pakistan—371 persons (4.65%)
Iran—236 persons (2.98%).

2.5 Balkan Route

2015 began recording a flow of illegal migration almost in parameters with the illegal migratory flows since 2014, but with the second semester, the findings of illegal border crossing have seen huge growth, reaching more than 12,000 people identified in an attempt to illegally cross the border within 24 h, reaching in October

Table 2 Illegal border crossing on South-Eastern route

		2015	2016	2017
Illegal crossing foreign citizens	Land border	936	1194	2840
	Black Sea	102	Data unavailable	537
Guides		211	140	251
Hidden in transportation means		317	197	683
Avoiding control/forged documents		143	396	457
Asylum requests (the requests regarding the grant of any form of protection by the Romanian state were submitted by third parties upon their discovery in the attempt of illegal border crossing) in the border points		255	Data unavailable	1600
Asylum requests at national level in Romania (Activity assessment of the General Immigration Inspectorate for 2017—sourcehttp://igi.mai.gov.ro/sites/default/files/evaluarea_activitatii_in_anul_2017.pdf, p. 10 (site accessed on January 18th, 2019)		1266	1886	4820
Illegal border crossing for smuggling purposes		437	352	253

Sources Romanian Border Police/Public information/Analysis of Romanian Border Police's activity for 2015, pp. 11–18, Analysis of Romanian Border Police's activity for 2016, pp. 5–7. https://www.politiadefrontiera.ro/ro/main/n-informaii-de-interes-public-analiza-activitatii-9/

2015 to over 577,000 people identified in attempted illegal border crossing from the Balkan states towards the Member States of the European Union (Frontex · Western Balkans Annual Risk Analysis 2016, p. 12).

In the last four months of 2015 the average of detections regarding illegal border crossing by persons who are not citizens of the Balkan countries was over 400,000 people on monthly basis (Frontex · Western Balkans Annual Risk Analysis 2016 "Figure 7. The migration observed in the Eastern Aegean pressure is later reflected in the Western Balkans", p. 14).

Due to the measures taken by the authorities in the Balkan states that were supported by the EU institutions, the illegal migration on the Balkan route recorded a deep loosening, reaching in 2016 to the identification of 279,282 people, and in 2017 to 26,979 people who tried to illegally cross the borders between the Balkan states (Table 3).

Between 2015 and 2017 most citizens discovered in attempted illegal border crossings, over 57% came from Syria, Afghanistan, Iraq, Pakistan and Iran (According to processed data from Frontex · Western Balkans Annual Risk Analysis 2017, pp. 46–47). If we refer to people who are not citizens of the Balkan states, citizens in Syria hold first place with more than 32% (764,500 persons), among those found in attempted illegal border crossing, followed by citizens in Afghanistan, about 15.4% (366,991 persons), citizens in Iraq at a rate of less than 7% (166,548 persons), citizens of Pakistan by 2% (48,016 persons) and Iran with 0.95% (22,804 persons). If

Table 3 Illegal border crossing on Balkan route

	2015	2016	2017
Illegal border crossing by citizens outside the Balkan space	2,081,366	279,282	26,979
Through border checkpoints	1140	1595	2775
Guides	1980	1155	763
Hidden in transportation means	372	560	958
Avoiding control/forged documents	931	855	1636

Source Frontex · Western Balkans Annual Risk Analysis 2017, "*Table 1. Overview of indicators*", p. 46

we relate to the Balkan states, we find that Albania with 28,928 persons representing 1.2% and Kosovo with 26,481 persons representing 1.1% stand out among the countries in the region whose residents were involved in illegal border transit (Frontex · Western Balkans Annual Risk Analysis 2017, p. 47 "*Table* 2. *Illegal border-crossings between BCPs*").

2.6 *Mediterranean Route*

On this migration route using false or forged documents is insignificant because the majority of illegal migrants try to reach Greece, Italy, Spain etc. using various vessels without bearing in mind the presentation at checkpoints (Table 4).

After the implementation of measures adopted after agreement between the EU and Turkey in 2017 it was recorded a serious decline in illegal border crossing on the eastern Mediterranean route, a trend that continued in the first half of 2018.

Table 4 Illegal border crossing on Mediterranean route

		2015	2016	2017	Sem. 1 2018
Illegal crossing at sea border	East Mediterranean route	873,179	174,605	34,732	14,084
	Central Mediterranean route	153,946	181,376	118,962	16,313
	West Mediterranean route	5740	8641	21,632	13,958
Facilitators	At sea border	1137	962	1032	186
	Inside the EU member states	4669	5199	4397	No data available

Between 2015 and 2017 the persons discovered in attempts of illegal border crossing on the Mediterranean route came from (data processed from Frontex · Risk Analysis for 2016, p. 16, Frontex · Risk Analysis for 2017, p. 18, Frontex · Risk Analysis for 2018, p. 18):

(a) East-Mediterranean route, citizens from Syria 54% (597,320 persons), citizens from Irak 11.4% (127,034 persons), citizens from Afghanistan 23.5% (260,740 persons);
(b) Central-Mediterranean route, citizens from Nigeria 17% (77,631 persons), citizens from Eritrea 9% (40,863 persons), citizens from Guinea 5.7% (26,264 persons);
(c) West-Mediterranean route, citizens from Algeria 17.3% (7031 persons), citizens from Morocco 14% (5637 persons), citizens from Côte d'Ivoire 12.3% (4991 persons), citizens from Guinea 10.3% (4175 persons).

3 Conclusions

Globalization has influenced the increase in volume of migration, has changed its social dynamics, created social networks to facilitate further migration flows. But situations of instability in various countries have boosted illegal migration, associated with an increase in crime, from illegal crossing the state border, to smuggling of migrants, fake identity, corruption and service offenses.

The immigration crisis management reflects a new way of handling safety at national and international level to ensure the best conditions of the security and justice area in Europe.

Illegal immigration remains the main problem that generates, in addition to the migration phenomenon which requires providing a form of humanitarian assistance, and the phenomenon of smuggling. Both affecting from economic point of view the society, while the opportunities offered by authorities in the surveillance and border control remain the spearhead of preventing, combating and investigating criminal activities at the border.

References

Analysis of Romanian Border Police, pp. 16–17 (2015)
Analysis of Romanian Border Police's Activity, p. 5 (2016)
Analysis of Romanian Border Police's Activity, pp. 11–18 (2015)
Frontex · Western Balkans Annual Risk Analysis, p. 12 (2016)
Frontex · Western Balkans Annual Risk Analysis, p. 14 (2016)
Frontex · Risk Analysis, p. 16 (2016)
Frontex · Risk Analysis, p. 18 (2017)
Frontex · Risk Analysis, p. 18 (2018)

Frontex · Western Balkans Annual Risk Analysis, pp. 46–47 (2017)
Frontex · Western Balkans Annual Risk Analysis, p. 47 (2017)
FRONTEX, Eastern Partnership, Risk Analysis Network Quarterly, Quarter 4 • October–December, pp. 10–16 (2016)
FRONTEX, Eastern Partnership, Risk Analysis Network Quarterly, Quarter 4 • October–December, pp. 9–12 (2017)
FRONTEX, Eastern Partnership, Risk Analysis Network Quarterly, Quarter 1 • January–March 2017, pp. 6–8, 11–14 (2017)
FRONTEX, Eastern Partnership, Risk Analysis Network Quarterly, Quarter 4 • October–December, p. 9 (2017)
FRONTEX, Eastern Partnership, Risk Analysis Network Quarterly, Quarter 1 • January–March, pp. 9–12 (2018)
FRONTEX, Eastern Partnership, Risk Analysis Network Quarterly, Quarter 3 • July–September, pp. 6, 10–13 (2016)
FRONTEX, Eastern Partnership, Risk Analysis Network Quarterly, Quarter 1 • January–March, pp. 6–7, 11–14 (2016)
FRONTEX, Eastern Partnership, Risk Analysis Network Quarterly, Quarter 2 • April–June, pp. 6–7, 12–15 (2016)
FRONTEX, Eastern Partnership, Risk Analysis Network Quarterly, Quarter 4 • October–December, pp. 6–7, 11–12, 15–16 (2016)
FRONTEX, Eastern Partnership, Risk Analysis Network Quarterly, Quarter 4 • October–December, pp. 6–7, 9–12 (2017)
FRONTEX, Eastern Partnership, Risk Analysis Network Quarterly, Quarter 2 • April–June, pp. 6–8, 11–14 (2017)
FRONTEX, Eastern Partnership Risk Analysis Network Quarterly Quarter 1, January–March, pp. 6–7, 9–12 (2018)
FRONTEX, Eastern European Borders Quarterly Quarter 1, January–March, p. 12 (2015)
FRONTEX, Eastern European Borders Quarterly Quarter 1, January–March, pp. 8–9, 18–21 (2015)
FRONTEX, Eastern European Borders Quarterly Quarter 2, April–June, pp. 8–9, 16–19 (2015)
FRONTEX, Eastern European Borders Quarterly Quarter 3, July–September, pp. 6–7, 11–14 (2015)
FRONTEX, Eastern European Borders Quarterly Quarter 4, October–December, pp. 11–14 (2015)
Geamănu, G.: Drept internaţional contemporan (Contemporary International Law). Editura Didactică şi Pedagogică, Bucureşti (1965)
Iancu, N.: Migraţia internaţională a forţei de muncă (International Migration of Manpower). Editura PRO Universitaria, Bucureşti (2013)
Iancu, N., Urziceanu, R.-M., Bădulescu, A., Iancu, E.-A., Simuţ, R.: The use of the gravity model in forecasting the flows of emigrants in UE's countries. Technol. Econ. Dev. Econ. **23**(2), 392–409 (2017). http://www.tandfonline.com/doi/abs/10.3846/20294913.2016.1213194
National Geographic Society: Human Migration Guide, pp. 6–8 (2005)
Risk Analysis Network Quarterly, Quarter 4 • October–December, p. 10 (2016)
Scăunaş, S.: Drept internaţional public (International Public Law), 2nd edn. C. H. Beck Publishing House (2007)
Stamate, G.: Frontiera de stat a României (Romania's State Border). Editura Militară, Bucureşti (1997)

Fuzzy Reasoning and Decisions in Primary School

Luciana Delli Rocili and Antonio Maturo

Abstract The work starts from the idea that the natural instrument of communication in a class is constituted by linguistic propositions, called fuzzy propositions, which are grammatically and syntactically correct statements, but which often can not be classified either completely true or completely false. The truth value of a linguistic proposition depends on the opinions of a decision maker that are conditioned by the circumstances that occur. In this order of ideas the linguistic proposition is called a fuzzy event. A problem of decisions under conditions of uncertainty can be seen as attributing truth values to fuzzy events for each of the states of nature. Using the classic logical operations on events we can define the logical reasoning methods. We consider also an extension to fuzzy events, highlighting the fact that this leads to having to make choices of a subjective nature, on which to open interesting debates in the classroom.

Keywords Linguistic propositions · Random events · Fuzzy events · Fuzzy reasoning · Decisions

1 Linguistic Propositions as a Natural Means of Communication in a Class

Classical logic, based on propositions that admit as truth values only true or false (Russell 1962), appears insufficient to make decisions in daily life. In fact, most of the propositions of spoken language, and therefore of the information we manage to obtain to make decisions, are nuanced, not susceptible to a clear classification between true propositions and false propositions (Zadeh 1965, 1975; Gentilhomme

L. Delli Rocili
Istituto Comprensivo Statale Pescara 5, via Gioberti No. 15, 65100 Pescara, Italy
e-mail: lucianadr@live.it

A. Maturo (✉)
Dipartimento di Architettura, Università G. d'Annunzio di Chieti-Pescara, Viale Pindaro, 42, 65127 Pescara, Italy
e-mail: antomato75@gmail.com

D. Flaut et al. (eds.), *Decision Making in Social Sciences: Between Traditions and Innovations*, Studies in Systems, Decision and Control 247,
https://doi.org/10.1007/978-3-030-30659-5_20

1968; Kaufmann 1975a, b; Fadini 1979; Klir and Yuan 1995; Ross 1995; Delli Rocili and Maturo 2013b, 2015; Maturo and Tofan 2016; Maturo and Porreca 2016). For example, propositions such as "Carlo is high" or "In the classroom there are many people" can be considered only partially true or partially false, with a degree of truth depending on the judgment of a decision maker who evaluates according to his parameters.

The statements of spoken language, to which a decision-maker can also attribute values of truth that are intermediate between true or false, are called *linguistic* or *fuzzy statements*.

A statement of classical logic, in which the possible truth values are only *true* or *false*, is said to be a *binary* or *crisp statement*. From a formal point of view, we can say that, in spoken language, a decision maker (who can be a listener, a teacher, a child, etc.) has at his disposal an ordered set of truth values V, having "false" as minimum and "true" as maximum.

A fuzzy statement with a set of values V can be interpreted as a question to which the decision-maker attributes as an element of truth an element of V (Delli Rocili and Maturo 2013a, 2015; Hošková-Mayerová and Maturo 2016, 2017, 2018a, b). The fuzzy statement is reduced to a crisp statement if it can take, as truth values, only the minimum and the maximum of V.

For example, in the school, after some questions to the student Pierino, the teacher, in the role of decision maker, must give a value of truth to the statement "Pierino is prepared". If the teacher were to reason according to the classical logic should put the vote "0" if the truth value is "false" and the vote "10" if the truth value is "true". But instead the teacher has a set of truth values corresponding to the numbers from 0 to 10. For example, a vote "7" means that the statement is more true than false, and a vote "5" that is more false than true. Normally it assumes "false" = 0, "true" = 1 and the intermediate truth values are calculated in proportion in the scale of values that goes from 0 to 1. So to the grade "7" corresponds the degree of truth "0.7".

In conclusion, in the usual communication between individuals, particularly in a class, we can not limit ourselves to statements that are only "true" or "false". Communication would be very poor, with little information and also rather boring.

Instead it is necessary to evaluate the information received, attributing to them a degree of truth based on our opinions and parameters of judgment.

Propositions such as "Carlo is good", "Giuseppe is tall", "Luigi knows how to play football well", "Pierino is a friend", "Giovanni is tall", "The teacher is nice" can not be ignored just because they can not be classified with a net "true" or "false", with a judgment of truth that depends on the various parameters of judgment of the decision maker.

The value of truth attributed to the various propositions influences the conclusions of a logical reasoning. The logical reasoning consists in passing from some propositions to others derived from them by means of some logical operations and in obtaining the truth values of the derived propositions. These truth values are fundamental for making decisions.

In the sphere of classical logic, the propositions deriving from others are often called theorems and logical reasoning serves substantially to prove or deny that the theorem is true. In the context of fuzzy logic, i.e. based on fuzzy propositions and their derivatives, the truth value obtained for a derivative proposition may be different from "true" or "false" and generally does not give rise to a definitive decision, but is a *decision support* element.

Moreover, in the fuzzy context, the type of logical reasoning to follow is also subject to evaluation, in the sense that there are choices to be made for the definition of logical operations that are:

- extension of those of classical logic;
- the most appropriate to the context.

The fuzzy logic has also revealed itself as a useful tool to deepen the sociometric investigations introduced by Moreno (1951, 1953). In fact, by introducing fuzzy relations and fuzzy sociometric matrices, it is possible to gather various new information on the social group constituted by a class (Hošková-Mayerová and Maturo 2016, 2017, 2018a, b).

A class is a complex system in which teaching must be integrated with an accurate socialization process (Delli Rocili and Maturo 2013a, 2017). The fuzzy reasoning allows to effectively model social systems, processes and deduce results that take into account both the classifications and the specificities of individual cases (Ragin 2000; Maturo et al. 2008; Maturo and Paone 2012).

In particular, fuzzy reasoning proves particularly useful for the treatment of data expressed in a qualitative manner or in conditions of semantic uncertainty, as highlighted in the text by Ragin (2000) and in the recent works of Fabrizio Maturo (2016) and Hošková-Mayerová (Maturo and Hošková-Mayerová 2017; Maturo and Maturo 2017).

2 Fuzzy Events and Decisions

A further complication in attributing a value of truth to propositions consists of uncertainty due to chance, that is, in the fact that the decision-maker's opinion depends on a set of circumstances that may occur.

In other words, the decision-maker often has partial information and his attribution of the degree of truth depends on the information he will have at a later time.

For example, the statement "Pierino is good at mathematics" has a degree of truth, assessed by the teacher who plays the role of decision maker, which depends on the answers that Pierino provides to a certain number of questions.

The statement "Carlo is good at football" has a value of truth, assigned by a coach as a decision maker, which depends on the way in which Carlo plays and integrates with the team in a set of games.

In order to make decisions under conditions of uncertainty, the concept of "state of nature" is introduced, analyzing the set of circumstances that may occur and which influence the evaluation of the decision maker.

The states of nature must be a partition of the certain event, that is, events that are not impossible, two by two incompatible and such that at least one of them must occur. In practice, the partition of the certain event in states of nature must be such that the decision maker, assuming the hypothesis that a state of nature occurs, must be able to attribute a value of truth to a proposition, consistent with its own criteria.

A crisp or fuzzy proposition is called an event (crisp or fuzzy) if its truth value is random, depending on the state of nature that occurs.

In decision theory, the possible choices of a decision maker are called alternatives. Let $A = \{A_1, A_2, \ldots, A_m\}$ be the set of alternatives.

In the presence of an objective O, a decision maker can consider, for each alternative A_i, the statement "A_i satisfies the goal O", giving score 1 if the goal is fully satisfied by the alternative A_i (true statement) and score 0 if he is not at all satisfied (false statement).

Since, in general, choosing an alternative, an intermediate situation occurs between total satisfaction and total dissatisfaction, the score in general must be a number between 0 and 1 and depends on the state of nature that occurs. Fixed the alternative A_i, the statement "A_i satisfies the objective O" is then a fuzzy and random statement, i.e. a fuzzy event.

Then, if we are in the presence of decisions under conditions of uncertainty, we must consider a set $S = \{S_1, S_2, \ldots, S_n\}$ of states of nature, and the degree of truth of the statement "A_i satisfies the objective O" depends on the state of nature that occurs. This statement is a fuzzy event called "fuzzy event associated" with the alternative A_i.

3 An Experimentation on Understanding and Classification of Propositions in Primary School

The experimentation was carried out in some summer schools of mathematics for teachers of the first cycle in Pescara and Pizzoferrato. In addition, an experimentation was also carried out among students In addition, an experiment was also carried out among the students of a fifth class of a primary school in Pescara, with appropriate assistance from the teacher.

The procedure followed for teachers who took the course is as follows.

First phase. The professor of the course assumes the role of coordinator of the experimentation and proposes to each teacher attending the course to write 10 sentences, possibly including, but not necessarily, a crisp statement, a fuzzy statement, an aleatory event, a fuzzy event, an unacceptable sentence as a statement linguistic.

Second phase. The coordinator, upon proposal of the teachers attending the course, forms a committee of 3 decision-makers, chosen among teachers participating in the course, who takes into consideration a sample of the sentences written by the group. The committee has the task of classifying sentences into four categories: non-random linguistic statements, crisp events, aleatory linguistic statements and non-linguistic statements. The coordinator stimulates a discussion on the classification of the whole group of the participant, leaving the final decision to the committee.

Third phase. The committee takes into consideration the non-random propositions, and, after a debate with the participants, divides these propositions into statements of the bivalent logic, in which the possible values of truth can only be "true" = 1 and "false" = 0, and fuzzy enunciations, in which there can be intermediate truth values. Subsequently the committee opens the discussion to establish the truth values of the various propositions. In order to obtain the truth values of the fuzzy statements, the Saaty method of the "pairwise comparison" described, for example, in Delli Rocili and Maturo (2017) is often used.

When commission members and participants do not agree on the truth value, it is given through a fuzzy number, consisting of an interval [a, b] contained in [0, 1], a point c inside [0, 1], and a function and a function φ: [a, b] $\rightarrow$ [0, 1], increasing in [a, c], decreasing in [c, a] and such that $\varphi(a) = 0$, $\varphi(c) = 0$, $\varphi(b) = 0$. For any $x \in [a, b]$, $\varphi(x)$ measures the "degree of trust" of the commission that the truth value is equal to x.

Fourth phase. The committee considers crisp events, and with the help of the participants, introduces a total pre-order relationship $\angle$ between such events, placing $A \angle B$ if A is judged "less likely" or "equally probable" than B. $\angle$ is a qualitative probability if it satisfies some coherence requirements (see e.g. Delli Rocili and Maturo 2013c). Participants are then invited to assign subjective probabilities p(E) to events compatible with the relation $\angle$, i.e. such that $A \angle B \Rightarrow p(A) \leq p(B)$. The committee checks the consistency of the probability assignment and the compatibility with the pre-order report.

Fifth phase. The commission considers the aleatory linguistic statements. Together with the participants works for a prediction of the various circumstances that may occur, which must be formalized as a partition of the certain event. The elements of this partition are the "states of nature". For each state of nature S, the linguistic propositions conditioned to S are non-random and their values of truth are attributed according to the criteria considered in the third phase. Instead the probabilities of the states of nature are assigned according to the criteria of the fourth phase.

Sixth phase. The coordinator, together with the committee, discusses the results obtained, the context, the criteria and the procedures with which we arrive at attributing the value of truth to the various statements. The coordinator opens the debate with all the participants.

The experimentation with the teachers lasted many hours. The analytical development of the various phases highlighted various difficulties in formulating and recognizing the various types of statements. Above all, it became clear that the application of theories carried out requires an interdisciplinary teaching between Mathematics and Linguistic Disciplines for a precise formulation and a full understanding of the meaning of the propositions.

4 An Educational Experimentation on Decision Problems in Conditions of Uncertainty

At the end of the day on which the experimentation described in Sect. 3 took place, the participants were invited to apply the concepts discussed to decision problems under conditions of uncertainty. To this end they have been offered the homework to formulate problems of decision under conditions of uncertainty taken from daily life. Two days later, in the next meeting for the summer school, there would be discussion for the most appropriate formulation of the proposed decision problems.

The experimentation also in this case can be divided into successive phases.

First phase. The coordinator begins by illustrating a theme proposed by the participants, for example that of improving the school system, and proposes to each teacher of the group to identify objectives and, in correspondence with each objective, to determine the states of nature.

Second phase. The coordinator forms a committee of 3 teachers to focus on one or two more relevant objectives. The relevance and correct formulation of the objectives is discussed by all the participants until an agreement is reached.

Third phase. For each of the objectives taken into consideration, the committee proposes to each teacher of the group to identify the states of nature and the possible alternatives.

Fourth phase. The coordinator, helped by the commission, after acquiring the opinion of the participants, attributes the probabilities to the states of nature.

Fifth phase. The commission asks the opinion of the participants for the assignment of the scores of the alternatives for each of the states of nature. The coordinator activates procedures to reach an agreement on these scores.

Sixth phase. All scores are reported to a numerical scale that has a minimum of 0 and a maximum of 1. The coordinator opens a discussion on the scores obtained and the interpretations in terms of fuzzy events.

Seventh phase. The coordinator and the commission establish criteria for aggregating scores, obtaining a final score for each alternative.

The coordinator opens the discussion on the results obtained, evaluating the relevance of the differences in the score between the various alternatives.

If the opinions of the participants are very different from each other, a procedure is activated to reach consensus.

5 The Foundations of Logical Reasoning with Crisp Events

Taking into account the possible randomness, it is convenient to introduce the logical operations on the events, as propositions of the binary logic with information that can be either total or partial. As a special case, if the information is total, operations on the statements of classical logic are obtained.

Table 1 Contrary Event

E	1	0
¬E	0	1

In classical logic, logical variables are generally used, i.e. symbols that can assume the values of truth: true or false.

In this paper we prefer to replace logical variables with events for a set of reasons, including:

(1) The random events can assume both values: *true* or *false*, and then they perform the same functions of the logical variables.
(2) Results are obtained directly in an environment that is broader than that of classical logic, in which partial information may exist.
(3) One can attribute the concrete meaning of a particular random event to each symbol used.
(4) The formulas of classical logic remain unaltered, simply replacing the word "binary variable" with the word "random event" and the word "tautology" with the word "certain event".

Finally, thanks to the concepts of "certain event partition" and "constituent", an event is represented in a very significant way and the link between logical and set-theoretical operations is very intuitive.

We indicate with "0" the value of truth "false" and with "1" the value "true". The first logical operation on the events to be considered is the negation, that at each event E associates another event ¬E (called the contrary of E) that is true when E is false and instead is false when E is true. It is a unary operation, that is done on a single event.

The concept can be presented in primary school with Table 1.

For the pair (E, ¬E) the following principles apply:

- *non-contradiction*: E and ¬E can not occur together;
- *excluded third*: at least one between E and ¬E occurs.

From the numerical point of view it can be observed that the sum of the elements of a column of Table 1 is always equal to 1, so we can write:

$$\mathrm{E} + \neg\mathrm{E} = 1 \tag{5.1}$$

One aspect to be explored in a class is the concept of equality. The equality between two propositions A and B, i.e. writing A = B, does not mean that the two propositions say the same thing. Equality only means that the two propositions have the same truth value, i.e. they are either both true or both false. This is a remarkable process of abstraction to highlight with numerous examples.

Table 2 Incompatible Events

A	1	0	0
B	0	1	0

In particular, it is worthwhile to show that $\neg(\neg E) = E$.

5.1 Some Basic Definitions on Events

The opportunity to define the relations between events is immediately presented, starting with the introduction of the incompatibility relationship.

Definition of incompatibility. Two events A and B are said to be incompatible if when one is true, the other is false.

In the form of a table, we obtain (Table 2):

From the numerical point of view it can be observed A and B are incompatible if and only if the sum of the elements of a column is always less than or equal to 1, that is:

$$A + B \leq 1 \tag{5.2}$$

Definition of partition of the certain event. A family of random events $\{A_1, A_2, \ldots, A_n\}$ is said to be a partition of the certain event if any two events in the family are incompatible and at least one of them must occur.

It is an extension to the case of any number of events of the principles of non-contradiction and excluded third: the latter is replaced by the more general fact that an event not belonging to the family is excluded. The condition that $\{A_1, A_2, \ldots, A_n\}$ is a partition of the certain event partizione dell'evento certo an be expressed numerically as an extension of (5.1) with the formula:

$$A_1 + A_2 + \cdots + A_n = 1 \tag{5.3}$$

5.2 Logical Operations Between Events

Binary logic operations between events associate a third event with each ordered pair of events. The most important are the disjunction "$\vee$", conjunction "$\wedge$" and implication "$\rightarrow$".

Disjunction or union operation. Given two events A and B, the disjunction or union between A and B, indicated by $A \cup B$ or $A \vee B$, is the event that is true if at least one of A and B is true.

From the numerical point of view there are various formulas that represent the disjunction, among which:

(maximum formula) $A \vee B = \max\{A, B\}$,
(algebraic formula) $A \vee B = A + B - AB$,
(limited sum formula) $A \vee B = \min\{A + B, 1\}$.

Conjunction or intersection operation. Given two events A and B, the conjunction or intersection between A and B, indicated by $A \cap B$ or $A \wedge B$, is the event that is true if it is true if both A and B are true. From the numerical point of view there are various formulas that represent the conjunction, including:

(minimum formula) $A \wedge B = \min\{A, B\}$,
(algebraic formula) $A \wedge B = AB$,
(limited sum formula) $A \wedge B = \max\{A + B - 1, 0\}$.

The operations "$\vee$" e "$\wedge$" are commutative and associative (as well as having other important algebraic properties). Furthermore, the formula is significant:

$$A \vee B + A \vee B = A + B \tag{5.4}$$

The most important operation for logical reasoning is implication, which leads to the "rules of inference" on which the correctness of logical reasoning is based.

The implication operation is, with respect to the union and the intersection, more difficult to understand in depth, but it is more efficient than the others in order to obtain the results of logic, being at the base of the syllogisms.

Implication operation. The implication $A \to B$ is defined as the true event if it is false A or if A and B are both true. From the numerical point of view there are various formulas that represent the implication. For example:

(formula of the union) $A \to B = \neg A \vee B$;
(maximum formula) $A \to B = \max\{\neg A, B\}$;
(minimum formula) $A \to B = \min\{1, 1 - A + B\}$;
(algebraic formula) $A \to B = (1 - A) + B - (1 - A)\,B = 1 - A + AB$.

5.3 *The Main Rules of Logical Reasoning*

The main rules of logical reasoning are related to binary and negation operations. They can (and perhaps should) be gradually and appropriately introduced into the first-cycle classroom to train students to correct reasoning and then be aware of making decisions. We propose a form of the various rules of inference easily extensible to the fuzzy case of the next paragraph, using the symbol $\leq$.

The Modus Ponens rule. If an event A is true and the implication $A \to B$ is true, then it is also true B. In symbols:

$$\text{(Modus Ponens Formula)} \quad (A \vee (A \to B)) \leq B. \tag{5.5}$$

For example, let A = "I drank the wine", $A \rightarrow B$ = "If I drink the wine I quench my thirst", B = "I quench my thirst".

The need for attention to the Modus Ponens rule is due to the fact that often, in the common life, the wrong conclusions are due to an incorrect application of the rule. For example, if they are true $A \rightarrow B$ = "If I drink wine I quench my thirst" and B = "I have quenched my thirst", I can not conclude that I drank wine, as one can quench one's thirst by drinking more liquids. The use of $\leq$ seems efficient to avoid logical errors. It can be shown that, even if $B = 1$ and $A \rightarrow B = 1$, it is not said that it is $A = 1$. If I have quenched my thirst, it is not said that I drank wine.

The Modus Tollens rule. If an event B is not true and the implication $A \rightarrow B$ is true, then it is not true A. In symbols:

$$\text{(Modus Tollens formula)} \quad (\neg B \wedge (A \rightarrow B)) \leq \neg A. \tag{5.6}$$

The Modus Ponens is an alternative to the Modus Tollens to deny an affirmation.

For example, if it is false B = "I have quenched my thirst" and is true $A \rightarrow B$ = "If I drink wine I quench my thirst", it can be concluded that I did not drink the wine. Again the use of the symbol $\leq$ seems efficient to show that if they are true $\neg A$ and $A \rightarrow B$ it is not said to be false B.

The third rule of inference is that which allows a chain of reasoning and therefore the demonstration of theorems.

The rule of Hypothetical Syllogism. If the implication $A \rightarrow B$ is true and the implication $B \rightarrow C$ is true, then the implication $A \rightarrow C$ is also true. In symbols:

$$\text{(Formula of the Hypothetical Syllogism)}$$
$$((A \rightarrow B) \wedge (B \rightarrow C)) \leq (A \rightarrow C). \tag{5.7}$$

For example: If it is true $A \rightarrow B$ = "If I drink wine I quench my thirst" and it is also true $B \rightarrow C$ = "If I disagree I feel good", then it can be concluded that the proposition is true $A \rightarrow C$ = "If I drink the wine I feel good". Also in this case the use of the symbol $\leq$ is useful to show that if the implication $A \rightarrow C$ is true and one of the two implications $A \rightarrow B$ and $B \rightarrow C$, it may occur that the other implication is not true.

6 The Game of Fuzzy Reasoning

At the level of primary school, fuzzy reasoning should be presented as a game in which we can attribute to each event a value belonging to the interval [0, 1], which represents "the degree of truth" of the proposition that characterizes the event, evaluated by the boys after an appropriate reflection assisted by the teacher.

As a first step, it is necessary to extend to the fuzzy case the definitions of a contrary event and of a partition of the certain event. A simple way is to impose that even in the fuzzy case the conditions (5.1) and (5.3) are valid.

So we can give the following definitions:

Contrary fuzzy event. Given a fuzzy event E, we define the *contrary* of E the fuzzy event

$$\neg E = 1 - E. \tag{6.1}$$

Fuzzy partition of the certain event. A family of fuzzy events $\{A_1, A_2, \ldots, A_n\}$ is said to be a partition of the certain event if

$$A1 + A2 + \ldots + An = 1. \tag{6.2}$$

The extension of the operation "union" to the fuzzy case involves a choice, since the formulas of the maximum, algebraic and the limited sum do not provide the same result (and various other formulas, called t-conorms, are possible).

Similarly, the extension of the operation "intersection" to the fuzzy case implies a choice between the formulas of the minimum, the algebraic and the limited sum (and various other formulas, called t-norms, are possible).

A constraint for the two choices is the fact that it continues to be valid relation (5.4). In this case we say that the choices are consistent. In the following we admit that consistency is verified.

As for the implication we can choose one of the four formulas of the union, of the maximum, of the minimum, the algebraic (and any other). The students will be able to experiment with the value of truth of logical proposition containing fuzzy events and fuzzy operations of union, intersection and implication, especially with concrete cases, and sometimes they will be surprised by the results obtained.

Once the operations of union, intersection and implication have been defined, we can extend the formulas (5.5)–(5.7) to the fuzzy reasoning.

For example, if I drank a moderate amount of wine the proposition A = "I drank the wine" could have a truth value of 0.5 and, if I am not completely sure that the wine then quenches the proposition $A \to B$ = "If I drink wine, I'm afraid" could have a truth value of 0.7.

Then the formula of the Modus Ponens $(A \wedge (A \to B)) \leq B$ allows to find a lower bound of the truth values that can assume B. For example, using the formula of the minimum for the intersection, we obtain $B \geq \min(0, 7; 0.5) = 0.5$.

7 Conclusion

The work and the proposed educational path aims to sensitize students to logical reasoning even if there are uncertain data. Another objective is to train young people to make decisions, to learn or to look for criteria to evaluate the correctness and the limits of their opinions and their deductions.

The possibility and the necessity to use values of truths intermediate between 0 and 1 empowers the boys to find logical criteria to attribute the values of truth.

The fact that in the fuzzy case the formulas of union and intersection do not lead to the same result induces the boys to reflect in depth on these operations, and to ask why choose one formula rather than another.

A separate discussion must be made for the implication operation. The understanding of this operation and the rules of Modus Ponens, Modus Tollens, Hypothetical Syllogism, quickly introduces the boy into the world of logical reasoning.

Surely it is not easy for the boys to understand the extent of the extension of the implication to the fuzzy case, nor is it easy to decide on which of the possible formulas to choose. However, in any case the boys can see such extensions as a game and are able, if trained, to comment on the results obtained.

In any case, the work carried out leads the students to a greater awareness in the use of the spoken language, to its links with the scientific language, to an interdisciplinary vision of the learning processes.

References

Delli Rocili, L., Maturo, A.: Teaching mathematics to children: social aspects, psychological problems and decision-making models. In: Soitu, D., Gavriluta, C., Maturo, A. (eds.) Interdisciplinary Approaches in Social Sciences. Editura Universitatii A.I. Cuza, Iasi, Romania (2013a)

Delli Rocili, L., Maturo, A.: Logica del certo e dell'incerto per la scuola primaria. Sci. Philos. **1**(1), 37–58 (2013b)

Delli Rocili, L., Maturo, A.: Probabilità e Statistica nella scuola primaria: esperienze didattiche e proposte. Sci. Philos. **1**(2), 49–78 (2013c)

Delli Rocili, L., Maturo, A.: Interdisciplinarity, logic of uncertainty and fuzzy logic in primary school. Sci. Philos. **3**(2), 11–26 (2015)

Delli Rocili, L., Maturo, A.: Social problems and decision making for teaching approaches and relationship management in an elementary school. In: Hošková-Mayerová, Š., et al. (eds.) Mathematical-Statistical Models and Qualitative Theories for Economic and Social Sciences. Series in Studies in Systems, Decision and Control, vol. 104. Springer, pp. 81–94 (2017)

Fadini, A.: Introduzione alla teoria degli insiemi sfocati. Liguori, Napoli (1979)

Gentilhomme, M.Y.: Les ensembles flous en linguistiques. Cahiers de linguistique theorique et appliquée, Bucarest **5**(47), 47–65 (1968)

Hošková-Mayerová, Š., Maturo, A.: An analysis of Social Relations and Social Group behaviors with fuzzy sets and hyperstructures. Int. J. Algebr. Hyperstruct. Appl. **2**(1), 91–99 (2016)

Hošková-Mayerová, Š., Maturo, A.: Fuzzy sets and algebraic hyperoperations to model interpersonal relations. In: Maturo, A., et al. (eds.) Recent Trends in Social Systems: Quantitative Theories and Quantitative Models Studies in Systems. Decision and Control, vol. 66. Springer, pp. 211–222 (2017)

Hošková-Mayerová, Š., Maturo, A.: Decision-making process using hyperstructures and fuzzy structures in social sciences. In: Collan, M., Kacprzyk, J. (eds.) Soft Computing Applications for Group Decision-Making and Consensus Modeling. Studies in Fuzziness and Soft Computing, vol. 357, pp. 103–112 (2018a)

Hošková-Mayerová, Š., Maturo, A.: Algebraic hyperstructures and social relations. Ital. J. Pure Appl. Math. 39, 701–709 (2018b)
Kaufmann, A.: Theory of Fuzzy Subsets, vol. I. Academic Press, New York (1975a)
Kaufmann, A.: Introduction a la théorie des sous-ensemble flous, vols. II & III. Masson, Paris (1975b)
Klir, G., Yuan, B.: Fuzzy Sets and Fuzzy Logic: Theory and Applications. Prentice Hall, Upper Saddle River, NJ (1995)
Maturo, F.: Dealing with randomness and vagueness in business and management sciences: the fuzzy-probabilistic approach as a tool for the study of statistical relationships between imprecise variables. Ratio Math. **30**(1), 45–58 (2016)
Maturo, F., Hošková-Mayerová, Š.: Fuzzy regression models and alternative operations for economic and social sciences. In: Recent Trends in Social Systems: Quantitative Theories and Quantitative Models. Series Studies in Systems, Decision and Control, vol. 66, pp. 235–247 (2017)
Maturo, A., Maturo, F.: Fuzzy events, fuzzy probability and applications in economic and social sciences. In: Maturo, A., et al. (eds.) Recent Trends in Social Systems: Quantitative Theories and Quantitative Models Studies in Systems. Decision and Control, vol. 66. Springer, pp. 223–233 (2017)
Maturo, A., Paone, F.: New processes of socialization in education: theoretical analysis, proposals for intervention and formalization with fuzzy models. Procedia: Soc. Behav. Sci. **46**, 4886–4893 (2012)
Maturo, A., Porreca, A.: Algebraic hyperstructures and fuzzy logic in the treatment of uncertainty. Sci. Philos. **4**(1), 31–42 (2016)
Maturo, A., Sciarra, E., Tofan, I.: A formalization of some aspects of the Social Organization by means of the fuzzy set theory. Ratio Sociol. **1**(2008), 5–20 (2008)
Maturo, A., Tofan, I.: Fuzziness Teorie e applicazioni. Aracne Editrice, Roma (2016)
Moreno, J.L.: Sociometry. Experimental Methods and the Science of Society. Beacon Press, New York (1951)
Moreno, J.L.: Who Shall Survive?. Beacon Press, New York (1953)
Ragin, C.C.: Fuzzy-Set Social Science. University Chicago Press, Chicago, USA (2000)
Ross, T.J.: Fuzzy Logic with Engineering Applications. McGraw-Hill, New York (1995)
Russell, B.: Introduzione alla filosofia matematica. Longanesi, Milano (1962)
Zadeh, L.A.: Fuzzy sets. Inf. Control **8**, 338–358 (1965)
Zadeh, L.A.: The concept of a linguistic variable and its application to approximate reasoning. Inf. Sci. **8**(Part I), 199–249 (1975); Part II, 301–357; **9**(Part III), 43–80 (1975)

Life Expectancy at Birth: A Causal Analysis of the Health Sector in Spain

Pedro Antonio Martín Cervantes, Nuria Rueda López and Salvador Cruz Rambaud

Abstract Life expectancy at birth has become the main indicator of the performance of a country's health system. From a causal point of view, there is a series of socio-economic indicators which affect life expectancy. The objective of this work is to determine the indicators which exhibit a unilateral or bilateral causal relationship with life expectancy, using the principal components analysis (PCA). Once selected these indicators, we will highlight, through the Dumitrescu-Hurlin version of the Granger causality test, those having a unidirectional causality relationship with life expectancy. Undoubtedly, our findings can provide suitable information which may be useful for the management of health resources from the point of view of maximizing social benefits.

Keywords Life expectancy at birth · Socio-economic indicators · Panel data · Principal component analysis · Granger causality test (Dumitrescu-Hurlin)

1 Introduction

Given its universal and decentralized character, the degree of accessibility and the level of primary care of the health sector in Spain have been always analyzed from the perspective of efficiency and its relation with some indicators, preferably of social and economic nature. In this way, if we had to choose a variable representative of the success of a health system, we think that this role should be played by the life expectancy of the population in which such system is implanted. Within this variable, life expectancy at birth is generally the most accepted health care outcome, although some scholars, such as Self and Grabowski (2003) and Caballer (2007),

P. A. Martín Cervantes · N. Rueda López · S. Cruz Rambaud (✉)
Departamento de Economía y Empresa, Universidad de Almería, Almería, Spain
e-mail: scruz@ual.es

P. A. Martín Cervantes
e-mail: pmc552@ual.es

N. Rueda López
e-mail: nrueda@ual.es

D. Flaut et al. (eds.), *Decision Making in Social Sciences: Between Traditions and Innovations*, Studies in Systems, Decision and Control 247,
https://doi.org/10.1007/978-3-030-30659-5_21

use life expectancy adjusted for disability, whilst, in other areas, life expectancy is considered at different ages (National Institute of Statistics—INE 2018). Logically, the interrelation between life expectancy and the chosen indicators must be carried out in a macroeconomic context because the evolution of life expectancy must be linked to general indicators of the health system (Urban and Hoskova-Mayerova 2017). Another question, quite different, is if we intend to carry out a microeconomic analysis, in which case it would be more difficult to identify the economic and operational management of a hospital or a health complex with its impact on the life expectancy of its insured people.

On the other hand, the existence of an increasing demand for health services, whether publicly or privately funded, justifies that the level of performance of the health system is put under evaluation. The growth of this demand is due, fundamentally, to the following three causes:

- The progressive aging of the population. In effect, Fig. 1 shows the current (2018) and prospective population pyramid (forecast for the year 2050), which clearly reflects the aging of the Spanish population.
- The increasing innovation, in absolute terms, of the medical technology sector. In effect, although within the budget resources addressed to R&D&I in the General State Budgets (GSB), the amount allocated to health research in the GSB 2018 was 270.85 million euros, compared to the 269.96 million in 2017, the percentage over the total was 4.14% in 2017 compared to 3.84% in 2018 (GSB 2018). On the other hand, Fig. 2 shows the evolution, during the period 1996–2016, of the number of operating devices, per 100,000 inhabitants, of computerized axial tomography (CAT), magnetic resonance imaging (MRI), hemodialysis and hemodynamics, according to the information extracted from the Ministry of Health, Consumer Affairs and Social Welfare (Ministerio de Sanidad, Consumo y Bienestar Social 2019).
- The greater presence of the so-called "Welfare State", which has raised mainly due to the social changes which are taking place in our society. In effect, it is necessary to bear in mind that, in the last 7 years (during the period 2011–2017), the percentage of the population with low social class has decreased minimally, staying at around 47%.

According to INE (2018), life expectancy is the average number of years which a person of a certain age would expect to live if the currently observed age-specific mortality pattern (mortality rates at each age) is maintained. In this work, we will use the life expectancy at birth because it is the most widely used indicator to make comparisons on the incidence of mortality in different populations and, based on this, on the health conditions and level of development of a population. In Western countries, life expectancy has experienced remarkable advances in the last century which has been achieved because of decreases in the probability of dying due to medical and technological advances, reduction in infant mortality rates, changes in nutritional habits and lifestyles, improvement in the levels of material living conditions and in education, as well as the population's access to health services (Svarcova et al. 2016). Although, in this work, we will focus on the variables related to the health sector,

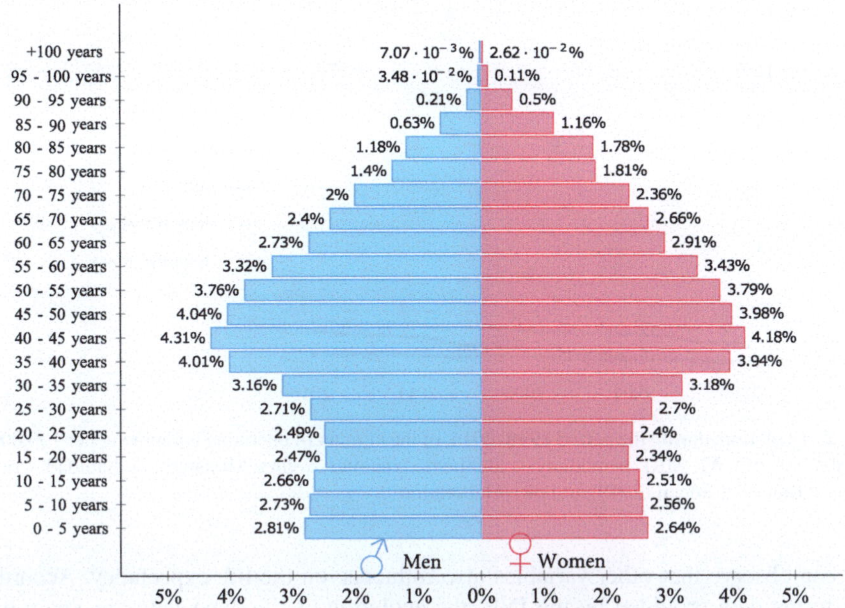

(a) Population Pyramid. Spain, 2018.

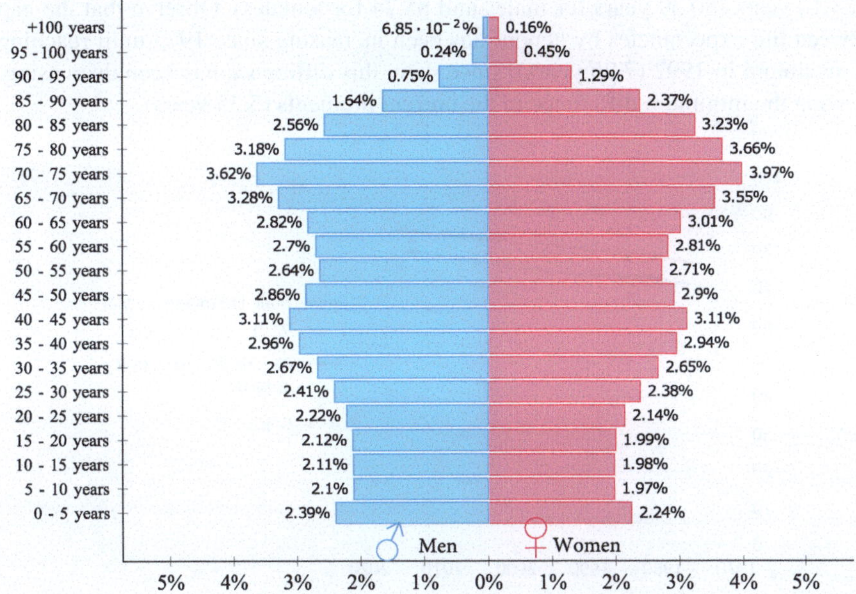

(b) Population Pyramid. Spain, 2050 (estimated).

Fig. 1 Pyramid of the Spanish population (2018 and estimated into 2050). *Source* INE (2018) and own elaboration

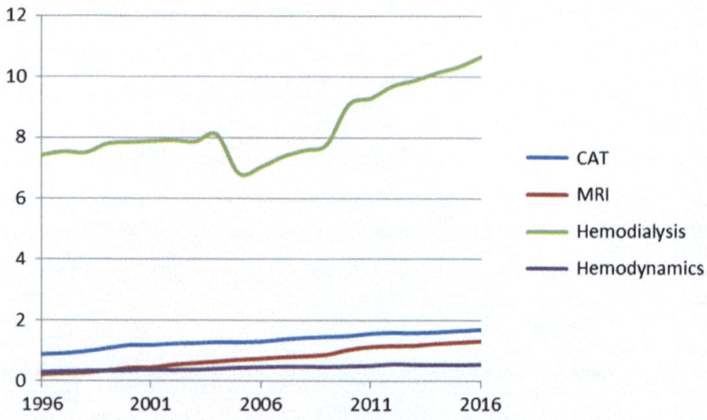

Fig. 2 Evolution, during the period 1996–2016, of the number of operating equipment, per 100,000 inhabitants, of CAT, MRI, hemodialysis and hemodynamics. *Source* Ministerio de Sanidad, Consumo y Bienestar Social (2019) and own elaboration

we can observe that other variables also influence on the life expectancy. According to the data provided by the INE, the evolution of life expectancy in Spain by gender has followed during the period 1975–2017 the tendency shown in Fig. 3. In effect, according to the provisional data for 2017, life expectancy at birth in Spain is 83.10 years (80.39 years for males and 85.74 for females). Observe that the gap between life expectancies by gender has been increasing since 1975 until reaching its maximum in 1992 (7.26 years); since then, this difference has been decreasing, showing the minimum difference in the current moments (5.35 years).

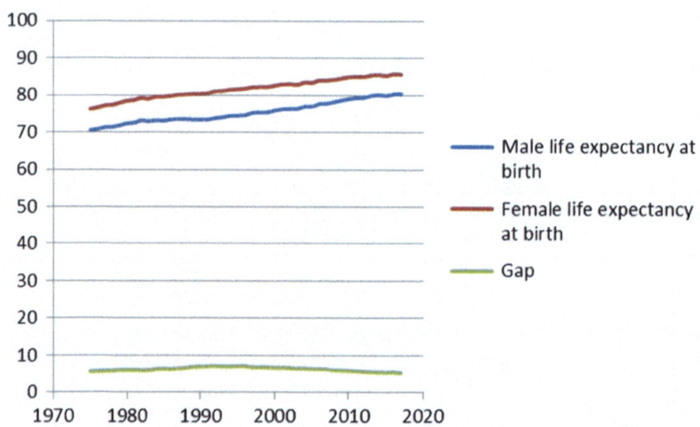

Fig. 3 Evolution of life expectancy at birth, by sex, during the period 1975–2017. *Source* INE (2018) and own elaboration

According to the data from the World Bank (2019), Spain is the fifth country or territory worldwide in terms of life expectancy at birth, behind Japan and Switzerland. However, taking into account the projection to 2040 made by this agency with demographic criteria, these positions are more or less maintained: Spain falls to the sixth position whilst Singapore appears in the fourth place (see Table 1).

In Fig. 4, we can observe the projection until 2065 of life expectancy at birth by sex, carried out by INE (2018) in Spain, where the convergence of life expectancies at birth for males and females must be highlighted (the gap tends to be zero).

At this point, it is necessary to compare these data with the results from Foreman et al. (2018) which place Spain in third place, behind Japan and Singapore, surpassing

Table 1 Life expectancies at birth according to countries or territories

Life expectancy at birth (2016)		Life expectancy at birth (projection to 2040)	
 Hong Kong SAR, PR China	84.23 years	 Hong Kong SAR, PR China	87.50 years
 Japan	83.98 years	 Macao SAR, PR China	87.35 years
 Macao SAR, PR China	83.85 years	 Japan	86.95 years
 Switzerland	82.90 years	 Singapore	86.53 years
 Spain	82.83 years	 Swizerland	86.50 years
 Singapore	82.80 years	 Spain	86.36 years

Source World Bank (2019) and own elaboration

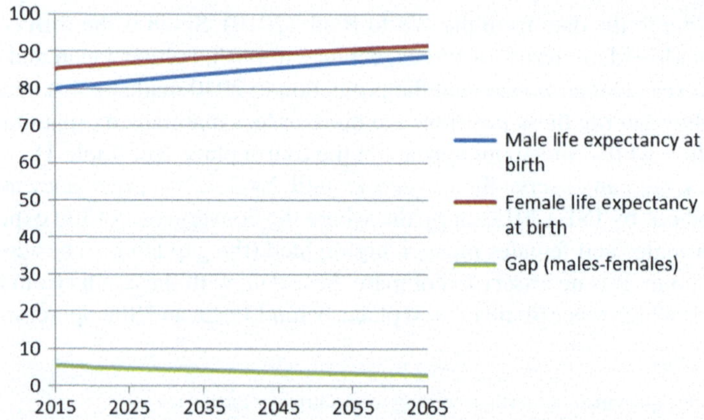

Fig. 4 Projection until 2065 of life expectancy at birth, by sex. *Source* INE (2018)

Switzerland in its projection to 2040. To do this, they analyze 250 causes of death and, using the data from the period 1990–2016, generate predictions for the period 2017–2040 where these countries show life expectancies over 85 years. The model incorporates changes in 65 risk factors and the underlying mortality rate for each cause which is a function of the per capita income, the educational level and the rate of fertility below 25 years.

The main objective of this work is to analyze the degree of performance of the Spanish health care system, by relating, in terms of causality, the life expectancy at birth of the Spanish population with the main socio-economic indicators. To do this, we will consider the following secondary objectives which will help to develop the primary objective:

- To identify the socio-economic variables which determine the success of the performance of the health care system.
- To analyze the significance of each variable selected for the proposed econometric model.
- To find the unidirectional causal relationships between the main indicators and the life expectancy at birth with the aim of contributing to the management of resources destined for the health care sector in Spain.

The methodology to be used in this work is mainly based on panel data with information disaggregated by autonomous communities, the PCA and the application of the Granger causality test, using the Dumitrescu-Hurlin approach (2012). Finally, the structure of this work is as follows. In Sect. 2, a review of the main theoretical and practical studies published to date on this topic is presented. In Sect. 3, we will propose a list, as complete and justified as possible, of the variables having an impact on the performance of the Spanish health system. Next, suitable econometric tools will provide us with the significance and, therefore, the contribution of each variable to the achievement of our health system. Finally, Sect. 4 presents the main conclusions and future lines of research of our work.

2 Literature Review

Auster et al. (1969) were the first scholars to analyze the health status of the population through an aggregate production function. Starting from this work, two lines of research have proliferated in the empirical literature. The first focuses on the use of macro-data, whilst the second, generally applied to clinical and epidemiological problems, analyzes the results in terms of health from an individual perspective and by using micro-data (on the latter, see for example, Wise 1997; Khuder 2001; Martínez-Sánchez et al. 2001; Kiuila and Mieszkowski 2007; Meara et al. 2008; Hansen and Strulik 2017). As indicated in the Introduction, in the present work we have opted for the first approach based on the use of macro-data. Among other advantages, this will allow us to analyze the total effect of the use of health care resources on the health status of citizens and, therefore, to obtain suitable economic policy implications.

Table 7 (see Appendix) includes a selection of the most impact papers, arranged chronologically, which differ not only in the country or groups of considered countries, the type of handled data and/or the analyzed time period, but also in the methodological strategy and the variable used to proxy the average health status of population. This review has been limited to those works which follow a macro approach and mainly focused on developed countries, since they agree in classifying the determinants of the health status of the population in three categories, namely socio-economic factors, health resources and lifestyle-related factors. Therefore, studies exclusively referred to developing countries has been excluded of this review due to the use of explanatory variables different from the former ones and more related with the degree of economic and social development, such as the access to drinking water, the degree of literacy, belonging to a specific ethnic group, the degree of urbanization, the geographic location, the access to health and the fertility (see, for example, Filmer and Pritchett 1997; Wang 2002; Aguayo and Lamelas 2003; Fayissa and Gutema 2005; Kabir 2008).

Table 7 shows that there are many papers referred to OECD countries and, to a lesser extent, referred to European countries or a combination of both groups. Most of these works use the mortality rate (at a specific age or in the childhood) and/or some dimension of life expectancy as proxy variables of the health status of the population. In this context, the life expectancy at birth predominates, although some works focus on the life expectancy at a specific age, for example 65 years, or differentiating by sex. Next, we are going to review the existing literature according to the criteria formerly outlined.

(a) With respect to the socio-economic factors, almost all works consider GDP or per capita income, and with less frequency some indicator of income distribution (Rodgers 1979; Berger and Messer 2002; Hill and Jorgenson 2018) or poverty (Crémieux et al. 1999). Unemployment is also analyzed in the works of McAvinchey (1988), Laporte (2004), Tapia-Granados (2005), Nixon and Ullman (2006), and Shin-Jong (2009), as a determinant of the health status of the population as it is a representative measure of the fluctuations of the economic cycle. Other "less traditional" macroeconomic variables used in this type of analysis are inflation and gross capital formation, both used by Monsef and Mehrjardi (2015).

Education is frequently included in these analyzes (Self and Grabowski 2003; Cutler and Lleras-Muney 2012; Ketenci and Murthy 2018, among others); as well as the level of occupation (Auster et al. 1969; Or 2000a, b; Hoskova-Mayerova 2016).

Among demographic variables, some works include the older population—generally the one that exceeds 60 or 65 years old—, as seen, for example, in Wolfe and Gabay (1987) and Hitiris and Posnett (1992).

Within the category of socioeconomic factors, some works include variables of institutional nature. Thus, Elola et al. (1995) analyze the relationship between different health systems and health indicators, and Grubaugh and Rexford (1994) include a dummy variable representative of the type of health system. On the other hand, Robalino et al. (2001) study the effect of fiscal decentralization and certain indicators such as the level of corruption and political rights, among others, on health status.

Recently, Jiang et al. (2018) enrich the available literature by incorporating into their study on the determinants of life expectancy, four dimensions of social development (healthcare, education, eco-environment and social harmony development).

Another factor which has been incorporated into this type of analysis is the pollution, generally represented by emissions of polluting substances (Or 2000a, b; Nixon and Ullmann 2006; Monsef and Mehrjardi 2015) or by an environmental quality indicator (Mariani et al. 2010). Recently, this interest has increased due to the importance of the environment which has been proposed to incorporate other variables, such as global warming, in this type of model. In this way, Ketenci and Murthy (2018) outline that we will have to wait for consistent measures of this phenomenon and for having a sufficient period of years.

The growing interconnection between different economies and societies, explains the recent emergence of a new line of research in this type of studies, by introducing globalization as a possible determinant of the health status (Vallejo et al. 2017). In this sense, Bergh and Nilsson (2010) consider up to three dimensions of globalization; namely: economic, social and political.

(b) In this type of literature, it is usual to use some expression of the health expenditure as a representative variable of health resources. In this way, we can consider either the total expenditure (Hitiris and Posnett 1992; Crémieux et al. 1999; Heijink et al. 2013, or the entire component of public health expenditure (Cochrane et al. 1978; Or 2000a, b; Berger and Messer 2002) or analyze public and private health expenditure separately (Linden and Ray 2017). Miller and Frech (2002), Shaw et al. (2005) and Crémieux et al. (2005) have incorporated other items such as the pharmaceutical expenditure and, later, Bradley et al (2011), Reynolds and Avendano (2017) and Van den Heuvel and Olaroiu (2017) consider various categories of social spending in order to assess their differential effect on health status.

Part of this literature also incorporated some proxy variables of health resources—mainly doctors and nurses (Crémieux et al. 1999; Or 2000b)—; and, less frequently, the number of beds (Cochrane et al. 1978; Nixon and Ullmann 2006; Van den Heuvel and Olaroiu 2017). Additionally, medical innovation (Lichtenberg 2000)—measured by the use of innovative medicines—and the requirement of having a medical prescription for the consumption of certain medications (Peltzman 1987) have been analyzed as conditioning factors of the health status.

(c) In order to capture the effects of health habits and lifestyle, in this type of literature it is usual to include other specific variables such as the consumption of tobacco and alcohol, and other variables representative of diet such as the intake of vegetables, fruits, sugar, butter, calories, fats, proteins, or even the level of obesity. Within this group of works, we can highlight the papers by Cochrane et al. (1978), Crémieux et al. (1999), Miller and Frech (2002), Shaw et al. (2005), Halicioglu (2011) and Heijink et al. (2013). Sometimes, the former variables are replaced by the expenditure made on these substances or foods (Crémieux et al. 2005).

Starting from this review, it is evident that some studies based on macro data to identify the determinants of health status, have been generalized at a country (or group of countries) level. However, those works carried out at the state, regional or municipal (subnational) level are scarcer, such as Auster et al. (1969), Thornton (2002), Hill and Jorgenson (2018), referred to North American states; Crémieux et al. (1999, 2005), Tapia-Granados (2005) and Jiang et al. (2018), focused on the provinces of Canada, Spain and China, respectively; and Soares (2007), for Brazilian municipalities.

Whereas there is some consensus in the existing literature when classifying the categories of factors which determine the health status into health resources, socioeconomic factors and factors referring to lifestyle, there is more controversy when choosing the methodology to be applied, although starting from this review (see Table 7), it is found that econometric models with panel data predominate. In this way, the paper by Joumard et al. (2008) can be very useful to specify the methodology (solving some methodological questions to be taken into account in the panel data models) and the explanatory variables.[1] On the contrary, other techniques used in the specialized literature are unit root testing and cointegration analysis; Overlapping Generations Models (OLG); Auto Regressive Distributed Lag (ARDL) approach to cointegration method; methodologies proposed by Ketenci and Murthy (2018); Mariani et al. (2010); and Halicioglu (2011), respectively.

At this point, our work aims to contribute a regional approach to the existing literature, by applying for the first time a novel methodology such as the Dumitrescu-Hurlin version of the Granger causality test to the study of the conditioning factors of health status. As reviewed and following the line of Joumard et al. (2008), a possible approach would be to apply panel data and considering the data from autonomous Spanish communities.

[1]Later, the paper by Joumard et al. (2010) extended the previous work by including some efficiency measures which will not be discussed in this chapter.

3 Empirical Analysis

According to the OECD (2014), there exists a close relationship between the rates of medical consultations and the life expectancy, being observed the highest consultation rates in those countries with the highest life expectancy. This simple correlation, however, does not necessarily imply causality since the totality of living standards can influence the consultation rates and the expected life expectancy. Arvas and Torusdağ (2017) confirm the previous statement in a study carried out in Turkey (using data from the period 1975–2015), where causal relationships (according to Granger) were analyzed among the health expenditure per capita, the number of doctors per thousand inhabitants and life expectancy in different age segments, and also among its variations. Finally, Young (2001) states that, despite its (spurious) positive correlation, the availability of medical specialists has a low impact on mortality rates, compared to the economic and social variables which have been used as control variables.

This section has been structured into four different subsections in order to obtain a representation, as detailed as possible, of the evolution and causality of the variable "life expectancy at birth" in the different autonomous communities during the time horizon from 2006 to 2016.

Figure 5 summarizes graphically the steps followed to implement the empirical analysis in this section.

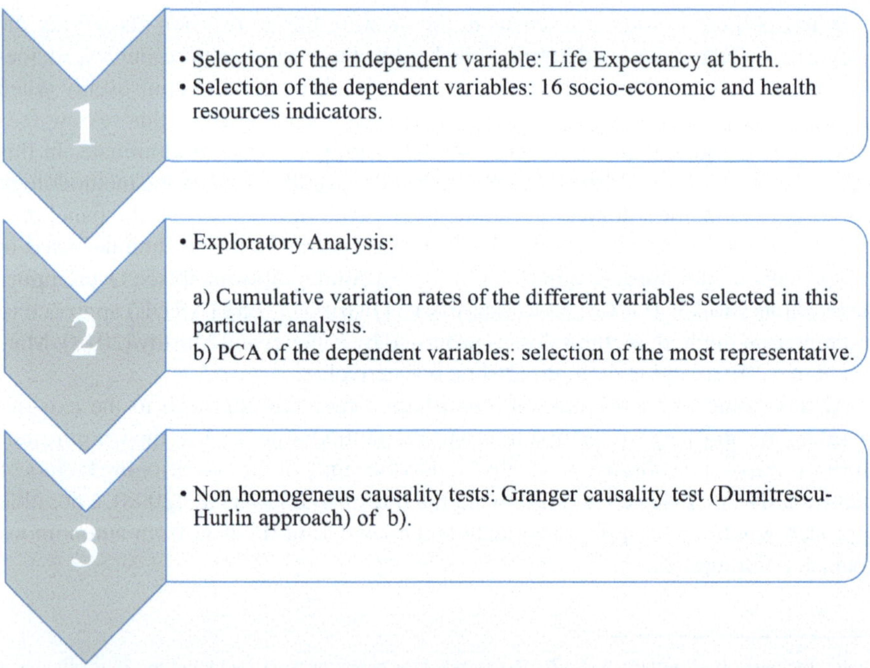

Fig. 5 Sequential development of Sect. 3. *Source* Own elaboration

3.1 Variables

In our analysis, we have used the panel data which include the variable "life expectancy at birth" and 16 other closely related socioeconomic variables, whose description is shown in Table 2, thus configuring a database composed by 3179 observations, corresponding to annual observations of the dependent and independent variables for each of the 17 autonomous communities.

In the following subsection of this chapter, we are going to carry out an exploratory analysis (EDA Methodology: Tukey 1977) based on the analysis of the annual average variation rates and the PCA, with the aim of defining the degree of interrelation between the 16 selected explanatory variables. This exploratory technique has facilitated us to identify the differences and analogies in the performance of these variables at an autonomous community level.

Later, according to the results of the PCA, most representative variables will be used for applying the Granger test according to Dumitrescu-Hurlin (DH) in which the prospective causal relationship between these and the dependent variable "life expectancy at birth" will be analyzed. In a beginning, the implementation of this test has to analytically verify the existence of a cause-effect relationship between the life expectancy and those dependent variables considered the most significant for the PCA.

With respect to the Granger test (DH), it is convenient to clarify that this is a relatively novel econometric methodology which, to the extent of our knowledge, has not yet been implemented to analyze health status in Spain, this work being the first in applying it. Probably, the same statement could be made, at a global level, regarding its use in the context of the public health care system.

3.2 Descriptive Analysis. Cumulative Variation Rates

The so-called annual cumulative variation rate (hereinafter, ACVR) of each variable $X_{i,j}$, calculated according to Eq. (1) as the geometric mean of the simple or arithmetic increments of each variable for the entire analyzed time horizon, gives us the opportunity to make a preliminary diagnosis of the evolution of "life expectancy at birth" and the explanatory variables listed in Table 2, broken down by regions.

$$\text{ACVR} = \sum_{j=1}^{k-1}\left[\left(\frac{X_{i+1,j}}{X_{i,j}}+1\right)\times\left(\frac{X_{i+2,j}}{X_{i+1,j}}+1\right)\times\cdots\times\left(\frac{X_{n,j}}{X_{n-1,j}}+1\right)\right]^{1/j} \quad (1)$$

The increase of life expectancy at birth close to 0.3% (see Table 3) justifies a positive outlook for population. In fact, with the exception of the autonomous community of Asturias, the final user of the health system shows a positive degree of satisfaction with respect to the quality of the offered services. The progressive increase in the life

Table 2 Description of the health resources, socio-economic and environmental variables

	Variables		
	Nomenclature	Short description	Source
Dependent variable	LEAB	Life expectancy at birth	Key indicators of the National Health System. Ministry of Health, Consumer Affairs and Social Welfare (Ministerio de Sanidad, Consumo y Bienestar Social, MSCBS)
Independent variables	CHRON	People with long-term disease or health problems (population percentage greater than or equal to 16 years)	Survey of Living Conditions. National Institute of Statistics (Instituto Nacional de Estadística, INE)
	PC_INCOM	Average net annual income per person and unit of consumption by autonomous communities (euros)	Survey of Living Conditions (INE)
	POVER	Poverty rate (%)	Key indicators of the National Health System (MSCBS)
	STUDI	Level of studies: population from 16 to 64 years with 2nd level of secondary and post-secondary, non-higher education (%)	Labor Force Statistics (INE)
	SENES	Senescence: population percentage greater than or equal to 65 years	Statistics of Continuous Register (INE)
	MED_SPE	Medical staff in specialized care per 1000 inhabitants	Key indicators of the National Health System (MSCBS)
	MED_PRI	Medical staff in primary care per 1000 inhabitants	Key indicators of the National Health System (MSCBS)
	NUR_SPE	Nursing staff in specialized care per 1000 inhabitants	Key indicators of the National Health System (MSCBS)

(continued)

Table 2 (continued)

	Variables		
	Nomenclature	Short description	Source
	NUR_PRI	Nursing staff in primary care per 1000 inhabitants	Key indicators of the National Health System (MSCBS)
	BEDS	Hospital beds per 1000 inhabitants	Key indicators of the National Health System (MSCBS)
	CAT	Technological endowment: computerized axial tomography (CAT) equipment per 100,000 inhabitants	Key indicators of the National Health System (MSCBS)
	NMR	Technological endowment: nuclear magnetic resonance (NMR) equipment per 100,000 inhabitants	Key indicators of the National Health System (MSCBS)
	POLL	Pollution: population suffering from pollution and other environmental problems (% of the total population)	Survey of Living Conditions (INE)
	POLL_PM10	Pollution: population (weighted average) affected by the annual average concentration of PM10 ($\mu g/m^3$)	INE Database (INE)
	UNEM	Unemployment rate by autonomous community	INE Database (INE)
	SATIS	Degree of citizens satisfaction with the operations of the public health system	Key indicators of the National Health System (MSCBS)

Source Own elaboration

expectancy may also be due, among other significant circumstances, to the decrease in some exogenous magnitudes which have a negative impact on health, such as the pollution rates at a generalized level (but not in Extremadura) or pollutants from high harmfulness like POLL_PM10. However, it is observed a continuous increment on pollution in the autonomous community of Murcia.

At a general level, it is confirmed that health authorities have chosen to increase their investment in cutting-edge technologies, a fact which can be verified by observing that CAT and NMR (endowment of CAT and NMR per 100,000 inhabitants) are two of the independent variables which show a greater increase after the variable

Table 3 Annual cumulative variation rates of the different variables selected for our analysis

C. A.	LEAB	CHRON	PC_INCOM	POVER	STUDI	SENES	MED_SPE	MED_PRI	NUR_SPE	NUR_PRI	BEDS	CAT	NMR	POLL	POLL_PM10	UNEM	SATIS
Andalucía	0,30%	2,11%	1,51%	1,50%	1,44%	1,32%	1,50%	0,00%	0,46%	0,17%	-1,56%	3,36%	6,36%	-6,58%	-2,23%	8,74%	0,52%
Aragón	0,29%	0,79%	1,95%	1,52%	2,18%	0,50%	0,52%	0,59%	2,16%	1,00%	-0,65%	2,51%	6,11%	-7,02%	-9,43%	10,49%	0,76%
Asturias	0,29%	1,79%	2,63%	0,30%	1,32%	1,15%	1,58%	0,81%	2,41%	0,30%	-1,01%	0,83%	8,19%	-1,71%	-4,73%	4,85%	-0,53%
Baleares	0,26%	2,70%	1,74%	1,70%	1,78%	1,26%	1,08%	0,33%	1,14%	0,62%	-0,32%	0,22%	5,76%	-4,25%	-4,14%	8,09%	0,40%
Canarias	0,39%	5,33%	1,58%	1,76%	1,67%	2,37%	1,82%	0,98%	1,88%	0,68%	-1,46%	1,18%	3,75%	-0,19%	-6,44%	8,15%	1,00%
Cantabria	0,32%	5,04%	1,19%	2,63%	1,62%	1,10%	0,46%	0,39%	1,15%	0,89%	-1,01%	2,17%	9,10%	-7,61%	-4,93%	8,18%	0,51%
Castilla y León	0,28%	-0,56%	2,46%	-3,24%	1,42%	0,75%	1,99%	0,09%	1,14%	0,23%	-1,48%	4,67%	11,18%	-12,94%	-6,00%	6,93%	0,77%
Castilla-La Mancha	0,28%	6,29%	1,84%	1,14%	2,35%	-0,11%	1,54%	0,50%	1,68%	0,56%	-1,00%	3,29%	0,92%	-7,02%	-5,26%	10,60%	0,08%
Cataluña	0,31%	1,23%	2,28%	-0,15%	-0,24%	1,30%	-0,30%	-0,14%	1,39%	-0,44%	-1,21%	3,72%	10,99%	-7,69%	-6,75%	8,51%	0,27%
Extremadura	0,27%	6,88%	2,66%	-2,08%	2,23%	0,48%	1,42%	0,46%	1,92%	0,48%	-0,54%	4,48%	1,96%	4,59%	-3,96%	8,14%	0,57%
Galicia	0,26%	3,39%	2,46%	-1,22%	1,26%	1,29%	1,16%	0,76%	1,48%	1,81%	-0,86%	1,20%	2,20%	-2,86%	-5,72%	7,53%	1,41%
Madrid	0,38%	3,71%	1,66%	4,34%	-0,51%	1,91%	2,46%	0,45%	1,14%	0,00%	-1,25%	2,56%	4,22%	-3,95%	-6,26%	8,50%	0,68%
Murcia	0,33%	1,76%	0,96%	0,53%	0,85%	1,18%	1,82%	1,76%	1,30%	1,73%	0,35%	2,05%	11,36%	-3,57%	1,26%	9,00%	0,98%
Navarra	0,22%	0,81%	1,22%	-1,53%	1,02%	1,08%	1,60%	0,52%	0,42%	0,56%	-0,78%	3,72%	4,06%	-8,53%	-9,63%	8,11%	1,01%
País Vasco	0,31%	3,30%	3,18%	-1,44%	0,99%	1,75%	3,64%	2,29%	5,38%	2,29%	-1,40%	2,62%	5,04%	-13,91%	-6,56%	5,89%	0,04%
La Rioja	0,34%	2,11%	2,66%	-6,17%	-0,70%	1,07%	1,46%	1,01%	0,65%	3,16%	0,35%	4,92%	2,60%	-8,62%	-5,51%	4,71%	0,37%
Valencia	0,32%	3,71%	0,49%	2,21%	1,17%	1,70%	2,44%	0,57%	1,06%	0,32%	-0,53%	1,89%	4,27%	-1,74%	-5,99%	8,49%	0,37%
Average period (2006-2016)	**0,30%**	**2,96%**	**1,91%**	**0,11%**	**1,17%**	**1,18%**	**1,54%**	**0,67%**	**1,57%**	**0,85%**	**-0,84%**	**2,67%**	**5,77%**	**-5,51%**	**-5,43%**	**7,94%**	**0,54%**

Source Own elaboration

UNEM, obviously influenced by the effect of the recent economic crisis. However, the negative economic situation has not been enough to decrease the number of nurses in specialized care (NUR_SPE) or in primary care (NUR_PRI) in all autonomous communities, with Catalonia being the only community that, in this way, shows a slight decrease. Of the same manner and except for Catalonia, the variation of variables MED_PRI and MED_SPE are also positive, being necessary to point out the effort made, in particular, by the Basque Public Health Entity (Osakidetza).

The cutback in the health sector has probably been most important for the provision of hospital beds (BEDS), which can be justified because health authorities have been forced to restructure and reorganize their corresponding hospital units. In effect, only the autonomous communities of La Rioja and Murcia exhibit positive increments of this item.

The slow increase in the variable "life expectancy at birth" has obviously been accompanied by a rise practically in unison of the variable SENES (with the unique exception of Castilla-La Mancha) and the same could be said about the variable CHRON which, only, decreases in Castilla-León. In any case, the increase in this subtype of diseases, taking into account the way in which CHRON has been quantified, should serve to denote the efficiency of health systems from a non-curative but palliative perspective. Hence, to a certain extent, it can be considered and even valued as a success of public health system.

Likewise, we can observe a generalized increase, at a national level, in per capita income, whilst the rate of poverty increased in a great part of the autonomous communities. Both extremes are not incompatible and, in fact, have an easy explanation. First of all, we must underline that the 2006–2007 biennium has been included in this analysis, and that in this period the harshness of the economic crisis had not yet affected the whole population. On the other hand, the supposed post-crisis economic improvement did not reach the entire population or did not materialize in a palpable improvement in the equidistribution of per capita income (Jurado and Pérez 2014).

On the other hand, it is noteworthy to point out that the variable "level of studies" (STUDI) has been increasing in almost all of autonomous communities, although it reflects negative growth rates in some specific geographical areas which, traditionally, have been Spanish areas of growing economy (for example, Madrid, Cataluña and La Rioja).

As observed in Fig. 13 (see Appendix), the variable "life expectancy at birth" has been increasing during the analyzed period since in the considered 11 years, the average life expectancy in Spain has increased more than 2 years, showing only a slight setback in the period 2014–2015. Moreover, it is observed that the behavior of this variable is not uniform in all autonomous communities, oscillating (in average values) between the minimum registered in Andalucía and the maximum value corresponding to the autonomous community of Madrid.

A priori, it seems that life expectancy at birth is heterogeneous and depends on the differences between the autonomous communities. In effect, one might think that there is a certain degree of correlation between the variables "per capita income" and "life expectancy" in all autonomous communities. However, Table 4 certifies that this

Table 4 Correlation coefficient of Pearson by autonomous communities (LEAB versus PC_INCOM and LEAB versus UNEM)

Autonomous community	ρ_{LEAB,PC_INCOM}	$\rho_{LEAB,UNEM}$
Andalucía (AND)	0.14126555	0.866003
Aragón (ARA)	0.652852855	0.7811751
Asturias (AST)	0.44765929	0.7942051
Baleares (BAL)	0.16751285	0.4544221
Canarias (CANA)	0.37959946	0.8414974
Cantabria (CANT)	0.077689567	0.8895257
Castilla-León (CL)	0.68040438	0.8800386
Castilla-La Mancha (CM)	0.34886274	0.9254192
Cataluña (CAT)	0.59132869	0.7719413
Extremadura (EXT)	0.50793988	0.9147664
Galicia (GAL)	0.83210593	0.9241171
Madrid (MAD)	0.585825479	0.8766449
Murcia (MUR)	−0.20180879	0.7998938
Navarra (NAV)	0.75345885	0.8593637
País Vasco (PVA)	0.6399369	0.8549346
La Rioja (LRI)	0.711399574	0.7129911
Valencia (VAL)	−0.07230718	0.7831451

Source Own elaboration

link is weak enough, if not negligible, in terms of the linear correlation coefficient of Pearson (denoted by ρ). In the same way, we precisely find the opposite effect, viz a strong linear relationship between the variable "life expectancy at birth" and the unemployment rate which, without being a general rule, is especially noteworthy in those autonomous communities with a lower per capita income (i.e., Castilla-La Mancha, Galicia or Extremadura).

In order to obtain a greater detail of the variables included in this work, Table 6 in Appendix lists their main descriptive statistics, by detailing, among other aspects, their coefficients of asymmetry and kurtosis, and the degree of compliance of the hypotheses of normality according to the Jarque-Bera test. These parameters have been disaggregated for Spain, for a community with a high per capita income (País Vasco) and for another with a significantly lower level of income per inhabitant (Andalucía). Figure 6, in which the original values of the variables have been rescaled[2] to carry out their joint comparison, offers a clear picture of the performance of each variable during the period 2006–2016 in each autonomous community.

[2]The average values of all variables included in Table 6 have been multiplied by three numerical constants: SATIS, POLL, POLL_PM10, UNEM, CHRON, STUDI, POVER and SENES by 10; BEDS, NMR, CAT, NUR_SPE, NUR_PRI, MED_SPE and MED_PRI, by 100; and, finally, PC_INCOM by 1/100.

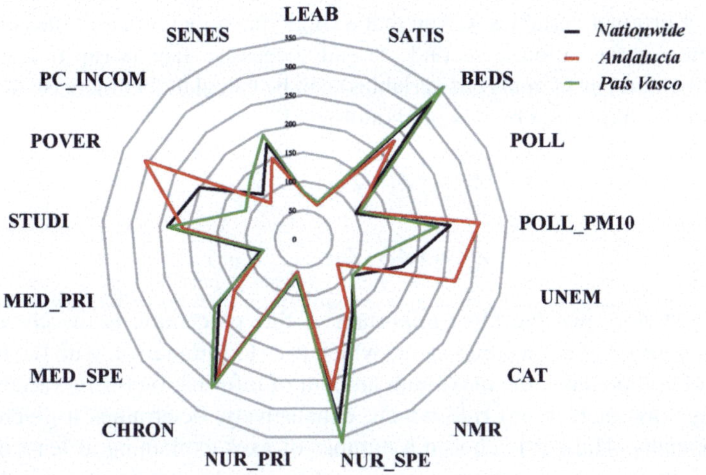

Fig. 6 Diagram of Kiviat of the considered variables, corresponding to the average values of Spain, Andalucía and País Vasco. *Source* Own elaboration

3.3 Principal Components Analysis

Starting from the high number of explanatory variables initially considered (16), the PCA (Pla 1986) gives the possibility to visualize, summarize and condense their information in a smaller number of new variables which, generically, are called "components".

The PCA is based on the following approach: If a variable is assumed to be a function of another or other ones, it should have to include some information considered as redundant (Zuur et al. 2007). Therefore, a priori, if p variables are strongly correlated, it is logical their replacement by a smaller number of them, without this fact supposing a significant loss of the information provided by those variables. Therefore, if Y_{ij} is the value of the ith observation ($i = 1, 2, \ldots, M$) corresponding to the variable j ($j = 1, 2, \ldots, N$), we propose as a starting hypothesis that the linear combination (2) will synthesize the values of the original variable Y_{ij} depending on some constants $c_{i,j}$ known as "charges":

$$Z_{i1} = c_{11}Y_{i1} + c_{12}Y_{i2} + \cdots + c_{1N}Y_{iN}. \quad (2)$$

Therefore, each linear combination, represented by a vector of dimension M, is named indistinctly main component[3] (3), axis or gradient, being considered that it summarizes the "most important" characteristics and the information of the N

[3]Observe that, for this reason and due to purely explanatory purposes, we have simplified the PCA implementation, for example, in which concerns the transformation of the original data (standardization) or the use of the matrix of correlations and covariances when determining the eigenvalues of a given dataset. With respect to this work, the correlation matrix has been implemented which allows rescaling the data, thus avoiding the problems derived from the use of variables with different

primary or original variables. Given that a single main factor determines only a part of the initial information, a second (4) and successive factors (up to a maximum equal to the number of analyzed variables) can be calculated to increase the level of information summarized by this methodology:

$$Z_1 = (Z_{11}, Z_{21}, \ldots, Z_{M1}) \tag{3}$$

$$Z_2 = (Z_{12}, Z_{22}, \ldots, Z_{M2}) \tag{4}$$

The next step, just the most important of this procedure, is the choice of the optimum number of main components which tries to summarize, with the minimum number of components, the maximum amount of information (or its variability).

In this way, there is no rule which, exhaustively, determines a given number of components. Thus, it is chosen a number of axes explaining at least over 80% of the information originally explained by the variables, or alternatively five main components (Kaiser 1961).

Another alternative is given by the "Scree Plot", by means of which the number of eigenvalues to be retained by the analyst is visually and intuitively selected. From this point, the PCA provides a number of techniques directly aimed at defining the degree of correlation between the variables and the common patterns described by the analyzed individuals. In this work, we have used punctually, for restrictions of space, only some of them (Figs. 7, 8, 9, 10 and 11).

In order to implement the PCA to the considered independent variables (see Table 2), we have transformed the original dataset into a new matrix which includes

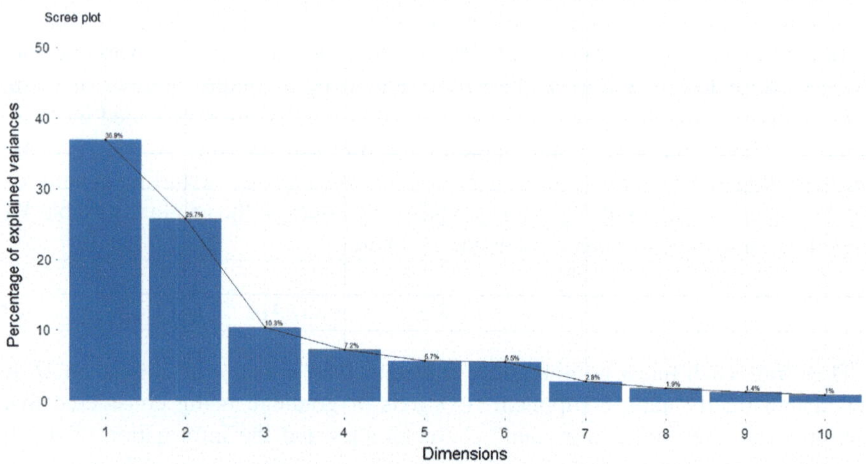

Fig. 7 PCA. Scree plot. *Source* Own elaboration

units of measure. In this way, overweighting some variables over other ones has been avoided when explaining the total variability of the dataset.

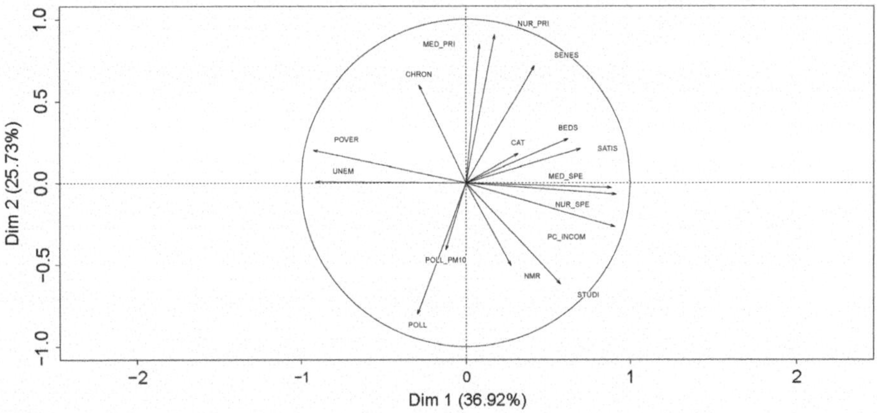

Fig. 8 PCA. *Biplot* of the analyzed variables. *Source* Own elaboration

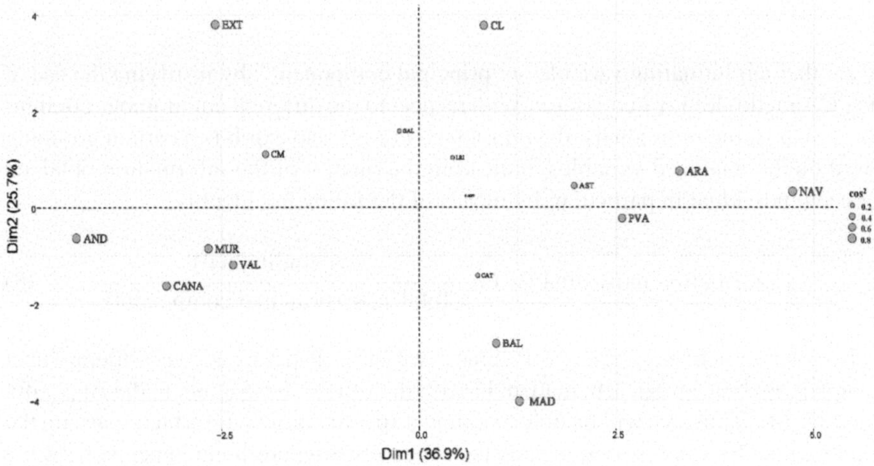

Fig. 9 PCA. *Biplot* of the analyzed individuals (I). *Source* Own elaboration

the average values of each variable during the period 2006–2016. This matrix is composed by 272 observations from 17 individuals (in this case, autonomous communities) and 16 variables (the average values of the aforementioned independent variables). Figure 7 shows that only the first two components are able to explain about 62.64% of the overall variability of data, by opting to select the first 5 components which show a greater explanatory power, namely 85.95%.

On the other hand, Figs. 8 and 9 represent, respectively, the *biplots* of both independent variables and individuals. In the first case, a high correlation is observed, which predicts a probable redundancy of the information included in the variables. Such are the cases of POVER and UNEM, MED_SPE and NUR_SPE, or BEDS and SATIS. This fact implies that the original variables can be conveniently substituted

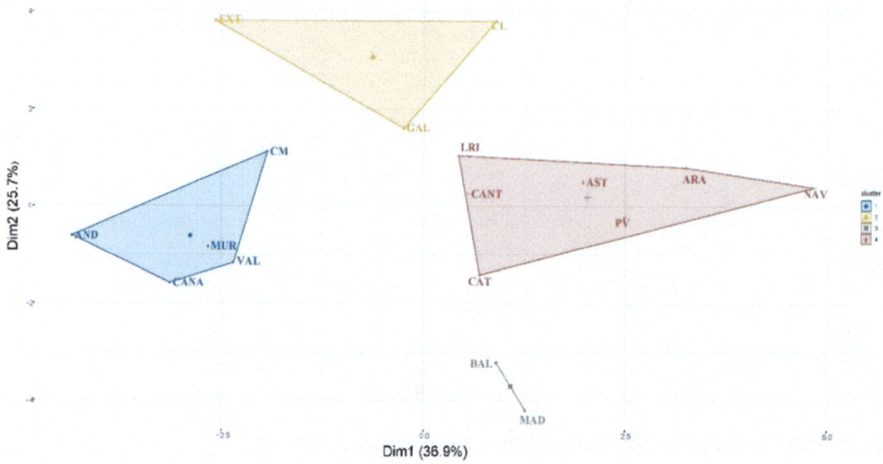

Fig. 10 PCA. *Biplot* of the analyzed individuals (II). Grouped by clusters. *Source* Own elaboration

by another agglutinating variable or "principal component", by justifying the use of the PCA methodology in this case. With respect to the different autonomous communities, that is, the individuals, the *biplot* (see Fig. 9) also exhibits a certain grouping based on the analyzed variables, indicating the quality of the information obtained for each individual in particular, by means of the following identity:

$$\text{Contribution individual } (\%) = \frac{\cos^2(\text{individual})}{\text{Total } \cos^2(\text{principal component})} \tag{5}$$

In order to achieve a greater accuracy, we have also carried out a hierarchical grouping by three levels, schematizable according to the dendogram in the Appendix (see Fig. 14). Thus, we will be able to obtain a specific basis with which to group the communities by establishing groups that, theoretically, are homogeneous between them and more heterogeneous with respect to the rest. Such a conglomeration of individuals according to their common characteristics is shown in Fig. 10.

Surprisingly, Fig. 10 shows that there is a certain geographical cohesion in the formed groups in such a degree that each of them can be perfectly assimilated to a particular area or region given, except the insular autonomous communities:

1. Southern and Central-Eastern area: Andalucía, Murcia, Castilla-La Mancha, Valencia and Canarias.
2. Western area: Extremadura, Castilla-León and Galicia.
3. Northern and North-Eastern area: Navarra, País Vasco, Aragón, Asturias, Cantabria, La Rioja and Cataluña.
4. Rest of communities: Madrid and Islas Baleares.

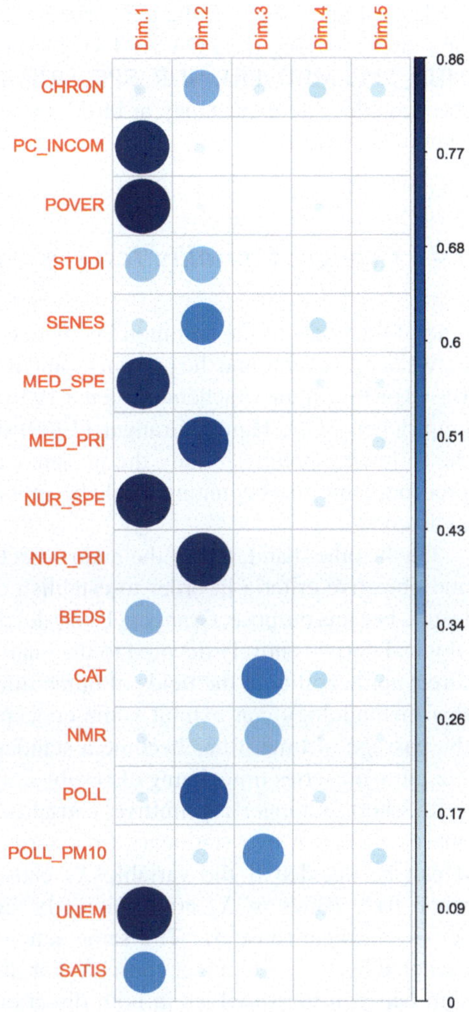

Fig. 11 PCA. Degree of correlation between the analyzed variables and the first five resulting dimensions. *Source* Own elaboration

We must insist on the fact that, except for the last group, everything seems to indicate that the clusters formed from a purely hierarchical grouping correspond to autonomous communities relatively close or with strong cultural ties between them, of course, this appreciation being always understood from a relative point of view.

The choice of variables which, to a greater extent, summarize the analyzed information can be done with different criteria: Either by extracting those variables which have a greater impact on the first two components or by choosing a predetermined number of them. For our analysis, we have opted to select those variables which present a greater degree of correlation with the first 5 components, as summarized in Fig. 11, in which correlations appear ordered by variable in a decreasing order.

According to Fig. 11, the ten variables which optimally synthesize the information after applying the PCA are CHRON, PC_INCOM, POVER, STUDI, SENES, MED_SPE, MED_PRI, NUR_SPE, NUR_PRI and BEDS, which will be linked to the variable "life expectancy at birth" by using the Granger causality test (version Dumitrescu-Hurlin).

3.4 Granger Causality Test (Version Dumitrescu-Hurlin)

One of the goals of the common law is to establish some cause-effect relationships between past events and the present moment which may be objectively generalizable. Based on this intuitive scheme (Wiener 1956), clearly influenced by the philosophical empiricism of D. Hume, Granger (1969) designed an econometric test based on the following premise: "only the present can be the ultimate cause of the past", a procedure that, at a beginning, could be applied to the joint analysis of life expectancy indicators.

On the other hand, one of the main objectives of Science is to implement rational and objective criteria in order to establish cause-effect relationships which can be tested. For this purpose, Granger (1969) developed a causality test which is a methodological corpus entirely devoted to the analysis of causality between time series, of direct applicability to the fields of humanities and experimental sciences. Although this methodology can exhibit some conceptual weaknesses (Maziarz 2015), over the passage of time it has become a standard widely accepted by the econometric literature on series time, being also subject to multiple revisions or extensions.

In short, Granger's primitive causality test starts from the following basic approach: Given two stationary time series X_t and Y_t, related by the model (6), it can be decided if the variable X_t causes Y_t in the sense of Granger (1969) when past values of X_t are sufficiently significant predictors of Y_t (denoted by $X_t \longrightarrow$ Granger causesY_t) The same scheme can be applied in the opposite way (denoted by $Y_t \longrightarrow$ Granger causesX_t) or under the premise that there is a relationship between the variables in both directions, in which case we would be talking about a bidirectional causality in the sense of Granger or "feed-back" (see López and Weber 2017).

Finally, determining the causality in Granger's sense in the model:

$$y_t = \alpha + \sum_{k=1}^{K} \beta_k y_{t-k} + \sum_{k=1}^{K} \gamma_k x_{t-k} + \varepsilon_t. \tag{6}$$

is subject to the implementation of a Fischer test, consisting in a hypothesis test for the following null hypothesis:

$$H_0 : \gamma_1 = \cdots = \gamma_K = 0. \tag{7}$$

One extension of the Granger causality test arises from its implementation to panel-type data, research initiated by Holtz-Eakin et al. (1988) when proposing the model with fixed effects (colloquially called *stacked pairwise Granger causality test*):

$$y_{i,t} = \alpha + \sum_{k=1}^{K} \beta_{ik} y_{t-k} + \sum_{k=1}^{K} \gamma_{ik} x_{t-k} + \varepsilon_{i,t} \quad (8)$$

and the hypothesis is null derived from it:

$$H_0 : \gamma_{i1} = \gamma_{i2} = \cdots = \gamma_{iK} = 0, \quad (9)$$

by establishing the equality between the coefficients corresponding to all individuals included in the panel, that is, throughout all cross observations.

In any case, according to Campo and Mendoza (2018), the previous way to conceptualize Granger causality, limited to panel-type data, presents some weaknesses related with the viability of applying this test which depends on the different sizes of N and T (Nickell 1981; Hurlin and Venet 2001) and on the significance of the null hypothesis (9) given that, according to Granger (2003), to consider that one variable causes another one in a panel of data, involves per se adding a null hypothesis to the one already proposed.

To solve this inconvenience, Dumitrescu and Hurlin (2012) designed a homogeneous non-causality test, derived from (8) in which a *balanced panel* composed by $X_{i,t}$ and $Y_{i,t}$, that is to say, the observations of two stationary variables corresponding to an individual i in a period t and with a number of delays K exactly the same for all individuals, it is possible to violate the "classical" assumption (heterogeneous causality test) by allowing coefficients γ_{ik}, being invariant in time, can differ between individuals, thus arriving at a new null and alternative hypotheses:

$$\text{Homogeneous non-causality test} \begin{cases} H_0 : \gamma_{i1} = \gamma_{i2} = \cdots = \gamma_{iK} = 0; \; i = 1, \ldots, N \\ H_1 : \begin{cases} \gamma_{i1} = \gamma_{i2} = \cdots = \gamma_{iK} = 0; \; i = 1, \ldots, N_1 \\ \gamma_{i1} \neq 0 \text{ or } \ldots \text{ or } \gamma_{iK} \neq 0; \; i = N_1, \ldots, N \end{cases} \end{cases} \quad (10)$$

Observe that the alternative hypothesis establishes the plausible causality between variables X_t and Y_t for some individuals, not for all of them, so rejecting H_0 does not necessarily imply the existence of causality between some individuals (López and Weber 2017). This hypothesis is well known as the homogeneous non-causality hypothesis and differs from the hypothesis (9) in that the alternative hypothesis allows the causality X_t to Y_t for some individuals, but not for everyone.

However, (10) presents some handicaps since a priori N_1 ($N_1 = 1, \ldots, N - 1$) is not known. If $N_1 = 0$, it is assumed the existence of causality for all individuals of the panel and, on the other hand, N_1 must be strictly less than N, because, otherwise, there would be no causality for all individuals and H_1 would be constrained to H_0.

Dumitrescu and Hurlin (2012) solve these drawbacks by calculating the standard regressions of the Granger causality for each individual in order to obtain the individual Wald statistics and, starting from it, determine the statistic $\overline{W}$ (average Wald or W-bar):

$$\overline{W} = W_{N,t}^{HNC} = \frac{1}{N}\sum_{i=1}^{N} W_{i,t},$$

where $W_{i,t}$ is the Wald statistic corresponding to the ith individual in t according to the null hypothesis $H_0 : \gamma_{i,t} = 0$. In the same way, taking into account the hypothesis that the different Wald statistics are independent and identically distributed for each individual, it can be demonstrated that the standardized statistics follow a standard normal distribution, when T and N tend to infinity:

$$\overline{Z} = (\overline{W} - K)\sqrt{\frac{N}{2K}} \xrightarrow[T,N\to\infty]{d} N(0, 1) \tag{11}$$

Moreover, given a fixed dimension of T, $T > 5 + 3K$, the standardized statistic $\tilde{Z}$ also follows a standard normal distribution:

$$\tilde{Z} = \frac{T - 3K - 3}{T - 3K - 1}(\overline{W} - K)\sqrt{\frac{N}{2K}\frac{T - 3K - 5}{T - 2K - 3}} \xrightarrow[N\to\infty]{d} N(0, 1). \tag{12}$$

This last observation conditions the number of delays to be selected when applying the homogeneous non-causality test (Dumitrescu-Hurlin), although originally, as López and Weber (2017) indicate, Dumitrescu and Hurlin (2012) did not specify a given or optimal number of delays, so a viable solution might be the use of estimators as the Akaike information or the Hannan-Quinn Bayesian criteria.

Table 5 Granger causality

Null hypothesis	W-Stat.	Zbar-Stat.	Prob.
BEDS does not homogeneously cause LEAB	1.92798	0.77748	0.4369
NUR_SPE does not homogeneously cause LEAB	2.22473	1.21445	0.2246
NUR_PRI does not homogeneously cause LEAB	2.00144	0.88564	0.3758
CHRON does not homogeneously cause LEAB	0.71378	−1.01049	0.3123
MED_SPE does not homogeneously cause LEAB	3.31763	2.82378	0.0047
MED_PRI does not homogeneously cause LEAB	2.40673	1.48244	0.1382
STUDI does not homogeneously cause LEAB	1.12455	−0.40561	0.685
POVER does not homogeneously cause LEAB	0.92645	−0.69733	0.4856
PC_INCOM does not homogeneously cause LEAB	2.7489	1.98631	0.047
SENES does not homogeneously cause LEAB	0.50179	−1.32265	0.186

Source Own elaboration

To conclude, testing the null hypothesis (10) is performed as a function of $\overline{Z}$ and $\tilde{Z}$, by comparing them with the standard critical values (see Table 5). Thus, the null hypothesis would be rejected when $\overline{Z}$ or $\tilde{Z}$ are greater than the standard critical values, taking into account that, for panel data in which N and T are relatively large, it is more convenient to test the null hypothesis by means of $\overline{Z}$, remaining the use of $\tilde{Z}$ for those cases in which N is relatively large and T is relatively small. In this case, we can conclude that there is Granger causality according to the version by Dumitrescu and Hurlin (2012).

4 Conclusions and Future Research

In this chapter, we have shown that, according to the Granger causality test for panel data (Dumitrescu-Hurlin version), the explanatory variables BEDS, PC_INCOM, MED_PRI, MED_SPE, NUR_PRI and NUR_SPE causes the life expectancy at birth (see Fig. 12). Of course, this result is very important for the design of health policies in Spain as it shows the main factors over which it is necessary to stress to increase the life expectancy at birth.

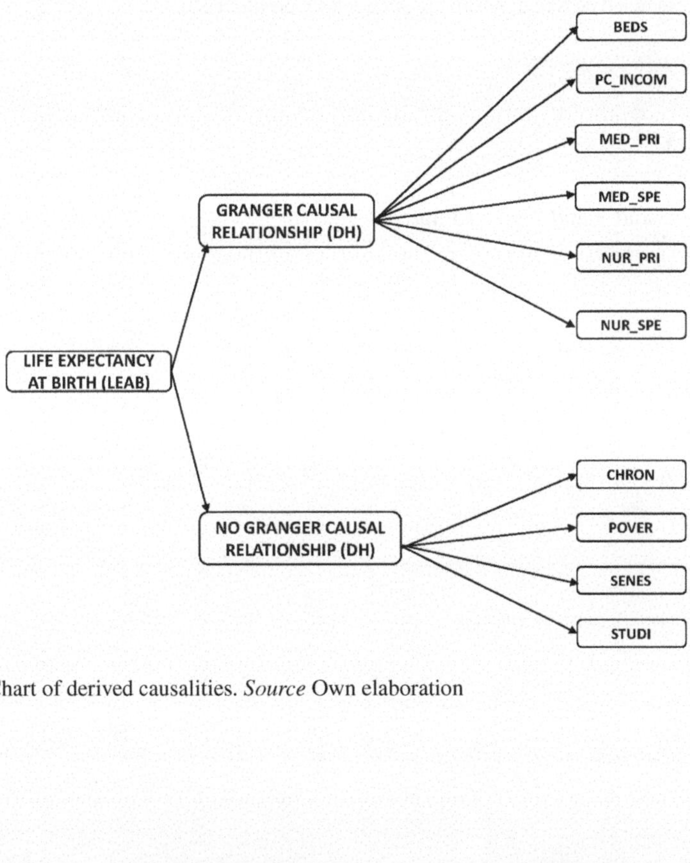

Fig. 12 Chart of derived causalities. *Source* Own elaboration

The main contribution of this work is twofold. On the one hand, this is the first time that the Granger causality test (Dumitrescu-Hurlin version) has been applied to a causal analysis within the health sector. This contribution becomes more relevant if we take into account the few existing works about the causality of life expectancy at birth in a country. On the other hand, we have to point out that another added value of this study is the use of panel data, which is completely necessary to deal with disaggregated data by autonomous communities.

However, the main concern of this work is the limited number of available observations, which has resulted in a reduction to one of the number of delays. Despite this limitation and in line with other authors, this chapter is a relevant contribution to the evaluation of the impact on health and a useful tool for its possible implementation in public health programs.

Finally and as future research, we intend to deepen this type of contrast by using a database with a longer time horizon and incorporating new possible explanatory factors of the life expectancy at birth.

Acknowledgements We would like to thank the authors of the Granger causality test (DH), E. I. Dumitrescu (Université Paris Nanterre) and C. Hurlin (Université d'Orléans) as well as S. Weber (Université de Neuchâtel) for their wise advices which, to a great extent, have been taken into account to write this work. The work presented in this paper has been supported by the Ministry of Economy and Competitiveness of Spain (research project No. DER2016-76053R: "La sostenibilidad del Sistema Nacional de Salud: reformas, estrategias y propuestas").

Appendix

See Tables 6 and 7 and Figs. 13 and 14.

Table 6 Descriptive statistics

Panel I: Spain								
Nation wide	LEAB	SATIS	BEDS	POLL	POLL_PM10	UNEM	NMR	CAT
Mean	82.463	6.524	3.239	9.963	25.332	17.583	1.056	1.581
Median	82.490	6.490	3.300	9.200	23.870	17.170	1.020	1.590
Maximum	85.220	7.550	4.350	23.800	54.330	36.260	1.950	2.190
Minimum	79.770	5.150	2.170	1.400	12.100	4.280	0.360	0.960
Std. Dev.	1.122	0.431	0.494	4.824	7.133	7.459	0.311	0.223
Skewness	−0.188	−0.239	−0.491	0.416	1.133	0.306	0.356	−0.274
Kurtosis	2.673	3.113	2.643	2.526	5.037	2.433	3.387	3.071
Jarque-Bera test	1.932	1.879	8.522	7.153	72.363	5.417	5.125	2.388
Probability	0.381	0.391	0.014	0.028	0.000	0.067	0.077	0.303
Sum	15,420.540	1220.060	605.780	1863.100	4737.010	3288.070	197.420	295.560
Sum Sq. Dev.	234.110	34.517	45.342	4328.295	9464.221	10,349.780	18.007	9.278
Observations	187	187	187	187	187	187	187	187

Panel I: Spain									
Nation wide	NUR_SPE	NUR_PRI	CHRON	MED_SPE	MED_PRI	STUDI	POVER	PC_INCOM	SENES
Mean	3.248	0.659	29.259	1.844	0.778	23.701	20.132	10,542.050	17.738
Median	3.160	0.660	28.800	1.790	0.760	23.700	18.100	10,493.000	17.800
Maximum	4.930	0.890	45.900	2.550	1.100	29.700	40.200	14,838.000	24.100

(continued)

Table 6 (continued)

Panel I: Spain

Nation wide	NUR_SPE	NUR_PRI	CHRON	MED_SPE	MED_PRI	STUDI	POVER	PC_INCOM	SENES
Minimum	2.530	0.470	16.100	1.370	0.590	16.200	5.300	6668.000	12.100
Std. Dev.	0.502	0.102	5.874	0.234	0.104	2.560	7.732	1880.223	2.904
Skewness	1.015	0.357	0.219	0.624	1.370	−0.336	0.440	0.271	0.223
Kurtosis	3.561	2.583	2.840	2.902	5.597	3.061	2.226	2.238	2.253
Jarque-Bera test	34.549	5.319	1.694	12.221	111.039	3.547	10.699	6.814	5.891
Probability	0.000	0.070	0.429	0.002	0.000	0.170	0.005	0.033	0.053
Sum	607.360	123.260	5471.500	344.740	145.520	4432.000	3764.700	1,971,364	3317
Sum Sq. Dev.	46.938	1.920	6417.751	10.219	2.026	1219.030	11,118.710	658,000,000	1568.440
Observations	187	187	187	187	187	187	187	187	187

Panel II: Andalucía

Andalucía	LEAB	SATIS	BEDS	POLL	POLL_PM10	UNEM	NMR	CAT
Mean	81.035	6.365	2.306	10.700	30.373	27.040	0.728	1.445
Median	81.160	6.320	2.270	10.000	29.180	28.250	0.710	1.510
Maximum	82.180	6.620	2.540	15.500	40.630	36.260	1.000	1.670
Minimum	79.770	6.070	2.170	7.400	24.970	12.220	0.540	1.200
Std. Dev.	0.882	0.157	0.137	2.814	4.729	8.113	0.158	0.148
Skewness	−0.252	0.083	0.633	0.752	0.919	−0.738	0.311	−0.297
Kurtosis	1.692	2.624	1.966	2.232	3.020	2.392	1.830	1.959
Jarque-Bera test	0.901	0.077	1.225	1.307	1.548	1.166	0.805	0.658
Probability	0.637	0.962	0.542	0.520	0.461	0.558	0.669	0.720

(continued)

Table 6 (continued)

Panel II: Andalucía

Andalucía	LEAB	SATIS	BEDS	POLL	POLL_PM10	UNEM	NMR	CAT
Sum	891.390	70.010	25.370	117.700	334.100	297.440	8.010	15.900
Sum Sq. Dev.	7.777	0.248	0.187	79.180	223.664	658.139	0.250	0.218
Observations	11	11	11	11	11	11	11	11

Panel II: Andalucía

Andalucía	NUR_SPE	NUR_PRI	CHRON	MED_SPE	MED_PRI	STUDI	POVER	PC_INCOM	SENES
Mean	2.661	0.577	28.455	1.498	0.727	20.927	30.682	8465.091	15.064
Median	2.630	0.580	29.100	1.540	0.730	21.100	29.500	8408.000	15.000
Maximum	2.910	0.610	33.500	1.590	0.750	22.600	35.700	9406.000	16.300
Minimum	2.570	0.530	23.100	1.370	0.710	19.500	27.900	7231.000	14.200
Std. Dev.	0.097	0.026	3.325	0.075	0.010	1.190	2.844	676.235	0.742
Skewness	1.696	−0.359	−0.289	−0.563	0.567	0.054	0.860	−0.277	0.367
Kurtosis	5.043	2.063	1.991	1.751	3.868	1.558	2.224	2.134	1.739
Jarque-Bera test	7.185	0.639	0.620	1.297	0.935	0.958	1.632	0.484	0.977
Probability	0.028	0.727	0.733	0.523	0.627	0.619	0.442	0.785	0.614
Sum	29.270	6.350	313.000	16.480	8.000	230.200	337.500	93,116	165.70
Sum Sq. Dev.	0.094	0.007	110.547	0.057	0.001	14.162	80.856	4,572,933	5.51
Observations	11	11	11	11	11	11	11	11	11

Panel III: País Vasco

País Vasco	LEAB	SATIS	BEDS	POLL	POLL_PM10	UNEM	NMR	CAT
Mean	82.883	6.784	3.580	10.782	23.179	12.046	1.051	1.687
Median	82.950	6.830	3.630	11.600	24.700	12.270	0.980	1.740
Maximum	83.930	6.960	3.800	16.100	30.110	16.600	1.430	1.890

(continued)

Table 6 (continued)

Panel III: País Vasco

País Vasco	LEAB	SATIS	BEDS	POLL	POLL_PM10	UNEM	NMR	CAT
Minimum	81.610	6.480	3.300	3.600	15.280	5.900	0.790	1.460
Std. Dev.	0.840	0.132	0.189	4.555	5.520	3.768	0.242	0.155
Skewness	−0.219	−0.993	−0.342	−0.347	−0.323	−0.254	0.361	−0.147
Kurtosis	1.654	3.635	1.481	1.720	1.537	1.946	1.629	1.479
Jarque-Bera test	0.919	1.992	1.272	0.971	1.172	0.627	1.100	1.101
Probability	0.632	0.369	0.529	0.615	0.557	0.731	0.577	0.577
Sum	911.710	74.620	39.380	118.600	254.970	132.510	11.560	18.560
Sum Sq. Dev.	7.057	0.173	0.357	207.436	304.681	141.948	0.587	0.239
Observations	11	11	11	11	11	11	11	11

Panel III: País Vasco

País Vasco	NUR_SPE	NUR_PRI	CHRON	MED_SPE	MED_PRI	STUDI	POVER	PC_INCOM	SENES
Mean	3.594	0.630	30.345	1.934	0.717	23.500	11.155	13,640.000	19.364
Median	3.840	0.640	29.300	2.110	0.780	23.600	10.500	14,046.000	19.200
Maximum	4.420	0.710	40.100	2.220	0.790	24.500	13.800	14,838.000	21.300
Minimum	2.570	0.550	23.500	1.490	0.610	22.200	9.000	10,493.000	17.900
Std. Dev.	0.745	0.062	5.307	0.279	0.078	0.832	1.491	1351.356	1.219
Skewness	−0.380	−0.054	0.453	−0.602	−0.335	−0.434	0.551	−1.584	0.281
Kurtosis	1.471	1.272	2.200	1.599	1.258	1.858	2.194	4.000	1.648
Jarque-Bera test	1.336	1.374	0.670	1.564	1.597	0.943	0.855	5.056	0.983
Probability	0.513	0.503	0.715	0.457	0.450	0.624	0.652	0.080	0.612
Sum	39.530	6.930	333.800	21.270	7.890	258.500	122.700	150,040	213
Sum Sq. Dev.	5.551	0.039	281.607	0.779	0.062	6.920	22.227	18,261,628	14.87
Observations	11	11	11	11	11	11	11	11	11

Source Own elaboration

Table 7 Empirical studies on the determinants of health status of population

Author (date)	Country	Period	Proxy of health status[a]	Methodology and data
Auster et al (1969)	US states	1967	M	Cross-section regressions
Cochrane et al. (1978)	18 developed countries	1970 (1969 or 1971)	MAG	Cross-section regressions
Rodgers (1979)	56 countries (developed and developing countries)	Not specified	LEAB and IM	Cross-section regressions
Wolfe and Gabay (1987)	22 OECD countries	1960, 1970, 1980	LEAB, LE 60 and IM	Linear structural relations for simultaneous models using cross-section data
Peltzman (1987)	22 middle-income countries	1970–1980	M	Cross-section regressions
McAvinchey (1988)	5 European countries	1960–1982	M	Almon Distributed Lag (ADL) approach using time series data
Hitiris and Posnett (1992)	28 OECD countries	1960–1987	M	Panel data regressions
Grubaugh and Rexford (1994)	12 OECD countries	1960–1987	IM	Panel data regressions
Elola et al (1995)	17 European countries	1990 or 1991	LE, PM and IM	Cross-section regressions
Crémieux et al (1999)	Canadian provinces	1978–1992	LEAB and IM	Panel data regressions
Or (2000a)	21 OECD countries	1970–1992	PM	Panel data regressions
Or (2000b)	21 OECD countries	1970–1995	LE 65, IM and PM	Panel data regressions
Miller and Frech (2000)	21 OECD countries	1996	LEAB, LE 40 and LE 60	Cross-section regressions
Lichtenberg (2000)	US	1960–1997	LEAB	Maximum likelihood estimates using time-series data

(continued)

Table 7 (continued)

Author (date)	Country	Period	Proxy of health status[a]	Methodology and data
Robalino et al (2001)	67 countries (OECD and less developed countries)	1970–1995	IM	Panel data regressions
Berger and Messer (2002)	20 OECD countries	1960–1992	M	Panel data regressions
Miller and Frech (2002)	18 OECD countries	1998–1999	LEAB, LE 40, LE 60, DALEB, DALE 60 and PM	Cross-section regressions
Thornton (2002)	US states	1990	AAM	Cross-section regressions
Self and Grabowski (2003)	191 countries (developed, middle income and less developed countries)	2000	DALEB	Cross-section regressions
Laporte (2004)	US	1948–1996	M	Error correction model (ECM) using time series data
Shaw et al (2005)	19 OECD countries	1980, 1985, 1990 and 1997	LE 40, LE 60 and LE 65	Cross-section regressions
Crémieux et al (2005)	A set of Canadian provinces	1975–1998	LEAB, LE 65 and IM	Panel data regressions
Tapia-Granados (2005)	50 Spanish provinces	1980–1997	M	Panel data regressions
Nixon and Ullmann (2006)	15 European countries	1980–1995	LEAB and IM	Panel data regressions
Soares (2007)	A set of Brazilian municipalities	1970–2000	LEAB and IM	Panel data regressions
Shin-Jong (2009)	8 Asia-Pacific countries	1976–2003	M	Panel data regressions
Bergh and Nilsson (2010)	92 countries (with different levels of development)	1970–2005	LEAB	Panel data regressions (regressions using panel-corrected standard errors—PCSE)

(continued)

Table 7 (continued)

Author (date)	Country	Period	Proxy of health status[a]	Methodology and data
Mariani et al (2010)	132 countries (with different levels of development)	2006	LEAB	Overlapping Generations Model using cross-section data
Halicioglu (2011)	Turkey	1965–2005	LEAB	Autoregressive-Distributed Lag Model (ARDL) to cointegration method using time series data
Bradley et al (2011)	30 OECD countries	1995–2005	LEAB, IM, LBW, MM and PYLL	Pooled cross-sectional analysis
Cutler and Lleras-Muney (2012)	61 countries (with different levels of development)	2004–2009	Others health behavior indicators[a]	Cross-section regressions
Heijink et al (2013)	14 developed countries	1996–2006	AM	Panel data regressions and macro-level cost-effectiveness analysis
Monsef and Mehrjardi (2015)	136 countries (with different levels of development)	2002–2010	LEAB	Panel data regressions
Linden and Ray (2017)	34 OECD countries	1970–2012	LEAB	Panel Vector Autoregressive (VAR) model using panel time series data
Reynolds and Avendano (2017)	20 OECD countries	1980–2010	LE	Panel data regressions

(continued)

Table 7 (continued)

Author (date)	Country	Period	Proxy of health status[a]	Methodology and data
Van den Heuvel and Olaroiu (2017)	31 European countries	2013	LEAB	Cross-section regressions
Ketency and Murthy (2018)	US	1960–2012	LEAB	Unit root testing and cointegration analysis using time series data
Jiang et al (2018)	31 Chinese provinces	2000 and 2010	LE	Cross-section regressions
Hill and Jorgenson (2018)	50 US states and the District of Columbia	2000, 2005 and 2010	LEAB	Panel data regressions

LE Life expectancy; *LEAB* Life expectancy at birth; *LE 40* Life expectancy at age 40; *LE 60* Life expectancy at age 60; *LE 65* Life expectancy at age 65; *M* Mortality rate; *MAG* Mortality at specific age; *IM* Infant mortality; *PM*: Premature mortality; *AAM* Age-adjusted mortality rate; *AM* Avoidable mortality; *MM* Maternal mortality; *LBW* Low birth weight; *DALEB* Disability-adjusted life expectancy at birth; *DALE 60* Disability-adjusted life expectancy at 60 years; *PYLL* Potential years of life lost
Source Own elaboration

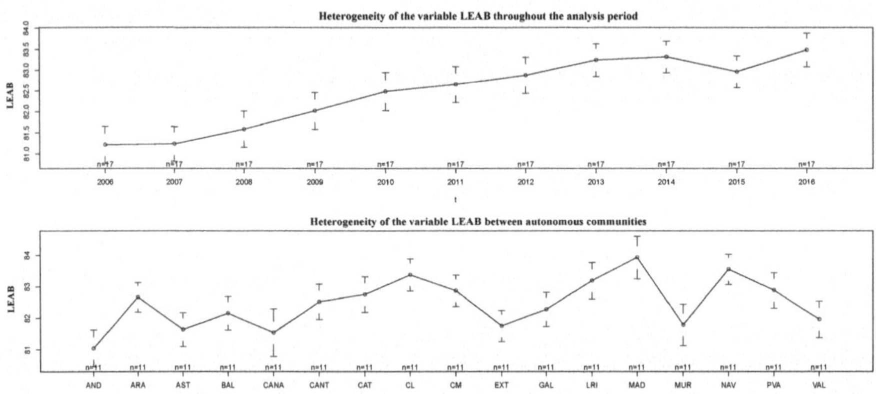

Fig. 13 Heterogeneity of variable "life expectancy at birth". *Source* Own elaboration

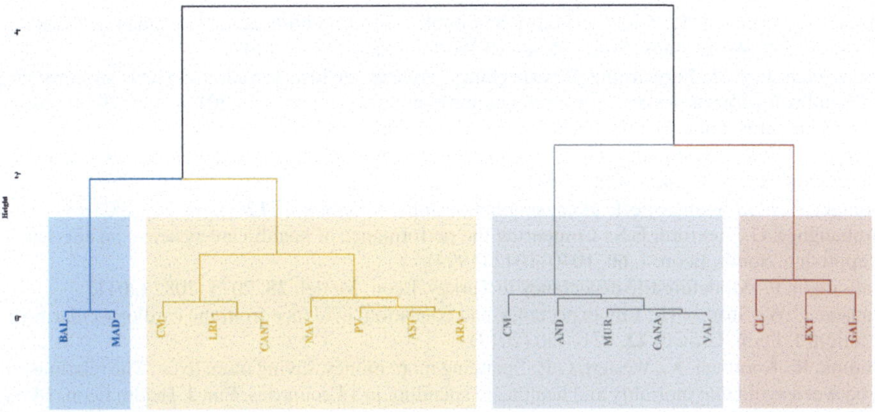

Fig. 14 Dendogram of clusters corresponding to the different autonomous communities. *Source* Own elaboration

References

Aguayo, E., Lamelas, N.: Factores determinantes de la esperanza de vida en los países del Mercosur, 1970–2000. Estudios Económicos de Desarrollo Internacional AEEADE **3**(2), 67–80 (2003)

Arvas, M.A., Torusdağ, M.: The link between health care expenditure and life expectancy: Turkey (1975–2015). Int. J. Hum. Soc. Sci. Invent. **6**(1), 59–64 (2017)

Auster, R., Levesoardln, I., Sarachek, S.: The production of health: an exploratory study. J. Hum. Resour. **4**, 411–436 (1969)

Berger, M., Messer, J.: Public financing of health expenditure, insurance, and health outcomes. Appl. Econ. **34**(17), 2105–2113 (2002)

Bergh, A., Nilsson, T.: Good for living? On the relationship between globalization and life expectancy. World Dev. **38**(9), 1191–1203 (2010)

Bradley, E.H., Elkins, B.R., Herrin, J., Elbel, B.: Health and social services expenditures: associations with health outcomes. BMJ Qual. Saf. **20**(10), 826–831 (2011)

Caballer, M.: Análisis y Evaluación Económica de Sistemas de Salud: Logro en Salud, Desempeño y Aspectos Organizativos. Tesis Doctoral de la Universidad Politécnica de Valencia (2007)

Campo, J., Mendoza, H.: Gasto público y crecimiento económico: un análisis regional para Colombia, 1984–2012. Lect. Econ. **88**, 77–108 (2018)

Cochrane, A.L., St Ledger, A.S., Moore, F.: Health service ‘input’ and mortality ‘output’ in developed countries. J. Epidemiol. Community Health **32**, 200–205 (1978)

Crémieux, P.Y., Ouellette, P., Pilon, C.: Health care spending as determinants of health outcomes. Health Econ. **8**, 627–639 (1999)

Crémieux, P.Y., Mieilleur, M.C., Ouellette, P., Petit, P., Zelder, P., Potvin, K.: Public and private pharmaceutical spending as determinants of health outcomes in Canada. Health Econ. **14**, 107–116 (2005)

Cutler, D.M., Lleras-Muney, A.: Education and health: international comparisons. NBER Working Paper 17738, pp. 1–30 (2012)

Dumitrescu, E.-I., Hurlin, C.: Testing for Granger non-causality in heterogeneous panels. Econ. Model. **29**(4), 1450–1460 (2012)

Elola, J., Daponte, A., Navarro, V.: Health indicators and the organisation of health care systems in Western Europe. Am. J. Public Health **85**(10), 1397–1401 (1995)

Fayissa, B., Gutema, P.: Estimating a health production function for sub-saharan Africa (SSA). Appl. Econ. **37**, 155–164 (2005)

Filmer, D., Pritchett, L.: Child mortality and public spending on health: how much does money matter? The World Bank, Policy Research Working Paper 1864 (1997)
Foreman, K.J., et al.: Forecasting life expectancy, years of life lost, and all-cause and cause-specific mortality for 250 causes of death: reference and alternative scenarios for 2016–40 for 195 countries and territories. Lancet (2018). https://doi.org/10.1016/s0140-6736(18)31694-5
Granger, C.W.J.: Investigating causal relations by econometric models and cross-spectral methods. Econometrica **37**(3), 424–438 (1969)
Granger, C.W.J.: Some aspects of causal relationships. J. Econom. **112**(1), 69–71 (2003)
Grubaugh, S.G., Rexford, E.S.: Comparing the performance of health-care systems: an alternative approach. South. Econ. J. **60**, 1030–1042 (1994)
Halicioglu, F.: Modeling life expectancy in Turkey. Econ. Model. **28**, 2075–2082 (2011)
Hansen, C.W., Strulik, H.: Life expectancy and education: evidence from the cardiovascular revolution. J. Econ. Growth **22**, 421–450 (2017)
Heijink, R., Koolman, X., Westert, G.P.: Spending more money, saving more lives? The relationship between avoidable mortality and healthcare spending in 14 countries. Eur. J. Health Econ. **14**(5), 527–538 (2013)
Hill, T.D., Jorgenson, A.: Bring out your dead!: a study of income inequality and life expectancy in the United States, 2000–2010. Health Place **49**, 1–6 (2018)
Hitiris, T., Posnett, J.: The determinants and effects of health expenditure in developed countries. J. Health Econ. **11**, 173–181 (1992)
Holtz-Eakin, D., Newey, W., Rosen, H.: Estimating vector autoregressions with panel data. Econometrica **56**(6), 1371–1395 (1988)
Hoskova-Mayerova, S.: Education and training in crisis management. Eur. Proc. Soc. Behav. Sci. EpSBS **XVI**, 849–856 (2016). https://doi.org/10.15405/epsbs.2016.11.87
Hurlin, C., Venet, B.: Granger causality tests in panel data models with fixed coefficients. Université Paris IX Dauphine. Cahier de Recherche EURISCO, No. 2001-09 (2001)
INE: Esperanza de vida a distintas edades por periodo. Brecha de género. España (2018). Available at: http://www.ine.es/jaxi/Tabla.htm?path=/t00/mujeres_hombres/tablas_1/l0/&file=d01005.px&L=0
Jiang, J., Luo, L., Xu, P., Wang, P.: How does social development influence life expectancy? A geographically weighted regression analysis in China. Public Health **163**, 95–104 (2018)
Joumard, I., André, C., Nicq, C., Chatal, O.: Health status determinants: lifestyle, environment, health care resources and efficiency. OECD Economics Department Working Papers, No. 627, OECD Publishing, Paris (2008). http://dx.doi.org/10.1787/240858500130
Joumard, I., André, C., Nicq, C.: Health care systems: efficiency and institutions. OECD Economics Department Working Papers, No. 769, OECD Publishing (2010). https://doi.org/10.1787/5kmfp51f5f9t-en
Jurado, A., Pérez, J.: Disparidades entre las comunidades españolas en el período 207-2012. FOESSA working paper, 2.9, pp. 1–35 (2014)
Kabir, M.: Determinants of life expectancy in developing countries. J. Dev. Areas **41**, 185–204 (2008)
Kaiser, H.F.: A note on Guttman's lower bound for the number of common factors. Br. J. Stat. Psychol. **14**, 1–2 (1961)
Ketenci, N., Murthy, V.N.R.: Some determinants of life expectancy in the United States: results from cointegration tests under structural breaks. J. Econ. Finance **42**, 508–525 (2018)
Khuder, S.A.: Effect of cigarette smoking on major histological types of lung cancer: a meta-analysis. Lung Cancer **31**, 139–148 (2001)
Kiuila, O., Mieszkowski, P.: The effects of income, education and age on health. Health Econ. **16**, 781–798 (2007)
Laporte, A.: Do economic cycles have a permanent effect on population health? Revisiting the Brener hypothesis. Health Econ. **13**, 767–779 (2004)
Lichtenberg, F.: Sources of U.S. longevity increase, 1960–1997. NBER Working Paper No. 8755 (2000)

Linden, M., Ray, D.: Life expectancy effects of public and private health expenditures in OECD countries 1970–2012: panel time series approach. Econ. Anal. Policy **56**, 101–113 (2017)
López, L., Weber, S.: Testing for Granger causality in panel data. IRENE Institute of Economic Research (17-03) (2017)
Mariani, F., Pérez-Barahona, A., Raffin, N.: Life expectancy and the environment. J. Econ. Dyn. Control **34**, 798–815 (2010)
Martínez-Sánchez, E., Gutiérrez-Fisac, J.L., Gispert, R., Regidor, E.: Educational differences in health expectancy in Madrid and Barcelona. Health Policy **55**, 227–231 (2001)
Maziarz, M.: Collectivism and economic development: a causal analysis. Socioeconomica **4**(8), 277–290 (2015)
McAvinchey, I.D.: A comparison of unemployment, income and mortality interaction for five European countries. Appl. Econ. **20**(4), 453–471 (1988)
Meara, E.R., Richards, S., Cutler, D.M.: The gap gets bigger: changes in mortality and life expectancy by education, 1981–2000. Health Aff. **27**, 350–360 (2008)
Miller, R.D., Frech, H.E.: Is there a link between pharmaceutical consumption and improved health in OECD countries? Pharmacoeconomics **18**(1), 33–45 (2000)
Miller, R.D., Frech, H.E.: The productivity of health care and pharmaceuticals: quality of life, cause. University of California, Santa Barbara, UCSB Departmental Working Paper, No. 12-02 (2002)
Ministerio de Sanidad, Consumo y Bienestar Social: Indicadores clave del Sistema Nacional de Salud (2019). Available at: http://inclasns.msssi.es/
Monsef, A., Mehrjardi, A.S.: Determinants of life expectancy: a panel data approach. Asian Econ. Financ. Rev. **5**(11), 1251–1257 (2015)
Nickell, S.: Biases in dynamic models with fixed effects. Econometrica **49**(6), 1417–1426 (1981)
Nixon, J., Ullmann, P.: The relationship between health care expenditure and health outcomes – evidence and caveats for a causal link. Eur. J. Health Econ. **7**(1), 7–19 (2006)
OECD/World Health Organization: Health at a Glance: Asia/Pacific 2014: Measuring Progress towards Universal Health Coverage. OECD Publishing (2014)
Or, Z.: Determinants of health outcomes in industrialised countries: a pooled, cross-country, time series analysis. OECD Economic Studies, No. 30, 2000/I (2000a)
Or, Z.: Exploring the effects of health care on mortality across OECD countries. OECD Labour Market and Social Policy, Occasional Paper No. 46 (2000b)
Peltzman, S.: Regulation and health: the case of mandatory prescriptions and an extension. Manag. Decis. Econ. **8**, 41–46 (1987)
Pla, L.E.: Análisis Multivariado: Métodos de Componentes Principales. Programa Regional de Desarrollo Científico y Tecnológico de la Secretaría General de la Organizaci6n de los Estados Americanos, Washington, DC (1986)
Reynolds, M.M., Avendano, M.: Social policy expenditures and life expectancy in high-income countries. Am. J. Prev. Med. **54**(1), 72–79 (2017)
Robalino, D.A., Oscar, F.P., Albertus, V.: Does fiscal decentralization improve health outcomes? Evidence from a cross-country analysis. World Bank Policy Research Working Paper Series 2565, pp. 1–14 (2001)
Rodgers, G.B.: Income and inequality as determinants of mortality: an international cross-sectional analysis. Popul. Stud. **33**, 343–351 (1979)
Self, S., Grabowski, R.: How effective is public health expenditure in improving overall health? A cross-country analysis. Appl. Econ. **35**, 835–845 (2003)
Shaw, J.W., Horrace, W.C., Vogelf, R.J.: The determinants of life expectancy: an analysis of the OECD health data. South. Econ. J. **71**(4), 768–783 (2005)
Shing-Jong, L.: Economic fluctuations and health outcome: a panel analysis of Asia-Pacific countries. Appl. Econ. **41**, 519–530 (2009)
Soares, R.R.: Health and the evolution of welfare across Brazilian municipalities. J. Dev. Econ. **84**, 590–608 (2007)
Svarcova, I., Hoskova-Mayerova S., Navratil, J.: Crisis management and education in health. Eur. Proc. Soc. Behav. Sci. EpSBS **XVI**, 255–261 (2016). https://doi.org/10.15405/epsbs.2016.11.26

Tapia-Granados, J.A.: Recessions and mortality in Spain, 1980–1997. Eur. J. Popul. **21**, 393–422 (2005)

Thornton, J.: Estimating a health production function for the US: Some new evidence. Applied Economics **34**(1), 59–62 (2002)

Tukey, J.W.: Exploratory Data Analysis. Addison-Wesley Series in Behavioral Science. Addison-Wesley, Reading (1977)

Urban, R., Hoskova-Mayerova, S.: Threat life cycle and its dynamics. Deturope **9**(2), 93–109 (2017)

Vallejo, A., Hoskova-Mayerova, S., Krahulec, J., Sarasola, J.L.: Risks associated with reality: how society views the current wave of migration: One common problem—two different solutions. Stud. Syst. Decis. Control **104**, 283–305 (2017). https://doi.org/10.1007/978-3-319-54819-7_19

Van den Heuvel, W.J.A., Olaroiu, M.: How important are health care expenditures for life expectancy? A comparative. European analysis. J. Am. Med. Dir. Assoc. **18**(3), 276.e9–276.e12 (2017)

Wang, L.: Health outcomes in poor countries: empirical findings from demographic and health surveys. World Bank Policy Research Papers, No. 2831 (2002)

Wiener, N.: The theory of prediction. In: Beckenbach, E.F. (ed.) Modern Mathematics for Engineers, chap. 8. McGraw-Hill, New York (1956)

Wise, R.A.: Changing smoking patterns and mortality from chronic obstructive pulmonary disease. Prev. Med. **26**, 418–421 (1997)

World Bank: Population estimates (2019). Available at: https://databank.worldbank.org/data/indicator/SP.DYN.LE00.IN/1ff4a498/Popular-Indicators

Wolfe, B.L., Gabay, M.: Health status and medical expenditures: more evidence of a link. Soc. Sci. Med. **25**, 883–888 (1987)

Young, F.W.: An explanation of the persistent doctor–mortality association. J. Epidemiol. Community Health **55**, 80–84 (2001)

Zuur, A.F., Ieno, E.N., Smith, G.M.: Principal components analysis and redundancy analysis. In: Samet, J., Tsiatis, A., Wong, W., Gail, M., Krickeberg, K. (eds.) Analysing Ecological Data, Statistics for Biology and Health, chap. 12. Springer, New York (2007)

Assessing Diversity and Inclusion in the Context of the U.S. Federal Department of Health and Human Services

Annamaria Porreca

Abstract In recent decades, the role of diversity in organizations has been much discussed. This is because, in the literature, insiders consider that diversity has a positive impact on multiple aspects. For example, in the corporate field, they suggest that a wide diversity of workforce is associated with higher levels of profitability. Nevertheless, at a theoretical level, not all scholars consider that diversity always has a positive impact on organizations. Because an excessive diversity can also lead to internal conflicts and may favor decision-making processes, not all researchers agree that diversity has a positive effect. Indeed, empirical studies also show conflicting results due to many problems. One of the main issue is how to measure diversity because numerous studies have adopted different indices. To solve this drawback, recent studies have proposed the use of diversity profiles in the organizational field. This research proposes an empirical application of this methodology in the organizational context; in particular, diversity profiles are applied for analysing diversity and inclusion of minorities in a United States Federal Agency, i.e. the Department of Health and Human Services.

Keywords Workforce diversity · Diversity profiles · Hill's numbers · Firms' heterogeneity · Diversity and inclusion

1 Introduction

Recently, the role of diversity in organizations has been much discussed because of its possible positive impact on performance (van Knippenberg and Schippers 2007; Harrison and Klein 2007; European Commission 2015; Ararat et al. 2015; Ruigrok et al. 2007; Shehata 2013). However, an excessive diversity can also lead to internal conflicts and may not favor decision-making processes; thus, not all researchers agree that diversity has a positive effect (Barako and Brown 2008; Hahn and Lasfer 2016).

A. Porreca (✉)
Department of Economics, "G. D' Annunzio" University of Chieti-Pescara, Chieti, Italy
e-mail: annamaria.porreca@unich.it

D. Flaut et al. (eds.), *Decision Making in Social Sciences: Between Traditions and Innovations*, Studies in Systems, Decision and Control 247,
https://doi.org/10.1007/978-3-030-30659-5_22

In this regard, empirical studies also found conflicting results. One of the main issue is how to measure diversity because numerous studies have adopted different indices (Harrison and Klein 2007). To solve this drawback, recent studies have proposed the use of diversity profiles in the organizational field (Maturo et al. 2017, 2018b).

Workforce diversity can be defined as a set of diverse individual that together help agencies pursue organizational objectives efficiently and effectively (Choi and Rainey 2010). During recent years, many worldwide countries have been promoting Institutional initiatives for stimulating diversity and minority inclusion in workplaces (European Commission 2015; OPM.gov 2017). Effectively, both European Union and particularly United States have promoted programs in this perspective.

The United States of America have always been rich in diversity where women and minority groups are a high percentage of total population, and for this reason it is necessary for organizations to increase diversity in the workforce to understand the better way to satisfy the needs of this slice of population (OPM.gov 2017). Public organizations promoted the managing of a diversified workforce, as a challenge for the organizational planning and also for enhancing social justice (Choi and Rainey 2010; Kellough 1990; Di Spalatro et al. 2017). But merely hiring a diverse workforce is not enough, it is necessary to make workplaces more inclusive to make every employee feel welcome and motivated to work (OPM.gov 2017).

The concept of workforce diversity theoretically concerns various aspects, e.g. race, gender, ethnicity, religion, personality, age, cognitive style, skill-set, and education. This study focuses on workforce diversity as ethnicity variety. Following the approach of Maturo et al. (2018b), diversity profiles are adopted for analysing U.S. Federal agencies data. However, in this context, Hill's numbers are considered instead of beta diversity profiles (Patil and Taillie 1979). Moreover, this research specifically focuses on the U.S. Department of Health and Human Services.

The U.S. Government instituted the Office of Diversity and Inclusion (ODI) which developed a Diversity and Inclusion (D&I) dashboard for agency use in workforce management (OPM.gov 2017). The D&I dashboard is a "*government specific tool created to provide agencies with demographic data about hiring, group attrition, employee inclusion perceptions, and overall accountability regarding D&I efforts*" (Cybercareers.gov 2016). Particularly, the Department of Health and Human Services' action plan aims to reduce Racial and Ethnic health disparities, for increasing federal involvement to create a more diverse healthcare workforce. The action plan promoted by the department is based on five major goals: transforming healthcare; strengthening the infrastructure and workforce of the nation's Health and Human Services; advancing Americans' health and well-being; promoting scientific knowledge and innovation; and upholding the accountability of HHS for making demonstrable progress (Koh et al. 2011; OPM.gov 2017). By these goals, the HHS Department brings the vision of a nation free of disparities in health and healthcare.

Despite a diverse workforce has several advantages, it comes with some disadvantages as well. An organization that recruits a diverse workforce promotes a culture of dignity and respect to avoid tension between employees. Contrary, organizations must find the right way to integrate minorities because the latter could create tensions at the workplace or leave the organization if their personal prejudices prevent them from working with colleagues from a different background.

The measurement of diversity within organizations is generally performed through the use of proportions, richness, Shannon-Wiener (Shannon 1948) and Blau-Simpson (Simpson 1949) indices. These indices express different aspects of variety, and thus, in certain conditions, they may show different results (Di Battista et al. 2017). It can therefore be said that there is no universally recognized way of assessing diversity within an organization. In recent years, many researchers have tried to find the best way to measure diversity, proposing various solutions. Specifically, this paper proposes Hill's diversity numbers, in accordance with the methodological FDA approach of other studies (e.g. Maturo 2018; Maturo et al. 2019), as a measure for monitoring the workforce diversity of the U.S. Department of Health and Human Services over time.

The remainder of this paper is structured as follows: Sect. 2 proposes the classical indices adopted for measuring diversity in organizations and Hill's numbers; Sect. 3 proposes an application to the workforce diversity of the U.S. Department of Health and Human Services using a real dataset; and finally, the conclusions are presented.

2 Materials and Methods

In the organizational field, the most common methods for measuring and monitoring workforce diversity are proportions, the Shannon-Wiener index (Shannon 1948), Simpson index (Simpson 1949), and the richness index. Also some variants of these indices has been proposed in the last decades, e.g. the exponential of the Shannon-Wiener index and the reciprocal of the Simpson formula. Moreover, some composite alternative have been suggested by scholars such as the approach of Maturo et al. (2018b). The latter exploits functional data analysis for considering simultaneously different aspects of diversity. Following this last perspective, this study proposes an application of this method with a little adjustment, i.e. Hill's numbers are used instead of beta diversity profiles.

Table 1 shows a brief summary of the above indicators of diversity. For simplicity reasons, proportions are not indicated in the table; effectively, they are simply the ratio of individuals belonging to a category (e.g. male or female) to the total number of subjects of the same category. Of course, this type of measurement is very easy to apply and understand but limited from a theoretical point of view. This is because this statistic does not consider the distribution of individuals among the different modalities.

Also the richness index is very easy and intuitive to apply. This consists in the number of modalities characterizing a statistical variable. The main limit of this index is that it is oversensitive to the number of categories, and thus it neglect the distribution among modalities such as proportions. The Shannon-Wiener and also the Simpson indices do not present this disadvantage because they take into account also the distribution and are not oversensitive to the number of categories. However,

Table 1 Most common diversity indices (k = number of ethnicities, f = relative frequencies, and q = Hill's numbers parameter (see Sect. 2))

Diversity Index	Equation
Richness	$D_1 = k$
Shannon	$D_2 = -\sum_{i=1}^{k} f_i \log(f_i)$
Exponential of Shannon Index	$D_3 = e^{-\sum_{i=1}^{k} f_i \log(f_i)}$
Simpson	$D_4 = 1 - \sum_{i=1}^{k} f_i^2$
Simpson Reciprocal Index	$D_5 = 1/\sum_{i=1}^{k} f_i^2$
Hill's numbers	$D_q = \sum_{i=1}^{k} f_i^{\frac{q}{1-q}}$

despite their properties, also these indices are not perfect. This is because diversity is a multivariate and multidimensional concept, and thus these two different indices focus their attention on different aspects of diversity. Particularly, the first one gives less attention to the dominant categories if compared with the second one. Therefore, no one of these indices could alone synthesize organizational diversity.

Only Hill's numbers, if implemented with functional data analysis (FDA), are able to consider richness, Shannon-Wiener, and Simpson indices simultaneously. Hill's numbers are a mathematically unified family of diversity indices differing among themselves only by an exponent q, which indicates their tendency to include or to ignore the relatively rarer categories (Hill 1973). The diversity number of order q is defined in Table 1 where f_i is the relative abundance of the category i in the workforce organizations.

Particularly, when $q = 0$, we get the richness index, if $q = 1$, we obtain the exponential of the Shannon-Wiener index, and finally for $q = 2$, we have the reciprocal of the Simpson's index. Increasing q, the Hill's function becomes less sensitive to the rarer categories (Hill 1973). For $q = -\infty$, we consider only the rarest category whereas, for $q = +\infty$, only the commonest one are considered. Hence, Hill's numbers can be computed for different tendencies in including or ignoring the relatively rarer category.

Following the approach of Maturo et al. (2018b), functional data analysis (e.g. Ramsay et al. 2009; Ferraty 2011; Aguilera et al. 2011; Maturo et al. 2018a; Fortuna et al. 2018) is used for considering simultaneously different aspects of diversity. However, Hill's numbers are considered instead of beta diversity profiles. In this regard, the method is more similar to that of Maturo and Di Battista (2018) but with the difference that the latter approach is originally considered in an ecological context. The latter proposes the contemporary use of FDA and Hill's numbers for building the diversity surface and related functional tools. Adopting the diversity surface, the volume under this function can be computed as follows (Maturo and Di Battista 2018):

$$\hat{V} = \int_0^4 \int_1^T D(q,t)\, dq\, dt - 4 \cdot (T - 1) \tag{1}$$

where $t = 1, 2, \ldots, T$ is the period in which we are monitoring workforce diversity. Hence, $0 \leq \hat{V} < +\infty$ because there is no theoretical limit to the number of categories which can compose the workplaces.

With a simple approximation using N points, the volume under the surface of the interval t can be expressed by

$$\hat{V}_t = \sum_{j=1}^{N-1} D(\alpha_j, \beta_i) \cdot (q_{j+1} - q_j) \cdot D_t - 4 \cdot D_t \tag{2}$$

where D_t is a fixed time unit, $\alpha_j = \frac{q_{j+1}+q_j}{2}$, and $\beta_i = \frac{t_{i+1}+t_i}{2}$ (Maturo and Di Battista 2018). Thus, following this perspective, the percentage change in the workforce diversity in the passage from one period t to the following $t + 1$ is given by:

$$\lambda_t = \frac{\hat{V}_{t+1} - \hat{V}_t}{\hat{V}_t} \cdot 100 \qquad for \qquad \hat{V}_t \neq 0 \tag{3}$$

3 Application and Results

The Diversity and Inclusion (D&I) dashboard was developed by the U.S. Government for agency use in workforce management (OPM.gov 2017). This because, in the USA, the importance and the role of workforce diversity has assumed a great relevance during the last decades. In this application, we focus only on the U.S. Federal Department of Health and Human Services' (HHS) whose action plan aims to reduce racial and ethnic health disparities, for increasing federal involvement to create a more diverse health care workforce. As described in the introduction, the plan has five major goals: transforming health care; strengthening the infrastructure and workforce of the nation's health and human services; advancing Americans' health and well-being; promoting scientific knowledge and innovation; and upholding the accountability of HHS for making demonstrable progress (Koh et al. 2011; OPM.gov 2017).

Therefore, given the importance that is given to diversity, proper measurement methods are needed for assessing workplace inclusion of minorities. In this context, we refer to the data collected by the Diversity and Inclusion (D&I) dashboard and adopt the approach described in Sect. 2 using ethnicity variety. Table 2 describes the workplace diversity occurred during the period 2010–2014. We observe that, over time, quite half of workers are White people whereas Black ones vary from 20.3 to 21.4%. Also American Indian-Alaskan Native people greatly contribute to HHS workforce. Despite these first considerations, it is not easy to understand, from the original data, if diversity increased or decreased over time.

Figures 1 and 2 represent the diversity surface of the workforce within the HHS department in the period from 2010 to 2014. We observe that the profile in 2010 has a higher curvature and is slightly lower than the profiles of the following years.

Table 2 Percentages of each ethnicity in the workforce of the HHS during a period of five years (2010–2014)

Year	Native Hawaiian-Pacific Islander (%)	American Indian-Alaskan Native (%)	Asian (%)	White (%)	Hispanic (%)	Black (%)	More than one race (%)
2010	0.1	17.9	7.7	50.5	3.3	20.3	0.2
2011	0.1	17.3	7.9	50.5	3.2	20.5	0.3
2012	0.1	17.0	8.2	50.0	3.4	20.7	0.6
2013	0.1	16.6	8.6	49.5	3.3	21.3	0.7
2014	0.1	16.6	8.9	49.0	3.3	21.4	0.7

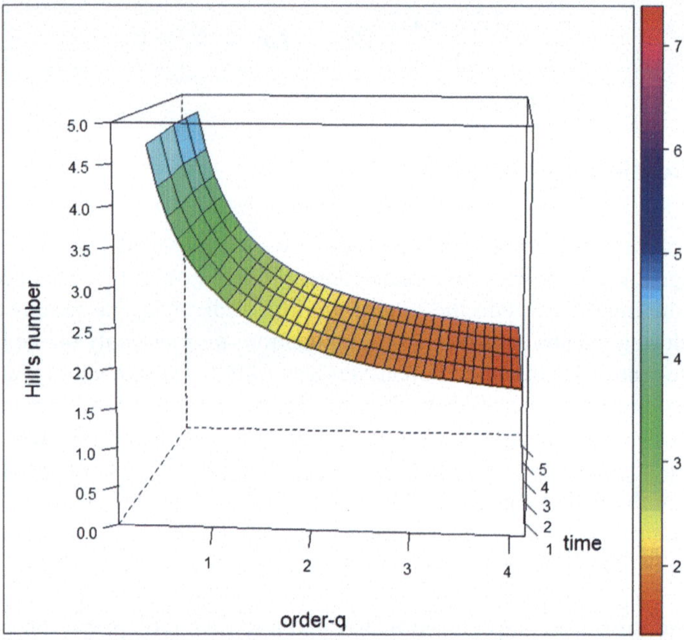

Fig. 1 Workforce diversity surface of HHS Department

We can see that, with increasing time, the profiles tend to become less curved and higher, indicating that diversity increases over time. Truly, however, it is not easy to grasp these small changes via the surface because the changes over time are almost imperceptible.

For this reason, we introduce the volume and the relative changes over time to detect these small differences in a more analytical way. Particularly, Table 3 illustrates the volume of each period computed using Eq. 2, and therefore of the time interval between one year and another. As we can see, over time, the diversity of the workforce

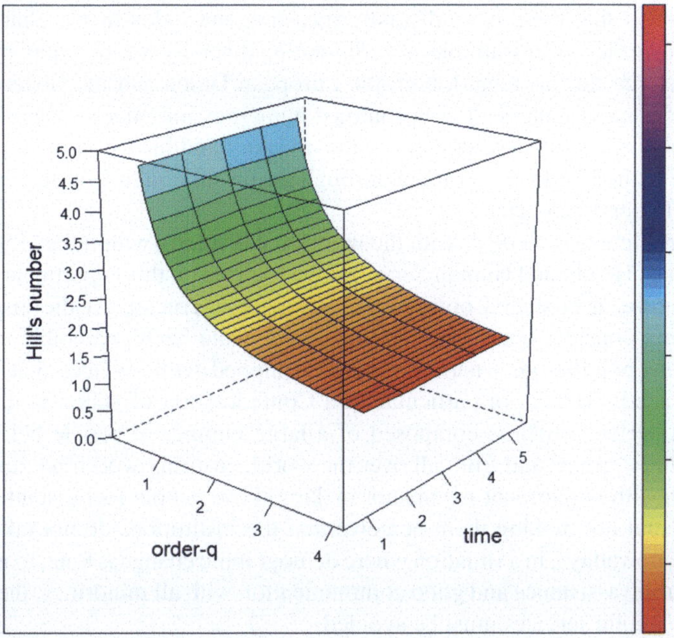

Fig. 2 Workforce diversity surface of HHS Department

Table 3 Workforce diversity volumes of HHS Department for each period (interval)

Volumes	$\hat{V}_t$
2010–2011	5.99
2011–2012	6.07
2012–2013	6.15
2013–2014	6.17

has an ever greater volume; in fact, in the first time interval the volume is equal to 5.99 while in the last period it is equal to 6.17.

Moreover, the relative changes of each period, computed using Eq. 3, are 0.0131, 0.0122, and 0.0036; note that the first period is not considered. Thus, the total change is 2.93% from 2010–2011 to 2013–2014.

4 Discussion and Conclusions

Recently, the role of diversity in organizations has been much discussed in the literature because researchers consider diversity as a positive element for multiple aspects of organizations' lives. Not all researchers agree that diversity always has positive consequences in organizations, and research results are also conflicting. One of the main reasons is that there is no diversity measurement index that is shared by all scholars. Nevertheless, institutions are stimulating diversity for including minorities in the workforce. In this regard, both the European Union and the Government of the United States of America have promoted numerous initiatives. One of the most famous initiatives is the development of the diversity and inclusion platform developed by the United States government through the promotion of workforce diversity in the U.S. Federal Agencies.

In this article, we have dealt with the workforce diversity within the U.S. Federal Department of Health and Human Services. We focused on this department because, in the literature, it is shared opinion that diversity is crucial. At the institutional level, insiders consider that, for providing an adequate health care able to serve a very different population, a nation must be equipped with services that meet the everybody needs. Hence, in particular in a Confederation of states as the United States of America, which is composed of a large number of people belonging to different ethnic groups and from all over the world, ensuring workforce diversity in the public health department means not making these people feel distant from the institutions and not making them perceive that the institutions do not take care of them. Thus, nowadays, in a situation where demographic change is king, to guarantee access to public assistance and good communication with all minorities, inequalities in access to public services must be avoided.

Moreover, the interest in subjective well-being has begun to play a crucial role in the political choices of nations. In effect, it is a shared opinion that the gross domestic product alone is not able to represent the evolution of nations' well-being but other indicators are necessary. Generally, the perception of people of their ability in accessing public services is an important indicator composing subjective well-being indices. Effectively, the access to public services is efficient only when all slices of the population are able to access it; however, this is possible only if population diversity corresponds to workforce diversity in Institutions.

Naturally, diversity is a multidimensional concept, and we can observe it from the perspectives, e.g. ethnicity, sex, experiences, languages, age, and many other characteristics. In this study, we have dealt with the diversity of ethnicity, and we tried to understand whether there was an increase in diversity across time in the U.S. Federal Department of Health and Human Services over a period 2010–2014. Specifically, we focused on measuring diversity with new functional tools recently introduced in the literature.

Our results showed that public intervention towards increasing diversity in the health sector had a small size. In fact, in the five-year period analyzed, there is an increase in diversity within the workforce. However, our findings also show that this increase was very low and that, to achieve evenness better reflecting the U.S. population, it will take a long time. However, we can certainly state that variety of ethnic groups within the workforce of the U.S. Federal Department of Health and Human Services is growing.

In the future, we hope that such initiatives will be taken also in other worldwide countries where the population is composed of different slices of socio-economic groups. Future research should focus on comparing diversity trends among different countries to understand who is achieving better results in this sense. At the same time, also the methodological research regarding diversity measures should continue for looking for new indices for monitoring it over time. Finally, we stress that this methodological approach could easily be extended to other context such as the study of threats dynamics (e.g. see Urban and Šarka Hošková-Mayerová 2017), multi-person decision-making problems (e.g. see Cavallo et al. 2015; Hošková-Mayerová et al. 2013), medical studies (e.g. Barassi et al. 2018, 2019), and also financial topics (see e.g. Longo et al. 2015; Rambaud and Pérez 2016; Olivieri et al. 2016).

References

Aguilera, A., Aguilera-Morillo, M.C., Escabias, M., Valderrama, M.: Penalized spline approaches for functional principal component logit regression. In: Contributions to Statistics, pp. 1–7. Physica-Verlag HD (2011). https://doi.org/10.1007/978-3-7908-2736-1_1

Ararat, M., Aksu, M., Tansel Cetin, A.: How board diversity affects firm performance in emerging markets: evidence on channels in controlled firms. Corp. Gov. (Oxford) **23**(2), 83–103 (2015). www.scopus.com

Barako, G., Brown, A.: Corporate social reporting and board representation: evidence from the Kenyan banking sector. J. Manag. Gov. **12**(4), 309–324 (2008). https://doi.org/10.1007/s10997-008-9053-x

Barassi, G., Bellomo, R.G., Di Iulio, A., Lococo, A., Porreca, A., Di Felice, P.A., Saggini, R.: Preoperative Rehabilitation in Lung Cancer Patients: Yoga Approach, pp. 1–11. Springer, New York, NY. https://doi.org/10.1007/5584_2018_186

Barassi, G., Bellomo, R.G., Porreca, A., Di Felice, P.A., Prosperi, L., Saggini, R.: Somato-visceral effects in the treatment of dysmenorrhea: neuromuscular manual therapy and standard pharmacological treatment. J. Altern. Complement. Med. **24**(3), 291–299 (2018). https://doi.org/10.1089/acm.2017.0182

Cavallo, B., D'Apuzzo, L., Squillante, M.: A multi-criteria decision making method for sustainable development of Naples port city-area. Qual. Quant. **49**(4), 1647–1659 (2015)

Choi, S., Rainey, H.G.: Managing diversity in U.S. Federal Agencies: effects of diversity and diversity management on employee perceptions of organizational performance. Public Admin. Rev., **70**(1), 109–121 (2010). https://doi.org/10.1111/j.1540-6210.2009.02115.x

Cybercareers.gov. Managerial & Executive Development (2016). https://www.cybercareers.gov/

Di Battista, T., Fortuna, F., Maturo, F.: BioFTF: an R package for biodiversity assessment with the functional data analysis approach. Ecol. Indic. **73**, 726–732 (2017). https://doi.org/10.1016/j.ecolind.2016.10.032

Di Spalatro, D., Maturo, F., Sicuro, L.: Inequalities in the Provinces of Abruzzo: a comparative study through the indices of deprivation and principal component analysis. In: Mathematical-Statistical Models and Qualitative Theories for Economic and Social Sciences, pp. 219–231. Springer (2017). https://doi.org/10.1007/978-3-319-54819-7_15

European Commission: Diversity within small and medium-sized enterprises—best practices and approaches for moving ahead. Brussels (2015)

Ferraty, F.: Recent Advances in Functional Data Analysis and Related Topics. Physica-Verlag HD (2011). https://doi.org/10.1007/978-3-7908-2736-1

Fortuna, F., Maturo, F., Di Battista, T.: Clustering functional data streams: unsupervised classification of soccer top players based on Google trends. Qual. Reliab. Eng. Int. 1–25 (2018). https://doi.org/10.1002/QRE-17-0561

Hahn, P., Lasfer, M.: Impact of foreign directors on board meeting frequency. Int. Rev. Financ. Anal. **46**, 295–308 (2016). https://doi.org/10.1016/j.irfa.2015.11.004

Harrison, D.A., Klein, K.J.: What's the difference? Diversity constructs as separation, variety, or disparity in organizations. Acad. Manag. Rev. **32**(4), 1199–1228 (2007). www.scopus.com
Hill, M.: Diversity and evenness: a unifying notation and its consequences. Ecology **54**, 427–432 (1973)
Hošková-Mayerová, Š., Talhofer, V., Hofmann, A.: Decision-Making Process with Respect to the Reliability of Geo-Database, pp. 179–194. Springer, Berlin, Heidelberg (2013). https://doi.org/10.1007/978-3-642-35635-3_15. ISBN 978-3-642-35635-3
Kellough, J.E.: Integration in the public workplace: determinants of minority and female employment in federal agencies. Public Admin. Rev. **50**(5), 557 (1990). https://doi.org/10.2307/976786
Koh, H.K., Graham, G., Glied, S.A.: Reducing racial and ethnic disparities: the action plan from the department of health and human services. Health Aff. **30**(10), 1822–1829 (2011). https://doi.org/10.1377/hlthaff.2011.0673
Longo, A., Squillante, M., Ventre, A., Ventre, V.: The intertemporal choice behaviour the role of emotions in a multi-agent decision problem. AAPP Atti della Accademia Peloritana dei Pericolanti, Classe di Scienze Fisiche, Matematiche e Naturali **93**(2), C2–C14 (2015)
Maturo, F.: Unsupervised classification of ecological communities ranked according to their biodiversity patterns via a functional principal component decomposition of Hill's numbers integral functions. Ecol. Indic. (2018). https://doi.org/10.1016/j.ecolind.2018.03.013
Maturo, F., Di Battista, T.: A functional approach to Hill's numbers for assessing changes in species variety of ecological communities over time. Ecol. Indic. **84**, 70–81 (2018). https://doi.org/10.1016/j.ecolind.2017.08.016. ISSN 1470-160X
Maturo, F., Balzanella, A., Di Battista, T.: Building statistical indicators of equitable and sustainable well-being in a functional framework. Soc. Ind. Res. (2019). https://doi.org/10.1007/s11205-019-02137-5
Maturo, F., Migliori, S., Paolone, F.: Do Institutional or Foreign Shareholders Influence National Board Diversity? Assessing Board Diversity Through Functional Data Analysis, pp. 199–217. Springer, Cham (2017). https://doi.org/10.1007/978-3-319-54819-7_14
Maturo, F., Fortuna, F., Battista, T.D.: Testing equality of functions across multiple experimental conditions for different ability levels in the IRT context: the case of the IPRASE TLT 2016 survey. Soc. Indic. Res. (2018a). https://doi.org/10.1007/s11205-018-1893-4
Maturo, F., Migliori, S., Paolone, F.: Measuring and monitoring diversity in organizations through functional instruments with an application to ethnic workforce diversity of the U.S. Federal agencies. Comput. Math. Organ. Theory **24**(2) (2018b). https://doi.org/10.1007/s10588-018-9267-7
Olivieri, M., Squillante, M., Ventre, V.: Information and intertemporal choices in multi-agent decision problems. Ratio Math. **31**(1), 3–24 (2016). ISSN 2282-8214
OPM.gov. Diversity & inclusion (2017). https://www.opm.gov/policy-data-oversight/diversity-and-inclusion/federal-workforce-at-a-glance/
Patil, G., Taillie, C.: An overview of diversity. In: Grassle, J., Patil, G., Smith, W., Taillie, C. (eds.) Ecological Diversity in Theory and Practice, pp. 23–48. International Co-operative Publishing House, Fairland, MD (1979)
Rambaud, S.C., Pérez, A.M.S.: Valuation of barrier options with the binomial pricing model. Ratio Math. **31**(1), 25–35 (2016). ISSN 2282-8214
Ramsay, J., Hooker, G., Graves, S.: Introduction to functional data analysis. In: Functional Data Analysis with R and MATLAB, pp. 1–19. Springer, New York (2009). https://doi.org/10.1007/978-0-387-98185-7_1
Ruigrok, W., Peck, S., Tacheva, S.: Nationality and gender diversity on Swiss corporate boards. Corp. Gov.: Int. Rev. **15**(4), 546–557 (2007). https://doi.org/10.1111/j.1467-8683.2007.00587.x
Shannon, C.: A mathematical theory of communication. Bell Syst. Tech. J. **27**, 379–423 (1948)
Shehata, N.F.: How could board diversity influence corporate disclosure? Corp. Board **9**(3) (2013). https://doi.org/10.22495/cbv9i3art4

Simpson, E.: Measurement of diversity. Nature **163**, 688 (1949)
Urban, R., Šarka Hošková-Mayerová, J.D.: Central Eur. J. Reg. Dev. Tour. **9**(2), 93–109 (2017)
van Knippenberg, D., Schippers, M.C.: Work group diversity. Annu. Rev. Psychol. **58**(1), 515–541 (2007). https://doi.org/10.1146/annurev.psych.58.110405.085546

Recent Social Perspectives Between Traditions and Innovations

Romanian Language, Between Tradition and Innovation

Cosmin Căprioară

Abstract This chapter proposes an outline of the general characteristics of the Romanian language and the way in which tradition blends with innovation in this Romance language. We are considering aspects related to the manner in which progress is manifested in language, similar to a permanent interplay between the necessity of keeping the fundamental coordinates of the linguistic system and the need to adapt communication to the new challenges of reality. Some of the most important landmarks in the evolution of Romanian language and certain manifestations of linguistic variation, mostly the diatopic one (dialectal) are also highlighted. In order to delve into the issue of the relation between tradition and innovation in Romanian, a few representative linguistic phenomena from the phonetic, morphosyntactic and semantic fields are discussed.

Keywords Romanian language · Vulgar latin · Romance languages · Archaism · Change · Innovation · Tradition · Dialectal/regional elements · Formal language · Linguistic variation

Abbreviations

adj.	adjective
cf.	*confer*
e.g.	*exempli gratia*
fem.	feminine
fig.	figure
Lat.	Latin
m.	map
masc.	masculine

C. Căprioară (✉)
The Department of Romanian Philology, Classical and Balkan Languages, Faculty of Letters, "Ovidius" University of Constanţa, Constanţa, Romania
e-mail: ct.caprioara@yahoo.fr

D. Flaut et al. (eds.), *Decision Making in Social Sciences: Between Traditions and Innovations*, Studies in Systems, Decision and Control 247,
https://doi.org/10.1007/978-3-030-30659-5_23

pl. plural
sg. singular
s.v. *sub verbo*

1 Introduction

This chapter investigates the issue of the relation between tradition and innovation in the linguistic field or the way in which change occurs in language, what factors it depends on and what influences it.

Any natural language is under continuous transformation and it is tightly connected to the history of the community who speaks it, to include here the era, the space, the environment, social organization etc., but also all the events in which that community participates in or to which it is a witness. Contacts with other languages can have deep consequences, which are embodied in different manifestations depending on the level of development of the communities they interact with (see Gafton 2010). The appearance of major changes in society can even favour the birth of a diastratic or diaphasic variants of language, and the synchrony variation is a *sine qua non* condition of variation in diachrony (see Glessgen 2014, p. 131 sqq., Bârlea 2000, p. 13 sqq.).

The evolution of a language is somewhat imperative and it is produced at the same time as the evolution of the society which uses it. We hereby understand through "evolution" its forging in time, a forging which generally involves progress, but which can also include moments of regress. Overall, an idiom is more developed in a certain moment than in the past stages of its evolution. If society undergoes a continuous regression, then language is also in permanent regression (the way it happened with classical Latin in the second century A.D.; see, for example, ILR I 2018, p. 73 sqq., Glessgen 2014, p. 447), but these cases are usually rare, because periods of regress are followed by periods of progress and viceversa.

As small as the daily changes of its speakers are, language evinces elements of progress too: a new word, a new meaning etc., and these are strictly dependent on the particularity of the categories of language: phonetics, morphosyntax, lexis and semantics (see Căprioară 2011).

Regarding this idea, Lucia Wald (2017, pp. 90–91) claimed that: "The ways in which the progress of language manifests itself are determined by the nature of its elements and the type of relation between them. Within language, several levels can be distinguished, each comprising an inventory of units with a certain organization. Hence, it can be concluded that the evolution of a language can be observed through the rise in the number of units, through the transformation in quality of their content and the strengthening of the systematic character of each level and of language as a whole.

We can admit that one of the criteria of linguistic progress is the rise in quantity, on condition that not every acquisition can be a considered a gain."

Just like A. Meillet, J. Vendryès (quoted in Wald) believes that the history of a language must be scrutinized in connection to the history of society, but he thinks that a direction of transformation cannot be deduced. To him, it is clear that, as long as human collectivities are not at the same level of development, their languages are also unequal. From the comparison between languages, he does not draw conclusions regarding the general stages of the history of languages, but claims that each language must be assessed following a special unit of measurement, in response to different needs (Wald 2017, p. 74).

Linguistic development does not occur, however, without a permanent attachment to its most stable elements, which is generally represented by morphology. In fact, the dynamic of the linguistic system is permanently at play between tradition and innovation.

2 General Characteristics of the Romanian Language

The Romanian language, just like any natural language, is subject to the general laws of progress. This can be displayed in various ways, in all the compartments of language, given the age and complexity of the manifestations of Romanian in time and space.

Romanian is an inflectional language from the family of Romance languages. It represents one of the descendants of vulgar Latin in the Oriental side of the former Roman Empire, more specifically, The Nether Danube and the Balkans (ILR I 2018, *passim*, Rosetti 1968, p. 77), being isolated early on from the rest of Romania and thus illustrating a particular evolution, unlike the other Romance languages. Paradoxically in fact, it enjoyed greater freedom than Latin, but at the same time, it remained more faithful to it.

The first linguist who acknowledged Romanian ("valachian") as a Romance language, categorizing it alongside Italian in the Oriental group, was Fr. Diez in his comparative grammar (*Grammatik der romanischen Sprachen*, Bonn, 1836–1843). Long before, however, in 1242, Spanish historian Roderigo Jimenez de Rada (who lived between 1180–1247) listed our language among Romance languages, in *Chronica Hispaniae ab origine prima* (ILR I 2018, p. 282; cf. Armbruster 1972, pp. 35–36). Later on, both on Romanian grounds, as well as European, references to the Romance character of Romanian would multiply (see ILR I, *passim*, Armbruster 1972, *passim*), an aspect which is unquestionable nowadays, although there were attempts of placing Romanian in the family of Slavic languages during the Soviet period.

Discussing the typology of Romance languages, M.-D. Glessgen (2014, p. 75) considered Romanian to be a language created both through distancing (from the Latin prototype), as well as development, reaching a literary variant (*Abstandssprache*—'*language by distance*', respectively *Ausbausprache*—'*language by development*', in the terminology of the specialist in German studies H. Kloss), and regarding its Romance character, he believed that this is "the partial effect of relatinisations and francisations, operated during its creation, in the 10th century, and which reduced the number of Slavonic elements" (2014, p. 126). It is well-known, however, that belonging to a linguistic group mainly depends on morphology, and Romanian morphology is almost completely Latin.

Regarding the new geography of the Romance space, Glessgen (*ibid.*, pp. 458–460) treats Romanian differently, as a valuable counterexample of the characteristics of the space adjacent to Central and Western Romania (being the only isolated Romance language, spoken by an orthodox population and influenced by Slavic).

Following a prestigious series of reserachers, among whom we can mention Sextil Puşcariu and Alexandru Niculescu, Eugeniu Coşeriu (2002) summarized the great particularizing traits of Romanian language, in his conference entitled "Romanian—a Romance language", held at the Linguistics Institute "I. Iordan—Al. Rosetti" of The Romanian Academy.

Thus, from a *genealogical* point of view, Romanian displayed a few important features:

- It is Latin (Neolatin) in all its inherited aspects or the aspects developed from the inherited aspects, all of which represent its essential structures.
- As part of the Oriental Romance world, it evinces the most "original coincidences" (conservative acts and common innovations) with Italian.
- It represents an autonomous, very unitary and homogenous unit, both through its specific Latin elements which were kept in all dialects or just some of them, and coincidences with other conservative areas of Romania, on the one hand, as well as through the original "divergence", attributed to Danubian Latin (through its specific substrate and the old Greek influence) and the later one, attributed to the Slavic superstrate and its various influences (Hungarian, Turkish, Neogreek), on the other hand.

From a *typological* point of view, Romanian is an inflectional language. It displays the same structure as all Romance languages, except for French, which is neither "analytical" nor "synthetic". In fact, Romance languages are "analytical" to a higher extent than Latin but less "analytical" than French. To be more specific, according to E. Coşeriu, the basic principle of this structural type is: internal determinations ("paradigmatic") for internal functions ("unrelational") (e.g.: gender and number) and external determinations (syntagmatic: periphrasis) for external functions ("relational") (e.g.: case). In other words, for example, Romanian keeps the inflections to express nominal grammatical categories (gender and number) while in order to indicate case values, prepositions are used, just like in the other Romance languages, but to a larger extent than Latin.

Finally, concerning the *spatial* criterion, secondary in history, Romanian is an autonomous language in the Romance linguistic space, an area which is opposed to the great Western Romance area, that is, all the other Romance languages. "It is precisely from this point of view that Romanian is now the only representative of the Oriental Latin world." Moreover, "Danubian Latin and pre-Romanian have contributed decisively to the construction of the league of Balkan linguistics; a great deal of 'Balkan lexical items' are also 'Romance lexical items'." (Coşeriu 2002, p. 117).

3 The Origin and the Diatopic Variation of the Romanian Language

The birth of Romance languages occurred, among other things, through the appearance of diatopic variants of vulgar Latin, in the context of the fall of the Western Roman Empire and the social and political reorganization of Europe, after the great migrations. It is possible that Romanian might have appeared as a particular diatopic variant of the vulgar Latin spoken at the Nether Danube ("Danubian Latin", Fischer 1985).

Even if specialists did not exclude the possibility of a dialectal diversification right at the core of Danubian Latin, the majority of Romanian linguists claim that, before the birth of the four dialects, Romanian, as a Romance language recently appeared (in the VII-th–VIII-th century A.D.) has gone through a unitary stage of evolution, called *străromână* (Puşcariu 1940, borrowing the German a calque *Urrumänisch*, used by his professors W. Meyer-Lübke şi G. Weigand), *primitive Romanian* (Densusianu 1901, Philippide 1928), *common Romanian* (Şiadbei 1932, Rosetti 1968, Vasiliu 1968, Caragiu Marioteanu 1975 etc.), *common primitive Romanian* (Macrea 1961, Gheţie 1994) or *proto-Romanian* (Coteanu 1969). As one can notice, even the name of this first stage of the evolution of Romanian language can be placed between tradition and innovation.

Romanian was formed both at the North and South of the Danube, on a vast territory, which comprised the Danubian provinces of the Roman Empire, that is, Moesia Superior and Inferior, Dardania, Pannonia Inferior. The Thracian, Dacian and Illyrian population here was profoundly Romanised, the more so as the regions at the South of the river were part of the Roman state ever since the first century B.C. and only Dacia was conquered at the beginning of the second century A.D., following the wars of 101–102 and 105–106 between Trajan and Decebalus. Extensive linguistic and archaeological evidence certify the continuity of Dacian and Romanic life and, later on, Romanian life at the North of the Danube, even though the period of Romanian ethnogenesis coincides with the era of the great migrations, and starting from the 7th century the Oriental romance world is almost completely isolated from the Western one, through the imposition of a Slavic belt.

The Romanian language is thus created through the superimposition of the Latin *strate* over the Dacian and Thracian one (*substrate*), with the particularizing contribution of a old common Slavic *superstrate* (the other Romance languages have a German superstrate, or Moorish, for the Ibero-Romance languages: Spanish, Portuguese). Many influences of adstrate were added: Greek, Hungarian, Turkish, modern Slavic, German, Latin-Romanic (mainly French), Anglo-American.

Just like any other language which has reached a mature stage of development, Romanian displays a few diatopic or geographic variants. The four great diatopic variants (dialects): *Daco-Romanian*, *Aromanian* (*Macedo-Romanian*), *Megleno-Romanian* and *Istro-Romanian*, which are variants of current Romanian, are spoken on a large area between the Northern Carpathians up to the south of the Balkan

Peninsula (Pind and Thessalia) and from North of the Adriatic Sea up to East of Dniester. Out of these, Istro-Romanian (The Istrian Peninsula, Croatia) has practically disappeared, and Megleno-Romanian (The Meglen Plain, Greece, Northern Macedonia) still has a small number of speakers (a few thousand). Aromanian (spoken over vast regions and in many cities of Greece, Albania, Northern Macedonia, Bulgaria, Serbia and, over the last century, Romania), with geographical variants named after the places or the Balkan regions where they come from (*fărşerot* from *Frasher*, *grămostean* from *Gramoste*, *muzăchear* from *Muzachia*, *pindean* from *Pind* etc.), created, in its turn, a rudimentary literary language, but since it did not benefit from a statal life it could not develop fully. Daco-Romanian is the most widespread (Romania, The Republic of Moldavia, Ukraine and other neighbouring states) and it also displays two major diatopic variants: North-Western and South-Eastern [or six subdialects: *bănăţean* from the name of region Banat (southwest of the area), *crişean* from Crişana (northwest), *maramureşean* from Maramureş (north), *transilvănean* from Transylvania (center), *moldovenesc* from Moldova (northeast) and *muntenesc* from Muntenia (south)] and it represents Romanian by excellence (due to the fact that it reached full development, creating even a literary variant, the neatest, highest, and unitary one).

Aromanian, Megleno-Romanian and Istro-Romanian are also called South Danubian dialects, not just because they are spoken South of the Danube, but also because the area where they were formed is situated at the South of the river. Since they did not benefit from favourable conditions of development and although they were often naturally subjected to innovation, especially through borrowings from the neighbouring languages, they preserved a great number of old phenomena or intermediary forms, outdated in Daco-Romanian and with a high value in the process of reconstruction of the stages which do not have any written evidence. It is well-known that, even if there is isolated lexical evidence in earlier Slavonic or Latin documents, the first known Romanian (Daco-Romanian) text is *Scrisoarea lui Neacşu din Câmpulung* [The Letter of Neacşu of Câmpulung], written in 1521.

All these factors: the conditions in which Romanian appeared, the manner of manifestation of the substrate, superstrate and adstrate influences, its geographical spreading and diversification etc., tinge of the way in which the relation between tradition and innovation is manifested. In certain stages, innovation will trudge its way and will be imposed at a slower pace and the same will happen to certain variants of language.

4 The Individuality of Romanian in Relation to Latin and to Other Romance Languages

In linguistic theory, the elements which have to do with tradition are called "conservationist phenomena" or "archaisms". They are thus considered in relation to the latest studies of language evolution and they are relative, as they must be linked to different variants of the language. For instance, a conservationist phenomenon of Aromanian can be evaluated as such in connection to Daco-Romanian, or at least literary Daco-Romanian, but not on its own.

The elements of tradition in the Romanian language can be best highlighted by referring to Latin, from which Romanian inherited almost entirely the morphological system and about 2000 words, the majority being part of the main lexical field, with a great derivational power (Sala 1988).

Hence, the neutral gender was preserved and reorganized formally, consistently, probably due to the conservation of specific plural inflections, *-a* (*ouă* 'eggs', dialectal, *pârauă* 'creeks, brooks') or *-ora* [**câmpură* > *câmpuri* 'fields', **timpură* > *timpuri* 'time (at plural)'], extended to the feminine. Italian also shows remnants of the neutral gender with the plural of certain words, which, grammatically speaking, are considered masculine: sg. *il braccio*/pl. *le braccia del uomo*, 'the arm/the arms of the man' (cf. sg. *il braccio*/pl. *i bracci del fiume* 'the arm/arms of the river') and Spanish uses the neutral unaccented pronoun *lo* with a so-called "neutral" value, as a synthetic repetition of a previous sentence.

Romanian partially preserves the case inflection of nouns: the direct case (the nominative), the oblique case (accusative-genitive-dative), for feminine singular, and a common case (nominative-accusative-genitive-dative), moreover, the vocative (*Doamne*, *Ioane*, *lupe*, *Petre* from *Domn* 'God', *Ion*, *lup* 'wolf', *Petru*), for masculine singular. The other Romance languages only have one case value for the noun.

The absolute superlative of adjectives and adverbs is formed with the help of the Lat. *forte* (in the literary language and Daco-Romanian as a dialect), while the rest of the Romance idioms use *multum* and *trans*.

Romanian has preserved all four Latin verbal conjugations [in *-āre*, *-ēre*, *-ĕre* and *-īre*: *cântare* (< Lat. *cantare*), *vedere* (< Lat. *videre*), *zicere* (< Lat. *dicere*), *venire* (< Lat. *uenire*)]. In the rest of the Romance languages there is a reduction to three conjugations, through the confusion between the second and the third conjugation.

From a lexical point of view, Romanian inherited about 100 Latin words which no other Romance language has kept. Among them: *adăpost* 'shelter', *aşterne* 'to strew', *cuţit* 'knife', *cuvânt* 'word', *dărâma* 'to break down', *farmec* 'charm', *lânced* 'sick, feeble', *lingură* 'spoon', *împărat* 'emperor', *întâi* 'first', *înţelege* 'to understand', *mare* (adj.) 'big', *mărgea* 'beads', *mângâia* 'to caress', *mire* 'bridegroom', *pădure* 'forest', *plăcintă* 'pie', *plânge* 'to cry' etc. (cf. CDDE s.v.).

Also by referring to Latin, as well as other Romance languages, innovative elements can be pinpointed, many of which exist as tendencies even from Vulgar Latin (see Niculescu 1965, Niculescu 1978, 1999, 2003, *passim*). On a phonetic level, Romanian (Daco-Romanian and partially Aromanian) have developed a timbre vowel [y] ([î]), unknown to the other Romance languages or Latin. At the same time, many phonetical shifts and cumulation of marks are used, which leads to hypercharacterization from a morphological point of view (for example, the difference between singular and plural in the phrases *acest frate bun*/*aceşti fraţi buni* 'this good brother/these good brothers').

From a morphological point of view, Romanian has a specific inflection, *-le* on the feminine plural of nouns and adjectives *stele* 'stars', *grele* 'hard'. The definite article is enclitic and attached to the theme of the word in Romanian (while in other Romance languages it is exclusively proclitic). Furthermore, Romanian has a possessive article (*al, a, ai, ale*) and a demonstrative one (*cel, cea, cei, cele*), that are missing in the rest of Romania, which entails consequences in the formation of the relative superlative (masc. *cel mai frumos*, fem. *cea mai frumoasă* 'the most beautiful' etc.) or in the structure of the possessive pronoun (masc. *al meu*, fem. *a mea* 'mine' etc.) or the ordinal numeral (masc. *al doilea*, fem. *a doua* 'the second' etc.).

Original forms and values of the pronoun arise: the personal pronoun *dânsul* 'he, it', the politeness pronoun displays two degrees of expressing politeness (sg. *dumneata* 'you polite'/sg. and pl. *dumneavoastră* 'you polite'), two case values of the reflexive pronoun (dative: *sie, îşi*; and accusative: *sine, se*), the demonstrative pronoun of identity [*acelaşi* '(the) same'] and difference [*celălalt* 'the other (one)'], formal innovation within the relative pronoun (*ceea ce* approximately 'what'), indefinite pronoun (*ceva* 'something', *cineva* 'someone', *orice* 'anything', *orişicine* 'anybody' etc.) and negative pronoun (*niciunul* 'none, neither', *nimic* 'nothing').

The cardinal numeral from 11 (*unsprezece*) to 19 (*nouăsprezece*) is constructed according to a "local" model, *unus super decem*, which could have entered Danubian Latin from the Thracian and Dacian substrate; the names for tens follow a Slavic or a substrate model, and for 100 a Slavic word is used (cf. ILR 2018, pp. 169, 435–437).

Regarding the verb, the short infinitive replaces the Latin infinitive in *-re* (*spunere* > (*a*) *spune* 'to say', *venire* > (*a*) *veni* 'to come'), which becomes a noun. The supine mode is recreated, built with a preposition (*de, la, pentru*) and the participle (*de văzut* 'to see', *la scris* 'to write', *pentru suflat* 'for blowing'). The subjunctive mode, which is formed with the conjunction *si*, not the lat. *que,* as is the case in the other Romance languages, keeps from Latin only the third person forms, borrowing the rest from the present indicative. The future is built analytically with the auxiliary *voleo,* whereas the other Romance languages use the auxiliaries *habeo*, *debeo* and *vado*. The pluperfect of the indicative takes on the forms of the pluperfect subjunctive (*venisem* 'I had come', *crezuseră* 'they had believed'), although, at the level of dialects, periphrastic romance forms are also attested: *s-o fost dus* 'se dusese' ('he/she had gone').

From a syntactic point of view, in the succession of two verbs, the second is usually in the subjunctive and not the infinitive as in other Romance languages (with the exception of the structure *a putea* + another verb: (eu) *pot spune* 'I can say'). This can increase the importance of the subordination mood of the subjunctive.

The direct object in the accusative and the indirect object in the dative are anticipated or rephrased through unaccented forms which correspond to the personal pronoun (**o** *văd* ***pe mama*** and ***pe mama*** **o** *văd* 'I see my mother', respectively, **îi** *vorbesc* ***mamei*** and ***mamei*** **îi** *vorbesc* 'I'm talking to my mother'). The personal direct object is formed with the preposition *pe.* This object is also formed with a preposition, *a*, in Spanish, Catalan and Italian dialects.

From a lexical point of view, Romanian preserves over 150 words from the Thracian and Dacian substrate which first made their way into Danubian Latin and were then treated as Latin terms: *baci* 'chief shepherd', *baltă* 'puddle', *balaur* 'dragon', *brânză* 'cheese', *cătun* 'hamlet, small village', *copac* 'tree', *mal* 'shore', *mazăre* 'green peas', *mânz* 'foal', *sâmbure* 'kernel', *ţap* 'he-goat' etc.

Some words of Latin origin changed their meanings: *aprinde* 'to ignite', *codru* 'forest', *greu* 'hard', *inimă* 'heart', *înţelept* 'wise', *tânăr* 'young', *vârtute* 'force, vigor' etc., the original meanings ('to learn', 'quarter', 'serious, grave', 'soul', 'educated', 'affectionate, gentle', respectively 'virtue') being kept in other Romance languages (see CDDE, s.v.).

The entire lexis was the predominant field of manifestation for the Slavic superstrate: *boală* 'illness', *bob* 'seed', *boier* 'boyar, nobleman', *cârn* 'snubby', *coasă* 'scythe', *denie* 'evening service during passion week', *leac* 'cure, medicine', *a iubi* 'to love', *mândru* 'proud', *a plăti* 'to pay', *slab* 'weak', *slujbă* 'celebration, job', *ucenic* 'apprentice', *veac* 'century' etc., as well as other influences, of a different nature than those which acted upon the other Romance languages (or which had a very weak impact on them): Greek (important for the Southern Italian dialects): *drum* 'road, way', *folos* 'benefit', *frică* 'fear', *pedepsi* 'to punish', *strachină* 'bowl, tureen' etc.; Hungarian: *fel* 'kind', *gazdă* 'host, owner', *marfă* 'merchandise', *meşter* 'artisan, skillful', *oraş* 'town', *şoim* 'hawk', *vamă* 'customs', *viteaz* 'brave' etc.; Turkish: *bursuc* 'badger', *chef* 'binge', *chiftea* 'rissole', *dud* 'mulberry tree', *fudul* 'arrogant', *halva* 'halvah', *odaie* 'room', *pilaf* 'pilaff', *tavan* 'ceiling' etc.; German: *bliţ* 'flash', *glaspapir* 'sand paper', *spital* 'hospital', *ştrudel* 'strudel' etc.; Bulgarian: *borcan* 'jar', *cârtiţă* 'groundhog' etc.; Ukrainian: *maştehă* 'step mother', *samcă* 'jinnee' etc. (see DA/DLR, DER, DEX, s.v.).

Also as innovative elements with a particularising character for Romanian we can mention many individual formations, created through all the internal means of vocabulary development, but mostly through derivation (the abundance of Romanian affixes, especially suffixes—over 50, is noticeable) and the rich synonymy (popular and literary, old and new, contending derivates with a common base).

It is also worth mentioning that Romanian has also proceeded to a renewal of the Romance character of its lexis (a process called *re-romanisation* or *re-latinisation*), initiated in the first part of the 19th century. The Latin-Romance neological component reaches according to some statistics (Sala 1989, s.v. *română*) about 40% of the current lexis.

5 Tradition and Innovation in Present-Day Spoken Romanian

The fields of tradition/innovation can also be highlighted through the discussion of aspects dealing with linguistic microsystems, which are an integral part of the diasystem of language.

The maps of dialectal atlases outline very well the relation between tradition and innovation in different compartments of language. Romanian has benefited ever since the beginning of the 20th century from this type of linguisic works, covering all areas: national, regional, synthesis (see Saramandu 2005, *Prefaţă*, in ALRR. *Sinteză*,

Caragiu Marioţeanu et al. 1977, *passim*). They are accompanied by many dialectal monographies, compendiums of dialectal texts, glossaries, dictionaries, treaties, which places Romanian dialectology at the top of world and European linguistic research.

I chose a few illustrative examples, which will be tackled below, using a few maps of the *Atlasul lingvistic român pe regiuni. Sinteză* (ALRR. *Sinteză*). This centralizes the linguistic data of the most recent Romanian regional atlases and offers a clarifying image especially of the "popular" aspect of Romanian. One must not forget that popular language is at the origin of the literary variant, so certain phenomena considered to be popular or dialectal can also be literary (for example, the word *zăpadă* 'snow', from the Daco-Romanian subdialect of Muntenia is a regionalism, but at the same time, it was selected as a term of the literary language, at the expense of *omăt* from the Moldavian subdialect or the older *nea* from the trans-Carpathian subdialects).

As m. 10 of ALRR. *Sinteză* I shows, **[í] in SUBŢIRE 'mince'** (see Fig. 1), the Daco-Romanian linguistic territory is marked by two different manifestations of a phonetic phenomenon: the preservation of accented [i] or its "velarisation" (that is, its transition from its prior palatal series to the central series) under the form of [î], depending on the "soft" or "hard" character of the preceding consonante [ts] ([ţ]). If [ţ] is pronounced soft, [i] remains the same, and if [ţ] has a tough character, then [i] becomes [î], which results in the forms *subţire*, respectively *subţâre*. (In

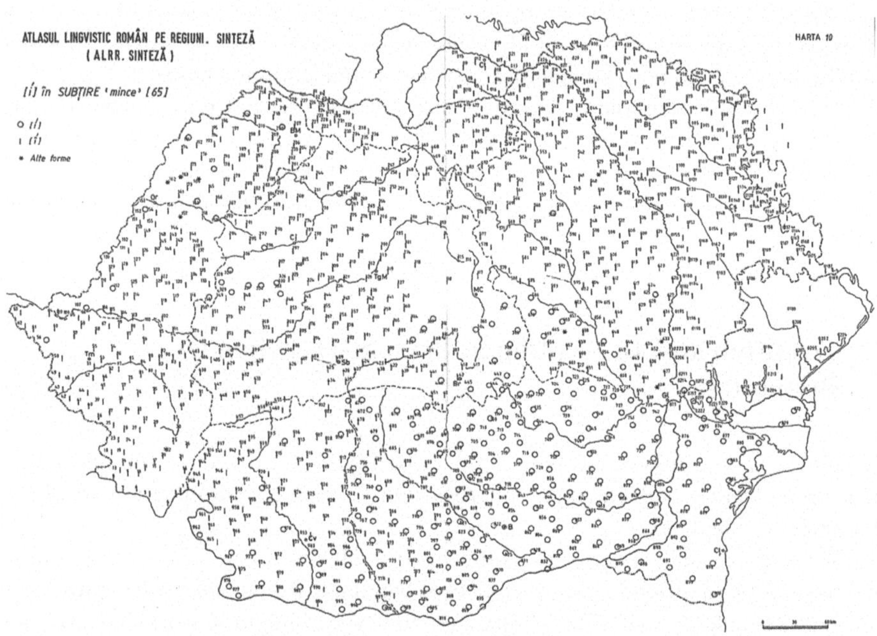

Fig. 1 The "soft" / "hard" character of the consonant [ts] in *subţire* (Source *Atlasul lingvistic român pe regiuni. Sinteză*, vol. I, EAR, Bucureşti, 2010, m. 10)

the current Romanian ortography, *â* and *î* represent two graphemes which render the same phoneme /y/, the first appearing just inside words, and the second at the absolute initial or at their end.)

At the origin of the Romanian term is the Latin *subtilis*, therefore the occurrence of the affricate [ţ], the rotacisation of the invervowel [l] and the velarisation of [i] represent innovations, whereas the use of the form *subţire* in some patois of the subdialect of Muntenia and Southern Transylvania, and then in the literary language represents manifestations of tradition and conservationist phenomena.

In order to stay om the same field of phonetics, we will take into consideration m. 133 of ALRR. *Sinteză* I, entitled **[z] in BUZĂ 'lèvre'** (see Fig. 2).

Buză represents a definite lexical item of the Thracian and Dacian substrate of Romanian, borrowed by Danubian Latin and inherited as such (see Brâncus 1983). Therefore, it illustrates a good example of how linguistic tradition is manifested. In its phonetic structure, it conserves in lateral areas of Daco-Romanian (in the subdialects bănăţean, maramureşean şi moldovenesc) the affricate consonant [dz], which is supposed to represent the actual original sound of the Geto-Dacian language. It must be remarked, however, that, throughout the largest part of Daco-Romanian [dz] was fricatised, turning into [z], a phenomenon which was already in use in the 16th century. The fricative form which resulted from this was preserved in the literary language.

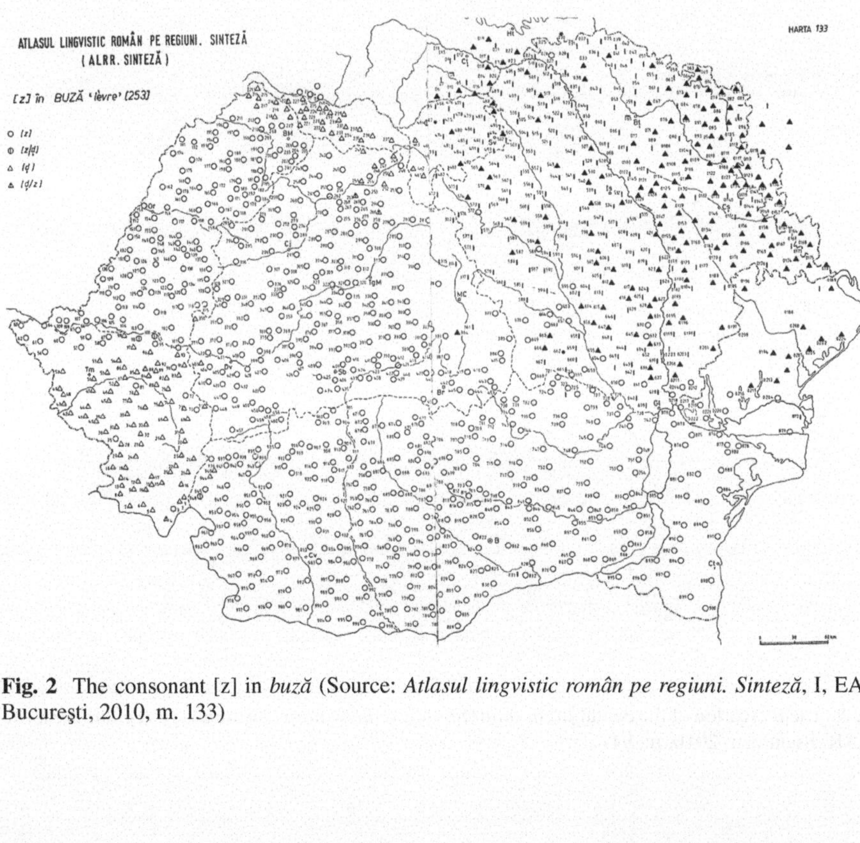

Fig. 2 The consonant [z] in *buză* (Source: *Atlasul lingvistic român pe regiuni. Sinteză*, I, EAR, Bucureşti, 2010, m. 133)

The phonetic compartment of language holds a special importance, because the transformations which occur here can have repercussions in morphosyntax, lexis and seemantics (for example, the dropping of the final consonants *-s* and *-m* with an inflectional role in Latin resulted in the Latin *densus* and *densum*, adjective forms of masculine and neutral, becoming homonyms (**densu*), which led to the weakening of the neutral gender as a whole. On the other hand, keeping the Latin inflection *-s* for the plural accusative of Latin nouns made it become a general and unique plural inflection in some Romance languages, such as Spanish.

A subtle play between tradition and innovation can be noticed in m. 94 of ALRR. *Sinteză.* I, which is entitled **SPUN 'je dis'** (see Fig. 3).

The map shows to what extent the dental [n] was recreated in the indicative present form, first person of the verb *a spune* from Daco-Romanian. This originates in the Latin *exponere* (DER, s.v. *spune*), with the same meaning. The popular form of the indicative present first person in Latin was *exponeo,* not *expono*, as in classical Latin (see DELL, s.v. *pōnō*), because the normal reflex in Romanian was (eu) *spui*(*u*), which can be explained through the palatization of the dental [n] after the *e-o* hiatus was eliminated and [e] became a semi-vowel (and the Latin word as pronounced in three syllables). But the Latin *expones* of second person singular also turned into *spui,* which led to the intolerable morphological homonymy (eu) *spui*—(tu) *spui.* As a result, an innovation was needed, which acted in the spirit of tradition: the old thematic consonant [n] was recreated in first person. The homonymy was resolved:

Fig. 3 The recreation of the dental [n] in *spun* (Source: *Atlasul lingvistic român pe regiuni. Sinteză*, I, EAR, Bucureşti, 2010, m. 94)

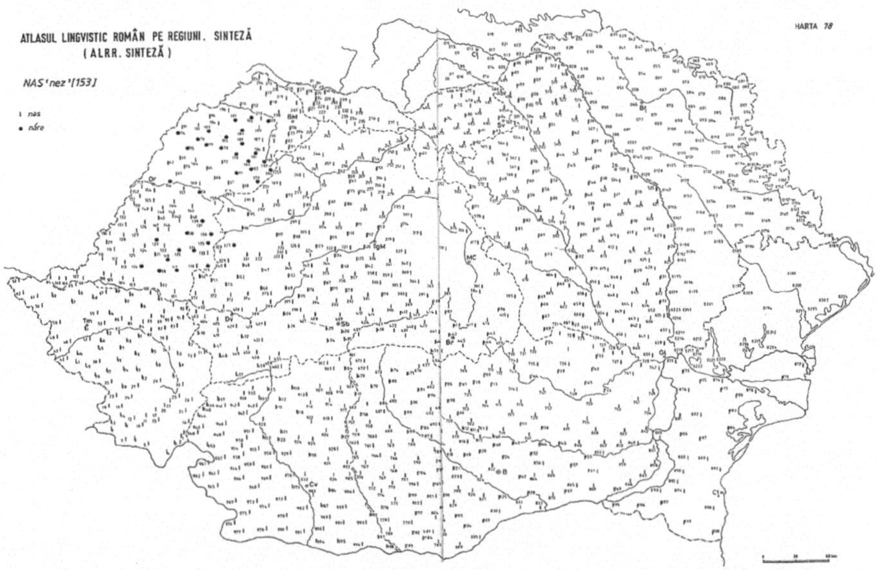

Fig. 4 Nas „nez" (Source: *Atlasul lingvistic român pe regiuni. Sinteză*, I, EAR, Bucureşti, 2010, m. 78)

(eu) *spun*—(tu) *spui*, and the literary language naturally preserved the most recent form. As it can be observed on the map, some dialects from the area of Moldova or Crişana attempted to resolve the homonymy differently, by eliminating [i]: (eu) *spuu*.

Maps 78 and 79 of ALRR. *Sinteză* I, entitled, **NAS 'nez'**, respectively **NARĂ 'narine'** (see Figs. 4 and 5) highlight the way in which tradition and innovation converge in the field of lexis and semantics. They must be read together, since the modifications undergone by a lexeme influence the use of the other.

Both words are inherited from Latin: *nas* comes from the Latin *nasus*, and *nară* (or the older *nare*) from the Latin *nare*(*m*), the accusative of *naris*, *-is*, without semantic modifications from the transition from Latin to Romanian (see CDDE, s.v., DER, s.v.).

However, as m. 78 indicates, in the area Crişana of Daco-Romanian *nare* replaced *nas*, and this may have happened ever since Danubian Latin. It has to do with a local innovation explicable through the metonymic character of the phenomenon: a part for the whole organ (*pars pro toto*).

The conservationism of the Crişana subdialect is acknowledged (see Teaha 1961), and in the lexical field it is noticeable through the preservation of Latin terms which were not in use or were lost in the other areas: *arină* 'sand' (from the Lat. *arena*), *brâncă* 'hand' (from the slang Lat. *branca*), *pedestru* 'orphan (someone) missing…' (from the Lat. *pedester*) etc.

If *nare* substitutes *nas*, the reality which it designates is preserved and it must carry a name, so, according to m. 79, other solutions can be found in the above-mentioned

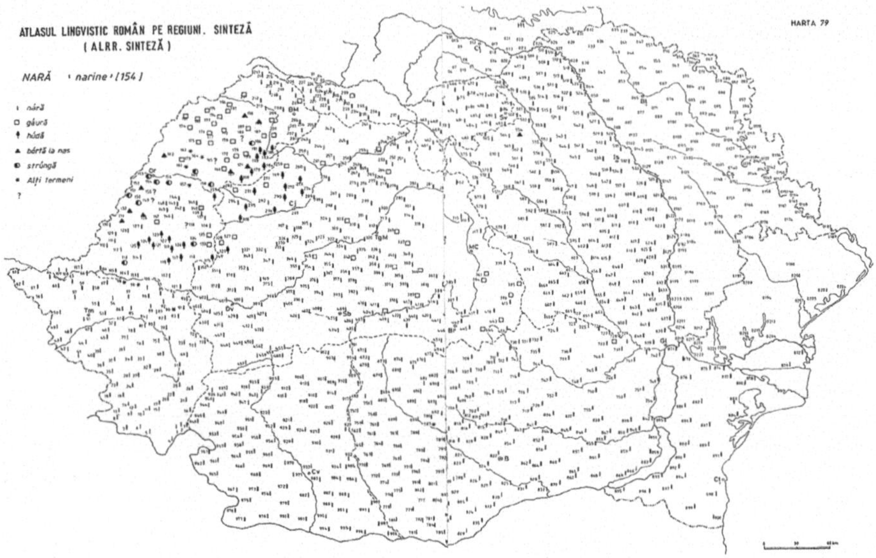

Fig. 5 *Nară* 'narine' (Source: *Atlasul lingvistic român pe regiuni. Sinteză*, vol. I, EAR, Bucureşti, 2010, m. 79)

area. Lexemes of different origins are used: *bortă*, *gaură*, *hudă* and *strungă* (all of them possibly from the substrate), with the same common meaning of 'emtpy space, cavity'.

Naturally, the literary Romanian language did not preserve the same marginal phenomenon, but it remains a perfect example for the transformations in meaning which occurred in vulgar Latin and for the synonymic competition in different areas in Romania.

6 Conclusions

In conclusion, Romanian is the Eastern most representative of the Romance family, with a powerful conservative character, but which does not exclude a big number of innovations, having been detached earlier from the common ancestry and with an evolution different in many ways, all of which made it crucial in the study of the other languages and furthered the legacy of spoken Latin.

The confrontation between old and new encompassed all the areas of language, but in a different way, at different times and intensities. The morphological compartment is the one resistant to change, but the transformations in the phonetic field have direct consequences here, as they do in lexis and semantics.

The relation between conservationism and change can only be followed in diachrony and it involves a system of reference, which is either the literary variant, or a different stage in language development.

Like any other natural language, Romanian subjected to the laws of progress and it developed continuously based on the continuous interplay between tradition and innovation. It was always about finding a balance between the pressure of the language diasystem and the necessity of change, which came not only from the succession of speakers' generations but also from the objective transformation of the surrounding reality.

References

DER = Al. Ciorănescu: Dicţionarul etimologic al limbii române. Editura SAECULUM I. O., Bucharest (2007)

ALRR. *Sinteză* = Saramandu, N. (coord.): Atlasul lingvistic român pe regiuni. Sinteză, vol. I. Editura Academiei, Bucharest (2005)

Armbruster, A.: Romănitatea românilor. Istoria unei idei. Editura Academiei, Bucureşti (1972)

Bârlea, Gh.: Introducere în studiul latinei creştine. Editura *Grai şi suflet—Cultura naţională*, Bucharest (2000)

Brâncuş, Gr.: Vocabularul autohton al limbii române. Editura Ştiinţifică şi Enciclopedică, Bucharest (1983)

Caragiu Marioţeanu, M.: Compendiu de dialectologie română nord- şi sud-dunăreană. Editura Ştiinţifică şi Enciclopedică, Bucharest (1975)

Caragiu Marioţeanu, M., Todoran, R., Ionescu-Ruxăndoiu, L., Rusu, V.: Dialectologie română. Editura Didactică şi Pedagogică, Bucharest (1977)

Căprioară, C.: Book review: Martin-Dietrich Glessgen, *Lingvistică romanică. Domenii şi metode în lingvistica franceză şi romanică*. Diacronia (www.diacronia.ro), 3/12 februarie 2016, A43 (pp. 1–7), Editura Revista „Diacronia", Iaşi (2016)

Căprioară, C.: Evoluţia limbii între conflict şi negociere. In A. Todi (coord.) *Filologie şi didactică*. Lucrările Colocviului Societăţii de Ştiinţe Filologice din România—Constanţa, 2011, pp. 70–75. Editura Universitară, Bucharest (2011)

CDDE = Candrea I.-A., Densusianu, Ov.: Dicţionarul etimologic al limbii române. Elementele latine (A-Putea). Librăriile Socec & Co., Bucharest (1907–1914)

Chivu, Gh.: La modernisation latino-romane du lexique roumain. Le modèle interne. Diversité et identité culturelle en Europe/Diversitate şi identitate culturală în Europa (DICE), XIII (2016), nr. 2, 7–18 (2016)

Coşeriu, E.: Limba română—limbă romanică. In In memoriam Eugeniu Coşeriu, from Fonetică şi dialectologie, XX–XXI, 143–165. Editura Academiei, Bucharest (2002)

Coteanu, I.: Morfologia numelui în protoromână (româna comună). Editura Academiei, Bucharest (1969)

DA/DLR = Academia Română: Dicţionarul limbii române, vol. I-XIX. Anastatic edition. Editura Academiei, Bucharest (2010)

DELL = Ernout, A., Meillet, A.: Dictionnaire étymologique de la langue latine. Histoire des mots. Ed. IV. Klincksieck, Paris (2011)

Densusianu, Ov.: Histoire de la langue roumaine. Ernest Leroux, Paris (1901)

DEX = Academia Română: Dicţionarul explicativ al limbii române. Editura Univers Enciclopedic, Bucharest (2016)

Fischer, I.: Latina dunăreană. Editura Ştiinţifică şi Enciclopedică, Bucharest (1985)

Gafton, Al.: Consecinţele profunde ale contactelor lingvistice. In Studii de limbă română. Omagiu profesorului Grigore Brâncuş, pp. 77–100. Editura Universităţii din Bucureşti, Bucharest (2010)
Gheţie, I.: Introducere în dialectologia istorică românească. Editura Academiei, Bucharest (1994)
Glessgen, M.-D.: Lingvistică romanică. Domenii şi metode în lingvistica franceză şi romanică. Preface and translation by Alexandru Gafton. Editura Polirom, Iaşi (2014) (Armand Colin, Paris, 2007)
Iliescu, M.: Româna din perspectivă romanică. Editura Academiei Române, Bucharest (2007)
ILR I 2018 = Sala, M., Ionescu-Ruxăndoiu, L. (coord.): Istoria limbii române. vol. I. Editura Univers Enciclopedic Gold, Bucharest (2018)
Macrea, D.: Probleme de lingvistică română. Editura Ştiinţifică, Bucharest (1961)
Niculescu, Al.: Individualitatea limbii române între limbile romanice, vol. I–II. Editura Ştiinţifică, Bucharest; vol. III-IV, Editura Clusium, Cluj-Napoca (1965, 1978, 1999, 2003)
Philippide, Al.: Originea românilor, vol. II: Ce spun limbile română şi albaneză. Editura Viaţa Românească, Iaşi (1928)
Puşcariu, S.: Limba română, vol. I: Privire generală. Fundaţia pentru literatură şi artă, Bucharest (1940)
Reinheimer Rîpeanu, S.: Lingvistică romanică. Fonetică. Morfosintaxă. Lexic. Editura All, Bucharest (2001)
Rosetti, Al.: Istoria limbii române de la origini până în secolul al XVII-lea. Editura pentru Literatură, Bucharest (1968)
Sala, M. (coord.): Vocabularul reprezentativ al limbilor romanice. Editura Ştiinţifică şi Enciclopedică, Bucharest (1988)
Sala, M. (coord.): Enciclopedia limbilor romanice. Editura Ştiinţifică şi Enciclopedică, Bucharest (1989)
Sala, M.: De la latină la română. Editura Univers Enciclopedic, Bucharest (1998)
Şiadbei, I.: Le Latin dans l'Empire d'Orient. Arhiva, XXXIX, 3–16 (1932)
Teaha, T.: Graiul de pe Valea Crişului Negru. Editura Academiei, Bucharest (1961)
Vasiliu, Em.: Fonologia istorică a dialectelor dacoromâne. Editura Academiei, Bucharest (1968)
Wald, L.: Progresul în limbă. Scurtă istorie a limbajului. Revised Second Edition. Text edited, preface and notes on the edition by Petre Gheorghe Bârlea. Editura Universităţii „Alexandru Ioan Cuza", Iaşi (2017)

Durkheim and Weber: Sociological Theory and Social Research

Vincenzo Corsi

In this chapter I analyse the method in the sociology of Émile Durkheim (1858–1917) and Max Weber (1864–1920). I identify in their work the basis of a research about the method that will have a fundamental value for sociology and for the empirical and theoretical development of the discipline in the 20th century. Durkheim and Weber are authors of works that contribute to the foundation of sociology, with regards to the objective and the method of the discipline.

Durkheim · Weber · Sociological theory · Methods of sociology

1 Introduction

In this chapter I analyze the method in the sociology of Émile Durkheim (1858–1917) and Max Weber (1864–1920). The work of the two sociologists identifies the theoretical foundation of the discipline and define its methodological developments.

Durkheim and Weber are authors of works that contribute to creating a methodological foundation to sociology. Durkheim starts from the positivist positions of the nineteenth century, Weber from German historicism. The conclusions reached by the two sociologists are different, but both are important for the methodology of sociological research.

The two sociologists operate between the end of the nineteenth century and the beginning of the twentieth. Durkheim operates within the nineteenth-century positivist tradition expressed by John Stuart Mill (1806–1873) and Herbert Spencer (1820–1903), on the English front, and by Auguste Comte (1798–1857) in the French tradition; he proposes an objectivist approach to the study of social phenomena, rooted in the idea of the applicability to the social sciences of the experimental

V. Corsi (✉)
Department of Business Administration and Management, University G. D'Annunzio of Chieti-Pescara, Viale Pindaro, 42, 65127 Pescara, Italy
e-mail: vincenzo.corsi@unich.it

D. Flaut et al. (eds.), *Decision Making in Social Sciences: Between Traditions and Innovations*, Studies in Systems, Decision and Control 247,
https://doi.org/10.1007/978-3-030-30659-5_24

method of physical and natural sciences. Weber works in the cultural context of German historicism and comes to the formalization of the object and method of sociology within the framework of an interpretivist paradigm of social action.

2 The Method in Sociology

The events that contribute to the birth of sociology must be sought in the eighteenth-century revolutions, which modify the structures and systems of economic production and the political reproduction of society. Sociology is the product of the crisis of social integration systems that has occurred in Europe since the eighteenth century following the French revolution and the British industrial revolution.

The new science of Sociology has the task of studying the laws of operation and change of society. The name of the discipline is due to Comte, who was concerned with building a science that can explain and predict social development. The cornerstones of the system are represented by the study of statics and social dynamics (Comte 1830–1842).

Historical-social events are a necessary condition, but not sufficient to determine the birth of sociology, we must also define the method by which the sociologist must study social phenomena. It is necessary to step back several centuries in order to understand the methodological aspects and go right back to Galileo Galilei (1564–1642) to find the methodological presuppositions on which modern science is based (Marradi 2007, p. 79). Galileo defines the characteristics of a scientific method suitable for the knowledge of nature (Amaturo 2012, p. 25). The aim of scientific work is the formulation of laws, valid in every time and place, able to explain the functioning of nature (Marradi 2007, p. 13).

Another important characteristic of a scientific work, which has consequences for the early developments of sociology, is the affirmation of the uniqueness of the method by which the sciences describe and explain the phenomena investigated, regardless of their physical or social nature. Science is understood as the discovery of laws according to a procedure of accumulation of knowledge.

Comte also promotes a positivist form of sociology within the framework of the acritical acceptance of the scientific method (Hughes and Sharrock 1997). Comte develops the idea of a sociology able to describe and explain social phenomena through the formulation of laws, and to foresee their change; as the method of physical and natural sciences has been able to arrive at the discovery of laws that govern nature, so sociology must work in search of general laws that govern the social world. In this position the origin of the discourse on the method must be identified, which is also a substantial basis for the birth of the science of society (Crespi 1985, p. 44).

For Comte the method is objective/experimental; social phenomena are related to each other in a constant relationship of cause and effect. Sociology must follow the following aspects of the process: observation of the facts and collection of data on a given phenomenon; formulation of hypotheses to explain the phenomenon; experimental verification; rejection of unverified laws.

The connection between sociology and the natural sciences is crucial; the techniques of the survey are the same as those of the natural sciences: observation must be guided by a preliminary theory; experimentation, even if partially possible in the context of human societies, remains a model valid also for the social; the comparison assumes the function of an important aspect in the application of the method (Smelser 1976, it. 1982, p. 40).

The positivist position contributes to give scientific basis to the discipline, albeit with all the limits posed by excessive confidence in the natural science method and its applicability to the study of society and the social behavior of people. The work of Émile Durkheim and Max Weber is within this scientific and cultural framework.

3 Émile Durkheim

The positivist sociology of Durkheim should be taken as one of the most significant starting points for the formation of the sociological method. The French sociologist anticipates the discourse in The Division of Social Work of 1893 and systematizes its own approach in 1895 with the publication of The Rules of the Sociological Method; in the study On Suicide in 1896 he provides empirical application. For Durkheim the knowledge of social phenomena must take place with the same method adopted by the physical and natural sciences, in order to arrive at the formulation of analogous laws in their form and capacity for explanation, to the laws of the natural sciences.

Durkheim is part of the naturalist positivism of the nineteenth century; his method involves observation, hypothesis formulation and comparison. From this methodological position are defined the epistemological bases of sociological analysis and the identification of the object that specifies the field of application of sociology: social facts (Durkheim 1895, it. 2001, p. 25).

The French sociologist precisely identifies the social fact, distinct from the generality of human events and behaviour, capable of characterizing sociology with respect to other social sciences in terms of the specificity of the object of study. The social fact is for Durkheim every way of behaving, more or less fixed, which is able to exert an external constraint on the individual. The social fact:

(a) is a way of behaving that exerts an external constraint on the individual to act in a certain way;
(b) imposes itself on the individual independently of their willingness to choose and can not be easily modified by the individual;
(c) has its own objectivity independent of the individual who observes it, who is conditioned by it in acting.

Social facts are not material entities, but have the same properties as things in the natural world (Corbetta 1999, p. 25). Social facts are regulated by laws based on the cause-effect relationship; they exist independently from observers, they are external to the individual, they pre-exist and condition their behaviour, they are coercive, because they impose themselves to the individual and are accompanied by sanctions or forms of social control, they have a general value, as they are valid for a large number of people. Social facts are therefore external to the human will, they condition it and limit it. The social facts must be explained by following this method: researching the cause that produces the social fact and identifying the function that the social fact plays within the social system. The cause must be sought in precedents, the function can be identified from its purpose, while the origin is given by the social environment within which the event takes place.

With the causal and functional imputation method of Durkheim, sociology assumes the character of a scientific explanation of social phenomena, which occurs through the determination of a cause-effect relationship between events, supplemented by the functional explanation. In this way Durkheim's program of methodology formalizes sociology as an autonomous science from the other sciences for the method and for the object (Durkheim 1895, it. 2001, p. 12).

The facts of which Durkheim speaks have a specificity not attributable to biological or psychological interpretation; the explanation is related to social determinants and takes place with reference to the socio-structural context in which they occur. The method is inductive, causal and functional; it rejects any reference to psychological or subjective dimensions or explanations; the goal is to establish cause-effect relationships between events, in order to reach general laws.

The methodology program of the French sociologist reveals cognitive needs derived from the specificity of the object; continuity with nineteenth-century positivism marks its ontological, epistemological and methodological characterization. During the twentieth century the nineteenth-century positivist paradigm, on which Durkheim's program of method is based, will have further developments and will be articulated in a progressive systematization that will see the emergence of post-positivism, in the matrix of Lazarsfeld, and of contemporary neo-positivism (Lazarsfeld 1958a, b). An undoubted influence on the sociology of the twentieth century should therefore be recognized in Durkheim.

4 The Method in Max Weber

Weber places the individual and social action at the center of sociological analysis. In defining the object and method of sociology he follows a different path from that of Durkheim. For the German sociologist social sciences are part of the sciences of the spirit, therefore there are epistemological and methodological limitations for the application of the cognitive procedures of natural sciences. According to Weber, in the study of society and social phenomena it is not possible to use mathematical-causal methods and it is not possible to arrive at the formulation of laws with general validity.

This position derives from the debate in Germany, between the late nineteenth and early twentieth century, which contrasts scientific and humanistic culture and aims to establish a precise distinction between the natural sciences and the sciences of the spirit, or historical social science. Weber's sociology is in line with the positions of Wilhelm Dilthey (1833–1911), Heinrich Rickert (1863–1936), and Karl Theodor Jaspers (1883–1969).

Weber takes part in the debate on the method from 1903; in 1904 he published the article The cognitive "objectivity" of social sciences and politics. The position is specified in the 1913 article, Some categories of comprising sociology, linked to a sociology based on the interpretation of human activity. Since 1909, Weber began working on Economy and Society, published posthumously in 1922, in which he defines the fundamental concepts of sociology and analyses relations with economy, law, politics and religion.

For Weber sociology, because of the specificity of the object of study, must operate with different methods from those of the natural sciences. Social phenomena are characterized by uniqueness, individuality, unrepeatability. Society itself is made up of individual actions, therefore in order to understand social phenomena we must start from the individuals and their actions. The sociologist must base their explanation not on external facts but on the understanding of the meaning of human-social action, identifying the motivations of the individual to act in a certain way (Weber 1922a, p. 241).

The object of the sociological inquiry is social action. Knowledge takes place through the understanding of the intention behind the action, which takes on typical forms in relation to the particular historical context in which the individual finds themselves. In Weber's sociology there are no social facts as in Durkheim's discourse, but only actions intended in a social sense. The researcher is called to investigate social actions; he must discover the meaning that the action assumes for the person who performs it (Cesareo 1993, p. 15).

For Weber, the sociologist, in order to understand society and social phenomena, must include in the analysis the individual and the subjective meaning of social actions, which must be studied with a comprehensive method, based on understanding the meaning, the subjective sense of individual action. In other words, sociology must understand, through an interpretative procedure, the social action of the individual and explain it causally in its course and its effects. The method consists of two moments: the understanding of the subjective sense of social action and the explanation of it. By understanding we mean referring to the meaning that action has for the person; to explain the process aimed at identifying the cause that pushes an individual to act in a certain way. Understanding is aimed at identifying the meaning, the motivation of action, while the explanation is aimed at describing the cause that is at the basis of acting and identifying any patterns.

The sociological explanation is achieved through researching and identifying the patterns of action, analysing the causes of a typical behavior. Consequently, the Weberian method consists of two concepts: understanding, and the ideal type. The first is explicit in the Weberian definition of comprehensive sociology, whose purpose is the reconstruction of the meanings that the individual confers to his social action; the second relates to the introduction of the concept of the ideal type concept. The ideal type consists in identifying the uniformity of social action; if one succeeds in understanding the reason and the cause of individual action, the ideal type allows the causal imputation of tendential uniformity of social action.

The ideal type is a framework of causal imputation, which does not represent reality, but serves as an interpretative framework within which reality is brought back, measured and compared. The construction of the ideal type is achieved through a process of separation and synthesis of elements of the observed reality (Weber 1922b).

In the research process the sociologist has an active role and is not a mere observer of social facts. They select the most significant empirical data and social problems that deserve to be investigated. Even the individual who performs an action plays a role in this process of knowledge. The human-social behaviour is grounded in the meanings that the person who acts socially gives to it; the explanation must be sought in the definition of the situation by the person who performs the social action and in the consequent process of understanding, on the part of the sociologist, of the sense attributed by the individual to the action.

For Weber, the collective and social structures do not exist independently of the individuals who compose them and who act in them socially.

With Weber, the object and method of sociology are defined in an individualistic sense. The sociologist, in order to understand social phenomena, must include in their analysis the individual and the subjective meanings that they attribute to their own action. The method cannot be that of the scientific-mathematical objectivity of the natural sciences, since social phenomena require a comprehensive method based on understanding the subjective meaning, the motivation, which lies at the base of human-social action; this happens through a methodological procedure that allows, by the interpretation of the causes, to a scientific knowledge of the social with a strong connotation of objectivity.

5 Conclusions

In sociology there are two traditions of thought and two different methodological approaches that originate from the works of Durkheim and Weber. The two positions clarify the differences both of the object and of the method of sociology.

For Durkheim social phenomena must be studied by identifying other social facts prior to the cause that determined them; the method is inductive, causal and functional. The paradigm of explanation is positivist, recognizes the independence of social facts and the conditioning role of individual behaviours; social facts must be studied with the same method of the natural sciences. With Durkheim and the causal and functional imputation method, sociology is characterized in terms of the analysis of causal relationships between variables, based on correlations between events, in a logic of cause and effect.

Weber places the individual and social action at the centre of his method; he does not recognize structures or objectivity outside the social action of individuals. The sociologist must seek and understand the subjective meanings of individual actions to explain social phenomena; the method is based on understanding the subjective sense and the motivation that underlies individual action.

While individual action in Durkheim's sociology depends on external constraints, in Weber social action is rooted within the individual himself, who acts on the basis of the meaning subjectively attributed to his action. From these two positions derive different methods of investigation aimed at treating objective variables in Durkheim, and subjective and motivational variables in Weber.

Durkheim and Weber represent important theoretical and methodological matrices for twentieth century sociology. The methodological positions of these two classical sociologists correspond to paradigms that refer to the duplicity of the object of sociological analysis, and which contrast a sociology of structure with a sociology of action, as well as holistic approaches to individualistic approaches.

The developments of the sociology of the twentieth century are to varying degrees marked by the theoretical and methodological positions of Durkheim and Weber, who have contributed to defining the object, paradigms and methods of twentieth century social research.

References

Amaturo, E.: Metodologia della ricerca sociale. Utet Università, De Agostini Scuola, Novara (2012)
Cesareo, V.: Sociologia. Teorie e problemi, Vita e Pensiero, Milano (1993)
Comte, A.: Corso di filosofia positiva, UTET. Torino, 1967 (1830–1842)
Corbetta, P.: Metodologia e tecniche della ricerca sociale, Il Mulino, Bologna (1999)
Crespi, F.: Le vie della sociologia, Il Mulino, Bologna (1985)
Durkheim, E.: Le regole del metodo sociologico, Edizioni di Comunità, Milano, 2001 (1895)
Hughes, J.A., e Sharrock, W.W.: The Philosophy of Social Research, Longman (1997)
Lazarsfeld, P.F.: Problemi di metodologia. In: Lazarsfeld, P.F., Metodologia e ricerca sociologica, Il Mulino, Bologna, 1967 (1958a)
Lazarsfeld, P.F.: Dai concetti agli indici. In: Lazarsfeld, P.F., Boudon, R., (edn.) L'analisi empirica nelle scienze sociali. volume I: Dai concetti agli indici empirici, Il Mulino, Bologna, 1969 (1958b)
Marradi, A.: Metodologia delle scienze sociali, Il Mulino, Bologna (2007)
Smelser, N.J.: La comparazione nelle scienze sociali, Il Mulino, Bologna, 1982 (1976)
Weber, M.: Economia e società. I. Teoria delle categorie sociologiche, Edizioni di Comunità, Torino, 1999 (1922a)
Weber, M.: Il metodo delle scienze storico-sociali, Einaudi, Torino, 1997 (1922b)

"Our Idea of School": Democracy Workshops in the Classroom

Fiorella Paone

Abstract The chapter derives from didactic activities experienced in some schools of the province of Pescara, highlighting its inclusivity and democracy. The pedagogical aim is to make students' point of view related to their school everyday life emerging in order to support their active role and participation to the cultural evolution process, in line with the latest perspectives of research of the movement Student Voice.

Keywords Inclusive pedagogy · Educational relationship · Responsability · Empathic listening · Scholastic pedagogy · Democratic school · Communicational barrieres

1 Introduction

The[1] chapter reflects on the importance which the emphatic listening of students' experiences can have in the process of cultural emancipation and self-awareness of the students themselves. Such an approach, indeed, is an important prerequisite of the pedagogical intentionality which aims to give space and voice to the opinions of school main characters, with the ambition to contribute to represent the nowadays school on the one hand, and to imagine future perspectives moving from the ideas, the needs and desires of the students on the other hand. In fact, starting from their words, it is possible to try to comprehend and guide the educational and training intervention in a scholastic well-being and equal opportunities-oriented direction, promoting educational paths of inclusion and active citizenship.

We agree, indeed, with the movement *Student Voice* results and its perspective of research based on the principle that (Cook-Starter 2013: 27–28): *there is something*

[1] In the chapter, all translations of students' quotes, Italian books' quotes, titles of books and professional organizations' names are to be taken as my own translation.

F. Paone (✉)
"G. D'Annunzio" University of Chieti, Chieti, Italy
e-mail: fiorella.paone@unich.it

D. Flaut et al. (eds.), *Decision Making in Social Sciences: Between Traditions and Innovations*, Studies in Systems, Decision and Control 247,
https://doi.org/10.1007/978-3-030-30659-5_25

deeply wrong in building and re-building a whole scholastic system without giving attention to people the system is made for (...) It is now time to consider the students and their rights to join both the critical analysis on education and the renovation of it. *Student voice* is the name of a pedagogic movement born in the '90s and internationally spread. Even if not so practised in Italy, it is deeply-rooted in UK, USA, Canada and Australia. It adopts a perspective of research that, even with different currents, it is based on the idea that voice of the students is to be recognised as a the powerful factor which changes school.

In order to show some tangible educational activity which implements that perspective, the "*Letters to School, listening to the real school and collaborative writing*" will be presented. The «Letters to School» project aims to give a start to a dialogue to reflect on the education's goals of a society which has to put the whole humanity at the centre rather than the performance. By involving teachers, students and workers in the sector students' contributions will be collected in order to write a cooperative text which portrays the current school reality and which can imagine the one of the future. In the organisers' intentions everyone can join the project sending a letter or a drawing, a video, any product which can express the needs of those who live the school everyday. It has been an activity launched at the beginning of the school season 2018/2019,[2] by the *Teachers Democratic Initiative Centre (CIDI)*[3] in Pescara and promoted by Curzio Maltese, Member of the European Parliament (MEP) and coordinator of the group in the Culture and Education Commission at the European Parliament.

This educational path of elaborating *Letters to school* invites the students to write both and individual and cooperative text which supports them to focus on the "ideal school".

By the analysis of the letters produced by the students involved, our research has the general aim, thus, to contribute and examine in depth the studies done within the *Student Voice* perspective, giving space to school ideas of the students in order to detect the characteristics of a teaching/learning path in line to their needs. More specifically, the aim is to critically analyse and focus on representations, feelings, expectations and wishes that students involved in the research have about their school experience, making the most meaningful traits emerge. Finally, the further aim is to define the pedagogical-didactic characteristics of a teaching/learning path able to give value to the students' experiences in an inclusion and socio-cultural promotion-oriented perspective.

---

[2]In the conference "The cancelled school?" ("La scuola cancellata?") held in Pescara the 6th September by the local CIDI in cooperation with the italian periodical *Insegnare*.

[3]As we can read from the Association's website CIDI is: *the Teachers Democratic Initiative Centre, an association of teachers from all levels of school and disciplines: its functions are promoting a teaching professionalism suitable to school needs, to implement and implementing reforms (...) Its first aim is contributing to build a more democratic school, with more culturally structured, nearer to girls and boys' interests (...) The Centre defines its democratic trait because it believes that considering culture as an emancipating and freedom mean is a democratic fact (art. 3 of Constitution). The first Centre was founded in Rome in 1972 (...) In 1979 national CIDI established with the aim to coordinate many local CIDI* (Available online at: http://www.cidi.it/site/cidi/cidi).

2 Methodology of the Research; Analyzing the School from Students' Point of View

The reflections presented derive from a research that, moving from the perspective of *Student Voice*, put the students' point of view as fundamental factor for the production of a pedagogical diagnosis on the current conditions of the school. Analysing students' experience starting from their own statements, in fact, could be interesting from a theoretical point of view at first, since could help to select the development line where its ordinary school life moves to.

From an applicative point of view, the critical reflection on the contents of the letter could help to draw the features of a pedagogical-didactic offer functional to the emerging needs of the classes.

A *case study*, thus, has been implemented and its has been about the documental materials produced by the students within the participation of schools of Pescara district into the initiative "*Letters to School*".

In order to give the research a start, it has been arranged a meeting with the whole of the teachers of the classes involved. The teachers have given all the information linked to times and contexts of the initiative "*Letters to School*" in their classes and they have been available to make all the materials produced by the students usable for the research's goal.

It has been done a documental analysis on the texts of the letters produced both in written and multimedial code. We want to highlight that, in accordance to what supposed by the documental observation, letters were elaborated independently from the researcher's action and with different aims from those of the research (Corbetta 1999, 437). This trait of independence from the research is a strength point since it means that the letters are not responsive, that is they are not influenced by the researcher-individual interaction, so avoiding potential distortions.

According to what the teachers declare during our meeting, we believe that proposing a educational path of elaborating a *Letter*, by which the students can express their point of view about their school experience, means moving the first step towards a student-centric system with students' hopes, feelings, troubles and potentialities.

It means giving deep relevance to their point of view and their role as main characters; it means overcoming the asymmetry of power which makes them passive actors and mere "containers to fill" that is rather the opposite of what school is made for, that is a school which teaches and guides to social participation, active citizenship and active use of the word, a school which makes the democratic pedagogical practices as tangible as possible.

An education based on solid democratic foundations should be a solid alternative to habitual meritocratic and selective approaches based on an obsessive evaluation attitude and based on discouraging methodological and cultural confusion, it should be an important point of reference building a school as a place where everyone can access the contents and the instruments of the culture, where social, economic, linguistic, cultural and individual differences does not turn into inequalities, a place where the full development of the person is supported in a holistic and emancipative perspective, according to Constitution. We specifically refer to art.3 od the Italian Constitution which states that: *All citizens have equal social dignity and are equal before the law, without distinction of sex, race, language, religion, political opinion, personal and social conditions. It is the duty of the Republic to remove those obstacles of an economic or social nature which constrain the freedom and equality of citizens, thereby impeding the full development of the human person and the effective participation of all workers in the political, economic and social organisation of the country.* (Available online at: https://www.senato.it/1025?sezione=118& articolo_numero_articolo=3).

We still need activities as that of "*Letters to School*" because the "Teacher" to whom students of Barbiana addressed still lives in the classroom (Cederna 2017: 53) with her authoritarian attitude and her melancholy for the strict school of the past and its failures, bad marks and strictness, a school where ordinary life avoid responsibility assumption by the students, a school where the weakest one are left behind, a school that turns into *hospital which recovers healthy persons and leaves behind the sick ones* (Scuola di Barbiana 1967: 15).

3 A Space of Dialogue Within a Plurality of Approaches

A first important fact emerges from the listening of the letters wrote by the students and from what told to me by the teachers about the didactic experience lived[4]: the suggested activity managed to build a communicative and relational context where students felt comfortable and motivated to express their feelings and ideas at school. This awareness confirms the main importance of taking care of the communicational dimension, which is an "invisible" prerequisite (Bonetta 2017) not sufficient but necessary, of the good result of each educational activity.

The opening of this space of dialogue among students, far from the attitude of judgements and marks but, on the contrary, making explicit that everyone's opinion would have been represented, turned on a process of responsibilisation which avoided a complaisant attitude towards adult's expectations on the one hand, and an oppositive one against school in general on the other hand.

Dealing with that, we can read in the letters:

- *We worked very hard (…) we reflected and we had a proactive attitude, we faced constructive exchanges*;
- *We got very involved in the task of the cooperative writing that has turned into a very beautiful experience.*

[4]Invited by local CIDI, four Schools of Pescara joined the invitation by Curzio Maltese involving teachers and students from 6 classes of middle and high school in the elaboration of a "Letter to school" during the October, November and December 2018. The schools, specifically, were: Liceo Sportivo dell'IIS "Volta", I.C. PESCARA 8,IIS "Emilio Alessandrini" Montesilvano- Pescara, IT "Tito Acerbo" of Pescara.

Every group worked on the topic "The school I wish" with the aim of writing a letter, but the results differ form one another in terms of texts produced. The students resorted to the written and spoken word, pictures, music, imagines in movement to express their feelings and thoughts. The language codes, therefore, were verbal and non-verbal depending on free choices of the students themselves who, thus, elected the most suitable medium and modality in tune with their needs and expressive inclinations, highlighting talents usually not appreciated in the ordinary educational activity.

Every teacher, besides, has organised the activities in different stages, but with different educational choices for each paths. Among the teachers, indeed, some has chosen the presentation of audio-visual inputs in order to create a climate of reflection on the topic (*Once in a Lifetime* film[5]), o narrative inputs (passages from *Letters to Teacher* of the Barbiana's School), others have chosen the individual and cooperative writing, another the *brainstorming*, or the collective debate, or the *circle time*, small group tasks or plenary discussions.

Anyway, teachers' stories makes emerging that each educational activity promoted shared non-classroom-taught lessons in order to create an effective contact and reduce the relational gap which, as the letters report, plays a negative influence on educational experiences and results. The active strategies, thus, were elected, non-direct-evaluative approaches able to promote an educational path of teaching/learning based on the dialogue and mutual emphatic listening (Hoskova-Mayerova, Maturo 2018, 2019). The teachers agree on such methodological choices as the key factor of the good results of the activity, methodological choices able to identify with the specific class context and able to combine with the specific characteristics of its main characters (teachers and students), indicating that to experiment never means reproducing pre-established practice, but professionally making methodological and content choices which moves from the acquaintance and respect of the other.

In this regard, we recall the words of Teacher Manzi, who declares: *If I should come back to school today, I would not do what I did till the last school day, because children change as well as we do* (Farnè 2011: 143).

4 The Words of the Students

Analysing the "letters", that is texts produced by the classes in the written or multimedia format, the most frequent terms are: discomfort, boredom, malaise, obligation, incomprehension, distance, hurdle.

The students complain the strict and irrelevant educational modalities as classroom-taught lesson, an impression supported by a number of data collected. The survey INVALSI (2016), for instance, reveals that traditional educational activities are yet prevailing, while activities of research in pair or in small groups and workshops are still uncommon.

[5] *Once in a Lifetime* (Les héritiers) is a 2014 film directed by Marie-Castille Mention-Schaar.

Even if the students do not have knowledge of these researches, they are aware of their own educational experience and in the letters they suggest solutions as having more activities and projects by external professionals or selecting such an American College-model.

Infact, in the letters we can read:

- *In USA and in other countries high-school students are allowed to choose some optional matters together with the obligatory ones and we think it is a very good option since the student is in condition to choose her/his own educational path responsibly and independently and we strong believe that the final results should satisfy our expectations in the end;*
- *We would like to have such a structure as the American college, where everyone gets her/his cabinet, more labs, a school canteen where having lunch all together, a broad courtyard and finally it would be great to have a classroom for each matter. We think that school should be both a source of education and a place where learning to live, fight and encounter with new people from the early age;*
- We would like to have a more structured education in terms of journeys, cooperative, conversation, graphics, informatics and arts workshops, with suitable tools and extra-school activities.

These solutions are sometimes naive and even dysfunctional, according to us, arguably linked to the most widespread and homologising media consumption—on which we would have to pay more attention—that makes the students themselves joining this process since they are the first observers of school dynamics, but with no cultural tools to elaborate their intuitions.

Moving from this alliance between students and teachers we can verify the hypothesis that the wish to turn the school-time in a set of projects corresponds with the need of a non-formal teaching methodology which foster the students' interests and move from those interests in order to expand their horizon and their cultural baggage. The wish to have "different teachers" with passion for their work, prepared to carry out self-examination and willing to grow up together with their students, in the researcher-teacher perspective[6] We could verify if the American college, usually seen through movies and fictions, represents to the students a school model able to personalise the educational paths, to make the students experience workshops methodology and different learning environment with motory activities, extended relaxed times which allow cooperative activities and interaction with different groups of peers and adults (Hoskova-Mayerova, Rosicka 2012).

[6]In the letters we can read:

– Nowadays we think that one of the problems of the school is the lack of teachers able to make us love matters they teach;

– We would like a school where everyone could express her/his opinions to build a more engaging classes together both with classmates and teachers in order to have a more fruitful and less hard lesson during school-time and so have less tasks at home. Such good communication would help us to have a more constant and effective working pace. In particular, we would like both the students and above all the teachers to open their minds and knowledge.

One thing for sure is that in the letters the students do not see explicitly these links and we are not so sure that we have adequately interpreted their wishes and needs, we argue that going deep in this matter is a right which a teacher has to grant. In the classroom projects of research and discovery are promoted in order to make the above mentioned perspectives tangible in line with what *Indicazioni Nazionali per il curricolo della scuola dell'infanzia e del primo ciclo di istruzione*[7] (MIUR 2012) suggest to even if it is left no space for such a purpose in the school.

In the letters girls and boys state that would the adults to recognise their autonomy:

- *what autonomy we talk about if our parents control everything we have to do once at home?*
- *school digital registers make us feel under the constant control of our parents.*

Students state that there should be suitable spaces and tools as well as free books:

- *we certify more and more that scholastic tools are not appropriate (...) in the compulsory school and books should be for free;*
- *cabinets for our school bags to be lighten would be very useful, a school courtyard where to study or spend our spare time during the breaks and eventually a more decorous and equipped site, starting from toilets.*

They tell they perceive a school more and more in trouble:

- *today society shows a particular helplessness to solve problems of school which seems to be in a deep and never-ending crisis it can't come out from.*

They tell they would have no school failure to exist and classrooms with lower number of students:

- *a solution could be having fifteen-student classrooms at maximum, in order to promote concentration both for students and teachers;*
- 15-*student classrooms would be more appropriate in order to follow each of them quietly and so personalise each educational interventions according to the student's needs;*
- *I wish a school with no failures, where everyone is given his/her time he/she needs to reach the goal.*

They state that everyone has special needs and labels are not fair:

- *each girl/boy has particular educational needs and nobody should be labelled;*
- *a school where nobody (teachers included) stops at prejudice and appearance, where nobody feels different.*

They tell that they wish a fair school which does not judge by mark-logic which they fail to understand, perceiving that those marks are used on a non-educational perspective:

---

[7]Currently the only regulation of reference for all Italian schools is "*Indicazioni Nazionali per il curricolo della scuola dell'infanzia e del primo ciclo*", entered into force with the D.M. 254 of 16 November 2012 (G.U. n. 30 of 5 February 2013).

- *we would like to have teachers with no preferences among students. We believe that motivations on marks should be given, so avoiding conflicts with teachers and knowing where and how to improve;*
- *we would like you [school] to be more evolved, open-minded and welcoming for each student without inequalities;*
- *we spent most of our lives at school and so we would like to feel good there;*
- *we wish a non-mark-identification school.*

They tell that teachers do not have respect and are not aware of their efforts, on the contrary teachers tend to undervalue:

- *there is no fair relationship with the teachers, since they only try to make us feel uncomfortable, undervaluing our abilities;*
- *teachers' indifference hurts us, they do not appreciate our efforts and so we are not gratified;*
- *the respect is very important matter to us, because where respect, there a comfortable climate.*

The Ocse-Pisa survey (2015) goes in that direction testifying they are not wrong: a test result is less brilliant if students perceive that the teachers think they have less competence than they actually have.

Besides, the teacher's evaluation influences a lot student's self-esteem producing states of anxiety which affect the performance.

A widespread state of anxiety is testified by the students reflections in the letters where they state that:

- *we live several days of anxiety and sadness*;
- *I started having problems in almost all the matters and each day was like hell (…) I was upset and scared for a failure*;
- *the school crisis arises as a discomfort for all people who deal with school*;
- *we would like to stay safe and comfortable without the recurring bullying, a phenomenon based on a series of abuses, humiliations, weak and strong psychological tortures by one or more boys in "bully gang"*;
- *I was really falling ill…I cried each days because of school duties*;
- *unfortunately we can't find comfort in the school for bullying episodes towards fragile and "defenceless" persons very often regarded as "different".*

Moreover, the above mentioned OCSE-PISA survey testifies that Italy, together with Portugal, ranks in first place in terms of anxiety, fear and worry perceived by the students also due to internal factors of the school. Among the 15-years olds' interviewed in the above-mentioned survey, in fact, 56% states to study with high level of stress, 77% to be very nervous and not able to do a task, 85% to be worried by getting bad marks, 70% to live into a high level on anxiety before a test, even if one has studied for it. The anxiety is linked to a decrease of performance.

In the letters we can read that the students feel obliged, bored:

- *I live school time deeply boring (…) teachers do not engage us effectively. This leads students to carelessness and confusion.*
- *School time seems never to pass. It's like torture.*

It is a fact that nowadays the school and its extra-scholastic proposal is less and less attractive to students, in opposition to what happened during Barbiana's period, where they contributed to their families' efforts as an alternative to school.

In the *Letters to Teacher* we can read, for instance, that students believe that (Scuola di Barbiana 1967: 10) *the school is even so better than the shit* which they would have had to shovel if they had to go to work to their fathers' stables.

Taking into account this cultural and social difference of the context, in which the school places, does not correspond to a "superficial" educational proposal, but it means keeping improving methodological abilities in order to make those proposals more effective, suiting them to the particular situation, beyond what such a "pedagogical automatism" manuals suggests.

In conclusion, we would like to highlight that in the letters students write that they feel as in prison:

- *the most difficult hurdle is to stay seated behind such small and uncomfortable desks suffering those obligations that make us feel as in prison;*
- *I can't understand how they can oblige us to shut up and immobile… we are young and alive;*
- *we have done nothing wrong, school should not be as a punishment.*

By the analysis of these words, as during Barbiana's times, there are still relations of power based on the strong opposition teacher-student, students who feel the teachers blind and deaf towards their needs:

- *thinking about school I don't like the couldn't-less-care attitude of some teachers who abuse their power for a psychological terrorism (…) How can we trust on such persons who praise their "higher" certificates but lacks in humanity?*
- *We would like teachers to understand and appreciate us, debate with them but they criticize us*

The teacher still acts the role of "school master", who *transmits* rather than *communicating* (Dolci 1988), who punishes rather than making them to feel responsible, who controls rather than promoting autonomy.

From the letters a self-critical attitude of the students towards their own behaviours that they judge unfair comes out:

- *we have responsibilities too because sometimes we don't respect teachers and schoolmates and we pay attention to reprimands only when converted in official suspension;*
- *I would like to talk about the wrong use of smartphone, because students spend a lot of time using it;*
- *what autonomy we talk about if we don't even care to note down school duties on our diaries due to the digital enhancement?*
- *right from the start the class has got an unsuitable attitude towards the school context, a mistake that specifically influenced the collective results of the students;*
- *at the end of the lessons the rubbish left on the floor is a very dirty sight.*

5 Conclusions

In conclusion, it seems that the *Letters of Barbiana* (1967) promoted democracy and social participation as well as today such educational participative practice near to class needs can activate virtuous processes. Processes able to involve both emotional and cognitive dimension of students and teachers and build communicational bridges which promote cooperation and fruitful contact, renewing motivation and passion to live the school and its activities.

Dialogue promotion means, indeed, educating to the emphatic listening and respect, it means to train to have opinions, argue, debate.

It means giving a place where focusing on our own ideas, turning experience into skills, getting in touch with our own feelings, beginning to host and work on them functionally to the cultural and human growth of our life project (Blezza 2018).

It means living that inter-subjective dimension which grants the word right to everyone, respecting the students, a dimension the cultural psychology considers as a prerequisite for any kind of learning process (Hoskova 2009).

Research has shown that:

- non-classroom-taught lesson educational paths help the students to meet challenges and become aware of their own ideas;
- class climate improves if the students have the space where express their opinions and feelings about the school experiences they live;
- non-direct-evaluative educational activities promote positive participation of the students;
- communicational distance between students and teachers is too often present in the school still and it is dysfunctional to inclusion and socio-cultural promotion good results which are general and final aim of the education system (Paone 2017; Hoskova-Mayerova 2016).

It is considered appropriate, thus, to follow-up students' inputs given in their letters, developing structured pedagogical strategies which support educational paths able to follow the direction that the students themselves have suggested, sometimes implicitly. The research highlights the importance of some pedagogical-didactic principles, such as cooperation, mutual identification, rules negotiation, clear goals sharing, clarification of criteria and modalities of working, respect for differences. These principles should convert into ordinary activities which make the students living in an authentic educating community.

We consider that this can renew their motivation, responsibility and enhance their gradual acquisition of languages, knowledge and tools to consciously face their learning path.

Taking this direction means to make the students matter for the school since in this way what school does concerns them.

These words could be heard as the echo of that *I Care* written on the walls of Barbiana, that *it is my concern (mi riguarda*) in opposition to the *Fascist 'I don't give a damn'* (*me ne frego fascista)* Scuola di Barbiana, 1967), the echo which encourages to address the pedagogical intentionality to take care of oneself and of the others.

References

Blezza, F.: Pedagogia Professionale, Che cos'è, quali strumenti impiega e come si esercita. Libreriauniversitaria, Padova (2018)

Bonetta, G.: *L'invisibile educativo. Pedagogia, inconscio e fisica quantistica*, Armando, Roma (2017)

Cederna, G.: (eds.) *Atlante dell'infanzia a rischio 2017. Lettera alla scuola*, Save the Children & Treccani (2017)

Cook-Starter, A.: *Legittimare il punto di vista degli studenti. Nella direzione della fiducia, del dialogo, del cambiamento in educazione*, In: Grion, V., Cook Starter, A., (eds.) *Student Voice. Prospettive internazionali e pratiche emergenti in Italia*, Guerini, Milano (2013)

Corbetta, P.: *La ricerca sociale: metodologia e tecniche,* Il Mulino, Bologna (1999)

di Barbiana, Scuola: Lettera ad una Professoressa. Barbiana, Libreria Editrice Fiorentina, Firenze (1967)

Dolci, D.: Dal trasmettere al comunicare. Non esiste comunicazione senza reciproco adattamento creativo, Sonda, Alessandria (1988)

Farnè, R.: *Alberto Manzi*. Bononia University Press, Bologna, L'avventura di un Maestro (2011)

Hoskova, S.: Experience with blended (distance) learning study materials. In: Hrubý, M., (eds.) Distance Learning, Simulation and Communication 2009, Proceedings, pp. 70–77. (2009). ISBN 978-80-7231-638-0

Hoskova-Mayerova, S.: Education and training in crisis management. The European Proceedings of Social & Behavioural Sciences EpSBS XVI, pp. 849–856 (2016). https://doi.org/10.15405/epsbs.2016.11.87

Hoskova-Mayerova, S., Maturo, A.: On some applications of fuzzy sets and algebraic hyperstructures for the management of teaching and relationships in schools. Models and Theories in Social Systems. Studies in system, Decision and Control **179**, pp. 343–353 (2018) Cham, Switzerland: Springer https://doi.org/10.1007/978-3-030-00084-4_19

Hoskova-Mayerova, S., Maturo, A.: On some applications of algebraic hyperstructures for the management of teaching and relationships in schools. Ital. J. Pure Appl. Math. **41**, 584–592 (2019)

Hoskova-Mayerova, S., Rosicka, Z.: Programmed learning. Procedia—Soc. Behav. Sci. **31**, 782–787 (2012). https://doi.org/10.1016/j.sbspro.2011.12.141

INVALSI. *I processi e il funzionamento delle scuole,* (2016) Available online at: http://www.invalsi.it/snv/docs/141016/Rapporto_Processi_2016.pdf

MIUR.: *Indicazioni Nazionali* per il curricolo della scuola dell'infanzia e del primo ciclo di istruzione,* (2012) Available online at: http://www.indicazioninazionali.it/wp-content/uploads/2018/08/Indicazioni_Annali_Definitivo.pdf

OCSE PISA 2015, *Students' well-being*, Available online at: https://www.oecd-ilibrary.org/docserver/9789264273856-en.pdfexpires=1549203686&id=id&accnameaccname=guest&checksum=5D73C62965D4483B57C054FDC3A676E5

Paone, F.: *Strategies for a sociological diagnosis of communicational environment of students.* In: Hoskova-Mayerova, S., Maturo, A., Kacprzyk, J., Soitu, D.T., (eds.) (a cura di), *Recent Trends in Social System: quantitative theorie and quantitative models,* Springer International Publishing, Zürich, 2017: 267–280

Aspects on the Evolution of the Press of Dobruja at the Turn of the 19th and 20th Centuries

Angela-Anca Dobre

Abstract The chapter deals with the development of the press of Dobruja in close connection with the stages of the social-political and economic evolution of the Romanian provinces returned to Romania in 1878. Previously there were very few local publications. Their number increases exponentially after the Union with Romania, the period from the turn of the 19th and 20th centuries, representing the peak of this phenomenon. The natural consequence of the growth and thematic diversification of the local press profile is the organization of journalism as a profession, a phenomenon that is part of the national efforts in this regard. The journalists from Dobruja, inspired by the effervescence of the national press, have also implemented journalistic principles at a time when the profession was in full organization and the definition of intellectual, aesthetic and moral landmarks was made. The period is marked by characteristic and memorable position in various national and especially local issues, articles and even newspapers integrating the work of high value publicists. Among them, we can mention: Ioan N. Roman, Petru Vulcan, Constantin N. Sarry, Constantin Moisil and others.

Keywords Press · Dobruja · Journalists · Social-political life · Modern age

1 Introduction

Following the War of Independence of Romania in 1877–1878, after four centuries of foreign rule, Dobruja joined with Romania. Transdanubian Romania, as politicians, intellectuals and journalists, called it enters the course of general development of the Romanian state in all its economic, political and social-cultural aspects.

The economic development of Romania after 1878 was a decisive factor in the general progress of the economy of Dobruja, having positive consequences on the social - political and cultural life of this region. It is not accidental that the nineties of the 19th century represent the period in which the most publications appear in

A.-A. Dobre (✉)
County Library "Ioan N. Roman", Constanţa, Romania
e-mail: ancaangeladobre@yahoo.com

D. Flaut et al. (eds.), *Decision Making in Social Sciences: Between Traditions and Innovations*, Studies in Systems, Decision and Control 247,
https://doi.org/10.1007/978-3-030-30659-5_26

Dobruja, the domains covered by the press are being diversified and important publicists like Ioan N. Roman, Constantin Brătescu, Constantin N. Sarry, Ioan Bănescu, Constantin Pariano, Petru Vulcan, Constantin Moisil are being recognized.

A direct consequence of Dobruja's return to Romania was the appearance of the local press and of the publications in general. In the modernization process of the Romanian society in Dobruja, in the creation of a modern social and political framework, the press played an important role. It had no traditions in Dobruja ruled by the Ottoman Empire, though in the "Voința Națională" of June 19, 1907, it was mentioned that the appearance in Sulina in 1875 of "Gh. Avgherinos Printing House" of the newspaper "Stravopodis", in Greek (Dumitru-Zamfir, Georgescu 1978, p. 326).

The first Dobrujan publication was "Steaua Dobrogei", a "local interest sheet", which appeared in Tulcea on June 22, 1879, under the guidance of the owner of "Aurora Dobrogei" Printing House, Basile Brănişteanu. He published in the newspaper's pages, his liberal principles: "He will be the faithful sentinel in the defense of all the interests of the people…" Steaua Dobrogei "will not be sold to boast nor criticize except when the facts based on evidence will provoke. The past testifies to this…" Moreover, concerned about the situation of Macedonian Romanians, Basile Brănişteanu publishes in the pages of his newspaper the status of the Macedonian-Romanian Culture Society (Lăpușan 2014, p. 37).

In Constanţa, the first modern newspaper and one of the most important in this space, was "Farul Constanţei". It appeared on 12 May 1880 and was the official newspaper of Constanţa County, publishing on its pages: ministerial circulars and prefect's decisions, briefings from the City Hall, general and local information (Dumitru-Zamfir 1985, p. 160). However, during it life he also approached other themes, underlining the concern of various editorial staffs for accuracy, detail and timeliness (Brătianu 2003, pp. 22–23), as shown by a description by Bucuţa in 1878: "…The message is seated, as it is with an official sheet, and information rich enough and vivid to make one of the good sources of documentation on the everyday life of the province".

It had four pages, divided into two columns and a slightly larger format than the "Monitorul Oficial". It seems that both the pattern and the content were of good quality, the information was rich and written in a balanced tone because it was an official publication. As a time of appearance, the "Farul Constanţei" stretches until 1888 and, with pauses, until 1913, having a weekly appearance. In all of its work, 2200 numbers have appeared, totaling more than 11,000 pages in various formats.

In 1900, the press from Dobruja was already very diversified; there were newspapers and magazines covering the most diverse areas, ranging from daily information to cultural, agricultural, veterinary or teaching magazines.

2 Professional Organization

Professional organization is the logical step in the evolution of the press and the profession of journalist.

An important moment in the history of Romanian media was recorded in November 1871, when the first Romanian press congress took place. Representatives of 13 newspapers adopted a resolution of 15 points. Besides paragraph X, where freedom of the press, freedom of speech, along with other fundamental freedoms were considered "the cardinal points on which our public edifice resides" (Petcu 2002, p. 260), all other provisions were more political than professional.

In 1883, the Society of Journalists was created, a moment evoked by Nicolae Iorga in terms of "the development of the press, the multiplication of the newspapers and the proof that journalism began to become a career, inspire some journalists (…) the determination to form a society of journalists" (Petcu 2002, p. 261).

Regarding the professional organization of journalists in Dobruja, it lacks until the First World War, although the diversification of publications and the numerical growth of the media ministers increased to 1912, which required a guild organization. The cause may be the small number of journalists, but especially the lack of the spirit of collegiality among them, the animosities that sometimes go through the printed rows or other reasons not necessarily of the profession but of the human nature.

Only in 1916 the journalist Peter Vulcan appealed to colleagues to organize a union press, which was not materialized because of the events of that summer, although on 5 April 1900 the first Union of Journalists in Bucharest was established. He advocated in his statute "the protection of the moral and material interests of journalists; the gathering of collegiality and professional solidarity between journalists; raising and maintaining the prestige and dignity of the corporation; creating an Aid House and boarding houses for union members" (Petcu 2002, p. 263).

Apart from the purpose of providing information and commentaries on current themes in the epoch, the Dobrujan press propagated ethical and moral principles that would help journalists in the exercise of their profession. They were concerned with raising the level of journalism, increasing the professional prestige and the profession itself, increasing the role and influence of the press in Dobruja and, without immediate result, the professional organization.

Like today, scandal news, a dubious quality press, often with gossip and slut, were being grown in the epoch. Quality journalists are organizing against this type of press and are campaigning for a press law. In 1931, the local newspaper "Dacia" published the article entitled "Law of the Press", in which it was said: "For our honor, for the defense of our unbridled existence, for the cult that we preserve the public and citizen's liberties, together with all our confreres, we picked up an honest and sincere voice of protest. But we can not forget one thing: that the respect of the written word has begun to cease to exist. The press has begun not to be the lab to cultivate ideas that are useful to the public. The press is no longer that noble nobility of words and ideas, it is no longer the altar whose servants have the duty to be considered priests of a cult to whom sacrileges can not be made. Because with the honest press, the press that defends it spreads, its ideas and beliefs with correct and knightly means, it has emerged as a sign of the times of irresponsibility and moral imbalance in which we live, another press: the revolver press, the blackmail press, the unscrupulous commercial business press, the pornographic press (Cheramidoglu 2011, p. 62).

On the need to regulate the profession of journalist, Alecu Demetriad (who wrote in many local publications such as "Cronica Dobrogeană", "Ziua", "Strălucitorul", etc.) wrote as reaction to the increase in the number of those who practiced the profession without any professional deontology or studies, about this profession of journalist: "Taken in the sense of a commercial business, the newspaper was the only industry that distributes the benefits among those who serve it according to the arbitrary criteria. For example, the newspaper's owner is always guaranteed to pay for the work he is making and the benefit of the capital invested in the business—otherwise he would leave the profession, the specialists who make the technical contribution to the newspaper, also receive their right-otherwise it would change the master-additionally they have a professional status that guarantees them the status of skilled craftsmen, sheltering them from the invasion of the seven craftsmen; remain the journalists, the editors, the reporters, the correspondents, etc., who when they do not work for the glory of the prophet of Nazareth, remain to gather the scraps falling from the tables of the previous categories and who have not at least the moral satisfaction to know that they are entitled by a title strengthened by law, and therefore sheltered by the wrath of those with many trades or those without any profession" (Demetriad 1934, p. 1).

The Constanta journalist thus expressed his support for the draft law made by the Journalists' Union in Bucharest, filed by Senator Al. Săndulescu, the president of this union, at the Ministry of Justice, to be analyzed. Demetriad was not aware of the bill other than articles published in the press, but expressed himself in his favor because he was pursuing: "the regulation of the profession of journalist" (Cheramidoglu 2011, p. 63). He justified his point of view by saying: "The group of journalists in Romania must be a select group, for journalism is a magistracy; so the sieve that will scrape the overproduction, the inflation of aspirants to the title of journalists, must be a fine sieve.

We are also opposed to enacting the principle of rights won. Because he was a charlatan and managed to live on the edge of the profession, by the time of the legislation of the journalist, cannot rely on the principle of rights won. We are requesting talent-based selection by virtue of a controllable activity of at least 10 years ago and a definite base at the entrance to the group. The title of journalist should be awarded after a rigorous examination of capacity, with a written and oral curriculum" (Demetriad 1934, p. 1).

From Demetriad's article, which is not singular in the local and nationalist landscape, we derive a general concern in defining and regulating the profession of journalist and putting it on a moral and ethical basis in addition to legal.

As a result of the interest shown in organizing the profession of journalist, in 1932 the Provincial Press Federation was established and held its third congress in Constanta on 29 and 30 September 1934. On this occasion, a newsletter was published which included the materials presented at the congress. Constantin Cheramidoglu points out that from the reading of the titles of the materials published in the bulletin the purpose of the society can be deduced: "Professional training for Federation members", "Provincial Press and Censorship", "Tourist Press and Propaganda" etc. The work of Al. D. Demetriad entitled "Contributions to the History of the Romanian

Press in Dobruja" was published separately (Dumitru-Zamfir 1985, p. 17). For it, the author said: "The scientist and the hurried reader are interested" and that it shows "the high degree of development our province has reached". This journalist printed the paper at his expense, in 500 copies distributed to all congressmen. He subsequently requested, according to an archive, that the City Hall purchase 25 copies at the price of 2500 lei (Cheramidoglu 2011, p. 63).

Over two years, the same federation of journalists continues protests against governmental tendencies to control the media and legislate without taking into account the interests of 2960 guild members. Their motion in this regard also appeared in the Constanta newspapers. For example, the newspaper "Dacia" published the article entitled "Protest against the law of the press" where the author likens the law with a "charging regime of the written word" which would not be successful in the time (article appears in 1936) than to sabotage the process of modern Romania's prosperity.

The government was required to take into account, when drafting a press law, and to the wishes of professional associations in the field (Cheramidoglu 2011, p. 63).

In the same newspaper "Dacia" was also published the speech of prosecutor I. Belcin (Ialomița Court) at the festive meeting of the Provincial Press Federation, in which he said that "the most holy thing of a civilized people is the freedom of the press" which must not be frozen than "moral and national commands, whether these commands are legalized or not". About provincial newspapers, he says that their writing is an expression of sincerity, although it does not always respect specific and technical rules and, through sincerity, can reach the hearts of the masses, towns and villages. Finally, the provincial press began to become "an honest correction of citizen's opinions, to work the only goal that the right journalist has in his heart; defending the interests of the nation and, above all, honoring the most valuable attribute of the soul: The Truth" (The speech of I. Belcin was published in "Dacia", An 24, no. 16, 30 June 1937, p. 1). The provincial press was an important factor within the national one, illustrated by the numerous articles devoted to this topic in various central and local publications.

3 Gazette Principles in the Press of Dobruja

"The school and the newspaper were born in Romania at the same time. For the same need. By the same wills. With the same people. We can say (…) that the press has emerged as a practical and lucrative efflorescence of culture. In our country, it was the nation's own alphabet. The first teachers of the Romanians were also her first journalists. At first it was the journalist…" (apud Cheramidoglu 2011, p. 63).

The comments on the purpose of press in Dobruja started from the local situation in this new Romanian province where, by laws and regulations (Rădulescu 1979, p. 296), the powers of the prefect were increased compared to those of the same state officials in the other counties of Romania (the situation will be perpetuated until 1912). This is the context that harmed the practice of journalists who criticized the abuses of the state administration. Because of that the authorities' suspended some publications (Lascu 1986, p. 68).

Dobruja has a special situation due to explosive economic development, until 1916, when the province, organically bound by the country, will become a pivot of national development. In Romania, civic education of citizens was more advanced than in Dobruja, where the principles of parliamentary democracy hardly penetrated.

The newspapers from Dobruja have mainly pursued the guidance of citizenship, necessary in the special context of Dobruja. It was necessary to move from the Ottoman administrative and legislative system to the modern Romanian (Romania had one of the most modern laws in Europe following the actions initiated by Alexandru Ioan Cuza, but especially by the founder of modern Romania, King Carol I), where the citizens' principles and liberties were sovereign.

In 1896 the newspaper "Constanţa" published an analysis of the local journalistic play and deplored the ephemeral of some publications. The reason why there were so many short or even single-edition newspapers was seen as the lack of support from the authorities to the press, combined with the lack of a political life comparable to that of the Old Kingdom. The article said that: "Over the Danube, where the press is not regarded as an annoying phenomenon, where the world began to learn the idea that journalism is useful" (Lascu 1986, p. 68).

Often, in the press of Dobruja appeared articles requiring an atmosphere of civilization, modernity and urbanism (It has also been seen as a form of civilization in a modern, progressive and fair world), both between competing publications and in the way of reflecting some problems and expressing the comments. This journalistic manner was seen as a way of doing career in areas such as politics or local government and was detrimental to the media and its image in general. The supporters of these opinions were Petru Vulcan or Constantin. C. Zamfiropol, who willingly generalized the situation to attract even more attention to it.

Among the most pertinent opinions on the role and purpose of the press in Dobruja, Nicolae Iorga, at a conference held in Constanța on November 6, 1911, stated "It can do miracles. Anyone looking for news, up to the bottom of the villages. Years will not pass, and you will see the lost Tartar in the middle of the new world, holding the diary in his hand… And if this sheet it strengthens a new consciousness, the sheet through which we are struggled and unhindered will be more superficial and more distracted. A good sheet must be local, definitely from Dobruja. It must be domestic and cultural, above all, indisputably fulfill a true and great service" (apud Lăpușan 2011, p. 137).

The efforts for progress and the desire to advance the inhabitants in Dobruja are noted and mentioned by Ioan N. Roman in his book "Dobruja and the Political Rights of its Residents": "Made up of people of origin, language and special faith, this population had to work, affirm and fight, in order to conquer as much land as possible and to ensure as much as possible its existence and settlement. It was and it is, from group to ethnical group and from man to man, a real struggle to compete,— and from these general efforts we have made the good progress we are speaking of" (Roman 1905, p. 128).

The newspaper "Dobrogea nouă" took over from the press-related ideas of the Romanian scientist in a series of articles about the press of Dobruja. An attempt to synthesize the stage of evolution of the press of Dobruja is also made by the teacher Petre Ştefănescu, who published his research results in the first issue of the "Romanian Gazette" of May 1, 1916 (Lascu 1986, p. 69). He made a panoramic view of the Dobruja publications until then, according to their orientations and themes.

The prints in which the Dobrujan newspapers are printed until 1916 are mostly local, in Tulcea or Constanta, but there are also occasions when the press of Dobruja is printed in Bucharest (Dumitru-Zamfir 1985, p. 327). The oldest Constanta printing house was the "Tipografia Română", which dates back to 1885, when the newspaper "Farul Constanţei" began printing here. She worked until 1915, when she published 13 newspapers. Other important printing presses were: "I. Georgescu Printing House", "Universal Printing" and "D. Nicolaescu Printing House".

The magazines of Dobruja published until 1916 aim at the cultural, moral and civic education of people from Dobruja, as it was written in the "New Magazine of Dobrogea" program, which appears monthly between 1901 and 1902 and was edited by a group of teachers. An important magazine for the constants was "Romania Mare" (1913–1916), whose program was perfectly illustrated by her name (Dumitru-Zamfir 1985, p. 327).

Until 1916, there were also 3 seasonal newspapers and 1 satirical newspaper ("Răcnetul Constanţei"), which reflected on their pages the activities of the summer season (Lascu 1986, p. 67).

The cultural circles whose work it reflects have edited three Constanta magazines. The most famous of these was "Ovidiu", mentioned above, and the other two were: "Cultura", between 1905 and 1906, with a special cover, made by the painter Marius Bunescu, who also signed poems and sketches, being edited by the Cultural Society with the same name and the "Society of Arts, Letters and Sports Society magazine" ("Revista Societății de Arte, Litere și Sport"), published between 1907 and 1910.

The nineties of the nineteenth century were marked by the appearance of two important publications in Constanta, longevity and outstanding in the local landscape: "Bulletin of the Chamber of Commerce and Industry" (1891–1944) and the newspaper "Constanța" (1891–1904, 1914), of a genuine publicist value. Until 1900, only 28 publications appeared in Constanta, followed by another 22 until 1910. Between 1911 and 1916, the press of Dobruja became enriched with another 40 new titles.

The German and Bulgarian occupation (Lascu 1986, p. 63) followed, which abruptly stopped the fulminant evolution of the press of Dobruja until 1916. During this period, only the papers edited and accepted by the administration of the two powers were allowed. It is the "Dobrutscha—Bote", a daily newspaper published between November 1916 and September 30, 1918, "Deutsche Etappen—Verwaltung in Dobrutscha" (the German Military Administration in Dobruja), which precedes the one before and appears between October 1916 and November 23, 1916, and the Bulgarian daily "Curierul Dobrogei" with an appearance from April 21, 1917 to September 30, 1918 (Lascu 1986, p. 70).

When looking at the statistics, the richest year in terms of new appearances was 1916, when 24 new newspapers were printed, among which we can mention: "Farul

Constanței", "Dobrogea Juna", "Conservatorul Constanței", "Buletinul oficial al Bursei", "Dacia", "Gazeta de Constanța", "Liberalul Constanței", "Romania Mare", "Șoimii Dobrogei", "Varda", "The Yearbook of the Muslim Seminary in Medgidia", "Tribuna Constanței" (Lascu 1986, p. 64).

As for longevity, there are great differences between the various Dobrujan publications of the time. Some have been published in a single issue, such as: "Curierul Băilor Constanţa-Techirghiol", "Ecoul", "Furnica", "Viitorul Constanţei", while others, few in number of 10 years: "Dobrogea Juna" which has a 40-year appearance between 1904 and 1944, the magazine "Ovidiu", which appeared between 1898 and 1910, with some interruptions, newspaper "Constanţa" (1891–1904, 1914), "Bulletin of the Chamber of Commerce and Industry" (25 years) and "Farul Constanţei" (36 years) (Lascu 1986, p. 65).

If we consider the character of the publications discussed, they are mainly newspapers—about 60 only in Constanța - then magazines, almanacs and yearbooks. As a frequency, most newspapers are weekly, with only a few being daily.

The first daily newspaper, expressly subtitled by its editors, was "Telegraful Constanței", whose subtitle is passed: "Everyday newspaper. Appears with night telegraph news. "Unfortunately, it had an ephemeral existence, of only 3 numbers, between June 26 and 30, 1913 (Aczel, s. a., p. 514).

Another great Constanta newspaper was "Romania Mare", from which 74 editions were published, between March 26 and June 26, 1915. The newspaper appeared in parallel with the magazine with the same name, from its publishers' wish to have a faster dispersion of mass information, especially as they propagated Romanian national ideas (Lascu 1986, p. 71). Both the newspaper and the magazine with the same title fought in their pages about the necessity of completing the Romanian national unity by publishing articles directed against the Austro-Hungarian and German spy, by morally supporting the Romanians in Transylvania and Bukovina and publishing original literary works of Romanians from occupied territories (After three months of publication, the newspaper changed its title to "Dacia", led by lawyer C. Irimescu, but maintains the same line of journalism. Another important daily was "Varda," which also changed its name to "Victoria" and militates for the same national ideal, so present in 1916, the year in which it appeared).

4 Conclusions

In the period up to Romania's entry into the First World War, the press of Dobruja was fully in agreement with the national issue, the dominant theme of the Romanian public debate, which is clear from the pages of some publications such as "Gazeta Dobrogei" or "Santinela Dobrogei".

As in other areas of the country in Dobruja has been an explosion of newspapers and magazines on varied themes, illustrating their political, social, economic and cultural activity, which intensifies as Dobruja integrates the Romanian political and social cultural framework. Until the summer of 1916, about 90 titles were published, most of them in Romanian, except for two Albanian publications, 4 in Turkish, 1 in Bulgarian and 3 in German (Aczel, s.. a., p. 639). In addition, there were publications dedicated to the presentation of budgets and reports that were not considered as part of the press.

In terms of subject matter, we can say that the press of Dobruja reflected, sometimes with a great deal of subjectivity, all current events but also the extraordinary, such as official visits or private politicians or cultural Romanian and foreign (Dobruja was visited by big names such as: Nicolae Iorga, I. L. Caragiale or Rabindranath Tagore) theatrical performances of some Romanian and foreign troops, the inauguration of some buildings such as the Constanţa Casino (In 1910, after a series of "adventures" related to changes in plan, of architects and budget, political changes were sometimes anecdotally reflected by local and even national press).

References

Aczel, I.: Publicaţii periodice dobrogene: 1878–1944, mss, s. n., 867 pp. Constanţa (s. a.)

Brătianu, T.: Scriitura şi genurile presei, 182 pp. Editura Fundaţiei „Andrei Şaguna", Constanța (2003)

Bucuţa, E.: Cincizeci de ani de presă dobrogeană. In:1878–1928. Dobrogea. Cincizeci de ani de viaţă românească, pp. 735–741. Editura Cultura Naţională, Bucureşti (1928)

Cheramidoglu, C.: Ziariști și ziaristică în Constanța interbelică. In: Lăpușan, Aurelia (coord.) Presa română interbelică între mitologizare și recuperare critică. Studii și cercetări de istorie a presei prezentate la cel de-al IV-lea Congres Național de Istorie a Presei, Constanța, 2011, pp. 62–66. Ovidius University Press, Constanța (2011)

Demetriad Al. D.: Contribuțiuni la istoricul presei românești din Dobrogea, s. n., 15 p. Constanța (1934)

Dumitru-Zamfir, C., Georgescu, O.: Introducere în istoricul presei dobrogene. In: Ionescu, Alexandru (coord.), Sion, Ion (coord), Stanciu, Marcel (coord.):Protecţia ecosistemelor, pp. 325–332 Constanţa (1978)

Dumitru-Zamfir, C.: Presa dobrogeană: (1879–1980): Bibliografie adnotată şi comentată, 392 pp. s.n., Constanţa (1985)

Lascu, S.: Apariţia presei în judeţul Constanţa şi dezvoltarea ei până la primul război mondial. Revista muzeelor şi monumentelor, nr. 3, pp. 63–71 (1986)

Lăpuşan, A.: O istorie a jurnalismului dobrogean, 264 pp. Editura Next Book, Constanţa (2014)

Lăpușan, A. (coord.): Presa română interbelică între mitologizare și recuperare critică. Studii și cercetări de istorie a presei prezentate la cel de-al IV-lea Congres Național de Istorie a Presei, Constanța, 2011, 290 pp. Ovidius University Press, Constanța (2011)

Petcu, M.: Istoria presei române, 294 pp. Editura Tritonic, Bucureşti (2002)

Rădulescu, A., Bitoleanu, I.: Istoria românilor dintre Dunăre şi Mare: Dobrogea, 440 pp. Editura Ştiinţifică şi Enciclopedică, Bucureşti (1979)

Roman, I.N.: Dobrogea şi drepturile politice ale locuitorilor ei, 159 pp. Tipografia „Ovidiu", Constanţa (1905)

Does Islamic Capital Asset Pricing Model Outperform Conventional Capital Asset Pricing Model?

Nousheen Tariq Bhutta, Biagio Simonetti and Viviana Ventre

Abstract This chapter aims to identify the difference between conventional and Islamic capital asset pricing model in order to ensure the efficient investment management of products and markets. Based on theoretical literature and critical and empirical analysis, it provides key justifications for why and how Islamic capital asset pricing model outperforms the conventional capital asset pricing model. The data has been extracted from 69 companies in the period 2000 to 2017, listed at Karachi stock exchange. It provides potential contribution towards the existing body of knowledge through incorporating risk and return perspective in capital market theory. Here, it has been found that Islamic investment is a low risk investment as compared to a conventional one. Moreover, Islamic lenders must share some proportional in profit and risk, which trade-off in risk shares, not in risk-return. It is suggested that Islamic investment strategy is more profitable than the other in case of direct sharing contract. Lastly, recommendations and policy reforms have been discussed for implementation of Islamic investment strategy.

Keywords Islamic capital asset pricing model · Risk management · Investment strategy

1 Introduction

A postulate shared in academics and in the practice of economic and financial choices is that people behave rationally and make rational decisions. Nevertheless rationality is on a concept not free from ambiguities and, also, as conceived in a classic

N. T. Bhutta
University of Sannio, Benevento, Italy

B. Simonetti
WSB University in Gdansk, Gdansk, Poland
e-mail: simonetti@unisannio.it

V. Ventre (✉)
National Institute of Geophysics and Volcanology (INGV), Naples, Italy
e-mail: viviana.ventre@unicampania.it

D. Flaut et al. (eds.), *Decision Making in Social Sciences: Between Traditions and Innovations*, Studies in Systems, Decision and Control 247,
https://doi.org/10.1007/978-3-030-30659-5_27

framework, it is not suitable to clarify some phenomena that are explained in the ambit of non-expected utility models: in other words, the only rationality cannot be recognized as the consistency motivation of the economic behavior. Indeed, as sustained and proved by many prominent scholars, economics is related to politics, in particular to ethics:

> The ethics-related tradition goes back at least to Aristotle. At the very beginning of *Nicomachean Ethics*, Aristotle relates the subject of economics to human ends, referring to its concern with wealth. [340 BC] The study of economics, though related immediately to the pursuit of wealth, is at deeper level linked up with other studies, involving the assessment and enhancement of more basic goals. (Sen 1987)

Financial system plays a very important role in the development of any country through exploiting opportunities for efficient allocation of resources. It also serves a financial intermediation to all users either borrowers or investor through channeling of funds in the whole economy. Therefore, efficient reallocation of resources should be required in order to facilitate the movement of money flow among all individuals who have plenty of financial resources to those who have a shortage of these resources.

A rational investor wants to maximize his return by minimizing risk, particularly risk averse. Due to this nature, he prefers to invest in a portfolio where he can efficiently diversify his risk and earn the maximum return (Markowitz 1952). Moreover, Markowitz proved the 'Efficient frontier' which means for given level of risk people always prefer investments having higher return. Later, Sharpe (1964) presented an efficient theory named the capital asset pricing model (CAPM), in order to determine required Rate of Return of a security or investment. Due to this significant innovation in finance theory, Sharpe achieved the Nobel Prize award in 1990. CAPM is a single factor model which specifies the risk and return for making investment decisions; and it is widely accepted throughout the world (Davis et al. 2000).

We will prove that dealing with ethical and historical principles in Islamic contexts may enhance profitability of financial operations.

However, under Islamic finance this mechanism works only through Profit & Loss sharing between investor and borrower at the time investment, instead of direct facilitation among agents in society. In an Islamic financial system each agent should be complemented to each other, which they bear equal risk and return in investment, which is the main milestone of Islamic finance. Some instruments of Islamic financial system, which are laid upon profit and loss sharing principle, have been presented in ensuing points.

- Musharakah (capital-capital partnership)
- Mudarabah capital-labor partnership
- Murabahah (capital equipment joint ventures)
- Al-Ijara (leasing)
- Al-Salam sale (forward sales).

These instruments/contracts depend upon equating the opportunity cost of capital through satisfying utility preferences of both parties in any Islamic society, which had been coined by Ibn-Taymia in his work. This principle yields the social benefit of both parties, which has not been measured in the neoclassical tradition (Lotfy 2005).

Due to corporate scandals accounting investors reinforce themselves ethical aspects towards making investment decisions and investment selection. This prompts the investors to adopt a financial system which provides social benefit in making investment decisions. Therefore, the concept of Islamic finance emerges from the establishment of Mit Ghamr Savings Bank of Egypt in 1963, which introduced Islamic banking to the investors (Sadaf and Andleeb 2014).

Moreover, Islamic banking has been practicing among all Muslim countries for a decade. In order to capture and retain investor many western countries also introduced the Islamic banking and offered Islamic products. In this essence, Dow Jones Islamic market index has been established by the USA which focus USA companies whose business are compatible with Islamic Shariah fundamentals. (Sadaf and Andleeb 2014).

The development in Islamic finance industry prompts us to investigate the Islamic fundamentals to investment decisions as compared to conventional finance. So, the main objective of this study is to provide the justification why and how the Islamic capital asset pricing model outperforms the conventional capital asset pricing model. Moreover, it provides potential contribution towards the existing body of knowledge through incorporating risk and return perspective in capital market theory.

Research Question:

Does the Islamic capital asset pricing model outperform the conventional capital asset pricing model?

The remainder of the present paper is organized as follows. The literature review is presented in Sect. 2. The methodology along with a theoretical framework is described in Sect. 3. Discussion is outlined in Sect. 4, concluding remarks along with future research directions in Sect. 5.

2 Conventional Capital Asset Pricing Model

Capital asset pricing model (CAPM) was firstly introduced by Treynor (1961), Sharpe (1964) and Lintner (1966). They defined risk is directly proportional to the return of security, exhibiting that high risk leads to higher return. Moreover, risk dimension comprises into two sub components, i.e. systematic and unsystematic. The former is uncontrollable, however; the latter can be diversified by efficient selection of assets within the portfolio. Mossin (1966) further extended the work of Markowitz on modern portfolio theory as well as diversification. Moreover, Black (1972) presented another dimension of CAPM, known as Black CAPM. This version was widely accepted all over the world.

Capital Assets Pricing Model comprises three components.

- Expected return on risk free asset,
- Beta, the sensitivity of security relative to market,
- Risk premium.

Risk free return basically concerns with the principle of the time value of money, which depicts that he would get surplus because of his risk and time factor. Generally, this rate is guaranteed and fixed. Beta measures the systematic risk, which is sensitivity of security relative to market. The risk premium is the economic profit over return of risk free asset (Sharpe 1964).

However, Roll (1977) criticized this model that without true market portfolio, cannot be applied. It may be attributable to measurement error because of using different proxies which gave different results. Black et al. (1972) investigated the impact of the capital asset pricing model of New York stock exchange through using time series data for the period of 1926 to 1966. He confirmed that beta is directly proportional to average return, which suggests that the investor can earn higher returns on highly risky investment. Moreover, Lau et al. (1974) applied the CAPM on Tokyo stock exchange through using the time series data from the period 1964 to 1969. They confirmed that the CAPM predicts the expected return of stocks accurately, which depicts efficient applicability of the CAPM model on Tokyo stock exchange. Similarly, Cheung and Wong (1992) tested the CAPM model on the Hong Kong stock market through using time series data for the period of 1980 to 1989, however, they found weak support towards the applicability of the CAPM model on the Hong Kong stock market. Raei and Mohammadi (2008) tested the 12 years data of 70 companies listed on the NASDAQ stock market for the period 1994 to 2005. They proved that CAPM assist in calculating the cost of capital.

After checking the applicability of the CAPM model on stock exchange of different countries, many researchers checked its validity among other analysis models. In this essence, Groenewold and Fraser (1997) compared the applicability of three models, namely capital asset pricing model, arbitrage pricing theory (APT) model and GARCH model through using time series data for the period 1983 to 1993. Their findings exhibit that the GARCH model and the APT model predicts the expected return on investment or assets accurately, however CAPM provide weak support in this scenario. Similarly, Fraser and Hamelink (2004) compared CAPM and GARCH models on the London stock exchange through using time series data for the period of 1975 to 1976. They confirmed that GARCH model well predicted the expected return instead of CAPM. Moreover, Eatzaz and Attiya (2008) also compared the capital asset pricing model with conditional multi risk factor model. The used data of 49 companies KSE for the period of 1993 to 2004. They reported multi risk factor model forecast more accurately instead of CAPM. Furthermore, Hanif and Bhatti (2010) investigated the CAPM model on 60 firms of Karachi stock exchange for the period from 2003 to 2008. They found that the validity of the CAPM model does not hold on KSE.

3 Islamic Capital Asset Pricing Model

Nowadays, great progress has been seen in both Islamic finance industry as well as Islamic finance theory. According to World Islamic Banking Competitiveness

Report 2013, Islamic banks have crossed the benchmark with the value of assets, \$1.8 trillion in 2013 over \$1.3 trillion in 2011 (Ernst and Young 2013). Therefore, Islamic banking is promoting all over of the world because it offers opportunities for both consumers and investors (Jomo 1992). There are three fundamentals which govern the Islamic financial system in any society which has been presented in the ensuing paragraphs.

- **Universal Complementarity**: It provides the social welfare to borrowers, investors, producers, government, consumers and other agents through equating the opportunity cost of capital because of absence of unified interest rate which serves as a benchmark in the economy (Selim 2008).
- **Abolition of Usury**: Like universal complementary fundamental, it also restricts the unified interest rate among all users in society. Therefore, borrowers and investors would share the profit and loss ratio, which ultimately depend on the real profitability of investment. Under the Islamic financial system, banks and other financial institution should abide this principle (Selim 2008).
- **Justice in Al-Hisba**: The Islamic financial society should introduce a judicial system, comprises of three components. The first is "Executive" branch which is executed through 'Al Hisba Institution' (regulatory committee); the second is "legislative" division which is applied through Al Shura (advisory committee); and the third is 'Judiciary' branch (independent). It provides a ruling body which ensures equal rights to all citizens, that they would receive suitable rewards in term of wages and rent based on education and entrepreneurship respectively (Selim 2008).

4 Methodology

The basic aim of our paper is to investigate whether Islamic capital asset pricing model outperform the conventional capital asset pricing model. We provide theoretical justification for the how it works. On the basis of above literature, our proposed model along with hypothesis has been presented as follows (Fig. 1):

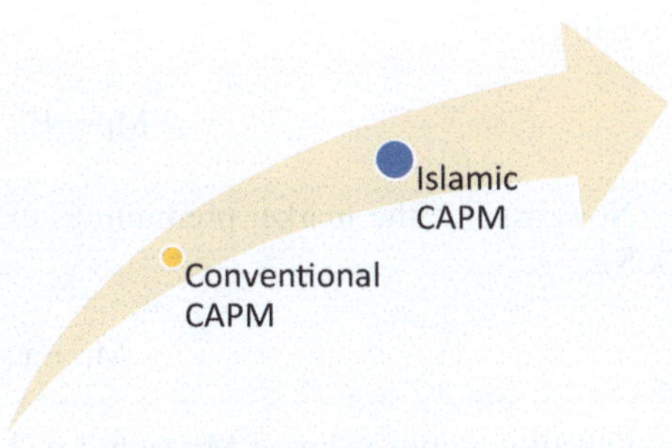

Fig. 1 How Islamic CAPM outperform the conventional CAPM

H1: Does Islamic capital asset pricing model outperform the conventional capital asset pricing model

Capital Asset pricing model depicts the asset behavior in terms of risk free interest rate in comparison of future rate of interest in exploiting opportunities. The main theory behind this model is net present value which was coined by Fischer (1930, 1965) as well as Hirschleifer (1958). However, it is criticized by Sharpe (1964) because both omit financial risk in the investment decision; even it is its key component. Later Lintner (1966); Fama and French (1992) and Markowitz (1997) further extended this model through inclusion of other possible variables. The most usable form of CAPM model is

$$R_C = R_f + \beta_1 (R_f - R_m). \tag{1}$$

This also can be written as

$$\beta_1 = \frac{Cov(R_c, R_m)}{Var R_m}, \tag{2}$$

where

R_c = Expected return on investment
R_m = Expected return on market
R_f = Expected return on risk free asset, which is guaranteed by government
$Cov(R_c, R_m)$ = covariance between return on market and investment
Var (R_m) = variance of returns of market
β_1 = sensitivity of investment related to market
$\beta_1 > 1$ implies to risky investment to the market
$\beta_1 < 1$ implies to low risky investment to the market
$\beta_1 = 1$ implies risky investment and $\beta_1 = 0$ implies risk free investment.

According to Islamic Abolition of Usury principle, we apply an interest free model to CAPM which is wholly based on the profit and loss sharing approach for investment decision. However, conventional CAPM comprises of risk free rate, expected rate of market, sensitivity of investment relative to market and expected return on investment (Selim 2008).

Market Premium = expected return of market − expected return of risk free investment

$$\text{Mp} = \text{R}_\text{m} - \text{R}_\text{f}$$

Now, assume the market premium as expected economic surplus of investment (λ S)

$$\text{Mp} = (\lambda\ \text{S}) \tag{3}$$

Substituting this value of Mp from Eq. 3 into Eq. 1, we get

$$R_c = R_f + \beta_1(\lambda\,S) \tag{4}$$

Now suppose the profit and loss contract between investor and lenders, in which γ is partnership of lender, $\gamma < 1$ = interest free partnership contributes to Islamic Rate of return

So we can say Islamic rate of return would be

$$R_I = \frac{(1-\gamma) + (\lambda\,S + \delta\,T)}{T}, \tag{5}$$

where R_I = Islamic rate of Return based on profit and loss sharing principle
$(1-\gamma)$ = investor partnership ratio based on profit and loss sharing principle
$(\lambda\,S)$ = Expected economic Surplus of Investment as given in eq. 3
$(\delta\,T)$ = Opportunity cost of expected economic Surplus of Investment
T = total value of investment.

According to our hypothesis Islamic CAPM should outperform the conventional CAPM

$$R_I > R_c \tag{6}$$

To prove this we assume some following conditions which has been cited (Selim 2008).

In order to implement the Islamic CAPM, we require that the lender's share of capital must be below the maximum threshold level, which means that they do maximum partnership which is inversely related to sensitivity of investment β and its high opportunity cost of capital α

So we can say

$$\gamma < \frac{1-\beta}{\alpha} \tag{7}$$

where the opportunity cost of capital of investment α is equal to change in expected economic surplus overopportunity cost of risk free asset.

$$\alpha = \frac{(\lambda\,S) - \delta\,T}{(\lambda\,S)} \tag{8}$$

Therefore, we can say that opportunity cost of capital of investment can be reduced through profit and sharing contract with $0 < \alpha < 1$.

Conditions

Selim (2008) presented the following conditions in his work:

(1) Both returns should be normalized

$$R_c = R_I,$$

(2) Then it implies that investor sharing $(1-\gamma)$ is equal to sensitivity of investment with expected economic surplus and opportunity cost of risk free assets,

$$(1-\gamma)=\beta\frac{(\lambda\,S)+\delta\,T}{(\lambda\,S)+\delta\,T} \tag{9}$$

which means sensitivity of investment under Islamic CAPM is less than 1 and $\gamma > 0$ which provides strong support as compared to conventional CAPM $\gamma = 0$, which means $\beta = 1$.

With Islamic partnership, the sensitivity of investment can be reduced, ultimately increase the expected required rate of return.

$$\gamma+\beta_I=1 \tag{10}$$

Thus, on the basis of above discussion, it can be conclude that Islamic CAPM should outperform the conventional CAPM i.e. $R_I > R_c$.

5 Data Collection

In order to test our hypothesis that Islamic CAPM outperforms conventional CAPM. For conducting analysis, monthly closing prices of 69 KSE Companies for the period of FY 2000–FY 2017 have been extracted from Brecorder. KSE 100 index is considered as market return and closing monthly prices are used to calculate Market return. The CPI and WPI data have been downloaded from Federal Bereau of Statistics of Pakistan and percentage return has been used. Moreover, the Risk free rate has been measured through monthly national saving rates.

For calculating the market return, portfolio return we used the following formula

$$\text{Return} = (\text{Ending price} - \text{beginning price})/\text{beginning price}$$

In order to conduct the regression analysis, we calculate conventional CAPM through the following equation by Hanif and Bhatti (2010)

$$R_p - R_f = \beta_o + \beta(R_m - R_f) + \varepsilon$$

For testing the Islamic CAPM, we calculate through the following equations with CPI proxy and WPI proxy by Hanif and Bhatti (2010)

$$R_p - N = \beta_o + \beta(R_m - N) + \varepsilon$$

Portfolio return has been calculated through this formula through giving equal weights to each stock

$$R_p = \sum WiPi$$

After findings returns six streams have been calculated as $R_p - R_f$ as excess return over portfolio, $R_m - R_f$ is market premium considering risk free rate and $R_p - N$ as market premium considering CPI and WPI.

6 Empirical Results

Results have been represented in Descriptive Statistics, Trend analysis and Regression analysis in the pursuing paragraphs. Regarding the descriptive analysis mean of CPI and WPI is higher than of RFR, portfolio return and market return. However, the standard deviation of CPI and WPI is higher than risk free rate, market return and portfolio return.

Descriptive Statistics

From the trend analysis of returns, it has been seen portfolio monthly return and market return are highly volatile, however risk free rate, CPI and WPI shows the moderate trend in Table 1 and Figs. 2 and 3.

The result of linear regression has been presented in following paragraph. As per findings, the R square for RFR proxy is 13.44%, which is lower than of CPI and WPI proxies. Moreover, the coefficient is also increased from 8 to 99% with significant P value at 0.000. The difference in R square and coefficient confirms that Islamic CAPM outperforms the conventional CAPM (Table 2).

These findings have significant contribution as the inflation rate becomes higher than Risk free rate. The first choice by any investor is the capital maintenance, which can be ensured through an inflation proxy, as traditionally it is considered as part of Risk free rate, however, it rose more than risk free rate. Here, we can conclude that investor are more concern to cover the inflation rate rather than risk free rate.

Table 1 Descriptive statistics

	R_p	R_m	CPI	WPI	R_f
Mean	0.044727	0.012814	4.057984	4.607403	0.102159
Standard Error	0.011527	0.000849	0.041793	0.05213	0.000199
Median	0.001038	0.019959	2.4	2.1	0.1056
Standard deviation	1.394485	0.102678	5.055776	6.306169	0.024016
Sample variance	1.944588	0.010543	25.56087	39.76777	0.000577
Kurtosis	6569.598	43.48737	5.21852	8.356411	−0.8822
Skewness	76.01625	−4.63281	2.052566	2.573749	−0.06687
Range	135	1.272666	26.6	37.7	0.0864
Sum	654.534	187.5039	59384.54	67424.74	1495
Count	14634	14633	14634	14634	14634

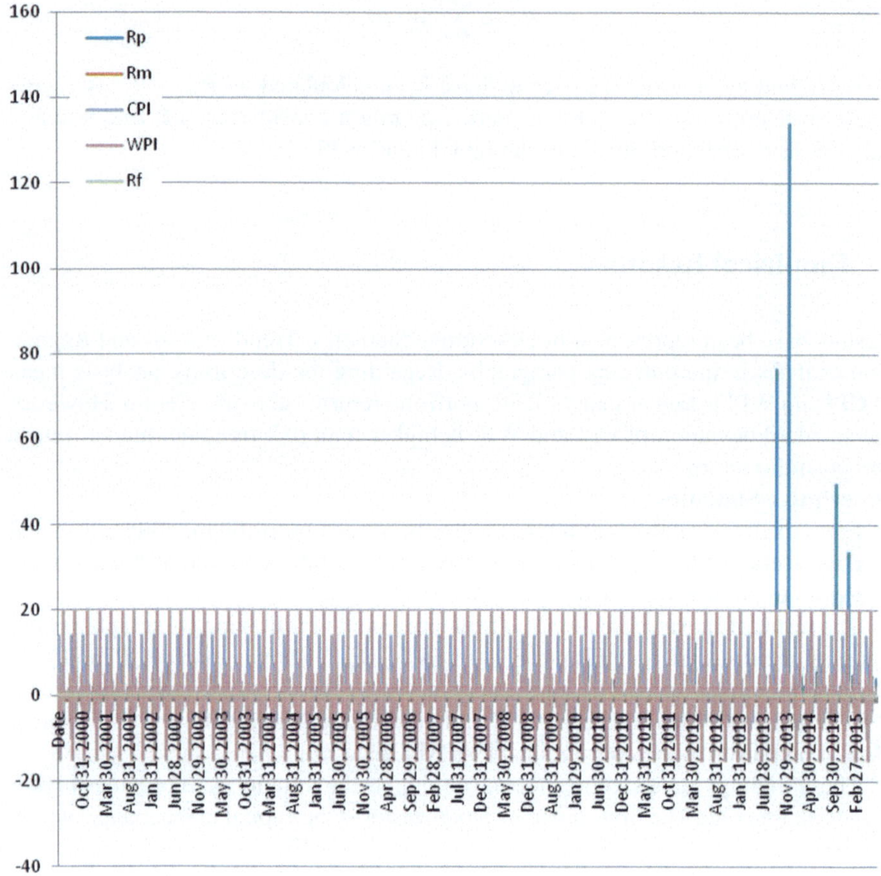

Fig. 2 Trend analysis of returns

7 Conclusions and Future Research Direction

The aim of this study is to compare the conventional Islamic capital asset pricing model and Islamic capital asset pricing model in order to ensure the efficient investment management of products and markets. Moreover, it provides basic justifications for why and how Islamic capital asset pricing model outperforms the conventional capital asset pricing model. The critical analysis shows that Islamic investment is low risk investment as compared to conventional investment. Moreover, Islamic lenders must share some proportional in profit and risk, which tradeoff in risk shares, not in risk-return. It is suggested that Islamic investment strategy is more optimal than the other one in case of direct sharing contract. Moreover, the empirical analysis confirms that Islamic CAPM outperforms conventional CAPM. The WPI proxy captures more significant result as compared to CPI and Risk free rate.

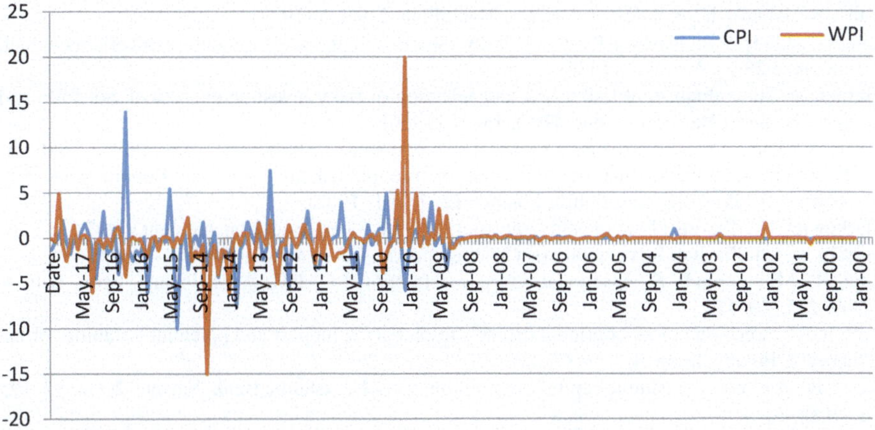

Fig. 3 Trend analysis of CPI and WPI returns

Table 2 Goodness of fit measures

Description	RFR proxy	CPI proxy	WPI proxy
R-Square	0.1344	0.9996	0.9997
Adjusted R-Square	0.1343	0.9996	0.9997
Coefficient	0.0807677	0.998128	0.9983457
P-Value	0.0000	0.000	0.000
F statistics	0.0000	0.0000	0.000

References

Black, F.: Capital market equilibrium with restricted borrowing. J. Bus. **45**, 444–455 (1972)

Black, F., Jensen, M., Scholes, M.: The capital asset pricing model: some empirical tests. In: Jensen, M. (ed.) Studies in the Theory of Capital Markets. Praeger, New York (1972)

Cheung, Y.L., Wong, K.T.: An assessment of risk and return: some empirical findings from Hong Kong stock exchange. Appl. Financ. Econ. **2**, 105–114 (1992)

Davis, J.L., Fama, E.F., French, K.R.: Characteristics, co variances and average returns: 1929 to 1997. J. Finance **55**, 389–406 (2000)

Eatzaz, Attiya, : Testing multifactor capital asset pricing model in case of Pakistani market. Int. Res. J. Finance Econ. **25**, 114–138 (2008)

Ernst, Young: World Islamic Banking Competitiveness Report 2013 (2013)

Fama, E., French, K.: The cross-section of expected stock returns. J. Finance **47**, 427–660 (1992)

Fischer, I.: The Theory of Interest. Augustus M. Kelly, New York, NY (1930)

Fischer, I.: The Theory of Interest (a Reprint), New York, NY (1965)

Fraser, P., Hamelink, F.: Time-varying betas and the cross sectional return risk relation: evidence from the UK. Eur. J. Finance **10**(4), 255–276 (2004)

Groenewold, N., Fraser, P.: Share prices and macroeconomics. J. Bus. Finance Account. **24**, 1367–1383 (1997)

Hanif, M., Bhatti, U.: Validity of capital assets pricing model: evidence from KSE-Pakistan. Eur. J. Econ. Finance Adm. Sci. **20**, 148–161 (2010)

Hirschleifer, J.: On the theory of optimal investment decision. J. Polit. Econ. **66** (1958)

Jomo, K.: Islamic Economic Alternatives. Macmillan, London (1992)
Lau, S.C., Quay, S.R., Ramsey, C.M.: The Tokyo stock exchange and the capital asset pricing model. J. Finance **29**(2), 507–514 (1974)
Lintner, J.: The valuation of risk assets and selection of risky investments in stock portfolio and capital budgets. Rev. Econ. Stat. **47**(1), 13–47 (1966)
Lotfy, A.: The Foundations Islamic Finance: A Comparative Study with Conventional Methods. The American University in Cairo, Research paper (unpublished), Economics Department (2005)
Markowitz, H.: Portfolio selection. J. Finance **12**, 71–91 (1952)
Markowitz, H.: Travels along the efficient frontier. Dow Jones Asset Management (1997)
Mossin, J.: Equilibrium in a capital asset pricing market. Econometrica **34**, 768–783 (1966)
Raei, R., Mohammadi, S.: Fractional return and fractional CAPM. Appl. Financ. Econ. Lett. **4**, 269–275 (2008)
Roll, R.W.: A critique of asset pricing theory's tests, Part 1: on past and potential testability of the theory. J. Financ. Econ. **4**, 129–176 (1977)
Sadaf, R., Andleeb, S.: Islamic capital asset pricing model. J. Islamic Bank. Finance **2**(1), 187–195 (2014)
Selim, T.H.: An Islamic capital asset pricing model. Humanomics **24**(2), 122–129 (2008)
Sen, K.: On Ethics and Economics. Blackwell, Oxford (1987)
Sharpe, W.: Capital asset prices: a theory of market equilibrium under conditions of risk. J. Finance, September (1964)
Treynor, J.L.: Toward a Theory of Market Value of Risky Assets. Unpublished manuscript, undated (1961)

Evaluative Aspects in Decision-Making Processes

E. Serena Sansevierio

Abstract In this work we deal with evaluative aspects in decision-making processes. Evaluative activity, however complex, precedes, accompanies and follows human actions. In fact, the cognitive outcomes which are reached with its progressive development, directs the action by positioning it towards desired directions and goals. The evaluative activity, contrary to what could be thought in a simplistic and practical way, is instead expressed through a series of operations that are distinct but mutually connected, all of great importance. To begin with, think of the attribution of value to facts, data, procedures, events or to some of their qualities in relation to the goals that the evaluator wants to pursue. It may happen that a fact deemed positive or negative in a context and in relation to certain purposes may not be considered as such if we are in the presence of other constraints and in view of other objectives. Therefore, the practice of evaluation, far from being standardizable, more often follows the law of the case by case and it is difficult to harness the attribution of value in rigid and pre-established schemes.

Keywords City · Representation · Images · Space · Time · Evaluation

1 Introduction

1.1 "Weak Town Planning": Paradigm of Contemporary Society

To introduce the central theme of this chapter, one can't ignore and not give the right place to what happened before the post-modern era in other disciplines such as philosophy and sociology, as they have much in common with that we call "weak urban planning".[1] "Weak thinking" is both a "metaphor" and a "paradox", because it

[1]Camagni 1986; Mela 1987; Mazza, 1986.

E. Serena Sanseviero (✉)
Università G. D'Annunzio, Chieti-Pescara, Italy
e-mail: sansevi@unich.it; serenasanseviero@alice.it

D. Flaut et al. (eds.), *Decision Making in Social Sciences: Between Traditions and Innovations*, Studies in Systems, Decision and Control 247,
https://doi.org/10.1007/978-3-030-30659-5_28

indicates a temporary but also contradictory situation; it does not give precise indications, but indicates multiple possibilities. "It is a experiment, an attempt to trace analysis, to move on the ground". With reference to the city, in recent years, post-modern philosophies have made it the object of attention and the transformations of metropolitan life habits have become a moment of confirmation and synthesis of hypotheses and theories. The "urban text" itself is "a linguistic construction", as a moment of production and communication. Among the interpretations that stress the linguistic structuring of the metropolis is that of Giovanni Vattimo who considers the city no longer as a system endowed with recognizable communicative signs but characterized by "disorder, phantasmagoria, seduction". "The contemporary metropolis is mirrored in language. It presents itself as a plot of expression, metaphors, proper names, proportional functions, times and verbal forms, disjunctions, implications. Nor is it an instructive analogy. The metropolis is indeed a linguistic formation, an environment constituted primarily by objectified speeches, predisposed codes, materialized grammars ... to orient oneself in a big city means to experience language".

In traditional urban policies, problems have a "simple" solution: they move in the great set of positive knowledge in which we identify the two subsets: the science of geographic—architectural settlements and the science of flows and links of a technical matrix - engineering.

This distinction, a bit reductive, with its different interpretations moves in an overview and is introduced only to highlight the inadequacy of different strands in positive knowledge to the characters of the new "complex" problems.

1.1.1 Weak Urbanism and the Birth of Complexity

The complexity born as an awareness that "the type of problems that new urban policies face aren't standardizable, like the problems faced by excellence in the strong disciplines. [...]. The use of procedural standards is not as useful as the proposal of physical standards".[2]

This awareness is not really new, but it can already be traced back to the '80s, when the weakening of the form of the nation-state, the crisis of large-scale territorial political units and the concept of territory as a single entity, now make the "gravitational" spatial paradigm that assumes a hierarchical structuring of space obsolete.

The territory is therefore to be considered structured as a series of "planetary systems" formed by smaller centers and "central" locations linked by a dense polycentric, multipolar network as the "friction of space" has decreased.[3]

"The new sciences of complexity and the increasingly heated debate among the philosophers of science, from which winning relativistic and local positions, seem

[2]M. Cremaschi, op. cit., p. 112.

[3]R. Gubert.

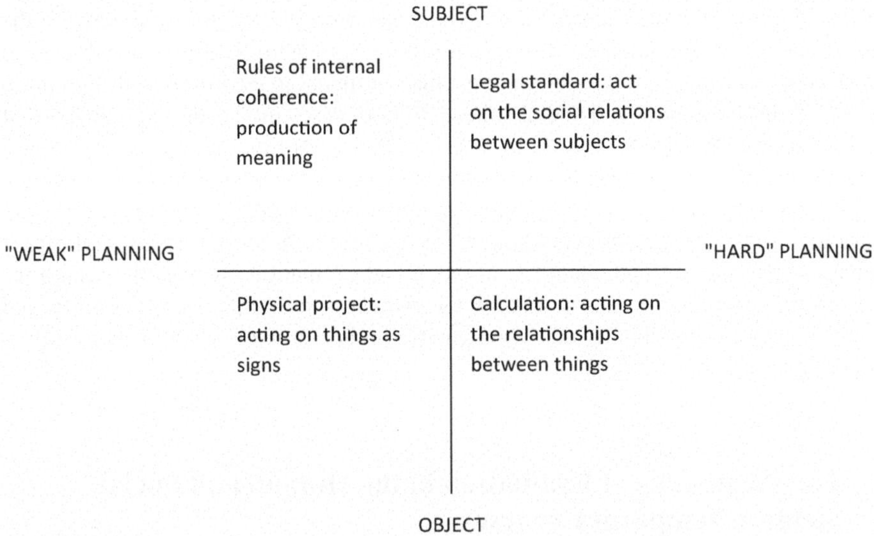

Fig. 1 Characteristics of the (landscape) plan. *Source* G. Dematteis, "Italian Geographical Rivista", n. 96, 1989, p. 452

to leave no room for the repetition of scientific paradigms with models of absolute validity".[4]

"... The local scientific approaches don't cooperate harmoniously with an image, a theory of knowledge and the universe, but on the contrary they intersect, overlap, ignore, contrast, integrate, split" (Fig. 1).

2 The Practice of Evaluation

Evaluative activity, however complex, precedes, accompanies and follows human actions. The evaluative activity, contrary to what could be thought in a simplistic and practical way, is instead expressed through a series of operations that are distinct but mutually connected, all of great importance. To begin with, think of the attribution of value to facts, data, procedures, events or to some of their qualities in relation to the goals that the evaluator wants to pursue. It may happen that a fact deemed positive or negative in a context and in relation to certain purposes may not be considered as such if we are in the presence of other constraints and in view of other objectives. Therefore, the practice of evaluation, far from being standardizable, more often follows the law of the case by case and it is difficult to harness the attribution of value in

[4]A.Petrillo "Current problems and possible solutions", COMPANY NOTES AND CONFLICT. Number 5/1995 URBAN QUESTIONS Southern Italy and Campania from ancient to contemporary—cap V.

rigid and pre-established schemes. The topic has been studied and conducted in the study with the related scientific-disciplinary skills both in the natural and anthropized environment, as can be the problem of evaluation linked to territorial or design practices both in the human and social sphere. In both cases, however, we realized that with the due differentiations of both weight and power of the various actors, and of instruments that can be used, the methodology remains the same whether we operate on the micro or on the macro-system. Grammar and evaluation syntax are formal rules that don't change with changing reference levels. A further transversal aspect emerges from the in-depth analysis: to talk about evaluation involves the search for interpretative methodologies that allow the evaluator to interpret lives, cultures and worlds that are immediately incomprehensible and cognitively foreign according to the thought of some authors.

3 The Paradigms of Evaluation in the Human and Social Field: A Temporal Excursus

Speaking about evaluation today at the time of cultural, territorial and individual differences entails the search for interpretative methodologies that allow those who have formative, social-assistance tasks (broadly ranging from welfare to housing) to interpret lives, cultures and worlds that are immediately incomprehensible and cognitively strangers. This consideration is valid in the social and relational sphere. In the socio-formative field, evaluative methods or practices are more valid than those that are more closely related to an interpretative-narrative paradigm which opposes and deviates from the decision-making one in which the evaluation assumes the role of scientifically and objectively basing the decision.

If the thought tends to be more and more ethnographic and negotiable, the evaluative processes tend to be more and more idiographic, they rely on narrative, interpretive and above all reflexive techniques. Oral metariflexion is also an extraordinary opportunity to teach students, for example, individually to reflect on their work and to develop their richer and more critical cognitive structure.

In philanthropy, evaluation is often used to test a specific theory of change: to see whether a grant-funded intervention has had the desired effect.

For decades, the prevailing answer has been that the research should have piloted new innovative programs, tested them to see if they work and, once a solution is found, let governments or others with larger resources finance the large-scale solution. It was an attractive approach to research: encouraged to continually experiment with new exciting ideas, each of which required only a small amount of funding and promised to promote an important step forward. Under this paradigm, the usual role of evaluation was to isolate the impact of a specific program, test its effect on a problem and make a judgment about its operation or not.

The only problem in the light of the initial considerations that this paradigm rarely—if ever—works. Social, political, territorial and planning problems are complex. They are created and perpetuated by a wide range of factors that are beyond the control of every single organization. Even if a promising new approach were discovered, the counterproductive political factions could block their adoption, more funding may not be available or the necessary management talent may be scarce.

4 Main Paradigms and Strategies for Evaluating Plans and Programs for the City and the Territory

A current interesting aspect in evaluation processes to support decision-making in the local area, with particular reference to public policies, is certainly "incomplete and uncertain knowledge of public problems".[5]

In fact, the latter "aren't observable data" and the plans take on a form such that in them such uncertainty can assume an exhaustive representation and/or solution. There is, however, a custom in planning to leave the definition of the problem in the background, while feeding the emergence of misunderstandings about the objectives and strategies to be followed.

This behavior or operational practice of planning tout-court generates an important consequence for the evaluation, as it obliges the evaluators (if called into question) to build ex-post the objectives, starting from the observed practices, or to limit themselves to impact analysis or released from the so-called goal-free goals.

"The latter are justifiable if they are constructed and outlined on a clear awareness, but often they have the effect of placing the evaluation "out of place" or of removing it excessively from the object".[6]

Really, planning and evaluation activities are to be considered closely connected. According to this line of reasoning it isn't possible to identify any urban plan without referring to the system of techniques that, more or less explicitly, have helped to orient the policies on the specific territorial area.

In order to propose an example we will talk about the evaluation of the possible negative effects produced by the realization of an energy infrastructure project (for example project for the installation of a wind power plant) in sites where good wind conditions usually coexist (profits for economic investment based on the project) and characteristics of naturalistic value. The use of natural resources has undergone a profound change during the years, and especially in the latter period; this is due to the evolution of economic development and the different needs of the user group.

In this regard, the problem around decision-making moment with respect to the complete phase of "construction of the quality of the environment" is often put forward, which is often presented to us in various ways.

---

[5]D.Patassini "Paradigms and strategies for evaluating plans, programs and policies," Urbanistica, n. 105, December 1995, pp. 50–59.

[6]Ibidem.

Sometimes this relationship is so inextricable that the choices of plan are not easily attributable to an autonomous evaluation moment. It's clear that the assessment is increasingly based on a set of practices to support the activities of the deliberative planner, so engaged in the public decision-making functions.

5 Rational Planning or "Hard Planning"

The hard planning consists of statutory planning laws, tools and institutions Soft planning is structured in processes of mutual learning, cooperation, negotiation and Coordination "Complex, overlapping, "tender" patchwork of activities, relationships and responsibility". Soft planning is "the preferred, indeed the only, realistic" model for soft spaces.

Today the strategic planning of space poses a certain danger of detachment in parallel systems for researchers: the planners in fact would face "a impossible choice between a legitimate rigidity of the statutory planning and an illegitimate strategic planning flexibility" (Balducci 2013).

"In rational planning the evaluation had an eminently function of optimizing the process to obtain a situation in which we love the relationship between the achievement of the objectives and the resource utilization, so it was at the beginning of the practice of the plan, In the opposite case of communicative planning, which appears to accompany the current system of the plan, the evaluation function consists of a continuous appreciation of the dialectical rhetoric that underlies the plan and, above all, of the debate on political programs".[7] Between the rational and the communicative planning "strategic" assumes a central position: the evaluation function consists in the analysis of the uncertainty of the process and the complexity of the decisions on the plan scenarios, while its main characteristics are the iterative adaptation and cyclical of the "plan assignments" package. It is therefore clear that evaluation is therefore a capillary component of the decision-making process, but is also to be understood as an "instrument of practical intelligence" with three problematic dimensions, which contribute to giving it a relative operational autonomy:

- as an orientation to action, as it originates within public action;
- as an objective point of view external to the system for understanding phenomena and processes;

[7] Strategic environmental assessment in spatial planning (La): New perspectives. Corrado Zoppi, Maurizio Garano Gangemi Editore spa.

- as a component of help associated with the decision that is realized through feedback devices[8] that allow verification of the decisions taken with appropriate criteria and tests.[9] (Bekesiene and Hoskova-Mayerova 2018)

Evaluation therefore participates in the processes of planning and management of resources in very different ways and forms, since it uses the knowledge of many disciplines and makes use of an extensive range of techniques, including a very wide-ranging area of study. Its importance tends to be emphasized for the role it plays within the decision-making process even if, in simple decision-making contexts, the evaluation function often ends up in a weighed comparison between the possible solutions. In more complex contexts, however, evaluation pervades the decision-making process by contributing to its structuring. It is precisely in this second case that we understand the absolute usefulness of the evaluation for the pursuit of objectives such as the efficient management of resources, the feasibility of forecasts, social equity and therefore, ultimately, the strengthening of the plan as an instrument. For the government of the territory.

6 An Example

Usually the public operator tends to ensure that the use of the environment, in its various meanings, urban, natural, etc., is compatible with what is already present on the territory and with uses not strictly natural, such as the complex of activities that could be started. It is stressed that this point is of great conflict and not easy to solve. Conflict is amplified since, even following a decision that seemed the best among those proposed, the costs and benefits are not distributed homogeneously either between the parties involved or between the intervention era and the remaining part of the territory. There is a need to try and recompose the conflicts between groups of people who propose different goals to the same common decision. The work proposes to highlight how the priority evaluation analyzes (called ex ante) can lead to the decision to realize or not the infrastructure (or in this case the wind farm). The point of arrival of the evaluation is the decision (heart of the decision-making process), which, at any level and especially when interventions concerning the territory are profiled, is proposed as the moment of reaching consensus, understood as the degree in which it is achieved maximum convergence towards choices of common satisfaction.

---

[8]Feedback devices can be: real (they activate organizational behaviors); artificial (simulate situations/ scenarios and use models of the described or explanatory type).

[9]Classification introduced by J. Leca "Sur le rôle de la connaissance dans the modernization of the age and the statute of the evaluation", Revue francaise d'administration publique n. 66, 1993 and taken up again by D. Patassini in note 2 of the cit.

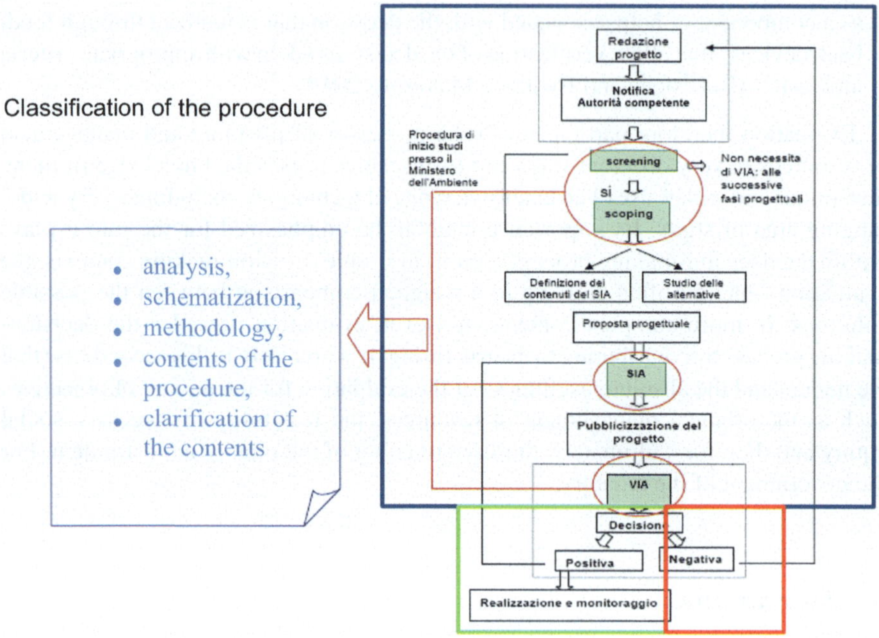

Fig. 2 The evaluation is the basis of the decision, if the evaluation proposes positive values for the impacts considered we arrive at the positive decision to build the infrastructure (green square) if, on the other hand, the evaluation concludes with negative values (and therefore the impacts are irreversible and negative for the territory) it is necessary to restart the design and remodel the intervention (red square)

In the proposed image we highlight the procedure for assessing the impacts of the proposed project which consists of several phases, as we said, linked together:

the evaluation is the basis of the decision, if the evaluation proposes positive values for the impacts considered (and therefore the latter are negligible and not negative for the territory) we arrive at the positive decision to build the infrastructure; if, on the other hand, the evaluation concludes with negative values (and therefore the impacts are irreversible and negative for the territory) it is necessary to restart the design and remodel the intervention (Fig. 2).

7 The New Role of Evaluation

The new role of evaluation can be summarized in three distinct functions:

- the rationalization function, that is to say, structuring the decision-making process;
- the legitimating function that affirms the principles of subsidiarity and responsibility;

– the participatory function.[10]

More authors point out that feasibility and equity are fundamental requirements for the success of the implementation tools; in relation to this it is also true that their achievement can't be separated from the disciplinary contribution of the economic-financial analysis, where the transforming factors directly or indirectly attributable to the environment assume an extremely important role. The feasibility of plans and projects can be more easily pursued if public and private finances intervene decisively, through forms of partnership between the interesting subjects. A significant example of this approach, which has become systematic in all the interventions desired and financed by the European Community, comes with the selective assignment of public financial contributions, which stimulates candidates (local administrations and public and private operators) to search for quality of the urban project.

It is precisely the theme of the quality of development, which must often be substantiated and supported ex ante, assessed a priori, to be the leitmotif of the interventions for the evaluation of plans and programs for the territory.

The theme and above all the application of certain evaluation procedures must therefore play a decisive role and must be conceived as an opportunity to compare scientific knowledge, technical know-how and common knowledge by putting together. Scientific knowledge is mainly exercised internally, in a self-referential way, thus delaying the transformation of new acquisitions into structured knowledge, useful to technical knowledge and therefore operatively manageable at the routine level. On the other hand, if the flow of new knowledge takes place in the most direct and immediate way, then the role of common knowledge also increases in the process of acquiring and structuring the concepts. In this regard, the application of evaluation techniques and the preparation of an evaluation study is a guarantee of whether the whole evaluation process is based on a shared vision of the objective-setting; the valuation procedures not referred to predetermined purposes inevitably remain within an absolute, purely technical, non-relativized and therefore not integrated logic (Maturo and Sanseviero 2004).

8 Conclusion

Finally they are two main lines on which the evaluation strategies are moving:

"Experimental type, effective in the presence of structured interactions";
"Non-experimental, advisable if the interactions between actors and the decision-making processes themselves are not sufficiently structured".

[10]Evaluation as a reference for planning—Speciale Ce.S.E.T.—XXXIII Study Meeting Study Center for Estimates and Territorial Economics edited by G. Marchi—University of Cagliari—Department of Land Engineering 2003 page 9.

"The experimental strategy makes use of" measures "and therefore uses quantitative techniques. The non-experimental strategy stimulates "opinions" with the help of techniques that are conventionally defined as qualitative".

These techniques are defined as such because "the analysis is not built through the direct observation of a group of variables by the decision maker, but by the detection of the opinions expressed by a group of subjects considered experts on the topic on which the decision is based".

It is clear that "planning activity makes use of both measures and opinions and for this it could activate several techniques simultaneously. In structured contexts, in which the main actors are recognizable, in which it is possible to decide what to decide over what to do and how to do it, the experimental strategy provides revisional analysis techniques, cost-benefits, analytical impact assessment and a heterogeneous range of indicators".

"In preliminary or not structured contexts, the decision-making process is preliminary and therefore the use of decision support techniques (Delphi, scenario writing, brainstorming) and methods of strategic choice becomes useful. Faced with this wealth of tools, the problem is not so much their positioning in the process of the plan, but rather how to give voice to the various evaluative demands placed by the plan (or alternative instruments) throughout their cycle".

An interesting approach to decision making process can be also found it the recent books, e.g. Flaut et al. (2018), Hoskova-Mayerova et al. (2017) or Ventre et al. (2013) or former papers e.g. Hoskova-Mayerova and Maturo (2018) or Hoskova-Mayerova et al. (2013).

References

Balducci, A., Mantysalo, R. (Eds): Urban Planning As a Trading Zone. Springer (2013)

Bekesiene, S., Hoskova-Mayerova, S.: Decision tree—based classification model for identification of effective leadership indicators in the Lithuania Army Forces. J. Math Fund. Sci. **50**(2), 121–141 (2018)

Flaut, C., Hoskova-Mayerova, S., Flaut, D. (eds.): Models and Theories in Social Systems, Studies in Systems, Decision and Control, 179, Springer International Publishing, 576p (2018). https://doi.org/10.1007/978-3-030-00084-4

Hoskova-Mayerova, S., Talhofer, V., Hofmann, A., Kubíček, P.: Mathematical model used in decision-making process with respect to the reliability of geodatabase, advanced dynamic modeling of economic and social systems. Ser.: Stud. Comput. Intell. **448**(2013), 143–162 (2013). https://doi.org/10.1007/978-3-642-32903-6_11

Hoskova-Mayerova, S., Maturo, A.: Decision-making process using hyperstructures and fuzzy structures in social sciences. In: Collan, M., Kacprzyk, J. (eds.) Soft Computing Applications for Group Decision-Making and Consensus Modeling. Studies in Fuzziness and Soft Computing, vol. 357. Springer, Cham (2018). https://doi.org/10.1007/978-3-319-60207-3_7

Hoskova-Mayerova, S., Maturo, F., Kacprzyk, J. (eds.): Mathematical-Statistical Models and Qualitative Theories for Economic and Social Sciences. Studies in Systems, Decision and Control, 104. Springer, Heidelberg, 442p (2017). https://doi.org/10.1007/978-3-319-54819-7

Maturo, A., Sanseviero, S.: A mathematical model for a complex evaluation and decision process. In: Franchino, R., Maturo, A., Ventre, A.G.S., Violano, A. (eds.) Strategies, Processes and Decision-Making Models for Environmental Management, pp. 228–237. Edizioni Goliardiche, Trieste (2004)

Ventre, A., Maturo, A., Hoskova-Mayerova, S., Kacprzyk, J. (eds.): Multicriteria and multiagent decision making with applications to economic and social sciences. In: Series: Studies in Fuzziness and Soft Computing, p. 315. ISSN 1434-9922, ISBN 978-3-642-35634-6. Springer, Heidelberg (2013)

Study on the Public Sector Personnel Motivation in the Caras-Severin County

Cornelia-Ecaterina Cornean

Abstract Personnel motivation represents one of the basic elements of human resources management, being at the same time one of the main research objective used by specialists in management, psychology, sociology etc. Human resources represent the engine of working activities, regardless of the public or the private sector these belong to, thus personnel satisfaction and motivation is adapted to each organization in order to lead optimum, efficient and rapid results. The research realized and presented in this chapter highlights the motivational factors with a high percentage used by public institution from the Caras-Severin County, which help managers in motivating persons to chose and keep a job in the public sector. The research methodology implies a quantitative factor, through the application of a questionnaire to a sample of 60 persons hired by two state institutions: a department of the Caras-Severin County Council and a national agency.

Keywords Human resources · Motivation · Public Sector

1 Introduction

In order to analyze personnel motivation in the public sector, the terms used should be briefly defined: public sector, management of human resources inside this sector and the personnel motivation term, an important component of the management of human resources (MHR).

Specialty literature offer a containing definitions; according to Lane, the public sector refers to institutions and organizations belonging to the state, which take decisions and apply consequences; to institutions that refer to consumption, investments and transfers realized by the government; and the production of goods (because the state is an owner, it can supply goods and hire work force, being nowadays the largest employer in the country) (Lane 1995: 15–17).

C.-E. Cornean (✉)
"Eftimie Murgu" University Resita, 1-4 Train Vuia Square, Resita, Caras-Severin County, Romania
e-mail: cornean.corneliae@gmail.com

D. Flaut et al. (eds.), *Decision Making in Social Sciences: Between Traditions and Innovations*, Studies in Systems, Decision and Control 247,
https://doi.org/10.1007/978-3-030-30659-5_29

The management of human resources was defined differently by the specialty literature, being in time defined as:

- being built of all activities referring to human resources;
- the ensemble of activities directed towards the efficient use of human capital in order to realize the objectives of the organization, while satisfying the employees' needs;
- being built of all decisions regarding relations between the management and the employees in order to make efficient the organizational activity;
- being formed of all measures regarding the recruitment, the selection, the activity, the rewarding, the satisfaction etc. realized during the employment period (Andreş 2009: 41).

Cole considers that the management of human resources represents an organizational activity that allows the most efficient use of employees in the reaching of organizational objectives, of the group and individual objectives (Cole 2000: 21–25).

The word motivation is most often used in the human resources domain. It comes from the Latin *motus* which signifies *to move, to influence and to affect.* Dictionaries describe motivation as being the "incentive to act" (Mitchell 1982: 81), while specialists add to the term a set of psychological processes that cause guidance and behavior modification so it can fulfill a goal (Robbins and Judge 2008: 209).

The explicative dictionary of the Romanian language defines motivation as being formed of all reasons and impulses (conscious or not) which determine a person to realize a certain action or to reach certain goals (www.dexonline.ro).

2 The Importance of Personnel Motivation in the Public Sector

Productivity represents the basis of management existence, thus managers are thriving to fulfill organizational objectives by developing efficiency. The performance of employees is one of the main factors contributing to production development (Mitchell 1982: 82). It has already been clarified that work motivation doesn't determine employee performance; it only influences him on how to realize his tasks (Rosicka and Hoskova-Mayerova 2013).

For a proper fulfillment of their tasks employees must possess necessary knowledge and abilities. Another condition is for them to understand what they are asked to do and be motivated in making efforts to fulfill their tasks. Environment is another important factor for the allowing of task fulfillment (Mitchell 1982: 83). Thus, a motivated employee can reach performances even if there is a lack of knowledge (Landy and Conte 2010: 365).

Unfortunately, it seems that a small number of managers are capable of motivating their employees, and this situation is worse in the public sector (Perry and Hondeghem 2008: 89). An explanation would be the lack of proper education and secondly, of functioning regulations established by the state management. In addition, the literature available at this point in Romania seems to be complicated and recommendations regarding their application even less numerous.

Motivation in the public sector seems difficult to apply, many factors that prevent motivational efforts being identified. Wright sustains that employees from the public domain are lazy and lethargic (Wright 2001: 560). In 2007 Buelens and Van den Broeck proved in their studies that employees from the public domain work less and are less devoted to their organization in comparison to the private sector (Wright 2001: 68). Still, this doesn't prove they are lazy, just that they desire to be "slaves" (Wright 2001: 69).

Secondly, as most of the population already knows, managers from the public sector don't possess the instruments necessary to motivate their personnel. One reason would be that firing people in the public sector is a difficult action and on the other side, they cannot offer substantial financial rewards without numerous approvals (Drobot et al. 2009).

Another factor is the employees' age. The retirement age being modified, some employees no longer face new demands and modifications. In addition, new employees no longer follow training periods so necessary to properly realize their tasks. Thus, the public activity is hardly fulfilled, while the large work volume overpasses the employees' capacities (Lupinca 2016: 37).

In addition, another factor is represented by the new trend of adapting to international demands and to those already applied in the private sector. Still the conditions in which this theory are applied are different for each sector, each sector functioning differently (Lupinca 2016: 38).

Frequent managerial changes, the results of changes in the state management, every four years, or even more often, lead to short term perspectives, to specific political agendas which make the application of managerial criteria a difficult task. Thus, the motivation of employees doesn't represent one of the most important aspects in the managers' careers, but their adaption to new values, missions and objectives that need to be fulfilled on a short term (Lupinca 2016: 38).

Another strong influence in the public system is represented by labor unions, which create new conditions in the employees' work contracts. These unions that are nearly extinct in the private sector "fight" more for an increase in productivity and for employees' satisfaction (Lupinca 2016: 38).

The transparency imposed to state systems becomes a negative characteristic when it comes to employees' safety. They cannot make their opinions heard, cannot take risks and nor they can implement innovations without being criticized by the general public (Lupinca 2016: 39)

Thus, the public sector managers' task is a difficult one; they must permanently adapt to leadership changes, without being able modify their own leadership methods, these also being imposed by the current government (Cornean 2014).

3 Study on the Public Sector Personnel Motivation in the Caras-Severin County

For the application of this research regarding the motivation of the public sector employee, the introduction of two different institutions, belonging to the state, was considered necessary: a state national agency—The National Agency of Cadastre and Land Registration (ANCPI), the manager of the Caras-Severin Office of Cadastre and Land Registration and a department that functions inside the Caras-Severin County Council—The Department for the contracting and monitoring of social services.

The purpose of this research is to underline the type of predominant motivation, intrinsic or extrinsic, in the choosing of a position in a state institution and if this result is compatible to the data obtained at an international level by specialty researchers. Numerous specialists classified motivation in large categories, still the most used one is the one based on intrinsic and extrinsic nature.

The hypothesis of this research starts from the difference observed in the specialty literature between the public and the private domains, underlining the idea that factors referring to the wellbeing, the prestige, the recognition of merits and the employee's happiness are more appreciated by employees in comparison to extrinsic factors, of a financial or professional nature, in choosing and keeping a this kind of job for a long time.

The hypothesis is: If the employees of the public sector are highly motivated by intrinsic factors then their loyalty is more developed.

The research method applied is a quantitative one, represented by the use of a questionnaire, applied to a sample of 60 subjects, employees of the two institutions mentions [15 from The Department for the contracting and monitoring of social services and 45 from the Caras-Severin Office of Cadastre and Land Registration (OCPI)].

The data obtained were analyzed with the help of the SPSS software (Statistical Package for the Social Sciences), separately for each institution and for both institutions together, in order to obtain a general image of motivation inside Caras-Severin county state institutions.

The complete results of this research were used in the Bachelor's diploma thesis *The motivation of the public sector employee* (Lupinca 2016), this chapter presents only the results obtained for the complete sample:

It is important to underline that three intrinsic motivational factors are chosen by the entire sample: safety at the working place in the public sector, the importance of the employees' daily activity and the fact that a professional development plan is available. On the other side, an intrinsic factor is rejected by all respondents: the idea de loyalty is rewarded in any way.

The same negative situation also appears for two extrinsic factors, also being rejected by all respondents: the accessibility of bonuses and the offer of rewards for the activity realized.

A percentage over 85% is observed for two intrinsic factors, 91.7% of the respondents considering that the authority, responsibility and the autonomy they dispose of represents an important factor in the choice of such a position, and 85% of the respondents chose the public sector because the tasks realized correspond to their faculty studies.

Percentages over 85% are observed for three extrinsic factors, the fact that salaries are established by law and that exact rules exist in this sector is considered an important motivational factor by 91.7% of the respondents, 85% sustain that the working environment is a nice one and 90% consider their schedule a satisfactory one.

Percentages over 70% are registered for other five intrinsic factors, the status obtained in a community is an important factor for 75% of the research participants; 73.3% of the respondents chose this domain due to a possibility of personal and professional development; 76.7% of the participants consider their work appreciated; 75% consider their aptitudes are well valued by the nature of their activities in the chosen sector; and 71.7% appreciate the equity and the clarity of the promotion process, possible due to the norms and standards imposed by law.

These percentages are also observed for extrinsic factors, 71.7% considering that the tasks realized inside their institution are attractive and 78.3% of the respondents sustain that a feedback is offered for the activity realized.

The other intrinsic factors introduced in this research were also chosen by a majority of respondents, 65% of the respondents consider their colleagues competent and the level of autonomy they dispose of through the position occupied in a state institution is a satisfactory one.

If we take into consideration the extrinsic factors, only two of them overpass 60% in the choice of positive values, thus 60% consider that the tasks to be fulfilled are clear and exact and 61.7% sustain that internal rules are easy to understand and apply, 58.3% of the respondents sustain that the salary is a transparent one, even if it is calculated according to regulations in force.

On the other side, two extrinsic factors make themselves noticed through negative values, 56.7% of the respondents sustaining that the choice of this working position is not influenced by the entourage; the second motivational factor with negative values refers to an internal communication of objectives which is underlined as being deficient by 60% of the employees.

According to the data presented, regarding the intrinsic factors introduced in the questionnaire applied it is obvious that these were chosen by a significant majority, 83.3% of the respondents sustaining intrinsic motivational factors. At the opposite point, for the extrinsic factors, positive values barely pass 50% of the respondents.

The data obtained confirms the conclusion that the motivation of an intrinsic nature, in the choice of a working position in the public sector, or at least for the two institutions belonging to the public sector, is stronger than that influenced by extrinsic factors.

In order to be able to offer a clearer image of a classification of intrinsic and of extrinsic factors, this part of this research is completed by a presentation of values, in a descending order, shaped as a table.

Thus it can be easily observed that intrinsic factors are chosen in higher percentages than extrinsic ones, some being chosen the entire group of respondents. The top choices are represented, as mentioned by the safety at the working place, the importance of the activity realized for the society and the existence of processional development. The second place is occupied by the responsibility induced to public sector employees and the autonomy of the tasks realized.

Thus, it can be stated that the results of the research applied in the Caras-Severin county, with the participation of employees from two state institutions, an institution dependent of the local administrative sector and an autonomous one, confirm what researches at an international level concluded, that the activity from the public sector, in comparison to the private sector, is mainly chosen based on intrinsic factors, connected to personal satisfaction (Table 1).

Table 1 Top intrinsic and extrinsic motivational factors

No.	Intrinsic motivational factors	Yes (%)	No (%)	Extrinsic motivational factors	Yes (%)	No (%)
1.	Safety of a working place	100	–	Salaries according to the law	91.7	8.3
2.	The importance of working tasks for the community	100	–	Working schedule	90	10
3.	The possibility of professional development	100	–	Working environment	85	15
4.	Authority, responsibility and autonomy	91.7	8.3	Feedback	78.3	21.7
5.	Correspondence with the diplomas obtained	85	15	Attractive tasks	71.7	28.3
6.	Work appreciation	76.7	23.3	Clear internal regulations	61.7	38.3
7.	Valuing of employees' abilities	75	25	Clear tasks	60	40
8.	Status and respect in the society	75	25	Salary transparency	58.3	41.7
9.	Personal and professional development	73.3	26.7	Objectives communication	40	60
10.	Equitable promotion plans	71.7	28.3	Influence of the entourage		56.7
11	Colleagues' competence	65	35	Bonuses accessibility	–	100
12	Satisfactory autonomy	65	35	Rewards based on the activity realized	–	100
13	Rewards for loyal employees	–	100			

Source Lupinca C.E 2016: 96

4 Conclusions

Numerous studies realized at the level of public or private organizations, use as reference internal and external factors, in the analysis of the predominant motivational type in the modification of employees' behavior and proved that intrinsic factors are necessary to be satisfied in order to obtain optimum results from employees.

The research presented in this chapter supplements the conclusions of specialty literature by confirming the idea that intrinsic factors are the one that lead to the efficient fulfillment of tasks among the employees from the Caras-Severin county public sector. In addition, a combination of the two types of factors, even if the intrinsic one reserved the top positions in the preferences of employees from the two participant institutions, contributes to the satisfaction of needs that encourage employees in the efficient realization of activities and to a loyalty towards the employer organization.

At present times, the practices of human resources management are oriented towards the identification of new methods through which the employees of the public sector may satisfy the demands of the society, while being motivated and satisfied with the work realized. The salaries continue to be calculated according to a unique salary system and the lack of financial resources still represents a barrier in the development of human resources practices regarding the recruitment and the selection processes, regarding the maintaining and the development of employees inside these institutions. The development of a strategy of human resources by public institutions is limited by the law, and the formulation of new ideas and innovations in this frame is impossible.

References

Andreş, S.: Managementul resurselor umane. Sinteze, teste, îndrumări (Management of Human Resources. Syntheses, tests, guidance). Eftimie Murgu, Resita (2009)

Cole, G.A.: Managementul personalului (Personnel management). CODECS, Bucharest (2000)

Cornean, C.E.: Motivaţia angajaţilor din sectorul public, Volume Optimizarea bunelor practici în consilierea integrativă educaţională (Motivation of Public Sector Employees, Volume "Optimization of Good Practice in Educational Integrative Counseling"). In: Ispas, C., Constantin, A.C. (eds.). PUBLIBANAT, Resita (2014)

Drobot, L., Petrovici, M., Rotaru, I., Cornean, C., Constantin, A.: Career orientation and reorientation. In: Katalinic, B. (ed.) Annals of DAAAM for 2009 & Proceedings of the 20th International DAAAM Symposium, Book Series: Annals of DAAAM and Proceedings, vol. 20, pp. 1443–1444 (2009). ISSN: 1726-9679, ISBN: 978-3-901509-70-4, WOS: 000282335600722

Landy, F.J., Conte J.M.: Work in the 21st Century: An Introduction to Industrial and Organizational Psychology, 3rd edn. McGraw Hill (2010)

Lane, J.-E.: The Public Sector: Concepts, Models, and Approaches. Sage Publications (1995)

Lupinca, C.E.: Bachelor's diploma thesis, Motivaţia angajaţilor din sectorul public (Motivation of public sector employees). "Babes-Bolyai" University of Cluj-Napoca, 85p (2016)

Mitchell, T.R.: Motivation: new directions for theory, research and practice. Acad. Manage. Rev. **7**(1), 80–88 (1982)

Perry, J.L., Hondeghem, A.: Motivation in Public Management: The Call of Public Service. Oxford University Press, Oxford (2008)

Robbins, S.P., Judge, T.A.: Organizational Behavior, 13th edn. Prentice-Hall (2008)

Rosicka, Z., Hoskova-Mayerova, S.: Motivation to study and work with talented students. Proc. Soc. Behav. Sci. **114**, 234–238 (2013). https://doi.org/10.1016/j.sbspro.2013.12.691
Wright, B.E.: Public Sector work motivation: review of current literature and a revised conceptual model. J. Public Adm. Theor. **11**(4), 559–586 (2001)

The Volkswagen Case: Analysis of a “Scandalous” Growth

Ferdinando Casolaro and Mario Cristiani

Abstract The purpose of this work is to study, through various regression models, the relationship between investment in Research and Development and the results achieved by the Volkswagen group in the years between 2001 and 2015, during which this company was at centre of a resounding fraud in the Automotive Industry, demonstrating as simple models such as those based on bivariate regression can be used to explain the decisions made by successful companies and also to predict the results.

Keywords Volkswagen · Research and Development · Gross Profit · Automotive Industry · Fraud · Regression

1 Introduction

Considering the Automotive Sector as a whole, it is impossible not to mention how investments in Research and Development have grown extraordinarily. According to a study by the Boston Consulting Group, Volkswagen, General Motors and Toyota were the companies of the Automotive Industry that in 2014 invested more in this field. Between 2007 and 2012 they increased their investments in Research and Development by 13% to 27.4 billion dollars, while in the same period their total turnover increased by 8% (Arcudi 2014).

In 2014 Volkswagen was reported, in a report drawn up by the European Commission, as the leading company for investment in Research. The complete ranking is in Table 1 (Ricciardi 2014).

F. Casolaro (✉)
Department of Architecture, University of Naples Federico II, Naples, Italy
e-mail: ferdinando.casolaro@unina.it; fcasolar@unisannio.it

M. Cristiani
Department of Law, Economics, Management and Quantitative Methods (DEMM), University of Sannio, Benevento, Italy
e-mail: mariocristiani31@gmail.com

D. Flaut et al. (eds.), *Decision Making in Social Sciences: Between Traditions and Innovations*, Studies in Systems, Decision and Control 247,
https://doi.org/10.1007/978-3-030-30659-5_30

Table 1 The top 50 companies for investments in R&D in the 2014 scoreboard and their rank change 2004–2014

Rank in 2014	Company	Country	R&D in 2013 (€M)	R&D intensity (%)	Rank change 2004–2014
1	VOLKSWAGEN	Germany	11743	6	Up 7
2	SAMSUNG ELECTRONICS	South Korea	10154.9	6.5	Up 31
3	MICROSOFT	US	8252.5	13.1	Up 10
4	INTEL	US	7694.1	20.1	Up 10
5	NOVARTIS	Switzerland	7173.5	17.1	Up 15
6	ROCHE	Switzerland	7076.2	18.6	Up 12
7	TOYOTA MOTOR	Japan	6269.9	8.2	Down 2
8	JOHNSON & JOHNSON	US	5933.6	11.5	Up 4
9	GOOGLE	US	5735.6	13.2	Up 173
10	DAIMLER	Germany	5379	4.6	Down 7
11	GENERAL MOTORS	US	5220.8	4.6	Down 5
12	MERCK US	US	5165	16.2	Up 17
13	BMW	Germany	4792	6.3	Up 15
14	SANOFI-AVENTIS	France	4757	14.4	Up 8
15	PFI ZER	US	4750.2	12.7	Down 13
16	ROBERT BOSCH	Germany	4653	10.1	Up 10
17	FORD MOTOR	US	4640.7	4.4	Down 16

(continued)

Table 1 (continued)

Rank in 2014	Company	Country	R&D in 2013 (€M)	R&D intensity (%)	Rank change 2004–2014
18	CISCO SYSTEMS	US	4563.8	13.4	Up 13
19	SIEMENS	Germany	4556	6	Down 15
20	HONDA MOTOR	Japan	4366.7	5.4	Down 4
21	GLAXOSMITHKLINE	UK	4154.3	13.1	Down 10
22	IBM	US	4088.9	5.7	Down 13
23	ELI LILLY	US	4010.8	23.9	Up 18
24	ORACLE	US	3735	13.5	Up 47
25	QUALCOMM	US	3601.6	20	Up 112
26	HUAWEI	China	3589.3	25.6	Up > 200
27	AIRBUS	The Netherlands	3581	6	Up 8
28	ERICSSON	Sweden	3484.8	13.6	Down 11
29	NOKIA	Finland	3456	14.7	Down 19
30	NISSAN MOTOR	Japan	3447.2	4.8	Up 4
31	GENERAL ELECTRIC	US	3444.3	3.3	Up 6
32	FIAT	Italy	3362	3.9	Up 12
33	PANASONIC	Japan	3297.2	6.2	Down 26

(continued)

Table 1 (continued)

Rank in 2014	Company	Country	R&D in 2013 (€M)	R&D intensity (%)	Rank change 2004–2014
34	BAYER	Germany	3259	8.1	Down 2
35	APPLE	US	3244.9	2.6	Up 120
36	SONY	Japan	3209.1	21.3	Down 21
37	ASTRAZENECA	UK	3202.8	17.2	Down 12
38	AMGEN	US	2960.6	21.9	Up 18
39	BOEHRINGER INGELHEIM	Germany	2743	19.5	Up 23
40	BRISTOL-MYERS SQUIBB	US	2705.4	22.8	Up 2
41	DENSO	Japan	2538.9	9	Up 12
42	HITACHI	Japan	2420	3.7	Down 18
43	ALCATEL-LUCENT	France	2374	16.4	Up 4
44	EMC	US	2355.2	14	Up 48
45	TAKEDA PHARMACEUTICAL	Japan	2352	20.2	Up 28
46	SAP	Germany	2282	13.6	Up 23
47	HEWLETT-PACKARD	US	2273.2	2.8	Down 24
48	TOSHIBA	Japan	2269.1	10	Down 18
49	LG ELECTRONICS	South Korea	2208.5	5.5	Up 61

Source The EU Industrial R&D Investment Scoreboards 2014 and 2004. European Commission, JRC/DG RTD

It is therefore evident that the research efforts were considered fundamental in the growth process that led, in those years, the German company to the top of the Automotive Industry. Toyota, which in 2013 was beaten by Volkswagen in regards to number of cars sold, is ranked seventh in this list. At the eleventh place there is General Motors, at the thirteenth BMW.

At this point, we can ask ourselves how much investment in Research has been important for the growth of Volkswagen in a glittering path inevitably stained by the scandal called "Dieselgate".

2 Methods and Results

Taking advantage of the data provided by the Volkswagen Annual Reports, we have identified the values of the expenses for Research and Development and the Gross Profit in the Automotive Division of the German company from 2001 to 2015 (Table 2).

The Gross Profit is one of the best suitable summary indicators to identify the health status of a company, to understand how it is and where it goes. With such a strong commitment to Research and Development by Volkswagen and the Automotive Industry in general, we have been concerned to develop a model that would allow us to identify the correlation between Research and Gross Profit.

Table 2 R&D investment and Gross Profit of VW from 2001 to 2015 (amounts in billions of euro)

Year	R&D	Gross Profit
2001	3.923	12.454
2002	4.371	11.958
2003	4.137	7.976
2004	4.164	8.046
2005	4.075	10.037
2006	4.240	11.596
2007	4.923	14.078
2008	5.926	14.737
2009	5.790	10.973
2010	6.257	18.060
2011	7.203	24.239
2012	9.515	30.664
2013	11.743	30.522
2014	13.120	31.226
2015	13.612	28.382

Source Volkswagen' Annual Reports 2001–2015

Table 3 Investments in R&D and Gross Profit as independent and dependent variables

x	y
3.923	11.958
4.371	7.976
4.137	8.046
4.164	10.037
4.075	11.596
4.240	14.078
4.923	14.737
5.926	10.973
5.790	18.060
6.257	24.239
7.203	30.664
9.515	30.522
11.743	31.226
13.120	28.382

However investments in Research and Development do not determine immediate results. They take time to produce their effects! Therefore the pairs of values to consider are not those of Table 2. By working on the observations, we then construct a new table with two columns and 14 rows (Table 3). In the first column, called x (independent variable), we insert the 14 values of expenses in Research and Development from year 2001 to year 2014. In the second, called y (dependent variable), we insert the 14 Gross Profit values in the Automotive Division from year 2002 to year 2015. In this way the values of the Gross Profit of the different years are associated with those of the expenses in Research and Development of the previous years.

Reported in a Cartesian system, the set of observations determines the scatterplot in Fig. 1.

From the scatterplot (Fig. 1) we can deduce the confirmation of the existence of a positive correlation between investment in Research and Development and Gross Profit. However we need a function to make forecasts on the future results of Volkswagen (in terms of Gross Profit) by forecasts concerning investment in Research and Development of the previous year.

Using the Microsoft Excel software, we have identified various regression functions starting from the available sample data and then we have selected the four that best approximate the functional relationship existing between the two variables.

The comparisons were made by the r-square index (coefficient of determination).

The first model is the linear regression model where, as can be seen from the equation as shown in Fig. 2 ($y = 2.5222x + 1.9318$), an increase in Research and Development expenses of one billion euros leads to an increase in Gross Profit of 2.52 billion (slope) in the following year. With a cost of 0 in R&D, in the following year the Gross Profit is 1.93 billion (intercept).

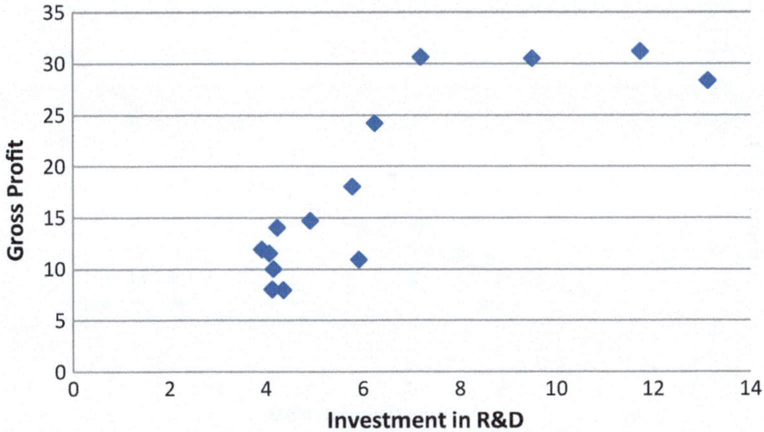

Fig. 1 The scatterplot of the VW case

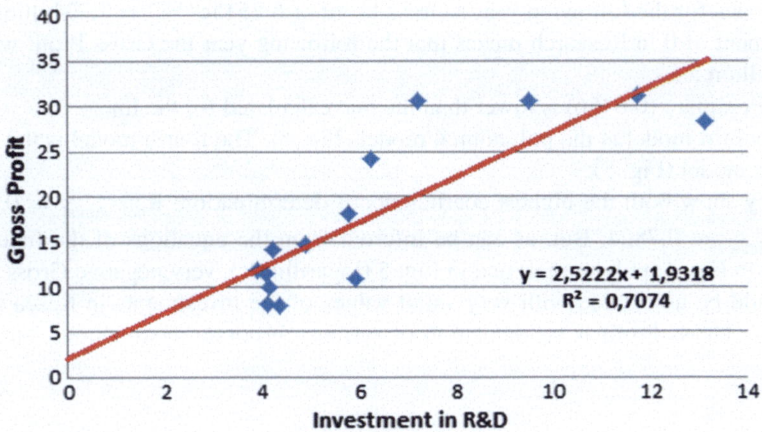

Fig. 2 The regression line in the VW case

Undoubtedly the contraction in Gross Profit seems to be excessive considering that investments in Research and Development are cumulative so, even in the case of an investment of 0 in one year, the company could still count on knowledge accumulated in previous years. Keeping up with competitors in terms of innovations, however, is fundamental in an industry such as the Automotive one in which companies are focusing on innovation to conquer substantial market shares.

The coefficient of determination $R^2 = 0.7074$ indicates a good ability of model 1 to explain the relationship existing between the variables considered (e.g. see Casolaro and Cristiani 2019).

The second model is the exponential regression model where, as can be deduced from the equation shown in Fig. 3, an investment of 1 billion euros in Research

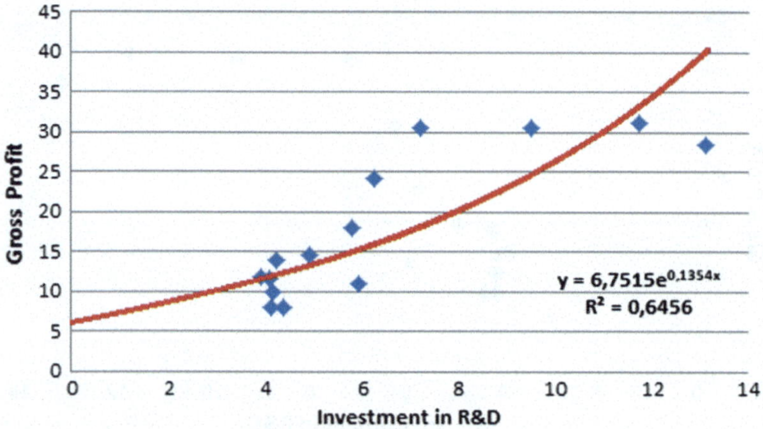

Fig. 3 The exponential function in the VW case

determines for the following year a Gross Profit of $6.7515e^{0.1354} = 7.73$ billion. An investment of 0 in Research means that the following year the Gross Profit will be 6.75 billion.

The r-square (0.6456) is lower than the one calculated for the line.

The third model is the polynomial model (Fig. 4). The fourth model is the logarithmic model (Fig. 5).

They show both the highest coefficients of determination: $R^2_{polynomial} = 0.8452$ $R^2_{logarithmic} = 0.7824$. But, as can be inferred from the equations of the functions shown in Fig. 4 (polynomial) and in Fig. 5 (logarithmic), very negative Gross Profits would be associated with very small values of the investments in Research. A negative Gross Profit is an indication of serious corporate dysfunctions, but very

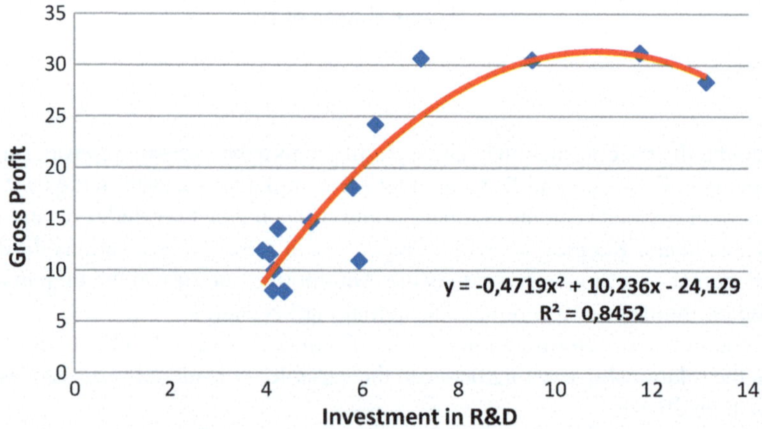

Fig. 4 The polynomial function in the VW case

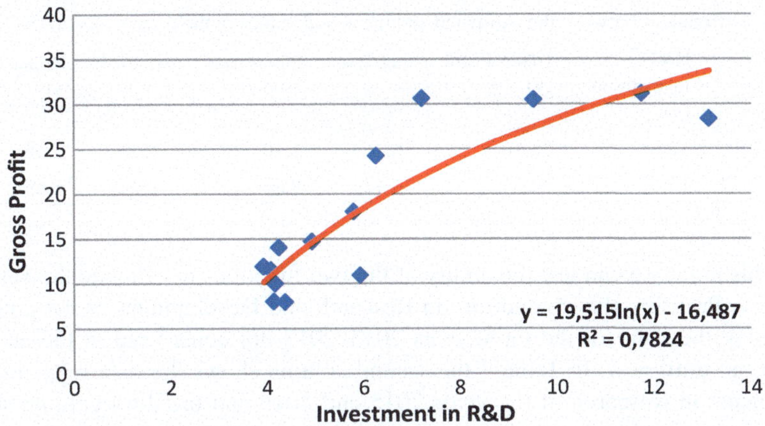

Fig. 5 The Logarithmic function in the VW case

negative Gross Profits are symptomatic of a real uneconomic approach to business management: something not justifiable even for a company in the start-up phase. For these reasons they are both to be excluded.

On the 18th September of 2015 Volkswagen received a notice of a violation of the Clean Air Act by the US Environmental Protection Agency (EPA). To pass the diesel engine emissions tests in the United States to obtain VW diesel-powered passenger cars approved for the US market, Volkswagen had manipulated the electronics of some diesel car models breaking the law and cheating millions of consumers. The scandal soon took on worldwide resonance and was named “Dieselgate”. The scandal hit in addition to the United States (with 480,000 vehicles) also Germany (with 2.2 million vehicles) and many other European countries and even China. VW admitted that the defeat device was installed on around 11 million cars worldwide, including eight million in Europe. Several public prosecutors and environmental protection agencies announced investigations into VW. After the outbreak of the Dieselgate in the United States the case spread in a growing number of countries. Investigations on the case have been opened in Italy, UK, France, Germany and even in China; meanwhile, around the world, politicians and environmental groups began to question the legitimacy of VW emissions testing (Bonato 2017; Otrisal et al. 2018).

How many of the millions of euros invested in Research and Development were used to design this highly sophisticated and fraudulent software? Answering this question is impossible.

The fact remains that, in any case, in the three years following the scandal the German company recorded absolute sales records and in 2016 it confirmed itself as a leading company in the Automotive Sector; at the same time it ranked first for investment in Research in the Eu R&D Scoreboard, a study promoted by the EU to evaluate the expenditure on Research incurred by the major industries on the world scene (www.truenumbers.it).

Table 4 Differences between the estimated and the actual values in the three years 2015–2017

Years	R&D investment (x)	Gross Profit (y)	$y_{stim\ (line)}$	$\|y_{estim\ (line)} - y\|$	$y_{estim\ (e)}$	$\|y_{estim\ (e)} - y\|$
2015–2016	13.61	35.16	36.26	**1.10**	42.64	**7.48**
2016–2017	13.67	36.34	36.42	**0.08**	42.99	**6.65**

At this point we can test the ability of the two functions to estimate the value of the Gross Profit by the expenditure in Research and Development of the previous year using the data for the three years 2015–2017 not considered in the analysis carried out until now. In Table 4 the second column shows the data related to the investments in Research of the years 2015 and 2016 and the third column shows the Gross Profit (Automotive Division) of the years 2016 and 2017 (Volkswagen' Annual Reports 2015–2017). The fourth and the sixth columns show the estimates of the Gross Profit for the years 2016 and 2017 obtained respectively for model 1 and model 2 by the Research investments of the years 2015 and 2016.

While the estimates obtained with the exponential function differ considerably from the results achieved, the Gross Profits estimated with the linear function in the years 2016 and 2017 are very close to the ones obtained by Volkswagen. The fifth column of the table shows the absolute values of the deviations for the line, the seventh the ones related to the exponential function.

3 Discussions and Conclusions

Taking advantage of Microsoft Excel software and based on our experience in business economics studies, we researched the regression functions that best approximated the relationship between Gross Profit and investments in Research and Development of the Volkswagen company, identifying them in the linear and exponential function.

The function that in terms of forecasting leads to the best performances is the linear one.

Regardless of the fraud made by the German company, investments in Research and Development were fundamental in obtaining results that allowed Volkswagen to make an extraordinary climb to the top of Automotive Industry, and the results of our work show as simple and undervalued models can be used to describe how decisions are made in big companies and also to predict their results: the line leads to a 97% accurate estimate in the 2016 and to a 99.8% accurate estimate in 2017!

The Dieselgate scandal has caused great bafflement even for the undisputed reputation of the brand and disappointment for the great indifference for relapses on health and the environment, but nevertheless there was little or no impact on the market that has kept rewarding a quality product, reliable and up to date, a sign of constant commitment and large investments in the Research and Development field.

It is possible to imagine improvements to the elaborate model. With longer time series available, the sums of research investments of three or four years could be considered as independent variables, or a multivariate regression model could be elaborated considering other independent variables such as investments in advertising, but we must point out that the strength of our model lies in its simplicity. At the same time, future developments may focus on more elaborated models such as fuzzy regression models (e.g. see Maturo 2016; Maturo and Hošková-Mayerová 2017), mixed regression estimators (e.g. see Kumar et al. 2017), nonparametric regression (e.g. see Vališ et al. 2017), and functional data analysis (see e.g. Ramsay et al. 2009; Febrero-Bande and de la Fuente 2012; Maturo and Hošková-Mayerová 2018; Fortuna et al. 2018).

References

Arcudi, S.: Settore auto: investimenti globali in ricerca e sviluppo a livelli record. Journal America **24**, 13-01-2014 (2014)

Bonato, V.: Il Dieselgate. Implicazioni giuridiche e cambiamento nel mercato dell'auto. Thesis for: Master Degree, Università degli Studi di Modena e Reggio Emilia (2017)

Casolaro, F., Cristiani, M.: Statistics for a Football Coach. Sci. Philos. **7**(1), 109–120 (2019). https://doi.org/10.23756/sp.v7i1.444

Febrero-Bande, M., de la Fuente, M.: Statistical computing in functional data analysis: the R package fda.usc. J. Stat. Softw. **51**(4), 1–28 (2012)

Fortuna, F., Maturo, F., Di Battista, T.: Clustering functional data streams: unsupervised classification of soccer top players based on Google trends. Qual. Reliab. Eng. Int. **34**(7), 1448–1460 (2018). https://doi.org/10.1002/qre.2333

Kumar, M., Bist, V., Kumar, M.: Some improved mixed regression estimators and their comparison when disturbance terms follow multivariate T-distribution. Ratio Mathematica, **32**, 3–19 (2017). http://doi.org/10.23755/rm.v32i0.330

Maturo, F., Hošková-Mayerová, Š.: Evaluating Journals Performance Over Time Using Functional Instruments (2018). https://doi.org/10.1063/1.5078474

Maturo, F., Hošková-Mayerová, Š.: Fuzzy Regression Models and Alternative Operations for Economic and Social Sciences (2017). https://doi.org/10.1007/978-3-319-40585-8_21

Maturo, F.: Dealing with randomness and vagueness in business and management sciences: the fuzzy-probabilistic approach as a tool for the study of statistical relationships between imprecise variables. Ratio Mathematica **30**(1), 45–58 (2016). http://doi.org/10.23755/rm.v30i1.8

Otrisal, P., Melicharik, Z., Svorc, L., Bungau, S., Virca, I., Barsan, G., Mosteanu, D.: Testing methods of assessment for the chemical resistance of insulating materials against the effect of selected acids. Materiale Plastice **55**(4), 545–551 (2018)

Ramsay, J., Hooker, G., Graves, S.: Introduction to Functional Data Analysis. Functional Data Analysis with R and MATLAB, pp. 1–19 (2009). https://doi.org/10.1007/978-0-387-98185-7_1

Ricciardi, R.: Investimenti: Volkswagen prima al mondo, Fiat al 32° posto. Journal Repubblica.it Economia & Finanza 4-12-2014

Vališ, D., Hasilová, K., Žák, L., Vintr, Z.: Mathematical modelling of soot particles in oil. Safety Reliab. Theor. Appl., 719–726 (2017). ISBN 978-1-138-62937-0

Volkswagen' Annual Reports 2001–2015

Volkswagen' Annual Reports 2015–2017

www.truenumbers.it. Volkswagen prima al mondo per investimenti in r&s, On 16 Febbraio 2018 In Futuro&Tech

Experience-Based Course of Social Generativity in Villaggio Celdit, Chieti. History of Challenge for Popular Quarter's Rebirth After Big Factory's Closure

Anna Crisante, Fabiola Nucci and Mario Perrucci

Abstract This chapter is dealing with **experience-based course of social generativity** in Villaggio Celdit in Chieti, its name due to "Cellulosa d'Italia CEL.D.IT.", later named "Cartiera CIR". The "Villaggio" is in fact a quarter built for working people of Fratelli Pomilio Paper-mill, factory born in 1938 after starting an advanced technique. In 70s, only the cellulose department had counting over 700 hands. But with wrong policy, factory closed in 2008, and the Villaggio transformed itself in a peripheral quarter, not only in the way of "distant from centre", but also in sense of "marginal", closed in its perimeter, in spite of nearness of college campus "G. D'Annunzio". In 2000, City Administration of Chieti determined to start local service offered to young, an aggregation space. It has allowed to lay, more than purposes and expectations of Administration and social operators, foundation stone to way of quarter's rebirth. It will also enounce the experience that, before social intervention, is people meeting, a workshop that rises and receives youth, associations, people, groups, organizations's requests, in a high potential quarter. It will demonstrate that social generativity is a course consisting to stay in the state of degradation to attain to know its real daily dimension and by running this inevitable bottleneck that one would gain intergenerational dimension, capable to "negotiate ties among past, present and future" (Magatti 2017).

Keywords Social generativity · Intergenerational experience-based course · Local workshop · Quarter's rebirth

A. Crisante
Via Vella n.75, Chieti, Italy
e-mail: annacrisante@yahoo.it

F. Nucci (✉)
Chieti Solidale srl, Via Mazzetti n.18, Chieti, Italy
e-mail: fabiolanucci@virgilio.it

M. Perrucci
Department of Letters, Arts and Social Science, Gabriele d'Annunzio University, Via dei Vestini n.31, Chieti, Italy
e-mail: m.perrucci@email.it

D. Flaut et al. (eds.), *Decision Making in Social Sciences: Between Traditions and Innovations*, Studies in Systems, Decision and Control 247,
https://doi.org/10.1007/978-3-030-30659-5_31

1 Introduction

Evolutionary paths of that periphery called to distinguish it from urban centre, depend on a lot of factors, and hold in three classes: demographical, social and economic one. First it will relate, by three ones, evolution of Villaggio Celdit, which has its history in its name.

Built in Chieti scalo to give hospitality to hands of factory called "Cellulosa d'Italia"—CELDIT in acronym –, the quarter Villaggio Celdit had passed through most important economic changes due to Italian industrial policies from last 30s. Pomilio brothers, manufactures of Chieti who opened that factory, had weighed heavy on Italian economy with their choices. Effects of Celdit's installation were considerable to local productiveness, promoting low city industrialization with others factories. At last 60s, thanks to Industrial Consortium and municipality, there were over forty establishments, a lot of big dimension, collecting five thousand workmen, being most industrialized area's supreme in Abruzzi. So, it's clear that our chapter starts from this company's history. And its heritage, local residents have got it in their own DNA. Successively, we examine the deep and generative action of last twenty years, from implementation of youth centre. Even before a social intervention it is the history of relationships testing to rejoin the present to a past of great social and cultural innovation. Celdit quarter was built in active context, actually peripheral in strict sense of marginal, closed in on itself watching very near hospital and university. These two important socio-economic realities refer to Celdit, not involving it, as it would surrounded by a rubber wall, in city springing up, while unrelentingly the up part, Chieti of ancient origin, dismisses low part, grown up fast, Chieti scalo. On the contrary, we will verify, experimental path in Villaggio Celdit takes at garret the obsolete and dichotomous opinion of periphery/centre, to prepare a rebirth able to slash the rubber wall of social and cultural isolation. By long stages the **paper quarter** preparing itself to be "place at the centre" of generative rebirth and therefore multiplier of central and meaningful actions.

2 Pulp Manufacturing Changes Chieti Scalo's History

2.1 Everything Starts from Pomilio's Family

Ernesto, Ottorino and Umberto Pomilio from Chieti, started to invest in chemical industry at the end of World War First beginning with electrochemical factory foundation in Naples, 1917. With production of sodium hydroxide and chlorine gas, the factory is part of war industry until 1918. In peacetime, having regard to the need of industrial reconversion, since 1920 studies about employment of chlorine gas in extraction of cellulose from annual plants initiated.

2.2 Pomilio Method Becomes Patent for Export

Cellulose is obtained especially from wood species, but also from fiber materials like esparto, reed and straw. In first decades of 20th Century it was used as raw material in paper industry, and in textiles (especially rayon), celluloid, cellophane, sorts of paints and plastics, as well as base of most part of explosives. Italy, by shortage of forests, imported cellulose, which was used in paper and rayon industries, trades in great expansion in 20s.

Umberto Pomilio, assisted by Francesco Giordani, formulated the theory of membrane electrolyzer, a modern instrument for electrolysis that named by two chemists and that will employed diffusely to product sodium hydroxide, last refine as chemical method to extract cellulose by chlorine and soda, those extracted, in turn, by electrolysis of salt. This proceeding was patented in Naples at "Elettrochimica Pomilio", so called "Pomilio Method". This had function to break vegetable tissue by multiple phases with dilute reagents. In the contrary of classic methods, like with sulphite and bisulfite, in those cellulose was submitted to high temperature with enormous expenditure of thermic energy, new method used principally electric energy to product reagents operating in cold or low temperature process. During first phase, raw material, cut and washed, was dealt by light alkaline solution. During the second one, the compound was processed by chlorine gas that, attacking lignonic unit, formed chlorine-lignone, easily to solve in an alkaline compound during the third phase. In 1922 Neapolitan factory started to product cellulose from wheat straw, stalk of hemp and esparto. Stalk of hemp was easy to be found, thanks to large hemp-growing around factory. Esparto came from Libya, from which frequently ships reached Naples. New product did not establish itself over classical cellulose, because the production was discontinuous and did not make sure economic benefits. Moreover, the production met with obstacles due to Italian papermakers's opinion that cellulose from annual plants was poorer to that from fir. Nevertheless, new method with chlorine gas attracted attention of Comit, that recently started to product cellulose with liquid chlorine in its establishment in Bussi sul Tirino (Pescara), still expensive in reagents, so without profits. In 1927 Neapolitan factory, with new society "Cellulosa cloro soda" (Ccs), was forced to stop its production because of impossibility to renew electricity supply a reasonable costs. This fact urged Umberto Pomilio to consider eventuality to sell his patent in South America. In fact, some entrepreneurs decided to contact Umberto Pomilio, the inventor of method for cellulose from wheat, in order that there was enormous quantity of wheat in Pampas. In 1929 it was formed the "S.A. Celulosa argentina" and to Umberto Pomilio, appointed technical director, some shares were offered in exchange to the use of patent. New establishment provided entirely paper-make, from transformation of straw in cellulose to realization of paper. This is the first industry in Argentina about cellulose from straw. Straw paper was suitable to make fine paper that in Argentina it was widespread. The home market absorbed all production «with an overcharge depending by excellent quality of products». Patent's renown spread and in 1932 Engineer Umberto Pomilio, at the head of a group of Italian technicians, went to Santiago in order to plan and direct the construction of an establishment that starts in June year after. The fact that also a State rich in forests like Chile would found an advantage from straw paper was «the proof of superior quality product» (Ostuni 2016).

In the meantime, the Ccs from Naples restarted activities strictly making soda, chlorine and derivatives. Later on, in 1933, to restore factory's economy, it had resolved to restart the cellulose department and, after new administrators announced, with the approval of the IRI management, especially Pomilio Method enjoyed to product cellulose, in the sense of forest preservation and without import making worse deficit of Italian trade balance. The consent to chlorine-gas-cellulose, with exploitation of national plants, came from necessity of war that needs a pressing quantity, but also *meridionalista* policy of Fascism. In those years the IRI wanted to concrete a «meridionalismo of action and pragmatism», turning to account agricultural and natural resources like salt and straw, plentiful presences in Southern Italy. IRI attention to Southern rendered on focus Pomilio's initiatives but its same entry as major shareholder induced Pomilio brothers to form a Corporation in 1934: "Sindacato Cellulosa Pomilio", which configured itself as a consulting company in the sphere of cellulose and paper, and was utilized to defend own interests in this context and to promote patents abroad.

2.3 *The New Establishment in Foggia,* Meridionalismo *of Pomilio Brothers and the Autarchy*

During 1933, Pomilio brothers, in the context of industrial development of the so called "Battaglia del grano" (Wheat Battle), planned a big establishment in the province of Foggia, really a barn of Italy, to make cellulose starting from wheat straw, and in addition a paper factory for integral cycle of production. In this case, it would reduce expenses on raw material from abroad. The establishment was considering in excellent position to supply with salt, since nearness to Royal saltern of Margherita di Savoia, only 55 km far. It was providing water from regional aqueduct and from new dedicated well. To the management of a complex establishment, it was created a specific company, the "Industria Cellulosa d'Italia" (Incedit). In this, the "Cellulosa cloro soda" by Naples was having preeminence «both for experience in management and technic and for the large share capital reserved to it» (Ostuni 2016). IRI was taking part at the company by "Cellulosa cloro soda", by means of an increase in share capital and election as Incedit CEO of Ottorino Pomilio.

«By law 1453 of June 13th, 1935, while works were starting to build up establishment, fascist government instituted an economic support to print, named "Ente nazionale per la cellulosa e per la carta". The "Ente" was a corporative institute. This resolution was in sense of autarchy, to protect the productive segment and to increase national cellulose production» (Ostuni 2016). The "Provveditorato Generale dello Stato", considering important paper consumption, deemed necessary to acquire the establishment through change of ownership to "Poligrafico dello Stato", in 1936. So, the establishment was nationalized. In 1937, the "Sindacato cellulosa Pomilio" stipulated with the Institute "Poligrafico" the exclusive right of chlorine-gas-cellulose.

By the opposite, there were interests of Italian papermakers, specially the "Società anonima Cartiere Burgo", founded by Luigi Burgo, that was oriented in press paper and that owned the national scene.

2.4 In 1938 CELDIT Constitution to Administrate New Establishment in Chieti Scalo

After left the Ccs share capital, Pomilio brothers became managers employed by IRI. When in 1938 CELDIT constituted, joint venture by IRI and Burgo, Ottorino Pomilio would nominate General Director. Company's purpose, in autarkic philosophy, was «to study best technical proceedings to economical cellulose production, as well as experimentation on lots of raw materials from which it is possible obtain cellulose». Chieti scalo was new geographical point to build up factory. «It is not impossible Chieti, Pomilio's home town, would select under pressure by Ottorino, staying in Management board. If this happen, the solution would be by policy, not by benefits and costs evaluation, so there was no interesting in industrial development in Abruzzi. Water resources proved be rather different than estimate. It needed to sink 30 m wells to satisfied production. By the plan, new establishment would work since 1939, on the contrary it started in April 1940, but it was working regularly at the end of July, a month after Italy's entrance into the war» (Ostuni 2016).

Chieti scalo site was 27 hectares extended and construction «continued by four years: in 1935 we can see terrain without buildings with first excavation works and next reinforced concrete plinths. In 1939, from other photo, we see industrial premises for straw stock. Magazine "Il Nuovo Abruzzo. Foglio d'ordine della Federazione dei fasci di combattimento di Chieti", nr. 15, Dec. 31st 1938, outlines the 48 Celdit worker's houses built up on 70 thousands metres squared, each with a garden and a yard of 450 m^2 dedicated only to two families in each building, like a happy place in the poor Pescara downriver valley» (D'Alessandro 1987). The workers get own house with a small levy from pay packet.

2.5 Chieti Scalo Development and Factory's First Difficulties

«In October 1938, Concezio Petrucci, one of the most important engineers, drew up the development plan for Chieti scalo to the order of Chieti municipality. Finished in 1939, the plan provided an industrial zone down Pescara river, railway sidings, a Celdit worker's newtown, three residential neighborhoods and the finalisation of Colonnetta street that would cut over 10 km distance from upper to lower town. The eruption of the military conflict partially stopped the plan» (Ferrandino and Napolitano 2014).

«During last five months in 1940, 13,903 quintals were made, rather little than expected average of 200,000 quintals per year, but it has to consider war effects about logistics and some inconveniences to the electrical and water systems. New conditions never let wildest expectations. Underperformance and high cost of straw depended on its origin: requisition of the military and little plots forced to source from Molise and Puglia, with expensive transport costs. Particular type of cellulose in Chieti aimed to maximal strength together with whiteness, though. An attempt was made to balance leaching and chlorination to obtain a bleaching not too aggressive. It could have a new raw material and make progress on quality» (Ostuni 2016). «In February 1940, it employed technicians and workforce, as we read in a "Relation about courses to technicians and workmen for establishment of Chieti", from "Istituto Industriale Luigi di Savoia, to train a number of students in technologies and cellulose industry. Two courses were instituted: to technics one and to workforce one. Later 43 units were employed» (D'Alessandro 1987). In 1941–42, 406 units were working in Chieti.

In 1943 the "Poligrafico dello Stato" dealt the acquisition of local factory, but the situation was still uncertain and at the beginning of October, the Germans occupied establishments and they started take machinery down, to send them in German, while those remained in Chieti they were damaged on 40%. Studies about restoring production plants started in 1945 yet, when IRI financed restructuring and transport of machinery from Treviso, but works walked slowly because of needs of electricity and means of transport: only in January 1948 factory recovered. In this period it produced 2500/3000 tons of fine quality cellulose, so thought to sell it in North America, England, France and Benelux. «Also a document of IRI inspectorate on July 1948 remarked that quality, thanks to the classic Pomilio method: manufacturing of *pergamina* paper makes supercalendered paper highly resistant and more brilliant and smooth» (Ostuni 2016). After some crises due to closures and re-openings, layoffs and resumptions, it decided for a paper industry that could absorb most part of cellulose of the local establishment. But it came only possible thanks to U.S. reconstruction founds (1948). Finally, in 1950, manufacturing of cellulose re-started and staff rehired only the strict number required. In the meantime, in the same year, a trolleybus will start: 8.5 km on the Colonnetta Street, connecting fastly the two parts of city, offered new transport to students and hands (Ferrandino and Napolitano 2014).

2.6 *Celdit's Complete Affirmation with Quality Product*

The 1951 had favourable conditions. The production program in the new paper industry listed two different type of paper: *pergamini*, with wider part of straw cellulose, and magazine and writing paper, or thin paper for accounting record books and packaging, with 25–75% straw cellulose. In August 1951, paper machine started work, so industrial program concerned about standard types: high quality paper with constant features and low price, so that beat the majors. The 1953 was rather a turning-point of the company: the year of the complete affirmation of paper type "Celdit", a special type of writing and for publishing paper, very thin, with enormous proportion of straw cellulose (60–75%), bleached fir cellulose and adhesives spiked. Then, paper production increased up to 14'000 tons and 70 million lire in 1956. At the end of the year, IRI approved definitively the planned extension of paper industry. At the beginning of 1960, after burnishing, the second paper machine started work, with great success: Celdit transformed itself from a chemical industry of cellulose into a paper industry. Qualitative performance of new machine was the best. It could not yet double the production but also vary it with coated and semi-coated paper, increasingly demanded. All types of Celdit were acclaimed on the market and prices rose. Years 1961 and 1962 were definitely the greatest for Celdit (Ostuni 2016).

2.7 *Industrial Disputes*

But starting from 1961 it happened a syndical battle, with some strikes to demand increase in wages. Demands appointed on incentive prize for hands. In September agreement was signed and it assigned to all hands the prize, part fixed and partly dependent on productiveness. Incentive prize's cost was 100 million lire, equal to more 15% of wages. That generated increase of the work hour from 378 to 454 lire between 1960 and 1961.

Between 1959 and 1964, municipality of Chieti was investing 150 million lire in industrial development of the Scalo, by purchasing land plots and constructing infrastructures. At the end of Sixties, with the activity of Industrial consortium and municipality, in Chieti scalo over 40 factories were active, a lot of big dimension, with five thousands workers employed.

From 1963, Celdit earnings began to get down, just like share dividends. To get down costs, the Management withdrew previous agreement in January 1965 and next month held four days lock-in, in reply to the strike decided by three trade unions of the confederation. The industrial dispute climaxed on March 11th, when establishment was occupied, and the Company retraced its footsteps. Some production sectors closed and so more layoffs. In 1966 started the third paper machine. Social unrest in 1967-68 reached a peak, with a new occupation of the site in March 1968, on reply to the umpteenth lock-in. In 1969 more disputes to abatement the "wage-cages".

2.8 From CELDIT to CIR

During October 1972, Celdit was incorporated in "Cartiere italiane". With this new group, named Cir ("Cartiere italiane riunite"), it reached the maximum of 700 employees. «In the Seventies, by political patronage, factory is transferred to the Fabocart group of Giovanni Fabbri and Carlo Bonelli, a real trust of paper market. The disastrous consequences were on employment and on production» (D'Alessandro 1987). In 1985 a special commissioner was nominated.

2.9 Closure of the Factory and Isolation of "Villaggio Celdit"

In 2008, the factory was demolished, when the transformation of Villaggio had already started. In 1974, indeed, lots of house-workers, bought by "Iacp", today "ATER" ("Aziende Territoriali per l'Edilizia Residenziale"), were tore down, while other buildings were constructing in their place. In this Villaggio, increasingly densely populated and less stores, in 1982, in Piazza San Pio X, a Cultural Centre was established there, transferred from via Torquato Scaraviglia, where it was in service since 1968. This Centre, managed by Government of Abruzzi, had got also a library of 15,000 books, Lida Buccella as responsible director. Then, always under the same administration, became "Agenzia per la Promozione Culturale", and 90s were the most successful period, thanks to good funding. There was a great level of enthusiasm and important public initiatives, from which a memorable outdoors play. To contrast the decision from the top, however, by a significant popular protest movement, the Centre closed in 2000 and it was transferred to upper city, in via della Liberazione. As a result of last earthquake, the Centre closed definitively, depriving city from rich book heritage that had stimulated the active and populated Villaggio.

3 Piazza San Pio X at the Heart of Villaggio Celdit: Social Generative Place "At the Point"

3.1 Youth Center: Generative Stone

As seen, the history of this quarter and its relationship spaces comes a long way, and has got own past—working-class neighborhood—that will go back, although not seems remember. Inevitably, it has to dovetail with its history to plan every possible future.

What is singularly impressive is that 2000, although before and next exist, marks a new start. Before: the closure of the Cultural Centre with its library; next: the opening of Youth Centre, by public administration's will, assessorship to social policies, like that want to restart from youth's instances. The library in Piazza San Pio X was a cultural node, of relationship and socialization. Became it an empty place, it meant take opportunity to city part that was just showing signals of social plain, because of rapid transformation around.

The opportunity of Plan of assistance for childhood and youth, as well known as "Law 285 of August 28th, 1997" (Provisions in progress for childhood and youth's rights and opportunities), offers to Municipality the finances to open the "Centro ragazzi 2000", a youth aggregation centre, by entrusting to voluntary association I.A.P.A., directed by sociologist Fabiola Nucci.

Starting from new generations, with an experience-based approach, as journalist Rita Gambescia reports in her report about centre activities, «is—by its nature—a process hic et nunc, always open to topics, which generates also for the staff new experience, tracing a history of community. In Celdit quarter, Centre staff came into contact with kids needing help to found **in group** resources to improve social relationship and cultural and value-based references. This was an educational process which has to interpenetrate youth's vibrant spheres». That meant promoting a regional coverage to connect these vibrant spheres «so, docked at many ports, round places recovering Centre, started floating», riding at generative anchor made «by people-to-people relationships dense with intentionality and significance». In the Preface to the report at Centre 10th anniversary, Eide Spedicato remembers this necessary interconnection, quoting Ivan Illich: «an inconvenient intellectual of the 70s, whom name is related to convivial society, a construction by social relationship far away from formalism and homologation to aim at renewed interpersonal and ambient nexus dense with intentionality and significance». Spedicato gives the best image of what professionalism we do, affirming dreamt about convivial society by Illich «although in a local microcosm, it seems came true» (Spedicato Iengo and Bongo 2015). The experience of "Centro ragazzi 2000" that still continue, 19 years later, and the thousands of transformations needed by intervening new generations, represents the generative stone to come out "Villaggio Celdit" rebirth.

3.2 Social Generativity

The concern for social generativity is clear from the sentence is prominently displayed to the heart of Villaggio, on circular stairwell: ***If we want construct together a mosaic with the thousands of its colours and infinite forms, first we imagine its empty set without colours any forms***. The concept of "generating" is carried on in sociology by Mauro Magatti: «It is a way of doing that assumes the presence of before, now and next, faced which we take responsibility to give ourselves, open to unknown». The generativity, so, is a conscious and voluntary action, respectful to past and actuality and opened to future, but ***you shall be imagined like an empty***

set without colours and forms. As already mentioned, generative way is necessarily undergoing uneasiness and degradation which in many moments come to generative requests. The perception of a loss, or eradication, or no sense has a lot in common with the Erikson's description of the stagnation in the seventh stadium—as adulthood—which he features to generating. Stagnation and generativity are interconnected, leaving opened gates to change which «knit actors by construction of relationship with development joint to social cohesion. Generativity is able to outline a sense direction not only self-referential but inter-subjectively» (Magatti and Giaccardi 2014) and intergenerational, to make ***a mosaic with a thousands of its colours and infinite forms***.

3.3 *The Bottleneck of Degradation:* Between Generativity and Stagnation

In ten years since "Centro ragazzi 2000" opened, the aim was to connect needs to resources in the quarter. We do not have yet the perception of a thousand of mosaic pieces, which shall grow up by generative approach. Next to foundation stone—the Centre—other services and events rose, stimulating factors of increasing neighborhood: the Intercultural Centre "Mondofamiglia", meeting place for people from different cultures; the "Ponte", project of job training offered to fragile boys; the "Banca del tempo", project to sustain auto- and inter-helping families in the quarter. In our idea, the spaces are first of all places, capable of producing collective thinking to solve and prevent problems, to join at events, meetings, conferences, workshops in education, culture and for social relationship. This great sense of unrest, soon would encounter stagnation during the commitment to "Azienda speciale multiservizi Chieti Solidale", now "Chieti Solidale srl" which in 2007 received social security and educative services of Chieti. The eriksonian stagnation resided in the perception of our powerlessness and frustration, so we consider fluctuating day by day experience over, and then other times imagine the spaces «as places which not exist yet, but able to mobilise social forces and project's skills». Without knowing it, we were tying all the loose in a generative texture that it guides us to reflect on this fact: even though our professionalism, on the contrary, in spite of it, fragilities were staying there and at the same time prepared to see the transformation of everyday context from space to ready-to-regeneration place. We had been in material space on stand-by to live in a place «potentially experienced and inhabited by social interactions, not only by a rest» (Marchesi and Marmo 2018). In this particular passage the transformation from space to place did not yet seem to us concrete, but achievable in every new start, in which there was what this travel was and will be.

3.4 The Relationship with Associations: New Age Network

Stagnation time, just told, signed one clear datum, the necessity to focus on the quarter with new energies and opportunities, to break the deadlock. A multitude of efforts always in the right way to reconnect the multiplicative experience of checkered last years. Unexpectedly, the occasion arose to a new start able to renew the speech and the lines of our memory as social operators. The "Piano sociale distrettuale – ambito n. 8", a sort of social master plan concerning all actions and services planned for a regional district (Chieti—Casalincontrada), was offering on a gold plate the striking change. Among many acts, provided entirely to Chieti Solidale the coordination of associationism world promoting healthy ageing. To this end—as in the Plan (Delibera 2017)—the Active ageing network is promoted to experience a methodology of inclusion for social prevention and a new idea about old age. It should be noted the name given to this, "Rete della nuova età" (New age network). It is obvious this is a transition period for the old age (thanks to advances in medicine, increased awareness of lifestyles, level of education higher than later, working context healthiest on average) that this could called new age.

The associations assented to a public notice to create a shared agenda of a variety of activities. Network was born in 2017, starting activities since January 2018: dynamic lessons about the use of new technologies; program of Mindfulness and psychosomatic and global wellness; program of shared lecture, later extended as a workshop of "walking lecture"; workshop of autobiographical and local history writing; workshop of mnemotechny and old-age-motricity; social gardens with cultural events and beautification of Piazza San Pio X. It is important dwell on methodology of this Network. The function of Chieti Solidale as coordination was determinant: ***methodological container*** of numerous local resources; ***sustain*** to new entry; ***bridge*** between people and resources; ***guarantor*** of labour's respect in the network; ***organizer*** of events, congress, meeting, workshops of all genres. The singular aspect in generative matter is in the nature become. As Fabio Folgheraiter (2011) says: «**key points of the network (nodes) are people not institutions**; crucial is the role of the **facilitator**, a sort of "relationship guide" which is not intended to deliver direct benefits, but to give supports to activate relationship they find appropriate solutions; **relationships quality** among people of a strict territory leads to common actions to equal needs». The texture of relationships was designing new scene of beginning and social regeneration, but at the same time it revealed a limit about the inclusion of the residents. Staying in this creation of relationships has to find right opportunity yet.

3.5 Inclusive Workshop and Network for Villaggio

This new opportunity offered thanks to Social Plan of the same district, which entrusted Chieti Solidale with the task of coordinating associations in perspective of a multifunction place, an **inclusive quarter workshop** which could assemble different functions, with everyone participation. You wanted to construct a dynamic and multifunctional place, intergenerational promoting, participation and inclusion oriented in, to retrain Celdit quarter. Plan's intuition was thinking about physical place in which all ages would interact, because the generative way would have to connect with past, present and future. Intergenerational dimension would give much value to the quarter's rebirth. The call of partnership for associations started the "Rete per il Villaggio" (Network for Villaggio), by ten associations. A number of activities and events has happened, with a view to involving residents, including the most vulnerable, by reaching out to them, which in that square staying everyday and seeing silently our care. It is meaningful the murder of the young Fausto Di Marco in October 2016, down the street to our square, and it was noticed the arrest of guy, a resident.

We started from the square, Piazza San Pio X, we invited associations and people to meet together and clean square with a view to plan party together. To each his own, for a common goal.

The square, heart of Villaggio Celdit, experimented in fieri socialization: staff, volunteers and citizens met starting from a material action. Coming back the concept of ecological relation: «the cohesion as generative mechanism to new social infrastructures capable to transforming spaces to places», as Paolo Venturi says, Chief of Italian Association for Cooperative and No profit Culture Promoting (A.I.C.CO.N.). But history told up to now is the so called "action of maieutic" by Venturi, to a regenerating process. This action «is not a master plan in which it arranged participation by instrumentations, but an action of maieutic, of speech, seeing at and shared with Economy of Inhabitants, so called by Giovanni Taneggi; a phase where it really matters informality and where **rituals and joys of cooperation** (R. Sennet) turning into generative mechanism of a process hardly to displayed by Gaant diagram» (Venturi 2017). There were a lot of moments "of maieutic" to focus on quarter and its potentiality. Chieti Solidale entry in the summer program "Giardino delle pubbliche letture", with the idea to expand events to peripheral areas, it was an important moment and seemed like a pretty big change by some respects. This summer program had already been operating «by cooperation between the different components of associationism, culture and institutions »—we read in 2018 report drawn up by organization of the "Giardino delle pubbliche letture". The objective was «promote, for the summertime, opportunities of socialization, city animation, knowledge, cultural development, interesting in books, by proposals of high standard responding to adult and young's wanting to participate, increasing social fabric in sympathy and care». This decision, about Chieti Solidale, was in two different ways, based on our thinking of operators. The first way: Celdit's square can get a new face with events, transforming itself from marginal place to cultural one (as we proudly read on a mural made by associations of Network: ***All we don't estimate culture is culture***). The second way: to absorb people from the square into activities, people sitting in that tiny club in front of us, apparently not interested by our strange operations. Boldly, we asked manager to take part at the events in program with a food court, homemade street food point. Affirmative response generated spirit of cooperation in him and his customers, which is become "**ritual and joy of cooperation**", connecting with us them. Another necessity came out: "to mend periphery and centre" —as Renzo Piano's definition suggests (Piano 2014) —by diffusing fissure from lower city to upper one, starting from Villaggio Celdit, which slowly but inevitably lead to a **place at the centre** of **generative rebirth** and so, grower of central and significant actions. They come back in us the ***mosaic pieces with a thousand of colours and infinite forms***, being aware that generative and experience-based way, by nature, is in process.

3.6 Urban Knitting: Women's Generative Capacity

This development and each possible answer depend of what Paolo Venturi names "telòs": «an action of social regeneration always implies a common action, having three particular traits: interdependence between participants; each person retain his own identity and responsibility in the common action; people support same awareness and tension to a common objective». The discussion on methods maintained by team determined that much longer still to be done. It has to detect and to channel resident's predilections and will. Above all, it was necessarily to find "stem cell"—as Venturi reminds us, which is able to spark a common action of social regeneration. The proposal by ladies group—some volunteers of mentioned Associations and women residents—for a urban knitting workshop named "Mettiamoci le mani, sferruzziamo per il Villaggio" (Let's put fingers, let's knitting for the Villaggio), was our **stem cell**. Women using knit decided "to reappropriate" places and street furniture collectively. Later, other women, not only resident, are inviting to join the convivial group to knit or crochet. It is suggested two projects, on the way of solidarity: first "Red bench". A bench in Piazza San Pio X coated with shades-of-red rags, sewed one to another. The work was opened on March 8th, 2019 on ending violence against women. Second one, named "Mettiamoci una pezza" (a phrase like "make up for it", "need to clean it up", containing the word "rag"), in expressing solidarity with L'Aquila in the 10th anniversary of April 6th earthquake. Rags will be a part of work designed by Animammersa association of L'Aquila. On this date, in Piazza San Pio X there will be an installation and an urban knitting day. These two projects the spark to new generative force. Women are makers not by tradition, because they take care making day-to-day, thanks to extraordinary ability to transform complex to simple situations; they want flexibility against the rigidity. They are generators, when they muster men, as in Villaggio, to clean up and free from neglect Piazza San Pio X together. It is interesting we should pause to reflect on group composition and its "preferences". Venturi defines them **sleeping asset**: «They are not unused real estate, neglected places, undeveloped common good, but *preferences of who inhabits* (inhabitants), of who wants re-inhabit (returners) and who finds meaning in orienting own powers and life plan far away (aliens). **Inhabitants**, **returners** and **aliens** are part of *prophetic minorities* on the basis of "stem cell" in regenerating patterns: such a key ingredient to encourage that "*social birth*" which is the precondition of entrepreneurial community and common business solutions able to start, by means of *multifunctional* and *multistakeholder* (well, hybrid), sustainable and durable patterns of regeneration» (Venturi 2017). This time we tell about is present. Women group of Villaggio, the inhabitants, returners and aliens, are strongly impregnating the generative seed able to renew the Celdit quarter's history, by joining its historic memory and its future.

4 Conclusions

The recurring layoffs and the subsequent closure of Celdit, followed by demolition, marked a shock of employment and values, in the strict signification of work and its dignity. The massive white factory which before dominated the valley, was not a peculiarity of landscape, torn down with its heritage of innovations. The rapid Scalo's development from agricultural and sometimes pastoral economy in an industrial one, outstripped the split between lower and upper Chieti, this last concerned with keep up its image of healthy city, but at the same time, lost when peasants free themselves from dependency on landowners, it could hurt the "invidia di vicinato", neighbour's envy (Bonomi 2010). But the "community of grudge" could be born to Scalo in the aftermath of the productive force lost, after the big factory closures, prosperity and stability failing.

Generative experimentation at the point of Villaggio Celdit we discussed gives us a quarter, albeit slowly, which is going to cut through the rubber wall, this made of its memories, grievances and indifference. Social agents, in communication and voluntary work, always set as facilitator for this rebirth. The paper quarter can set up social incubator for a more open and supportive community.

References

Bonomi, A.: Sotto la pelle dello Stato. Rancore, cura, operosità, Feltrinelli, Milano (2010)

D'Alessandro, M.: Appunti per una storia. In: Iezzi, U. (ed.) Il Villaggio della Fabbrica di Papà. Edizioni Materiali, Chieti (1987)

Delibera: to: Consiglio Regionale, Abruzzo, Delibera Consiliare del 29 settembre 2017 n. 315 Piano Sociale Regionale 2016/2018, Approvazione Piano Sociale Distrettuale – Ambito Sociale n. 8 "Chieti" (2017)

Ferrandino, V., Napolitano, M.R.: Storia d'impresa e imprese storiche. Una visione diacronica. Franco Angeli, Milano (2014)

Folgheraiter, F.: Fondamenti di metodologia relazionale. Centro Studi Erickson, Trento (2011)

Magatti, M., Giaccardi, C.: Generativi di tutto il mondo unitevi! Un manifesto per la società dei liberi. Feltrinelli, Milano (2014)

Magatti, M. (ed.): Social generativity. A relational paradigm for social change. Routledge, Abingdon-on-Thames (2017)

Marchesi, A., Marmo, M.: L'immaginazione, energia che apre traiettorie di futuro. Animazione sociale 8 (Dec., 2018)

Ostuni, A.: La cellulosa italiana. Promozione privata e intervento statale attraverso le iniziative della famiglia Pomilio. 1917–1963, Dissertation, Pisa University, A.A. 2015–16 (2016)
Piano, R.: Il rammendo delle periferie, Il Sole 24ore, Jan. 30, 2014, 25 (2014)
Spedicato Iengo, E., Bongo, G.: Società artificiale. Dal consumismo alla convivialità. Franco Angeli, Milano (2015)
Venturi, P.: Rigenerazione come trasformazione dell'esistente. Archivio Generatività sociale 6 (Dec. 2017)

Selected Aspects of Barrier Materials Assessment as a Part of the Reaction on Threats and Risks Connected with CBRN Problems

Pavel Otřísal and Šárka Hošková-Mayerová

Abstract The protection of civilians, military professionals, including specialists of the Czech Armed Forces chemical corps, specialists of the Czech Republic Fire Rescue Service chemical service and, last but not least, employees of industrial enterprises against toxic substances is a long-term priority of the responsible state administration bodies. The priorities of these state bodies are focused on protection of the civilian population and industrial workers against the effects of toxic substances and related material equipment. The current security environment is characterized by a range of threats, including the proliferation of weapons of mass destruction and the possibility of releasing industrial hazardous substances such as toxic industrial materials into the environment. It is therefore possible to talk about the phenomenon of CBRN (C-chemical, B-biological, R-radiological, N-nuclear) threats that are expressed by the existence or use of nuclear, biological or chemical means against the civilian population. Critical Infrastructure means both production and non-production systems, the malfunctioning of which would have serious implications for security, economy and the preservation of the essential scope of other essential functions of the state in crisis situations. The risk of the CBRN threat is determined by the degree of severity expressed by the constantly expanding range of entities that have or are seeking to obtain them. The aim of the chapter is to notice that barrier materials used in the branch of individual protection must be assessed in accordance with those characteristics which are connected with resistivity against the permeation of toxic compounds.

Keywords Barrier material · Permeation · Diffusion · Porous material · Isolative material · Fick's law

---

P. Otřísal (✉)
NBC Defence Institute, University of Defence, Sidliste Vita Nejedleho, 682 01 Vyskov, Czech Republic
e-mail: pavel.otrisal@unob.cz

Š. Hošková-Mayerová
Department of Mathematics and Physics, Faculty of Military Technology, University of Defence, Kounicova 65, 662 10 Brno, Czech Republic
e-mail: sarka.mayerova@unob.cz

D. Flaut et al. (eds.), *Decision Making in Social Sciences: Between Traditions and Innovations*, Studies in Systems, Decision and Control 247,
https://doi.org/10.1007/978-3-030-30659-5_32

1 Introduction

The protection of civilians, military professionals, including specialists of the Czech Armed Forces chemical corps, specialists of the Czech Republic Fire Rescue Service chemical service and, last but not least, employees of industrial enterprises against toxic substances is a long-term priority of the responsible state administration bodies (Švarc 2015). The priorities of these state bodies are focused on protection of the civilian population and industrial workers against the effects of toxic substances and related material equipment in case of emergency situations (Hoskova-Mayerova 2018). Somewhat in a narrower way the focus is put on the protection of the Fire Rescue Service chemical service' specialists and military professionals, including the Czech Armed Forces chemical corps' specialists and other special forces. The members of the Fire Rescue Service and specialists of the Czech Armed Forces chemical corps are indispensable for the protection of civilians. Their specialized training and specific skills play a very important role not only in their own crisis response process in all aspects of these activities, but also in the organization of work at individual workplaces, the operation of which has to be safeguarded for a long time (Hoskova-Mayerova 2015).

The current security environment is characterized by a range of threats, including the proliferation of weapons of mass destruction and the possibility of releasing industrial hazardous substances such as toxic industrial materials into the environment. It is therefore possible to talk about the phenomenon of CBRN (C-chemical, B-biological, R-radiological, N-nuclear) threats that are expressed by the existence or use of nuclear, biological or chemical means against the civilian population. Critical Infrastructure means both production and non-production systems, the malfunctioning of which would have serious implications for security, economy and the preservation of the essential scope of other essential functions of the state in crisis situations. The risk of the CBRN threat is determined by the degree of severity expressed by the constantly expanding range of entities that have or are seeking to obtain them. Furthermore, these substances introduce a high risk represented by the possibility of controlling a small number of resources by individuals or non-state groups. Last but not least, they also bring extraordinary dangers posed by terrorism in a number of regions of the world, or the real threat of using CBRN funds for extortion. All these aspects lead to the need for the armed forces to be permanently and precisely prepared to respond adequately to the threats and risks associated with CBRN threats (Procházka and Procházková 2018; Rybanský 2009). Besides basic material, which is the equipment of every soldier, the specialized components of the Czech Armed Forces must be equipped with means to ensure the possibility of long-term stay in a contaminated environment in carrying out tasks within a military operation (Drozd 2018; Stodola 2017), reconnaissance, detecting, identifying, decontaminating and removing the consequences of using weapons of mass destruction and releases of industrial hazardous substances into the environment. The Czech Armed Forces chemical corps, as a kind of the force, is designed to fulfill the most complex tasks and measures to CBRN Defence and chemical support in military operations. One of its tasks is to provide cooperation in the fulfillment of tasks of the integrated rescue system in the consequence management of natural disasters in the territory of the Czech Republic. To accomplish these tasks, the specialists of the Czech Armed Forces chemical corps must not only be trained and educated, but also equipped with such protective material that will ensure their long-term functionality in the environment contaminated by high toxic chemical, radioactive and biological substances. It is very important to realize that the number of highly professional forces is always limited and it is only very difficult to replace the forces that are adequate.

2 Use of Barrier Materials

Barrier materials that are used to produce individual protective equipment and some simple means of collective protection have to meet a number of requirements. One of the most important is resistance against permeation of toxic substances in the form of liquids, aerosols, gases and vapors. This protection must be absolute, thus the toxic compound must not pass through the barrier materials in any concentration. In practice, we talk about the state where the toxic compound is at its maximal concentration on one side of the barrier material and on the other is at its zero concentration. This state of affairs must be long-lasting, unchanging, and reassured by lightweight, thin, flexible and affordable polymeric materials that are easy to process and exhibit long stability in store. The complex of these requirements forces scientists and developers to think about problems how current materials describe, how they can be compared and how to find unifying evaluation criteria for their choice and improvement (Blaha 2018). The need for a unique criterion for evaluation is a very interesting and useful option for the end-user. On the basis of the presented and commented data, the end-user can make a decision as to which individual protective equipment is the one in which he will invest his funds from their point of its fro example life cycle properties (Urban and Hoskova-Mayerova 2017).

2.1 Former Use of Parameters Designated for Barrier Materials Assessment

The evaluation of the protective properties of the barrier materials was previously carried out essentially based on a single parameter, which was the breakthrough time. This time was understood as a time period from the beginning of the contamination (application of the toxic substance) to one side of the barrier material until a visible (color) change was observed on the other side. This change was defined by a certain concentration of toxic substance. However, it depended on the observer's experience and capabilities to what extent he was able to recognize the correct time that matched the requirements for ending the measurement and specifying the correct value of the breakthrough time, which in the case of chemical warfare agent was always related to the so-called threshold dose. In some cases this value can be used in a scope of their simulants (Otrisal 2018a). Methods that used color changes in the indicator layers did not allow any other description of the properties of the material than the breakthrough time. A very significant problem arose at a time when security threats turned away from the use of chemical weapons in the form of a classical weapon of mass destruction. We began to talk about the real possibilities of releases of toxic industrial chemicals, and thus the necessity to provide quality and long-term protection against their effects (Petrea 2018). The threshold dose for these substances is very difficult to determine and data that characterize it are largely absent or absent at all. It has therefore to be noted that simple colorimetric methods have to be replaced by automatic or semi-automatic methods which will provide more data than the value of the breakthrough time. These methods can use a huge number of various sensors for detection of chemicals that have quite universal possibilities (Švorc et al. 2018).

3 Permeation Through Non-porous Polymeric Materials and Its Math Specification

Permeation through a non-porous polymeric membrane runs during its exposition with gas and liquid phase of the toxic compounds. During gases permeation diffusion coefficients are independent on concentration of permeant in the membrane while during permeation of gas and liquids is this in contrary as the result of swelling of the material with permeated permeant. Opening polymeric membrane cases that the absolute permeation flow of the permeant through the membrane can be about two or three orders higher than in case of foul gases. That is why that in case of gas and liquids permeation the diffusion coefficient D which is exponentially dependent on concentration of the permeant in the membrane. For description of this process has been proposed a lot of theories from them the most frequently used is the theory based on a solubility-diffusion mechanism of the flow. This one is described in lots of publication, for example in Bernauer (2012).

Permeability as a transport property of non-porous polymeric systems reclines in the transportation of the mass through the membrane owing to gradient of temperature, pressure or concentration, eventually owing to other outer field of force. The mechanism of this transport depends mostly on the character of the membrane and permeated compounds. The model of the solubility-diffusion flow presumes that the transport consists from several subsequent steps. This model can use similar principles as modelling landscape changes or optimalization of the vehicle routes (Svatonova and Rybanský 2014; Rybanský 2014; Rybanský et al. 2016). Firstly, it comes to the sorption of the compound on the surface of the membrane and subsequently to its sorption in the membrane (Otrisal 2018c; Prikryl 2018). Secondly, it comes to diffusion of the compound through the membrane and after that to desorption from the surface on the second side of the membrane. A lot of authors described quantitatively this permeability with the help of the Fick's diffusion law and Henry's law about solubility. Further work and late studies of the structure of macromolecule compounds have contributed to next clarification of this mechanism and determination of dependence of the permeability of the polymer on their composition, structure phase state, orientation and other properties. For example in the scope of polymers in a semi crystallic state it is assumed that to transport of the compound comes only in the amorphous phase whereas the crystallic phase is impermeable for permeated compounds.

The assessment of the protective properties of barrier materials used in anti-gas protection is described in terms of resistance to permeation of toxic substances by several parameters. One of the basic parameters is a diffusion coefficient. This can be determined experimentally or using software tools that calculate it automatically based on generated permeation data. According to current standards, various methods are used based on a variety of physical principles. The methods themselves, however, allow the gain of information on the increase in the observed value. The diffusion coefficient value is recalculated and usually refers to a particular permeation measurement stage. However, some modern methods allow it to be determined on-line at any stage of measurement, which is used to describe the protective properties of barrier materials at different stages of their testing. However, the basic theoretical aspects of diffusion, based on the generally accepted mathematical approaches presented in the text below, must always be respected.

If we consider the polymer as the solvent within diffusion of the compounds through polymeric materials and the diffusing compound as a dissolved component it is possible to describe this diffusion process with only one parameter, thus with the diffusion coefficient D, for which it determines the density of the mass flow in accordance with the 1st and 2nd Fick's law. In this way distributive diffusion coefficient is generally dependent on temperature, pressure and concentration. If diffusion coefficient is the function not only of concentration but also even time, it deals with non-fick's diffusion which is not conducted with Ficks' laws. Suitable mathematical methods which will not be described here in the detail way have been elaborated for description of diffusion processes.

If gas or vapours in the polymeric membrane sorbs within creation of a very diluted solution it is possible to consider this system as the ideal one and in accordance with Henry's law the concentration c of gases or vapours in the membrane can be expressed as the linear function of outer balance pressure. The coefficient of permeability P is then a product of the diffusion coefficient D and the coefficient of solubility S which characterizes sorption. Generally all three coefficients are the function of the temperature and pressure. The coefficient of permeability P determines the amount of the compound which permeates through the unit of area of the membrane in the unit of time within the unit gradient of the pressure in standard conditions. The coefficient of solubility S determines the amount (weight) of the gas or vapours of dissolved toxic compounds in the unit volume of the solvent (material of the membrane) within the unit partial pressure and the particular temperature in standard conditions. The relationship between these variables is given by the equation:

$$P = D.S. \tag{1}$$

During exposition of the polymeric membrane with the liquid phase of the permeant it is also possible to describe the sorption in the polymer in accordance with Henry's law with an analogical formula. In accordance with this formula the concentration of the compound in the polymer is determined with the product of the concentration of the compound in the liquid phase on the surface of the membrane and with non-dimensional product of the solubility S. This product of the solubility has in the particular temperature a constant value and it expresses a ratio of concentration of the compound in and out of the membrane. Within the transport of the compound in polymers, however, the sorption of gases or vapors is not conducted by Henry's law but it is affected even with the course of an isotherm expressing the dependence of the balance concentration of the toxic compound sorbed in the polymer on the partial pressure with constant temperature.

Higher concentration of toxic compounds is created on the polymer surface. This one causes molecule migration through polymer in the direction of the concentration gradient with the aim to level concentration not only in inner and outer polymer materials areas but also on its underside side. This migration is sometimes called right diffusion. Permeation duration respectively molecules transportation through a non-porous membrane is schematically introduced in Fig. 1.

Explanations to Fig. 1:

1. diffusion either in gaseous or liquid phase;
2. chemical compound solution in material of the membrane;
3. diffusion through the membrane;
4. transportation into either gaseous or liquid phase;
5. diffusion from the membrane in liquid and gaseous environment.

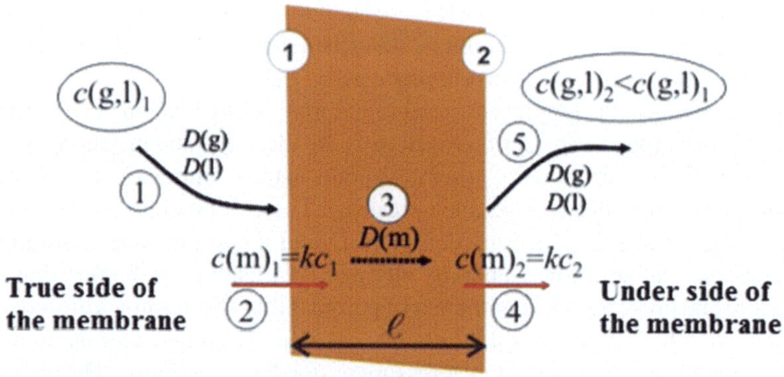

Fig. 1 Duration of non-porous membrane molecule transportation

Material flux j through non-porous membrane is possible to introduce with an mathematical formula:

$$j = -D(\mathrm{m})\frac{\Delta c(\mathrm{m})}{\ell} = -D(m)k\frac{\Delta c(\mathrm{g, l})}{\ell} = P\frac{\Delta c(\mathrm{g, l})}{\ell}, \tag{2}$$

where k is a distributive constant of a particular compound between a relevant phase and the membrane, P is coefficient of permeability, Δc introduces a concentration difference between both right and underside of the non-porous membrane and $D(m)$ is coefficient of diffusion of compound m which takes part in the process of membrane transportation with thickness ℓ.

In case of porous membrane, molecule transportation is due to toxic compound's coefficient of diffusion. Material flux j will be dependent on the concentration difference on membrane both right and underside and its thickness.

$$j = -D(g)\frac{\Delta c}{\ell}$$
$$D(g) \gg P \tag{3}$$

It is possible to formulate permeation process in accordance with picture 1 by the formula:

$$Q_p = PA\frac{\Delta c}{\Delta \ell}.t, \tag{4}$$

where Q expresses an amount of permeation toxic compound, A is the area of the polymer membrane by which permeation properties are researched, Δc is difference of concentration on the polymeric membrane right and underside, $\Delta \ell$ is the thickness of the polymer material, t is time of permeation and P is coefficient of permeability.

3.1 Process of Polymer Solution and Swelling

The process of solid polymer transportation into the liquid phase within temperatures under so called glass transition temperature (T_g) is composed of two following events:

- swelling and
- self-solution.

The process of swelling has a non-stationary character and its result is a swelling surface layer. For the swelling surface layer it is characteristic that from the protective properties' point of view does not fulfill a task of a totally close membrane.

The swelling rate always increases with higher temperature. The swelling process can be either exothermal or endothermic in accordance with polarity of both dissolution reagent and swelling polymer. During exothermal swelling decreases a balance degree of swelling. On the other hand, during endothermic processes increases.

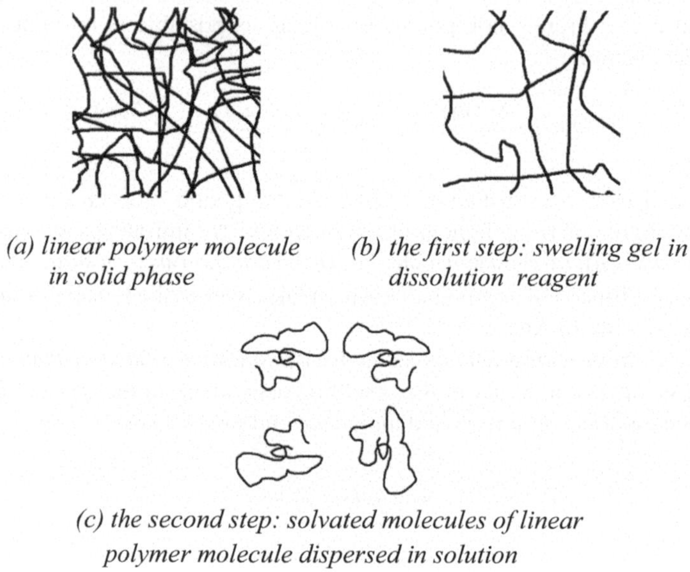

Fig. 2 Process of linear polymer swelling and solution (see steps a, b, c)

Macháčková (2008) claims that processes in SSL are very difficult. The rate of swelling and solution of the linear polymer is schematically shown in the Fig. 2.

During solution of the linear polymer comes to mixing of polymer with dissolution reagent. A lot of factors exist, however, the most important one is speed of diffusion in relation to dissolution reagent and researched polymer. This speed depends on both a polymer chemical structure and the dissolution reagent chemical structure. Mutual interactions between them play a very important role depending up temperature of solution and molecular weight of the dissolution reagent.

Within polymer material it is possible to summarize the swelling process into a scheme Fig. 3. Its practical usage touches not only researched polymer material but also the QCM detector polymer layer.

3.2 Evaluation of Permeation Characteristics with the Help of the Diffusion Coefficient and the Coefficient of Permeability

The issue of permeation thus the transportation of molecules of substances through biological materials under the influence of concentration gradient without the excitement has already devoted to Thomas Graham in 1829. Then he watched the gradual supercharging of pigs' bladders with the help of detection of escaping CO_2 (Dhingra 1997). At that time, the author observed significant phenomena that characterize the

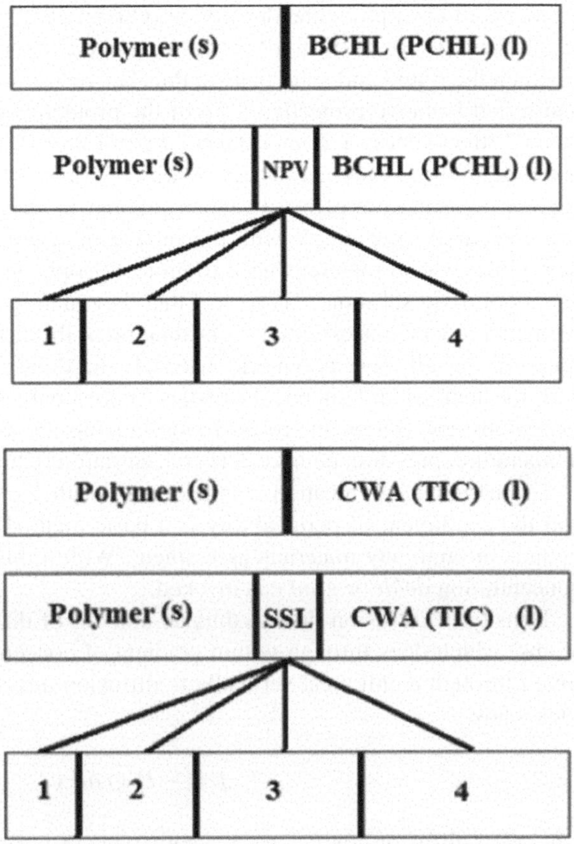

Fig. 3 Creation of distributed layers during swelling and solution processes. 1-infiltrated layer, 2-rubber layer, 3-layer of gel, 4-solution polymer layer with the solid phase, liquid phase (CWA—chemical warfare agent, TIC—toxic industrial material, SSL—swelling and solution layer)

permeation of gases. It was noted that it is possible to obtain various transport parameters and constants of barrier materials that do not depend on temperature, pressure and type of permeated gases. Knowledge of basic material properties allows the application of established methods of study of protective properties of constructive materials used to produce garments of an isolative type, which is a more accurate description of transportation of mass important. Moreover, the knowledge of those properties can help student to understand permeation processes within the educational process and their applications in a lot branches where physical protection is required (Badea et al. 2017; Barsan 2018; Svarcova et al. 2016).

The basic and characteristic process by which can penetration of toxic of substances through a barrier layer of protective clothing be describe is the diffusion, which is closely linked to the permeation process. Slabotínský (2006) for the case of isolative garments without ventilation states that "through intact and tight garment can penetrate only chemicals by the slowest process, thus by diffusion". It does not take place, however, as the first process, but it builds on the sorption processes, including absorption, adsorption and processes which are usually accompanied by a process of exsorption and desorption (Otrisal 2018b; Chovancová 1988; Romano and Lukey 2010). Diffusion belongs to among adsorption processes taking place between the liquid and solid phases, thus between test chemicals (toxic substances) and tested isolative protective folio of the protective garment (Mărieș 2017). Efficiency, effectiveness and importance of adsorption, however, always depend on the pore size of the researched surface on which it runs. The surface of the polymer folio formed by a mixture of butyl rubber polymer is insignificant in terms of porosity and compared for example with the surface of activated carbon porosity is almost negligible. For all materials used to protect people against the effects of gases and vapors of toxic substances is typical that they manifested a significant influence of permeation, which is essential for evaluation of the airtight of barrier materials. The properties of all these polymeric materials (hardness; porosity; permeability; thermal, mechanical and chemical resistance) obviously depend on the manufacturing conditions and can be improved by optimizing the producing parameters such as temperature, pressure, components concentration (Chira 2015).

Evaluation of permeation characteristics with the help of diffusion coefficient and the coefficient of permeability is a basic method for permeation of any compounds through any materials assessment. Within this affect so called gradient of concentration *dc/dt* or *grad c* is invoked.

Density of diffusion flow *J*, thus an amount of diffused compound (number of atoms) which does through within gradient of concentration *grad c* in a period of time *t* through a unit area, vertically to diffusion direction, is expressed by the first Fick's law.

$$\vec{J} = -D\ grad\ c, \tag{5}$$

where *D* is diffusion coefficient. A negative sign in the formula (4) expressed diffusion duration in a direction of concentration decrease.

Diffusion coefficient D ($m^2\ s^{-1}$) of a compound A in a mixture A, B is defined in accordance with the same law. In conformity with this law is the diffusion flow of molecules A through the flat in a place *x* proportional to gradient of concentration of compound A in a place *x* in one-dimensional direction: (Meissner 1987; Ditl 2008; Urban 2007; Lile 2015).

$$\vec{J_A} = -D\frac{dc_A}{dx},\ \left[\text{kmol (A) m}^{-2}\,\text{s}^{-1}\right] \tag{6}$$

respectively

$$\vec{J_A} = -D\frac{d\rho_A}{dx},\ \left[\text{kg (A) m}^{-2}\,\text{s}^{-1}\right] \tag{6}$$

where c_A is concentration of the compound *A* in the place *x* and ρ_A je density of the compound *A*.

The first Fick law is presented very often in a form:

$$J = F = -D\Delta c, \quad (7)$$

where F is a permeation rate. Fick law can be applied in a breakthrough steady rate of toxic compound through barrier material. The steady rate is reached in the time when concentration with the time stays constant and the rate does not change (Beňová 2019; Hoskova-Mayerova 2016).

4 Conclusion

The evaluation of the affectivity of barrier materials to capture toxic substances or to create an effective and long-lasting barrier against their permeation by diffusion coefficients and other experimentally detected values is a very important issue. It enables the very effective evaluation of the relationship between a toxic substance represented by a chemical warfare agent or the toxic industrial chemical and the barrier material. Each component of this relationship has its own specific characteristics that affect the resulting breakthrough time value very significantly. As a simplified example, it is possible to mention the different concentration of the used test chemical and the different thickness of the used barrier material. The influences, that fundamentally affect the resulting breakthrough time value and therefore the practical applicability of the barrier material to produce a particular type of individual protection equipment, are very numerous. All must be taken into account, assessed in the context of the experimental conditions, the used test methods and the observer's experience. The employment of semiautomatic and automatic methods to detect other permeation characteristics rather than just the breakthrough time will significantly contribute to the observer's error, and will allow a description of the permeation process of the toxic substance through the studied barrier material at various measurement stages.

Acknowledgements The work presented in this paper has been supported by the project for "Development of Basic and Applied Research developed in the long term by the Departments of Theoretical and Applied Bases FMT [Project code: "VYZKUMFVT (DZRO K-217)"] supported by the Ministry of Defence of the Czech Republic.

References

Badea, D., Bârsan, G., Moşteanu, D., Virca, I.: Using E-learning technology as a support for educational activities in the filed of critical infrastructure security. In: 13th International Conference eLearning and Software for Education "Could Technology Support learning Efficiency?" (2017). https://doi.org/10.12753/2066-026x-17-003

Barsan, G., Mares, J., Margin, C., Mosteanu, D., Carutasu, V.: Coal radioactivity and the impact of the mine exploitation on the environment in the sarmasag quarry. J. Env. Protect. Ecol. **19**(3), 1249–1260 (2018)

Beňová, P., Hošková-Mayerová, Š., Navrátil, J.: Terrorist attacks on selected soft targets. J. Secur. Sustain. Issues **8**(3), 453–471 (2019). https://doi.org/10.9770/jssi.2019.8.3(13)

Bernauer, B. et al.: Membránové procesy. (Membrane Processes), 293p. Publikace VŠCHT Praha (2012)

Blaha, M., Potužák, L., Šilinger, K.: Linear and angular issues in perspective artillery fire control system. In: 22nd International Conference on Circuits, Systems, Communications and Computers, Jan 2018, MATEC Web of Conferences, vol. 210(Spec. ed. 1), 02055, 5p (2018). https://doi.org/10.1051/matecconf/201821002055

Chira, D., Mărieș, G.H.R.E., Bungau, C.: The influence of subsequent pressure of (HDPE), (PMMA), (PC + ABS) on some mechanical properties of items obtained through injection. Rev. Mater. Plast. **52**(4), 572–577 (2015)

Chovancová, L.: Chemická technika: difúzní operace. (Chemical Technique: Diffusion Operations), 1st edn., 84p. Havlíčkův Brod, Fragment (1988). ISBN 80-7200-214-7

Dhingra, S.S: Mixed gas transport study through polymeric membranes: a novel technique. [Dissertation work], 172p. Virginia, Blacksburg (1997)

Ditl, P.: Difúzně separační procesy. (Diffusion and Separation Processes). [Textbook], 3rd edn, 231p. České vysoké učení technické v Praze, Praha (2008). ISBN 978-80-01-03-972-4

Drozd, J., Stodola P., Křišťálová D., Kozůbek, J.: Experiments with the UAS reconnaissance model in the real environment. In: Modelling and Simulation for Autonomous Systems, pp. 340–349. Springer International Publishing, Cham (2018)

Hoskova-Mayerova, S.: Education and training in crisis management. In: The European Proceedings of Social & Behavioural Sciences (EpSBS), XVI, pp. 849–856 (2016). https://doi.org/10.15405/epsbs.2016.11.87

Hoskova-Mayerova, S.: Emergency assessment in case of hazardous substance leakage at Czech Republic freight rail transport in 2008–2016. In: Haugen, S., Barros, A., van Gulijk, C., Kongsvik, T., Vinnem, J.E. (eds.) Safety and Reliability—Safe Societies in a Changing World Proceedings of ESREL 2018, pp. 1381–1386. CRC Press, London (2018). https://doi.org/10.1201/9781351174664-174

Hoskova-Mayerova, S.: Geospatial data reliability, their use in crisis situations. In: Knowledge Based Organization International Conference, vol. 21(3), pp. 694–698, De Gruyter open (2015). https://doi.org/10.1515/kbo-2015-0117

Lile, I.E., Freiman, P.C., Hosszu, T., Vasca, E., Vasca, V., Bungau, S., Vaida, L.: A subsidiary physical research of glass ionomers. Rev. Mater. Plast. **52**(2), 175–179 (2015)

Macháčková, A.: Difúzní proces rozpouštění polymerů a jeho využití v materiálovém inženýrství. (Diffusion Process of Dilution of Polymers and its Usage in Material Engineering). [Shorten Version of Habilitation Thesis], 1st edn., 26p. VUTIUM, Brno (2008). ISBN 978-80-214-3557-5

Mărieș, G.H.R.E., Chira, D., Bungau, C., Costea, T., Moldovan, L.: Determining the influence of the processing temperature by injection and of the subsequent pressure on the surface's hardness and indentation modulus of the products made of HDPE, PMMA, PC + ABS through nanoindentation—G-series basic hardness modulus at a depth method. Rev. Mater. Plast. **54**(2), 214–220 (2017)

Meissner, B., Zilvar, V.: Fyzika polymerů: Struktura a vlastnosti polymerních materiálů. (Physics of Polymers: Structure and Properties of Polymeric Materials), 1st edn., 306p. SNTL—Nakladatelství technické literatury, Praha (1987)

Otrisal, P., Florus, S., Barsan, G., Mosteanu, D.: Employment of simulants for testing constructive materials designed for body surface isolative protection in relation to chemical warfare agents. Rev. Chim. **69**(2), 300–304 (2018a)

Otrisal, P., Melicharik, Z., Svorc, L., Bungau, S., Virca, I., Barsan, G., Mosteanu, D.: Testing methods of assessment for the chemical resistance of insulating materials against the effect of selected acids. Rev. Mater. Plast. **55**(4), 545–551 (2018b)

Otrisal, P., Obsel, V., Buk, J., Svorc, L.: Preparation of filtration sorptive materials from nanofibers, bicofibers, and textile adsorbents without binders employment. Nanomaterials **8**(8), 564p (2018c). https://doi.org/10.3390/nano8080564

Petrea, N., Ginghina, R., Pretorian, A., Petre, R., Barsan, G., Otrisal, P., Mosteanu, D.: Experimental survey regarding the dangerous chemical compounds from military polygons that affect the military health and the environment. Rev. Chim. **69**(7), 1640–1644 (2018)

Prikryl, R., Otrisal, P., Obsel, V., Svorc, L., Karkalic, R., Buk, J.: Protective properties of a microstructure composed of barrier nanostructured organics and SiO_x layers deposited on a polymer matrix. Nanomaterials **8**(9), 679 (2018). https://doi.org/10.3390/nano8090679

Procházka, J., Procházková, D.: Problems of mobile risks in territory. In: Safety and Reliability—Safe Societies in a Changing World, p. 1783. Taylor & Francis Group, London (2018). ISBN 978-0-8153-8682-7

Romano, A.J., Lukey, E.B., Salem, H.: Chemical Warfare Agents: Chemistry, Pharmacology, Toxicology and Therapeutics, 2nd edn., 723p. Taylor & Francis Group, New York (2010). ISBN 1-4200-4661-6

Rybanský, M.: Modelling of the optimal vehicle route in terrain in emergency situations using GIS data. In: IOP Conference Series-Earth and Environmental Science, vol. 18. Article Nr: 012131 (2014). https://doi.org/10.1088/1755-1315/18/1/012131

Rybanský, M., Vala M.: Relief impact on transport, In: ICMT'09—International Conference on Military Technologies 2009, 9pp. Brno, Czech Republic (2009)

Rybanský, M., Brenova, M., Cermak, J., et al.: Vegetation structure determination using LIDAR data and the forest growth parameters. In: IOP Conference Series-Earth and Environmental Science, vol. 37. Article Number: UNSP 012031 (2016)

Slabotinský, J., Brádka, S.: Ochrana osob při chemickém a biologickém nebezpečí. (Protection People within Chemical and Biological Danger), 1st edn., 109p. Sdružení požárního a bezpečnostního inženýrství v Ostravě, Ostrava (2006). ISBN 80-86634-93-0

Stodola, P.: Improvement in the model of cooperative aerial reconnaissance used in the tactical decision support system. J. Def. Model. Simul. **14**(4), 483–492 (2017)

Švarc, J., Binar, T., Šilinger, K., Sedlaříková, M.: The assessment of a material life cycle based on an analysis of fracture surfaces after fatigue failure. In: 16th International Conference on Advanced Batteries, Accumulators and Fuel Cells, ABAF 2015, pp. 177–185. Electrochemical Society Inc. USA (2015)

Svatonova, H., Rybanský, M.: Visualization of landscape changes and threatening environmental processes using a digital landscape model. In: IOP Conference Series-Earth and Environmental Science, vol. 18. Article Nr: 012018 (2014). https://doi.org/10.1088/1755-1315/18/1/012018

Svarcova, I., Hoskova-Mayerova, S., Navratil, J.: Crisis management and education in health. Eur. Proc. Soc. Behav. Sci. (EpSBS) **XVI**, 255–261 (2016). https://doi.org/10.15405/epsbs.2016.11.26

Švorc, Ľ., Haššo, M., Sarakhman, O., Kianičková, K., Stanković, D.M., Otřísal, P.: A progressive electrochemical sensor for food quality control: reliable determination of theobromine in chocolate products using a miniaturized boron-doped diamond electrode. Microchem. J. **142**(November), 297–304 (2018). https://doi.org/10.1016/j.microc.2018.07.007

Urban, P.: Modeling of physical-chemical processes within transport in polymer Gel studie. Doctoral thesis summary, 35p. Univerzita Tomáše Bati, Zlín (2007). ISBN: 978-80-7318-636-4

Urban, R., Hoskova-Mayerova, S.: Threat life cycle and its dynamics. Deturope **9**(2), 93–109 (2017)

The manufacturer's authorised representative in the EU is Springer
Nature Customer Service Centre GmbH, Europaplatz 3, 69115 Heidelberg,
Germany. If you have any concerns regarding our products, please
contact ProductSafety@springernature.com

Printed and bound by CPI Group (UK) Ltd, Croydon, CR0 4YY
25/09/2026
02222851-0012